Synchrotron Radiation Theory And Its Development

In memory of I M Ternov

World Scientific Series on Synchrotron Radiation Techniques and Applications

---

Published

Vol. 1 Synchrotron Radiation Sources — A Primer
H. Winick

Vol. 2 X-ray Absorption Fine Structure (XAFS) for Catalysts & Surfaces
ed. Y. Iwasawa

Vol. 3 Compact Synchrotron Light Sources
E. Weihreter

Vol. 4 Novel Radiation Sources Using Relativistic Electrons
P. Rullhusen, X. Artru & P. Dhez

Forthcoming

Photoelectron Spectroscopy of Solids and Surfaces
K. C. Prince

Insertion Devices for Synchrotron Radiation
F. Ciocci, G. Dattoli & A. Renieri

X-ray Microscopy
ed. G. R. Morrison

Medical Applications of Synchrotron Radiation
eds. Al Thompson, B. Thomlinson & M. Ando

Synchrotron Radiation Studies of Non-Crystalline Biology Systems
J. Bordas & A. Svensson

Series on Synchrotron Radiation Techniques and Applications – Vol. 5

Synchrotron Radiation Theory And Its Development

In memory of I M Ternov

Editor

Vladimir A Bordovitsyn

Tomsk State University, Tomsk, Russia

World Scientific

Singapore • New Jersey • London • Hong Kong

Published by

World Scientific Publishing Co. Pte. Ltd.

P O Box 128, Farrer Road, Singapore 912805

USA office: Suite 1B, 1060 Main Street, River Edge, NJ 07661

UK office: 57 Shelton Street, Covent Garden, London WC2H 9HE

British Library Cataloguing-in-Publication Data
A catalogue record for this book is available from the British Library.

SYNCHROTRON RADIATION THEORY AND ITS DEVELOPMENT

ISBN 981-02-3156-3

Printed in Singapore.

Contents

Preface

Synchrotron radiation (SR) is an example of a thoroughly studied physical phenomenon in the sense that, at the moment, theory and experiment are in perfect agreement. The unique properties of SR (i.e. a continuous spectrum extending up to the X-ray region; great intensity; and a high degree of polarization) as well as the possibility of exact theoretical verification have resulted in extensive theoretical and technical applications of SR. This radiation opens up vast and new prospects of the investigation in experiments with polarized beams of particles in high energy physics, in solid state physics (photoelectron spectroscopy, crystallography, X-ray luminescence, etc.), chemistry (observation of reaction evolution), biology (study of DNA molecule structure), medicine (angiography, blood filtration), geology (elemental analysis), ecology (analysis of atmospheric aerosols, soil and water), and creates new advanced technologies (microelectronics, micromechanics, constructing new composite materials), etc.[1–4] (see also Ref.[5])

Scientific research related to SR has already passed into the field of engineering and technical applications where great experience has been accumulated. Unfortunately, purely theoretical investigations of SR have been somewhat left aside. As a matter of fact, the SR theory is at the juncture of classical, semiclassical and quantum electrodynamics and can serve as an excellent testing ground for clearing up relations between old and new physical theories. The energy of electrons (tens of GeV) associated with SR is distinguished by the fact that it is there that the individual properties of the electron begin to manifest themselves best of all thus enabling one to study the fundamental properties of elementary particles (spin, mass, intrinsic magnetic moment) which under nonrelativistic conditions are hidden under the cover of the wave function.

Mechanism of SR and the inverse Compton effect (the scattering of soft photons by relativistic electrons) have been extensively considered as a source for the production of X-ray in cosmic objects and in the galactic background radiation[6]. Besides, the methods worked out due to the development of the SR theory allow investigations in the behavior and properties of a substance under extreme conditions (superstrong magnetic fields and superhigh energies), which is important for many astrophysical applications of the SR theory[7]. The inverse Compton effect is also used in laboratory studies as a source of polarized high-energy gamma rays[8] and for express analysis of polarization effects in SR.[9]

The idea of writing this book with the purpose of systematizing the known results in the SR theory as well as filling the gap in the scientific literature on its latest development originated from I.M. Ternov and the author of this preface. A similar book of collected articles entitled "Synchrotron Radiation", edited by A.A. Sokolov and I.M. Ternov, was first published in Russian more than thirty years ago[10], and it has already become a rare book. It contains the fundamentals of classical and quantum theories of SR developed by the scientists from the Physical Department of Moscow State University and from Tomsk Polytechnical Institute and Tomsk State University. Later, the basic results of the above research were used in the well-known monographs written by A.A. Sokolov and I.M. Ternov.[11]

While discussing the outline of the book I.M. Ternov, in one of his letters to me, suggested to take his recent article entitled "Synchrotron Radiation", published in the journal "Uspekhi Fiz. Nauk" (in English — "Physics-Uspekhi") in 1995[12] as a basis of introduction. Unfortunately, I.M. Ternov's tragic death prevented him from actively participating in the publication of this new book on SR which is written by his disciples. The book is dedicated in fond memory to I.M. Ternov, an outstanding scientist, who is one of the founders of the modern SR theory and of a well-known scientific school of theoretical physics.

In order to show I.M. Ternov's unique style of reasoning and his way of thinking it was decided, by way of introduction, to quote some parts from the above-mentioned article in "Uspekhi". In that work he set forth the history of the origin and development of the SR theory, the discovery of radiative selfpolarization of relativistic electrons (Sokolov-Ternov effect) as well as its recent achievements.

The next three chapters in this book deal with a systematic description of the classical SR theory (V.A. Bordovitsyn and V.S. Gushchina), polarization properties of relativistic electrons (V.A. Bordovitsyn and V.S. Gushchina) and quantum theory of SR (V.G. Bagrov). Note that along with the well-known results one can find additional interesting information there which, for some reason, was not included into the earlier publications.

Also presented here are the latest results[13,14] concerning relativistic μ-radiation (i.e. the radiation of the electron intrinsic magnetic moment, or spin light) as well as mixed $e\mu$-radiation (V.A. Bordovitsyn and V.S. Gushchina). It should be noted that the mixed $e\mu$-radiation was first observed by G.N. Kulipanov's group in Novosibirsk on the basis of the Sokolov–Ternov effect.[15] Thereby, it was proven that both the electron intrinsic magnetic moment radiation and the spin properties of the relativistic electron are impossible to ignore in the SR theory.

Recently, new radiation devices with relativistic electrons have been developed, namely, undulators and wigglers. These are magnetic apparatus with a sign-variable magnetic field installed in the rectilinear sections between the bending magnets of electron storage rings. The above devices yield radiation which, according to I.M. Ternov, is "genetically related" to SR differing from the latter either in greater brightness due to a periodic increase in the angle of deflection of electrons in the magnetic field (a wiggler), or in stimulation of radiation at certain frequencies caused by the interference of radiation at periodic electron motion with a small deflection angle in the magnetic field (an undulator). The radiation is considered in detail by V.Ya. Epp in the chapter "Undulator Radiation".

The next chapter deals with the problems of relativistic radiation related to the electron path length in a homogeneous magnetic field (V.G. Bagrov and V.Ya. Epp). It turned out that when radiation occurs in uniform magnetic field, its properties, particularly its spectrum, greatly depend on the length of the arc of the circle that an electron moves along. It is only after a great number of revolutions in a homogeneous magnetic field that one can get an ideal SR spectrum.

A separate chapter is devoted to electron radiation under extreme conditions in superstrong magnetic fields (O.F. Dorofeev, V.Ch. Zhukovskii and A.V. Borisov). Interest in SR problems in superstrong fields arose due to investigation of physical processes occurring close by the pulsar in outer space where magnitudes of magnetic fields can be as large as those of the order of Schwinger's magnetic field ($4.41 \cdot 10^{13}$ Oe). At present nobody doubts the fact that SR "works" actively in cosmos.

The last chapter of the book is concerned with a wide area of application of the SR theory in astrophysics (G.S. Bisnovatyi-Kogan).

The group of authors presented here consists of mature scientists who have made appreciable contributions to the development of the SR theory. So it was the editor's task, besides discussing the contents and structure of the book, to retain the original styles of its authors. I beg the reader to forgive me, but for this reason, there is no single notation system in the book. To some extent this drawback has been compensated for by the detailed index of the terms used which the reader can find at the end of the book.

I hope our work will be useful for those who are interested in the problems of SR — one of the most wonderful discoveries in physics in the 20th Century.

I would like to thank Dr. V.I. Podkatov for his kind attention and support of this work, V.G. Bulenok and V.V. Kruglikov for drawing the graphs, and L.E. Trifonova for correcting the translation.

V.A. Bordovitsyn

References

1. E.E. Koch, ed. *Handbook on Synchrotron Radiation* (North Holland, Amsterdam, 1983).
2. I.M. Ternov, V.V. Mikhailin and V.R. Khalilov, *Synchrotron Radiation and its Application* (Harwood Acad. Publ., Chur, Switzerland, 1986).

3. H. Winick, ed. *Synchrotron Radiation Sources: A Primer. In: Series on Synchrotron Radiation Techniques and Applications*, **v. 1**. (World Scientific, 1995; Plenum, New York, 1980).
4. P. Rullhusen, X. Artru and P. Dhes, *Novel Radiation Sources Using Relativistic Electrons. In: Series on Synchrotron Radiation Techniques and Applications*, **v. 4**. (World Scientific, 1998).
5. Synchrotron Radiation. Proc. 11-th Nat. SR Conf. (SR-96), Novosibirsk, Russia, July, 9-12, 1996. Ed. G.N. Kulipanov *Nucl. lnstrum. and Meth. Phys. Res.* **A 405** (1998) 191.
6. V.L. Ginzburg, S.I. Syrovatskii, *The Origin of Cosmic Rays* (Oxford, Pergamon Press, 1964).
7. V.L. Ginzburg, *Teoreticheskaya Fizika i Astrofizika* (Moskva, Nauka, 1981).
8. O.F. Kulikov *Trudy Fiz. Inst. Akad. Nauk (FIAN)* **80** (1965) 3.
9. DESY Polarized Beams at PETRA, *CERN Courier* N20 (1980) 5.
10. A.A. Sokolov and I.M. Ternov, eds. *Sinkhrotronnoe Izluchenie* (Nauka, Moscow, 1966).
11. A.A. Sokolov, I.M. Ternov, *Synchrotron Radiation* (Akademie-Verlag, Berlin; Pergamon Press, New York, 1968); *Radiation from Relativistic Electrons* (American Inst. of Physics, New York, 1986).
12. I.M. Ternov, *Physics - Uspekhi* **38** (1995) 409.
13. V.A. Bordovitsyn, I.M. Ternov, V.G. Bagrov, *Physics – Uspekhi* **38** (1995) 1039
14. G.N. Kulipanov, A.E. Bondar, V.A. Bordovitsyn, V.A. Bordovitsyn and V.S. Gushchina, see in Ref.[5].
15. V.N. Korchuganov, G.N. Kulipanov, N.A. Mezentsev, et al., Preprint INP 77-83 (Novosibirsk, Institute of Nuclear Physics, 1977).

Life is real — life is earnest —
And the grave is not its goal:
Dust thou art, to dust returnest,
Was not spoken of the soul.

H.W. Longfellow

I. M. Ternov

Introduction[a]

In 1947, for the first time in history, Floyd Haber, a young staff member in the laboratory of Professor Pollock, observed radiation emitted by electrons as they moved circularly in the magnetic field of the chamber of an accelerator. This occurred during the adjustment of a cyclic accelerator — a synchrotron which accelerated electrons up to 70 MeV [2] (see also Ref.[2]). The radiation was observed as a bright luminous patch on the background of the chamber of the synchrotron. It was clearly visible in the daylight. In this way, 'electronic light' was experimentally seen for the first time as radiation emitted by relativistic electrons having a large centripetal acceleration. The radiation was called synchrotron radiation (SR)[b]since it was observed for the first time in a synchrotron.

It is hard to overestimate the importance of SR in our days, and the interest is growing incessantly since the radiation features a rare combination of fundamental properties and offers important scientific and technical applications.

It was sheer accident that the SR was observed: the opaque metallized cover of the chamber was removed to perform an adjustment and this allowed the light to be seen outside the chamber.

The discovery and first observations of synchrotron radiation were unexpected; its properties seemed mysterious and unusual at the initial stage of investigations. However, a number of theoretical studies on the emission of a relativistic accelerating electron had been carried out long before the experiment described above.

The first steps in this direction were taken by A. Liénard (1898) and O. Heaviside (1902)[4]. They extended the familiar Larmor formula for the plane power of a nonrelativistic electron,

$$W = -\frac{\partial E}{\partial t} = \frac{2e^2\dot{\boldsymbol{v}}^2}{3c^3}, \tag{0.1}$$

to a high-velocity particle. In modern notation it takes the form

$$W = \frac{2}{3}\frac{e^2}{m^2c^3}\left(\frac{dp_\mu}{d\tau}\frac{dp_\mu}{d\tau}\right) = \frac{2}{3}\frac{e^2\gamma^6}{c}\left[\dot{\boldsymbol{\beta}}^2 - [\boldsymbol{\beta}\dot{\boldsymbol{\beta}}]^2\right]$$

[a]From the article of I.M. Ternov, *Synchrotron Radiation* published in *Physics - Uspekhi* **38** (1995) 409.

[b]There is another name in the literature — the magnetic bremsstrahlung. This term is common in astrophysical problems (see Ref.[3])

(p_μ is the four-dimensional impulse, $d\tau = dt/\gamma$ is the intrinsic time, $\gamma = E/mc^2$, and $\boldsymbol{\beta} = \boldsymbol{v}/c$). Liénard turned his attention to the fast growth of losses in the energy of an electron describing a circle ($\boldsymbol{\beta} \perp \dot{\boldsymbol{\beta}}$) of radius R

$$W = \frac{2}{3}\frac{e^2c}{R^2}\beta^4\gamma^4. \tag{0.2}$$

The growth was proportional to the fourth power of the energy.

Subsequently G.A. Schott (1907) made an interesting detailed study of the radiation an electron emits as it follows a circular path[5]. Schott's objective was to explain the discrete nature of atomic spectra. Based on early models of the atom and especially on the Saturnian model in which electrons in an atom move about the positive charge in circles similar to the rings of Saturn, G.A. Schott made an attempt to calculate the spectrum and the distribution of spatial radiation of electrons in an atom by the strict methods of classical electrodynamics. He reckoned the spectral theory to be the most important issue of the theory of matter since he believed it to be the way to the working model of the atom.

The consistency and elegance of the Schott theory are admirable. However, his attempts to explain the atomic radiation within the scope of classical physics failed and for this reason Schott's work had been only of academic interest for 40 years and was actually forgotten. Their relevance was discovered in new circumstances 40 years later when the issue of an emitting charge moving in a macroscopic trajectory arose.

Of primary interest was the emission of accelerating electrons in a magnetic field. In 1939 I.Ya. Pomeranchuk established the radiative 'ceiling' for the energy of electrons in his attempt to determine the maximal energy the cosmic charged particles could possess at the Earth's surface due to radiative losses in the Earth's magnetic field[6]. By means of this estimation, the maximal energy was then predicted for a betatron — an induction accelerator in which electrons move in a magnetic field which builds up in time and is virtually homogeneous along the trajectory of the particle (D.D. lvanenko and I.Ya. Pomeranchuk)[7]. The existence of radiative losses in the energy of an electron in the magnetic field of an accelerator was soon verified in experiments conducted by J.P. Blewett (1946, Ref.[8]). He found that electrons moved in decreasing orbits as their energies increased: the particles moved in a converging spiral and ceased to accelerate because of a loss in energy to radiation (note that the energy, radius of the orbit, and magnetic field strength are related by the equation $\beta E = eHR$, see Ref.[22]).

Blewett's experiments could be considered to be a proof of the actual existence of radiation from relativistic charges and this radiation could even be called betatron radiation. However, attempts to visually observe this radiation were not a success: the search for radiation in the microwave range (dipole radiation) was a total failure. This exceptional situation when the energy losses of electrons were clearly observed while the radiation itself was elusive showed that large radiative losses alone did not explain the fundamental features of this extraordinary phenomenon.

Having studied theoretically the spectral distribution of the radiation power emitted by a circularly moving relativistic electron, L.A. Artsimovich and I.Ya. Pomeranchuk (1945, Ref.[9]; see also Ref.[10]) found out that the maximal power fell not on the fundamental frequency (as would be the case for dipole radiation), but on its higher harmonics: $\omega \sim \omega_0\gamma^3$. For electrons of an energy of $80 - 100\ MeV$, the radiation ought to be observed not in the microwave range but in the radiation range of higher multifields, i.e., in the visible range. This was revealed in an experiment on the synchrotron in the USA[2,1].

It was shown in Ref.[9] that the angular distribution of the power of synchrotron radiation is highly anisotropic, i.e. it is concentrated in a slender cone of an angle $\delta\psi \sim 1/\gamma$ in the orbital plane of revolution of the electron and is directed forward in parallel to its motion. The theoretical study of the coherence of radiation [9] (this is of special interest for the radiation of a cluster of electrons in a betatron, when they fill almost the whole orbit) showed that the coherence could manifest itself at the lowest frequencies only because of fluctuations of the current density in a beam for $\gamma \gg 1$ — far from the maximum of the spectral distribution of power.

Thus, the qualitative description of the properties of synchrotron radiation were known before it was observed for the first time. However, as it is noted above, it was discovered by sheer accident.

The discovery of electronic light in the synchrotron stimulated further investigations of the SR properties and, first of all, analysis of the spectral and angular distribution of the radiation power. This was a complicated problem since Schott's formulas [5] were inconvenient for describing the radiation spectrum of a relativistic electron when it involved higher harmonics of the frequency of the circular revolution of the electron. The conventional approach of expanding the series in terms of multifields was not applicable to the analysis of the radiation. Then the asymptotic problem in the radiation spectrum of a relativistic electron arose for a large relativistic factor of $\gamma \gg 1$.

V.V. Vladimirskii [11] successfully applied the Airy functions, which had been studied thoroughly by V.A. Fock, to describe the radiation spectrum of an electron moving in a magnetic field. The asymptotical formulas for the spectral composition of synchrotron radiation were independently determined in several theoretical studies [11–15]. They opened a possibility of experimental verification. The experiments demonstrated good agreement with the theory in the visible range[16], the vacuum ultraviolet range[17], and the X-ray range[18].

The classical theory of SR was then contributed to by investigations of the polarization features of SR[19]. It was established, for example, that SR is elliptically polarized in general and it is linearly polarized when observed in the direction close to the orbital plane of revolution. The first observations of linear polarization were made in the initial studies [1] but the polarization features of SR were investigated in detail by the staff of the Physical Department of the MSU (Moscow State University) on the synchrotron in the FIAN (Physical Institute of the Academy of Sciences) and showed definite agreement with the theory [20] (see also Ref.[21]).

Numerous investigations shaped the classical theory of synchrotron radiation to perfection, and the theory was included in a variety of monographs [3,22–25] and courses[14,26]. Synchrotron radiation became important for astrophysics, in the analysis of nonthermal cosmic radiation. The Swedish scientists H. Alfven and M. Herlofson suggested in 1950 [27] that the nonthermal radiation of our galaxy could be explained through the mechanism of SR. In Russia this problem was addressed at the same time by V.L. Ginzburg[28] and I.S. Shklovskii[29]. The recognition of the importance of synchrotron radiation, the development of its theory, and experiments stipulated the outstanding advances in radioastronomy.

During the last few years the problem has acquired a new and important feature — synchrotron radiation is used extensively in scientific research. In parallel with our widening knowledge of the nature of the phenomenon, the objective of accelerators and storage units of electrons has recently changed — they have become a major source of synchrotron radiation which has taken up an independent position in experimental physics.

The physics of undulator radiation, i.e. the radiation of relativistic electrons as they move in a periodic outer field, is of utmost importance in experimental applications of SR. Undulator radiation, which V.L. Ginzburg predicted in 1947 [30], has the same origin as SR and is similar to it in many aspects. It has attracted considerable attention lately.

Although the theory of SR seemed to be complete, it turned out that the electronic light possessed a variety of fine and interesting properties which the classical theory of an accelerating charge did not describe: the physical nature of SR turned out to be richer and the quantum theory had to be applied for its comprehensive description[22,24,25].

* * *

It proved worthwhile to develop the quantum theory of synchrotron radiation on the basis of quantum relativistic mechanics and quantum electrodynamics, applying the so-called 'method of exact solutions'[24,25]. In this case the wave function which describes the quantum state of an electron obeys the Dirac equation:

$$i\hbar\frac{\partial\Psi}{\partial t} = \left[c(\boldsymbol{\alpha}\hat{\boldsymbol{P}}) + \rho_3 mc^2\right]\Psi,$$

where

$$\hat{\boldsymbol{P}} = -i\hbar\nabla - \frac{e}{c}(\boldsymbol{A}^{\rm ext} + \boldsymbol{A}^{\rm qu}),$$

refers to the outer magnetic field which is considered exactly, and $\boldsymbol{A}^{\rm qu}$ to the quantum transverse radiation field. Here $\boldsymbol{\alpha}$ and ρ_3 are the Dirac matrices of four rows, and the wave function Ψ has four components (for details see Ref.[24]). Processes in which an electron in the bound state is involved when it interacts with the radiation field are considered by using the perturbation theory. In the perturbation theory, all expansions are made in terms of the full system of the exact solutions to the Dirac equation with an outer field (the Furry representation[31]).

Such an approach made it possible to predict and reveal a variety of new physical features of SR: quantum fluctuations of the trajectory of an electron when it moves in a cyclic accelerator and storage rings, the radiative polarization effect for electrons and positrons, the peculiar features of synchrotron radiation in strong and very strong magnetic fields, and a number of others (see Refs [24,25,32]).

Omitting the details of solving the Dirac equation (it is described in detail in Refs [24,25]) I want only to emphasize that the quantum state of an electron in a magnetic field $\boldsymbol{H} = (0,\ 0,\ H)$ is specified by the set of four quantum numbers: n is the energy number, s is the radial number, k_3 is the projection of the impulse onto the direction of the field , and ζ is the projection of the spin onto the direction of the field. In this case, the energy of an electron takes the form:

$$E = (m^2c^2 + c^2\hbar^2k_3^2 + 2e_0Hc\hbar n)^{1/2}. \tag{0.3}$$

If an electron executes a macroscopic motion, then the energy number s takes a very large value and the energy spectrum is continuous. In the nonrelativistic approximation, Eq. (0.3) describes the familiar Landau levels and also involves the kinetic energy of the motion of an electron along the field. Note that the energy spectrum given by Eq. (0.3) is degenerate with respect to the spin and to the radial number s. Taking into consideration the anomalous magnetic moment eliminates the spin degeneration and the energy spectrum takes the form (see Ref.[33])

$$E = mc^2\left(\left(\frac{\hbar k_3}{mc}\right)^2 + \left\{\left[1+\frac{H}{H_0}(2n+\zeta+1)\right]^{1/2} + \zeta\frac{H}{H_0}\frac{\mu-\mu_0}{2\mu_0}\right\}^2\right)^{1/2},$$

where

$$\mu = -\mu_0\left(1+\frac{e^2}{2\pi\hbar c}\right), \quad \mu_0 = \frac{e_0\hbar}{2mc}, \quad H_0 = \frac{m^2c^3}{e_0\hbar}.$$

Here H_0 is the Schwinger magnetic field. The ground state (n=0, k_3=0) [see Eq. (0.3)] corresponds to the spin oriented in the opposite direction of the magnetic field ($\zeta = -1$). The quantum numbers n and s are related to the radius of the orbit of an electron and the quadratic fluctuation of the radius:

$$\langle r^2\rangle = \frac{n+s+1/2}{\gamma_0}, \quad \langle r^2\rangle - \langle r\rangle^2 = \frac{s+1/2}{2\gamma_0}, \quad \gamma_0 = \frac{e_0H}{2c\hbar}.$$

The electrons are localized near the orbital plane of revolution for an extremely strong magnetic field since

$$R = \sqrt{\frac{n}{\gamma_0}} = \frac{\hbar}{mc}\sqrt{\frac{nH_0}{H}},$$

and the radius of the orbit is of the order of the Compton wavelength in weakly excited states as $H \to H_0$.

Restricting myself to the above remarks I move on to discuss synchrotron radiation. By the use of rather conventional methods [24,34], the expression for the synchrotron radiation power can be obtained in the form

$$W = \frac{e^2 c}{2\pi} \sum_{\nu, s', \zeta'} \int d^3k \, \delta(k - k_{nn'}) \Phi I^2_{ss'}(x), \tag{0.4}$$

where

$$ck_{nn'} = \frac{E_n - E_{n'}}{\hbar}, \quad x = \frac{k^2 \sin^2\theta}{4\gamma_0}, \quad \nu = n - n'.$$

It is taken into consideration in summing over k'_3 that the component of the impulse parallel to the field is preserved. The radial factor $I^2_{ss'}(x)$ appearing in the formula for the radiation power is a Laguerre function, which is related to the Laguerre polinomials $Q^{s-s'}_{s'}(x)$ by the equation

$$I_{ss'} = \frac{1}{\sqrt{s!s'!}} \exp\left(-\frac{x}{2}\right) x^{(s-s')/2} Q^{s-s'}_{s'}(x). \tag{0.5}$$

The function Φ depends on the elements of the Dirac matrix and is expressed in terms of the Laguerre functions $I_{nn'}$, which are approximated by McDonald functions $K_{1/3}$ by analogy with the classical theory of SR (see Refs[24,25]). Thus, the integral under the summation sign in Eq. (0.4) is the power of radiation electron emit in transitions $n \to n'$, $s \to s'$ with the spin flip $\zeta \to \zeta'$.

Making necessary manipulations, integrating with respect to angles and summing over the polarization states of the radiation field, we find the following expression for the spectral distribution of the synchrotron radiation power[24]

$$W = W^{cl} \frac{9\sqrt{3}}{16\pi} \sum_{s'} \int_0^\infty \frac{y dy}{(1+\xi y)^4} \, I^2_{ss'}(x) F(y), \tag{0.6}$$

where

$$\begin{aligned} F &= \frac{1+\zeta\zeta'}{2} \left[2(1+\xi y) \int_y^\infty K_{5/3}(x) dx + \frac{1}{2} \xi^2 y^2 K_{2/3}(y) \right. \\ &- \left. \zeta(2+\xi y) \xi y K_{1/3}(y) \right] + \frac{1-\zeta\zeta'}{2} \xi^2 y^2 \left[K_{2/3}(y) + \zeta K_{1/3}(y) \right]. \end{aligned}$$

Here the argument x of the function $I^2_{ss'}$ takes the form $\nu = n - n'$ upon changing the summation over y for integration with respect to the variable

$$x = \frac{\xi_1 y^2}{(1+\xi y)^2}.$$

It follows from Eq. (0.6) that the probability of spontaneous transitions of an electron is a function of two parameters[c]

$$\xi = \frac{3}{2}\frac{H}{H_0}\frac{E}{mc^2} = \frac{3}{2}\chi = \frac{3}{2}\left(\frac{E}{E_{1/2}}\right)^2, \quad E_{1/2} = mc^2\left(\frac{mcR}{\hbar}\right)^{1/2},$$
$$\xi_1 = \frac{9}{8}\frac{H}{H_0}\left(\frac{E}{mc^2}\right)^4 = \frac{9}{8}\left(\frac{E}{E_{1/5}}\right)^2, \quad E_{1/5} = mc^2\left(\frac{mcR}{\hbar}\right)^{1/5}.$$

Now I want only to note that the SR power depends solely on one invariant parameter χ since the sum over the radial quantum number

$$\sum_{s'} I_{ss'}^2(x) = 1$$

is equal to unity. Thus, the limits in which the classical theory was predicted to be applicable[11] were strictly borne out by the quantum theory. Note that this refers, however, to the SR power only.

Expression Eq. (0.6) is exact: it allows any values of the parameter $\xi = 3\chi/2$, including $\chi \gg 1$, which are realizable in the physics of neutron stars where the magnetic field strength is close to the critical H_0. In addition, the formula for the radiation power involves the contribution of radiation which is accompanied by the flip of the spin (spin-flip transitions) when $\zeta' = -\zeta$. As follows from Eq. (0.6), the probabilities of such transitions are proportional to the square of the Plank constant $\hbar^2$.

Since the SR power depends explicitly on the orientation of the spin of an electron, the radiation accompanied by the flip of a spin affects the orientation of the spin and stimulates the directed process of polarization of the electron beam. In addition, there is a spin dependence in the formula for the SR power, which enters the terms which are independent of the flip of a spin. This is not only of theoretical interest. In fact, given a small invariant parameter ξ it follows from Eq. (0.6) that the spectral distribution of the SR power of a polarized electron beam takes the form

$$W^{\text{pol}} = W^{\text{cl}}\frac{9\sqrt{3}}{8\pi}\int_0^\infty \left[(1-3\xi y)\int_y^\infty K_{5/3}(x)dx - \zeta\xi y K_{1/3}(y)\right] y dy. \tag{0.7}$$

Note that V.A. Bordovitsyn[35] made a large advance in interpreting quantum corrections to the classical expression for the SR power. He showed, in particular, that the quantum correction in Eq. (0.7) involves contributions from the interference of the radiation from electron, charge, and the spin magnetic moment of electron.

Thus, this offers a new possibility for visual observation of the polarization characteristics of an electron beam by determining the SR power at a fixed spectral

[c]The probability of spontaneous transitions can be found by dividing the integrand in Eq. (0.6) by the energy of the quantum of the electromagnetic field $\hbar\omega$

frequency[36]. The experiment which was performed in the Institute of Nuclear Physics, Siberian Branch of USSR Academy of Science[36], can be considered as a first visual observation of radiation which is directly associated with electron spin.

Further, the expression for the SR power can be found by summing up over the polarization states of an electron

$$W = W^{\mathrm{cl}} \frac{9\sqrt{3}}{8\pi} \int_0^\infty \frac{y dy}{(1+\xi y)^3} \left[\int_y^\infty K_{5/3}(x) dx + \frac{\xi^2 y^2}{1+\xi y} K_{2/3}(y) \right]. \tag{0.8}$$

The SR power which is uniformly applicable for any values of the parameter ξ was calculated exactly by V.G. Bagrov[37]. Here I cite two limiting cases in the form of asymptotic expansions. First, let us consider the case of small invariant parameter of $\xi \ll 1$, i.e., the quantum corrections to the classical radiation formula are

$$W = W^{\mathrm{cl}} \left[1 - \frac{55\sqrt{3}}{24} \xi + \frac{64}{3} \xi^2 + \dots \right]. \tag{0.9}$$

This correction was found up to a linear term in Ref.[38] and was corroborated by J. Schwinger for spinless particles[39].

In the second limiting case of $\xi \gg 1$ (high energies, extremely large magnetic fields), the formula for the SR power differs drastically from the classical one (the ultraquantum limit)[24,40]

$$W^{\mathrm{uq}} = W^{\mathrm{cl}} \left[\frac{2^{8/3}}{9} \Gamma \left(\frac{2}{3} \right) \right] \xi^{-4/3}, \quad \xi \gg 1. \tag{0.10}$$

In the ultraquantum limit the principal term of the radiation power is of quantum nature. Therefore transition to the classical approximation is impossible.

It is characteristic that in the ultrarelativistic limit the spectrum is terminated at the frequency $\omega_{max}^{\mathrm{uq}} = E/\hbar \ll \omega_{max}^{\mathrm{cl}}$, and does not reach the classical critical frequency

$$\omega_{max}^{\mathrm{cl}} = \frac{c}{R} \left(\frac{E}{mc^2} \right)^3 . \tag{0.11}$$

This follows immediately from the formula for the classical frequency of radiation (see Ref.[34]):

$$\omega_{max} = \frac{3}{2} \omega_0 \gamma^3 \frac{y_{max}}{1 + \xi\, y_{max}}, \quad y_{max} \sim 1$$

Of some interest is the radiation of weakly excited electrons (low-energy levels) in a strong magnetic field. It is peculiar for such a problem that the energy spectrum of an electron is discrete ('quantising' magnetic field). If electrons move perpendicularly to the field ($k_3 = 0$), the energy

$$E = mc^2 \left(1 + 2n \frac{H}{H_0} \right)^{1/2}$$

takes essentially discrete values for $n \sim 1, 2, \ldots$

In this case, only numerical methods are applicable. It turns out that the probability of spontaneous transitions depends no longer on the orientation of the spin of an electron and the probability of a transition with a change in the spin orientation is the same as that with no change of polarization.

The radiation powers of the σ and π components of the linear polarization take the forms $W_\sigma = 0.742\ W$ and $W_\pi = 0.258\ W$, where

$$W = 0.453 \left(\frac{H_0}{H}\right)^2 W^{cl}$$

(see Ref.[41]). Thus, the expression for the radiation power W differs from the classical formula by the invariant factor

$$f = \frac{1}{4}\frac{1}{H_0^2}\, H_{\mu\nu}H^{\mu\nu} = \left(\frac{H}{H_0}\right)^2 .$$

This result does not coincide with those obtained not only in the classical theory, but also in the ultraquantum case of an excited electron moving in a very strong field.

* * *

The quantum theory of SR [22,24] helps to explain the discrete nature of the radiation and its influence on the trajectory of the particle (the recoil effect). As the theory shows[42,43], this influence manifests itself in the quantum widening of the trajectory of an electron — the particle is involved in a peculiar Brownian movement and quantum fluctuations of the trajectory are macroscopic in nature. The latter fact proved to be important in the engineering of accelerator design and storage of electrons.

The quantum theory also made it possible to investigate the SR emitted by a polarized electron and the contribution of the spin of a particle to the radiation power. The analysis of the spin evolution during synchrotron radiation revealed the effect of the polarization of radiation of electrons and positrons in storage rings[44]. This effect is of special interest in connection with the problem of how to create a beam of relativistic particles with an oriented spin.

The effect is that the spins of particles are oriented in the same way under the influence of synchrotron radiation when they circulate in storage rings for a long time. This effect of radiative polarization was predicted by the author [45] and strictly established together with A.A. Sokolov with the use of the exact solutions to the Dirac equations [46] (see also Refs [34,47]).

The probability of quantum transitions accompanied by spin flips can be calculated with Eq. (0.6). The calculation shows that the probability still depends on the spin orientation and upon integration with respect to angles and to the spectrum:

$$w^{\uparrow\downarrow} = \frac{1}{2\tau}\left(1 + \zeta\,\frac{8\sqrt{3}}{15}\right), \tag{0.12}$$

where the polarization time τ has the form

$$\tau = \frac{8\sqrt{3}}{15}\,\frac{\hbar^2}{mce^2}\left(\frac{mc^2}{E}\right)^2\left(\frac{H_0}{H}\right)^3 . \qquad (0.13)$$

It follows that radiation induces electrons to transit predominantly in states with a spin oriented in the direction opposite to that of the magnetic field [45]. Positrons have the opposite spin orientation. States with a predominant spin orientation match the minimal potential energy of particles, the magnetic moment of which is $\boldsymbol{\mu} = -(e_0\hbar/2mc)\boldsymbol{\zeta}$ in the magnetic field $U = (-e/\mid e\mid)\boldsymbol{\mu H}$.

Omitting the details of the kinetics of the polarization process (see Refs [46,34] and turning our attention to an ensemble of particles, we can characterize the polarization of a beam of particles by the average value $\zeta(t) = \rangle\zeta\langle$, bearing in mind that the electron goes into the combined state as a result of interaction with the electromagnetic field. Then

$$\frac{d}{dt}\,\zeta(t) = \sum_{\zeta'}(\zeta' - \zeta)\,w^{\uparrow\downarrow} = -2\sum_{\zeta}\zeta w^{\uparrow\downarrow},$$

whence it follows that[d]

$$\zeta(t) = -\frac{8\sqrt{3}}{15}\left[1 - \exp\left(-\frac{t}{\tau}\right)\right] . \qquad (0.14)$$

The extreme degree of polarization is $P(\infty) = 8\sqrt{3}/15 = 0.924$ (for $t \gg \tau$).

The estimate for the polarization time shows that the effect of radiative polarization is accessible for observation in magnetic fields which are typical of accelerators only if the particles circulate in the magnetic field for a long time (about 1 hour). Storage rings provide a means for this possibility. Although there are effects depolarizing the beam in an actual storage ring, the effect of radiative polarization exists and provides a unique capability for creating polarization beams of high-energy electrons and positrons. The effect of radiative polarization was experimentally observed in the USSR, France, Germany, USA, Japan, and Switzerland in storage rings with electrons of energy 1–50 GeV (see Ref. [47]).

It should be noted that storage rings, in which there was a possibility of compensating for radiative energy losses, became a unique laboratory for studying quantum effects, since electrons could circulate for tens of hours under such conditions, with the average energy remaining constant. The quantum effects in synchrotrons were also verified experimentally. Thus, now both classical and quantum theories of SR are complete and reliable.

* * *

For a long time, synchrotron radiation was considered to be a nuisance in the operation of a cyclic accelerator. The reason was that it set a radiative 'ceiling' to the operation of a betatron[7]. Radiative losses of energy imposed a fundamental

[d]This result was later obtained by V.N. Baier and V.M. Katkov[48] and by J. Schwinger[49].

restriction on the inductive method of acceleration of electrons. Therefore, a new acceleration technique was adopted upon the discovery of automatic phase stabilization (Wecksler, McMillan), i.e. a synchrotron in which energy losses were compensated for. However, even early studies of the properties of SR [1,16–18] attracted the attention of experimentalists and soon aroused an interest in SR as a new source of radiation.

In the 1960s the first laboratories of synchrotron radiation appeared with the object of finding an application for SR in the physical experiment. This special attention to the new source of radiation was due to its peculiar properties: a wide spectral range of electromagnetic waves from the infrared radiation to the X-ray radiation; sharp collimination because of which the brightness of radiation was very high; high power; and natural polarization typical of this source. Of great importance was the fact that all the properties of SR were fully described theoretically. This made it possible to calculate its characteristics with a high degree of precision.

The 1980s were marked by a vigorous growth of a number of investigations in which SR was used, and by advances in new types of specialized sources of radiation. During those years there appeared scientific research centers of synchrotron radiation. They were equipped with sources of SR, free electron lasers (FELs), wigglers, systems of undulators, and auxiliary facilities built in the chamber of a storage ring. In Russia, centers of synchrotron radiation were established in the Institute of Nuclear Physics in Novosibirsk, in Lebedev Physical Institute (FIAN), and in Kurchatov Institute of Atomic Energy. Today scientists in many countries of the world such as the USA, FRG, Italy, Japan, the United Kingdom, France, Switzerland and others are engaged in research using synchrotron radiation.

The successful application of SR in the physical experiment has had a pronounced effect on the progress of physics of atoms and molecules and also on solid state physics. It is impossible to cover all these issues in our review in any detail since they constitute a vast problem, which is in itself of interest. Fortunately, there is no need for it since the issues related to applications of SR in the physical experiment are considered in detail in special literature. So, for example, the review of Gurdling was devoted in particular to the application of SR in atomic spectroscopy (see Ref.[50]). E.E. Koch and B. Sonntag reviewed applications of SR in molecular spectroscopy, D. Ling reviewed the spectroscopy of solid bodies (see Ref.[50]). Applications of SR in studies of the luminescence of crystals were considered in the monograph [25] and, finally, the review of R. Haensel is devoted to applications of SR in studies of optical properties of alkali halide compounds (see Ref.[51] and also Refs [52–55]).

Note that SR is virtually a unique source of high-intensity radiation in the range of $200 - 500\ A$. In the short-wavelength range of the vacuum ultraviolet radiation and in the soft X-ray range, the power of the radiation emitted by electrons of the energy of several GeV exceeds the power of radiation available from X-ray tubes by several orders of magnitude[25].

Of special importance is the application of SR in experiments in the soft X-ray range of radiation, in which its power exceeds several times that of all other sources of X-ray radiation. It should be added that SR has an advantage over other sources since it allows for continuous adjustment of the wavelength of radiation, especially for application of long-wavelength X-ray radiation.

This peculiar feature of SR opened a possibility for its application in biology, in the study of structures of biopolymers. The reduction of the exposure time, and the prevention of the object of investigation from being destroyed because of a much smaller radiative load make SR indispensable in studies of biological structures (see Ref.[25]).

The recent years have witnessed successful application of SR in medicine, particularly in angiography by X-ray techniques. There is a possibility of obtaining more information when a smaller radiative load is applied to a patient. In 1986 H. Winick conducted angiographic inspection of a man the Stanford Laboratory of SR (earlier such inspections were conducted on animals only) (see Refs [25,56,57]).

SR has been applied in microlithograpy for obtaining elements of microschemes used in modern semiconductor devices. The unique properties of SR, e.g. sharp directivity and large power in the X-ray range, make it possible to improve the quality of elements of microschemes and obtain new elements in microelectronics (see Refs [25,58]).

I will now draw the reader's attention to new possibilities of experiments centered around the direct visual observation of 'electronic light' [25,59], when one observes an electron beam passing through an accelerating cycle or moving in a storage ring. As noted above, an outstanding success was the experimental examination of the dynamics of betatron oscillations of electrons in the presence of forces of radiative damping and quantum fluctuations[60,61].

The basis for visual observation of the dynamics of an electron beam is the high-speed photography of synchrotron radiation emitted by an electron beam followed by the processing of photographs. High-speed photography of an electron beam for the purposes of analysis of its dynamics was first performed by Pollock's group on the synchrotron 'General Electric-70 MeV' [1], and then this technique was developed by Ado [62] and also in a series of studies performed under the supervision of F.A. Korolev on the synchrotrons 'FIAN-280 MeV' and 'FIAN-680 MeV' [63].

The analysis of photographs obtained in the experiments mentioned above restored the full history of evolution of betatron oscillations under the action of forces of radiative damping and quantum fluctuations[60,63]. I would like to emphasize that the source of information on the motion of an electron was the particle itself through emitted electromagnetic waves — 'luminous electron' (for details, see Ref.[25]).

In conclusion, I shall describe briefly ways in which SR sources can be improved. One of the crucial problems is how the brightness of a source can be increased. The term brightness is understood to refer to the number of photons which are emitted in one second from a unit area of an extended source into a unit solid angle. One way of increasing the brightness is to create a storage ring of small emittance which is a characteristic of a beam of particles, and is given by $\varepsilon = \pi\sigma\theta$, where σ is the Gaussian size of the beam in meters, and θ is the angle of the cone of radiation in radians. In advanced modern sources, small emittance is achieved through strong focusing of a beam of particles, combined with systems of permanent magnets of multiperiodic undulators (note that the angular size of the cone of radiation is the quantity $\delta\theta \sim 1/(\gamma\sqrt{N})$ for an undulator made up of N sections of magnets).

Further, it is clear that the brightness of a source depends primarily on the radiation power. Of interest in this context are coherent bunches of electrons clustered at distances less than the wavelength of their radiation. As noted in the early work [10], in this case, coherence would make it possible to increase the radiation drastically since the bunch of electrons behaves like an effective charge $e_{\text{eff}} = N_e e$ (N_e is the number of electrons in the bunch).

The creation of such coherent bunches is a very complicated problem, even in the microwave range. So far as the possibility of clustering electrons in bunches of the size of the order of the optical wavelength was concerned, the difficulties seemed to be insurmountable.

At present it is established that in the theory of free electron lasers, which considers the interaction of an electron beam in an undulator with an electromagnetic wave, there is a mechanism of self-modulation of an electron beam. Electrons are clustered in the longitudinal direction and form coherent bunches of length of the order of the optical wavelength.

The mechanism of self-modulation of an electron beam is similar to some extent to the clustering of electrons in a synchrotron under the combined action of the leading magnetic field of the accelerator and the vortical high-frequency field accelerating the particles (the self-modulation principle of Wecksler and McMillan). As a result of their action, the electron beam is divided into bunches the length of which is a function of parameters of the high-frequency electric field.

It is interesting that particles are clustered in the free electron laser even if an outer electromagnetic wave is absent, with the 'trigger' wave of spontaneous radiation playing its role. The self-amplification of spontaneous radiation is one of the important properties of the undulator. When electrons pass through a large-length undulator, there is no interference initially and the total radiation power is proportional to the number of particles $W^{\text{total}} = N_e W$. Then the clustering mechanism comes into play, and the radiation of a cluster of electrons becomes coherent. The spontaneous radiation amplifies itself and the radiation power is now proportional to N_e^2 because of the interference: $W^{\text{total}} = N_e^2 W$. This self-amplification phenomenon provides the grounds for a special 'strong' source of radiation — a large-length undulator.

The coherence problems of synchrotron radiation have recently attracted the attention not only of theoreticians but also of experimentalists[64–67].

I would like to say that synchrotron radiation has not only won recognition in physical experiments of the present time but will still be an experimental tool in the future.

In *The Thousand and One Nights*, there is a story about a boy named Aladdin and a magic lamp. The boy found a magic lamp and, when he rubbed it slightly with a pinch of sand, there appeared an enormous genie and said, "I am at your disposal, I am your slave". The electronic light burst out of the chamber of an accelerator in

1947 and since then synchrotron radiation, like the genie in Aladdin's magic lamp, has shown the way to knowledge in various fields of science.

Bibliography

1. F.R. Elder, A.M. Gurewitsch, R.V. Langmuir, H.C. Pollock, *Phys. Rev.* **71** (1947) 829.
2. *Am. J. Phys.* **51** (1983) 278;
3. V.L. Ginzburg, *Theoretical Physics and Astrophysics* (Pergamon, Oxford, 1969).
4. A. Liénard, *L'Eclairage Electr.* **16** (27) (1898) 5; O. Heaviside, *Nature* **67** (1723) (1902) 6.
5. G.A. Schott, *Ann. Phys.* **24** (1907) 635; *Electromagnetic Radiation* (Cambridge, 1912).
6. I.Ya. Pomeranchuk, *Zh. Eksp. Tear. Fiz.* **9** (1939) 915.
7. D.D. lvanenko, I.Ya. Pomeranchuk, *Dokl. Akad. Nauk SSSR* **44** (1944) 343.
8. J.P. Blewett, *Phys. Rev.* **69** (1946) 87.
9. L.A. Artsimovich, I.Ya. Pomeranchuk, *Zh. Eksp. Teor. Fiz.* **16** (1946) 370.
10. V.L. Ginzburg, *Izv. Akad. Nauk SSSR* **11** (1947) 165; V.L. Ginzburg, S.I. Syrovatskii, *The Origin of Cosmic Rays* (Oxford, Pergamon Press, 1964).
11. V.V. Vladimirskii, *Zh. Eksp. Teor. Fiz.* **18** (1948) 392.
12. A.A. Sokolov, *Vestn. Mosk. Univ.* 4 (1947) 77.
13. D.D. Ivanenko, A.A. Sokolov. *Dokl. Akad. Nauk SSSR* **59** (1948) 1551.
14. L.D. Landau, E.M. Lifshitz, *The Classical Theory of Fields* (Oxford, Pergamon Press, 1980).
15. J. Schwinger, *Phys. Rev.* **75** (1949) 1912.
16. Yu.M. Ado, P.A. Cherenkov, *Dokl. Akad. Nauk SSSR* **110** (1956) 35 [*Sov. Phys. Dokl.* **1** (1957) 517].
17. D. Tombulian, P. Hartman, *Phys. Rev.* **102** (1956) 423.
18. G. Bathov, E. Freitag, R. Haensel, *J Appl. Phys.* **37** (1966) 3449.
19. A.A. Sokolov, I.M. Ternov, *Zh. Eksp. Teor. Fiz.* **31** (1956) 373 [*Sov. Phys. JETP* **4** (1957) 396].
20. F.A. Korolev, E.N. Akimov, V.S. Markov, O.F. Kulikov, *Dokl. Akad. Nauk SSSR* **110** (1956) 542 [*Sov. Phys. Dokl.* **1** (1957) 568].
21. P. Joos, *Phys. Rev. Lett.* **4** (1960) 558.
22. A.A. Sokolov, I.M. Ternov, *Synchrotron Radiaton* (Berlin, Akademie-Verlag; New York, Pergamon Press, 1968).
23. D.D. Ivanenko, A.A. Sokolov, *Klassicheskaya Teoriya Polya* (Classical Field Theory) (Moscow-Leningrad, Gostekhizdat, 1951).
24. A.A. Sokolov, I.M. Ternov, *Synchrotron Radiation* (Akademie-Verlag, Berlin; Pergamon Press, New York, 1968); *Radiation from Relativistic Electrons* (American Inst. of Physics, New York, 1986).
25. I.M. Ternov, V.V. Mikhailin and V.R. Khalilov, *Synchrotron Radiation and its*

Application (Harwood Acad. Publ., Chur, Switzerland, 1986).
26. A.A. Sokolov, I.M. Ternov, V.Ch. Zhukovskii, A.V. Borisov, *Kvantovaya Elektrodinamika* (Quantum Electrodynamics) (Moscow, Moscow State University, 1983).
27. H. Alfvén, M. Herlofson, *Phys. Rev.* **78** (1950) 616.
28. V.L. Ginzburg, *Dokl. Akad. Nauk. SSSR* **76** (1951) 377.
29. I.S. Shklovskii, *Dokl. Akad. Nauk. SSSR* **90** (1953) 6.
30. V.L. Ginzburg, *Izv. Akad. Nauk SSSR* **11** (1947) 165;
31. W.H. Furry, *Phys. Rev.* **81** (1951) 115.
32. I.M. Ternov, O.F. Dorofeev, *Fiz. Elem. Chastits At. Yadra* **25** 1 (1994) 5.
33. I.M. Ternov, V.G. Bagrov, V.Ch. Zhukovskii, *Vestn. Mosk. Univ., Ser. Fiz. Astr.* 1 (1966) 30.
34. A.A. Sokolov, I.M. Ternov, V.G. Bagrov, R.A. Rzaev, in Synchrotron Radiation (Moscow, Nauka, 1966) p. 72 [in Russian].
35. V.A. Bordovitsyn, Dissertation of Doctor of Phys. and Math. Sci. (Moscow, Tomsk, 1983); I.M. Ternov, V.A. Bordovitsyn, *Vestn. Mosk. Univ., Ser. Fiz. Astr.* **24** 5 (1983) 69; **28** 2 (1987) 21.
36. V.N. Korchuganov et al., Preprint Inst. Yad. Fiz. Sib. Otd. Akad. Nauk SSSR No 77-83 (Novosibirsk, 1977); S.A. Belomestnykh et al. *Nucl. lnstr. Meth.* **227** (1984) 173.
37. V.G. Bagrov, *Izv. Vuz. Fiz.* 5 (1965) 121.
38. A.A. Sokolov, N.P. Klepikov, I.M. Ternov, *Zh. Eksp. Teor. Fiz.* **24** (1953) 249.
39. J. Schwinger, *Proc. Nat. Acad. Sci.* **40** (1954) 132.
40. N.P. Klepikov, *Zh. Eksp. Teor. Fiz.* **26** (1954) 19.
41. I.M. Ternov, V.G. Bagrov, O.F. Dorofeev, *Izv. Vyssh. Uchebn. Zaved., Fiz.* N10 (1968) 97
42. A.A. Sokolov, I.M. Ternov, *Zh. Eksp. Tear. Fiz.* **25** (1953) 698.
43. A.A. Sokolov, I.M. Ternov, *Dokl. Akad. Nauk SSSR* **92** (1953) 537.
44. A.A. Sokolov, I.M. Ternov, in *Tr. Mezhd. Konf. po Uskoritelyam Visokoi Energii* (Proc. International Confer. on High-Energy Accelerators) (Moscow, Gosatomizelat, 1964) p. 921.
45. I.M. Ternov. Dissertation of Doctor of Phys. and Math. Sci. (Moscow, 1961).
46. I.M. Ternov, *Fiz. Elem. Chastits At. Yadra* **17** (1986) 884 [*Sov. J. Part. Nucl.* **17** (1986) 389]; A.A. Sokolov, I.M. Ternov, *Dokl Akad. Nauk SSSR* **153** (1963) 1052 [*Sov. Phys. Dokl* **8** (1964) 90].
47. J. Le Duff et al., in *Proc. of All-Union Symposium on Accelerators of Charged Particles* (Moscow, 2-4 October 1962) (Moscow, Nauka, 1973) p. 371; U. Camerini et. al., *Phys. Rev.* **D12** (1975) 1855; R.F. Switters et. al., *Phys. Rev. Lett.* **35** (1975) 1320; *CERN Courier* 20 (1980); A. Blondel, in *Proc. 9th Int. Symp. (Bonn, FRG, September 1990)* Vol. I, p. 128.
48. V.N. Baier, V.M. Katkov, *Zh. Eksp. Teor. Fiz.* **53** (1967) 1478 [*Sov. Phys. JETP* **26** (1968) 854].
49. J. Schwinger, W. Tsai, *Phys. Rev* **D9** (1974) 1843
50. C. Gurdling, E. Koch, B. Sonntag, D. Ling in *Synchrotron Radiation. Techniques*

and Application. Ed. by C. Kunz, with contribution by K. Colding, W. Gudat, E.E. Koch et al. (Springer, Berlin, 1979).
51. R. Haensel in *DESY – Bericht* F41 – 69/2.
52. G.N. Kulipanov, A.N. Skrinskii, *Usp. Fiz. Nauk* **122** (1977) 369 [*Sov. Phys. Usp.* **20** (1977) 559]; G.N. Kulipanov. Dissertation of Doctor of Phys. and Math. Sci (Novosibirsk, 1994).
53. P. Barnes, *Phys. Chem. Sol.* **52** (1991) 1299.
54. M. Krause, *Bull. Am. Phys. Soc.* **38** (1993) 911.
55. J.B. Hastings et al. *Phys. Rev. Lett.* **66** (1991) 770.
56. *Synchrotron Radiation Research* (Eds H. Winick, S. Doniach) (New York, Plenum Press, 1980).
57. *Handbook of Synchrotron Radiation* ed. E.E. Koch (Amsterdam, North-Holland, 1983).
58. A.P. Alekhin et al. *Elektron. Promyshl.* 3 (1992) 19.
59. A.A. Varfolomeev, *Lazery na Svobodnykh Elektronakh i Perspektivy lkh Razvitiya* (Free-Electron Lasers and Prospects for Their Development) (Moscow, Inst. Atomn. Energ. im. I.V. Kurchatova, 1980).
60. F.A. Korolev, et al. *Opt. Spektrosk.* **24** (1968) 316.
61. O.F. Kulikov, *Tr. Fiz. Inst. Akad. Nauk* **80** (1975) 3.
62. Yu.M. Ado, *Zh. Eksp. Teor. Fiz.* **31** (1956) 533 [*Sov. Phys. JETP* **4** (1957) 437].
63. F.A. Korolev, A.G. Ershov, O.F. Kulikov in *Uskoritel' Elektronov na 680 MEV* (680 MeV Electron Accelerator) (Moscow, Gosatomizdat, 1962) p. 75.
64. N.P. Klepikov, I.M. Ternov, *Izv. Vyssh. Uchehn. Zaved., Ser. Fiz.* **33** 3 (1990) 9.
65. K. Ichi et al. *Phys. Rev.* **A43** (1991) 5597.
66. Y. Schibata et al. *Phys. Rev.* **A44** (1991) 3449.
67. K.J. Kim, A. Sessler, *Science* **250** (4977) (1990) 88.

Chapter 1

Classical Theory of Synchrotron Radiation

V.A. Bordovitsyn, V.S. Gushchina
Physics Faculty, Tomsk State University
634050 Tomsk, GSP-14, Russia[a]

1.1. Electromagnetic Fields of Relativistic Charge

The field of a point charge on arbitrary motion is calculated on the basis of Liénard-Wiechert potentials. Usually these potentials are obtained by either simple covariant generalization of Coulomb potential[1] or consecutive integration of retarded potentials[1−4]. Derivation of Liénard-Wiechert potentials by means of Lorentz transformations[5] is less known. Let us consider the latter method in more detail.

For description of the electromagnetic field created by a moving charge we introduce a system of coordinates shown in Fig. 1.1. According to this figure, the charge position is determined by the radius-vector $\boldsymbol{r}(t)$ which is related to the radius-vector of the observation point $\boldsymbol{R}$ by the formula

$$\boldsymbol{r}(t) = \boldsymbol{R} - \tilde{\boldsymbol{r}}(t),$$

or in the four-dimensional form

$$r^{\mu}(t) = R^{\mu} - \tilde{r}^{\mu}(t),$$

where the light four-vector $\tilde{r}^{\mu}(t)$ has the following components[b]

$$\tilde{r}^{\mu}(t) = (c(\tilde{t} - t), \boldsymbol{R} - \boldsymbol{r}(t)) = (\tilde{r}(t), \tilde{\boldsymbol{r}}(t)), \quad \tilde{r}_{\mu}(t)\tilde{r}^{\mu}(t) = 0. \tag{1.1}$$

[a]E-mail: bord@urania.tomsk.su
[b]We use a metrics in which $g^{\mu\nu} = diag(-1, 1, 1, 1)$.

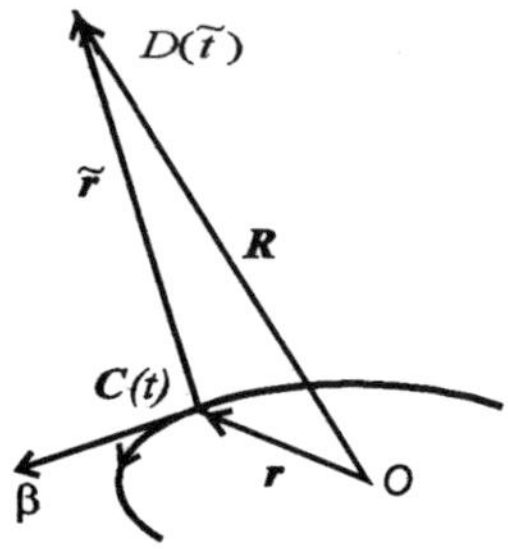

Fig. 1.1. "Charge-observer" coordinate system: O is the origin of the coordinates, C is the charge position at the moment of emission, D is the detector of radiation at moment $\tilde{t}$.

In the general case, the Lorentz transformations for $\tilde{r}^{\mu}(t)$ from the rest system of the charge (the primed system) to the laboratory one have the form

$$\begin{aligned} \tilde{\boldsymbol{r}}'(\tau) &= \tilde{\boldsymbol{r}}(t) - \gamma\boldsymbol{\beta}\tilde{r}^0(t) + \frac{\gamma^2}{\gamma+1}\boldsymbol{\beta}(\boldsymbol{\beta}\tilde{\boldsymbol{r}}(t)), \\ \tilde{r}'(\tau) &= \gamma[1-(\boldsymbol{n}\boldsymbol{\beta})]\tilde{r}^0(t), \end{aligned} \tag{1.2}$$

where $\boldsymbol{\beta} = \boldsymbol{u}/c$, $\boldsymbol{u}$ is the velocity of a charge, c is the speed of light, $\gamma = 1/\sqrt{1-\beta^2}$, τ is the proper time. Unit vector $\boldsymbol{n} = \tilde{\boldsymbol{r}}(t)/\tilde{r}^0(t)$ according to Eq. (1.2) is transformed by the rule

$$\boldsymbol{n}'(\tau) = \frac{\boldsymbol{n} - \gamma\boldsymbol{\beta} + \frac{\gamma^2}{\gamma+1}\boldsymbol{\beta}(\boldsymbol{\beta}\boldsymbol{n})}{\gamma[1-(\boldsymbol{n}\boldsymbol{\beta})]}, \quad \boldsymbol{n}'^2 = 1, \tag{1.3}$$

where all values in the right-hand side of expression are determined at moment t.

Eqs (1.2) and (1.3) together with the Lorentz transformations for the fields solve completely the stated problem. Really, in the system of rest only the scalar potential $\varphi' = e/\tilde{r}'(\tau)$ is a nonzero one. Therefore

$$\begin{aligned} \varphi(\tilde{t}) &= \gamma\varphi'(\tau) = \frac{e}{[1-(\boldsymbol{n}\boldsymbol{\beta})]\tilde{r}(t)}, \\ \boldsymbol{A}(\tilde{t}) &= \gamma\boldsymbol{\beta}\varphi'(\tau) = \frac{e\boldsymbol{\beta}}{[1-(\boldsymbol{n}\boldsymbol{\beta})]\tilde{r}(t)}. \end{aligned} \tag{1.4}$$

These are Liénard-Wiechert potentials[6,7].

The expressions for fields of an arbitrarily moving charge in the rest system have the form[c]

$$\boldsymbol{E}'(\tau) = e\frac{\boldsymbol{n}'}{\tilde{r}'^2} + e\frac{[\boldsymbol{n}'[\boldsymbol{n}'\boldsymbol{a}']]}{c^2\tilde{r}'}, \quad \boldsymbol{H}'(\tau) = -e\frac{[\boldsymbol{n}'\boldsymbol{a}']}{c^2\tilde{r}'}. \tag{1.5}$$

The terms containing the particle acceleration $\boldsymbol{a}'$ are responsible for dipole radiation of a charge (see Ref.[1], § 67).

[c]We denote the vector product by square brackets.

In the laboratory system we have the fields

$$\begin{aligned} \boldsymbol{E}(\tilde{t}) &= \gamma(\boldsymbol{E}' - [\boldsymbol{\beta}\boldsymbol{H}']) - \frac{\gamma^2}{\gamma+1}\boldsymbol{\beta}(\boldsymbol{\beta}\boldsymbol{E}'), \\ \boldsymbol{H}(\tilde{t}) &= \gamma(\boldsymbol{H}' + [\boldsymbol{\beta}\boldsymbol{E}']) - \frac{\gamma^2}{\gamma+1}\boldsymbol{\beta}(\boldsymbol{\beta}\boldsymbol{H}'). \end{aligned}$$

Taking into account Eqs (1.2) and (1.3) and also

$$\begin{aligned} \boldsymbol{a}' &= \gamma^2\left[\boldsymbol{a} + \frac{\gamma^2}{\gamma+1}\boldsymbol{\beta}(\boldsymbol{\beta}\boldsymbol{a})\right], \\ [\boldsymbol{n}'\boldsymbol{a}'] &= \frac{\gamma^2}{1-(\boldsymbol{n}\boldsymbol{\beta})}\left\{[(\boldsymbol{n}-\boldsymbol{\beta})\boldsymbol{a}] - \frac{\gamma}{\gamma+1}\boldsymbol{\beta}(\boldsymbol{\beta}[\boldsymbol{n}\boldsymbol{a}])\right\}, \end{aligned} \tag{1.6}$$

after rather simple but unwieldy calculations we obtain

$$\boldsymbol{E}(\tilde{t}) = \frac{e}{\tilde{r}^2}\frac{(\boldsymbol{n}-\boldsymbol{\beta})}{\gamma^2[1-(\boldsymbol{n}\boldsymbol{\beta})]^3} + \frac{e}{\tilde{r}}\frac{[\boldsymbol{n}[(\boldsymbol{n}-\boldsymbol{\beta})\boldsymbol{a}]]}{c^2[1-(\boldsymbol{n}\boldsymbol{\beta})]}, \quad \boldsymbol{H} = [\boldsymbol{n}\boldsymbol{E}], \tag{1.7}$$

that coincides with the well-known expressions for field intensities of the arbitrarily moving relativistic charge[1].

In conclusion we note that the given method is of general purpose. This method allows obtaining electromagnetic fields for a more difficult relativistic case of particles possessing, for example, an electric or magnetic dipole moment (see Chapter 4).

The Liénard-Wiechert potentials are united in the four-dimensional vector

$$A^\mu = -\frac{ev^\mu}{\tilde{r}_\rho v^\rho}, \tag{1.8}$$

where $v^\mu = dr^\mu/d\tau = \gamma(c, \boldsymbol{\beta})$ is a four-vector of the charge velocity,

$$\tilde{r}_\rho v^\rho = -c\gamma[1-(\boldsymbol{n}\boldsymbol{\beta})]\tilde{r}(t).$$

The field intensities given by Eq. (1.7) are united in the electromagnetic field tensor, which for radiation[d] can be defined as following

$$H^{\mu\nu} = \widetilde{D}^\mu A^\nu - \widetilde{D}^\nu A^\mu, \tag{1.9}$$

where $A^\mu = (\varphi, \boldsymbol{A})$ are potentials of Eq. (1.8) and the generalized derivative $\widetilde{D}^\mu$ is calculated at the observation point with the radius-vector R^μ and takes into account the retardation of radiation of the moving charge

$$\widetilde{D}^\mu = D^\mu + \frac{d}{d\tau}\cdot\left(\frac{\partial\tau}{\partial R_\mu}\right). \tag{1.10}$$

[d]The definition of radiation quantities will be given more precisely further.

The first term in the right-hand side of Eq. (1.10) gives the values of potentials at the fixed moment $\tilde{t}$ which is related to the trajectory time t by Eq. (1.1), i.e.

$$D^\mu = \frac{\partial}{\partial R_\mu} = \left.\frac{\partial}{\partial \tilde{r}_\mu}\right|_{t=const} = \tilde{\partial}^\mu.$$

The second term in Eq. (1.10) takes into account the instantaneous variation of the charge position.

To find $\partial\tau/\partial R_\mu$ we calculate the derivative

$$\widetilde{D}^\mu \tilde{r}_\rho \tilde{r}^\rho = 0.$$

Hence it follows that

$$\frac{\partial\tau}{\partial R_\mu} = \frac{\tilde{r}^\mu}{\tilde{r}_\rho v^\rho}$$

and thus

$$\widetilde{D}^\mu = \tilde{\partial}^\mu + \frac{\tilde{r}^\mu}{\tilde{r}_\rho v^\rho}\frac{d}{d\tau}. \tag{1.11}$$

It is possible to show that the Lorentz condition in terms of the generalized derivative $\widetilde{D}^\mu$ for the Liénard-Wiechert potentials takes the form

$$\widetilde{D}_\mu A^\mu = 0. \tag{1.12}$$

The substitution of potentials of Eq. (1.8) into the electromagnetic field tensor given by Eq. (1.9) yields

$$H^{\mu\nu} = -e\frac{\tilde{r}^{[\mu}w^{\nu]}}{(\tilde{r}_\rho v^\rho)^2} + e(c^2 + \tilde{r}_\alpha w^\alpha)\frac{\tilde{r}^{[\mu}v^{\nu]}}{(\tilde{r}_\rho v^\rho)^3}, \tag{1.13}$$

where w^μ is the four-vector of the charge acceleration

$$w^\mu = dv^\mu/d\tau = \gamma^2(\gamma^2(\boldsymbol{\beta a}), \boldsymbol{a} + \gamma^2\boldsymbol{\beta}(\boldsymbol{\beta a})), \quad v_\mu w^\mu = 0,$$

and the square brackets mean an antisymmetric combination

$$a^{[\mu}b^{\nu]} = a^\mu b^\nu - a^\nu b^\mu.$$

The field intensities can be obtained if one takes into account that in the used metrics

$$H^{\mu\nu} = \begin{vmatrix} 0 & E_x & E_y & E_z \\ -E_x & 0 & H_z & -H_y \\ -E_y & -H_z & 0 & H_x \\ -E_z & H_y & -H_x & 0 \end{vmatrix} = (-\boldsymbol{E}, \boldsymbol{H}).$$

Then from Eq. (1.13) we obtain the formulas coinciding with Eq. (1.7).

One can show that the tensor $H^{\mu\nu}$ satisfies the Maxwell equations

$$\widetilde{D}^{\rho} H^{\mu\nu} + \widetilde{D}^{\nu} H^{\rho\mu} + \widetilde{D}^{\mu} H^{\nu\rho} = 0, \quad \widetilde{D}_{\mu} H^{\mu\nu} = 0 \tag{1.14}$$

for a pure electromagnetic field.

The components of the tensor $H^{\mu\nu}$ consist of two absolutely different terms

$$H^{\mu\nu} = \overline{H}^{\mu\nu} + \widetilde{H}^{\mu\nu},$$

$$\overline{H}^{\mu\nu} = c^2 \frac{e}{\widetilde{r}^2} \frac{n^{[\mu} v^{\nu]}}{(n_\rho v^\rho)^3}, \quad \widetilde{H}^{\mu\nu} = -\frac{e}{\widetilde{r}} \left[\frac{n^{[\mu} w^{\nu]}}{(n_\rho v^\rho)^2} - n_\alpha w^\alpha \frac{n^{[\mu} v^{\nu]}}{(n_\rho v^\rho)^3} \right], \tag{1.15}$$

where $n^\mu = (1, \boldsymbol{n})$. At long distances from the charge the components of $\widetilde{H}^{\mu\nu}$ related to acceleration dominate. They satisfy all properties of the plane electromagnetic field

$$\widetilde{\boldsymbol{H}} = [\boldsymbol{n}\widetilde{\boldsymbol{E}}], \quad (\boldsymbol{n}\widetilde{\boldsymbol{E}}) = (\boldsymbol{n}\widetilde{\boldsymbol{H}}) = (\widetilde{\boldsymbol{E}}\widetilde{\boldsymbol{H}}) = 0. \tag{1.16}$$

At shorter distances the components of $\overline{H}^{\mu\nu}$, which do not depend on acceleration, dominate. For these components one of the conditions of Eq. (1.16) is not realized: $(\boldsymbol{n}\overline{\boldsymbol{E}}) \neq 0$. One can show that both $\overline{H}^{\mu\nu}$ and $\widetilde{H}^{\mu\nu}$ satisfy the first pair of Maxwell equations. At the same time, the second pair of Maxwell equations is not fulfilled for them separately. Nevertheless, the field $\widetilde{H}^{\mu\nu}$ is usually related to the electromagnetic radiation of the charge. Further, we shall consider this problem in more detail.

1.2. The Wave Zone and Covariant Definition of Radiation

The space region where radiation field dominates is called a wave zone[8]. It has been shown in Refs[9,10] that the wave zone of a charge moving at a high velocity has a strong space anisotropy. Its general tendency is to approach the charge as the velocity increases. In any direction where the radiation disappears the wave zone goes to infinity.

As we shall be interested mainly in radiation, it is interesting to work out a special mathematical procedure for calculations of radiation fields in the wave zone.

First we note that in the wave zone the generalized derivative $\widetilde{D}^\mu$ transforms into the form

$$\widetilde{D}^\mu \approx \frac{\widetilde{r}^\mu}{\widetilde{r}_\rho v^\rho} \frac{d}{d\tau}, \tag{1.17}$$

because the first term $\widetilde{\partial}^\mu$ in Eq. (1.11) raises the falloff degree of the field after acting on Liénard-Wiechert potentials. The Lorentz condition of Eq. (1.12) in the wave zone takes the form

$$\widetilde{r}_\mu \overset{\circ}{\widetilde{A}}{}^\mu \approx 0, \tag{1.18}$$

where

$$\overset{\circ}{\widetilde{A}}{}^\mu = \widetilde{\left(\frac{dA^\mu}{d\tau}\right)} = -e \left[\frac{w^\mu}{\widetilde{r}_\rho v^\rho} - \frac{v^\mu \widetilde{r}_\alpha w^\alpha}{(\widetilde{r}_\rho v^\rho)^2} \right]. \tag{1.19}$$

The known relations in the case of the Lorentz gauge of potentials follow from Eq. (1.18)

$$\overset{\circ}{\tilde{\varphi}} \approx (\boldsymbol{n}\, \overset{\circ}{\widetilde{\boldsymbol{A}}}), \quad \tilde{\varphi} \approx (\boldsymbol{n}\widetilde{\boldsymbol{A}}),$$

which are easily verified if one takes into account that

$$\frac{d\boldsymbol{n}}{d\tau} = \frac{1}{\tilde{r}} c\gamma(\boldsymbol{n} - \boldsymbol{\beta}).$$

In the wave zone the electromagnetic field tensor $\widetilde{H}^{\mu\nu}$ can be written in the form

$$\widetilde{H}^{\mu\nu} \approx \frac{n^{[\mu}\, \overset{\circ}{\widetilde{A}}{}^{\nu]}}{n_\rho v^\rho}. \tag{1.20}$$

Further, it is expedient to proceed from the derivative with respect to the proper time to the derivative with respect to the time of the observer. Taking the zero component in Eq. (1.17) we get

$$\frac{d}{d\tau} = -\frac{1}{c} n_\rho v^\rho \frac{d}{d\tilde{t}} = [1 - (\boldsymbol{n}\boldsymbol{\beta})] \frac{\partial}{\partial \tilde{t}}\,.$$

Now we have

$$\widetilde{\boldsymbol{E}} \approx \frac{1}{c}[\boldsymbol{n}[\boldsymbol{n}\, \overset{\circ}{\widetilde{\boldsymbol{A}}}]], \quad \widetilde{\boldsymbol{H}} \approx [\boldsymbol{n}\widetilde{\boldsymbol{E}}],$$

where $\overset{\circ}{\widetilde{\boldsymbol{A}}} = \partial\widetilde{\boldsymbol{A}}/\partial\tilde{t}$. These are the well-known relations for the plane electromagnetic wave[1]. We note also that

$$\widetilde{E}^2 = \frac{\overset{\circ}{\widetilde{A}}_\lambda \overset{\circ}{\widetilde{A}}{}^\lambda}{(n_\rho v^\rho)^2} = \widetilde{H}^2. \tag{1.21}$$

The problem of definition of electromagnetic radiation has a long history. Nevertheless, the completed definition does not exist so far. Detailed discussions can be found in Refs[11–17]. Here we give the covariant definition of radiation satisfying our requirements.

In the field theory the momentum and force density are connected by the relation

$$\mathcal{P}^\mu = \frac{1}{c}\int \mathcal{F}^\mu d\Upsilon,$$

where $d\Upsilon = cdtdV$ is the four-dimensional volume element.

For the electromagnetic field

$$\mathcal{F}^\mu = -\partial_\nu \mathcal{P}^{\mu\nu}$$

with the tensor of energy-momentum density

$$\mathcal{P}^{\mu\nu} = -\frac{1}{4\pi}\left(H^{\mu\lambda}H_\lambda{}^\nu + \frac{1}{4}\, g^{\mu\nu} H_{\alpha\beta}H^{\alpha\beta}\right). \tag{1.22}$$

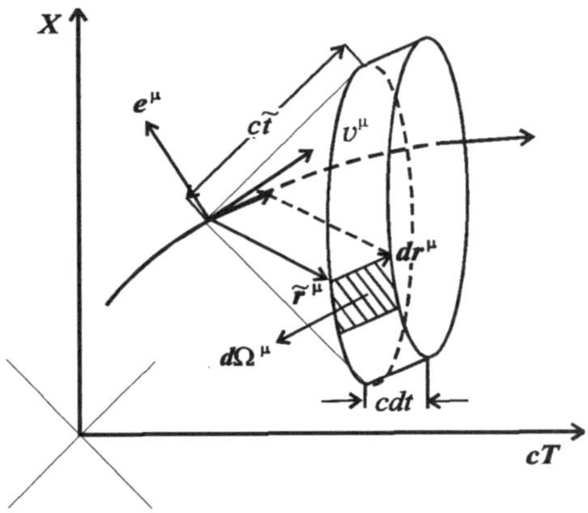

Fig. 1.2. The world line and light cones of radiating charge. Radiation emitted during time dt passes through the timelike surface of the form of a narrow cylindric strip.

Or in components

$$\mathcal{P}^{\mu\nu} = \begin{pmatrix} \mathcal{W} & \vdots & \frac{1}{c}\boldsymbol{S} \\ \cdots & \cdots & \cdots \\ \frac{1}{c}\boldsymbol{S} & \vdots & \mathcal{P}_{ij} \end{pmatrix},$$

where

$$\mathcal{W} = \frac{E^2 + H^2}{8\pi},$$

is the density of energy,

$$\boldsymbol{S} = \frac{c}{4\pi}[\boldsymbol{EH}]$$

is the Poynting vector, and

$$\mathcal{P}_{ij} = -\frac{1}{4\pi}\left[E_iE_j + H_iH_j - \frac{1}{2}\delta_{ij}(E^2 + H^2)\right].$$

Momentum $\widetilde{\mathcal{P}}^\mu = \left(\frac{1}{c}\widetilde{\mathcal{E}}, \widetilde{\boldsymbol{P}}\right)$ emitted by the particle in the form of radiation is determined by the expression

$$\widetilde{\mathcal{P}}^\mu = \frac{1}{c}\oint \widetilde{\mathcal{P}}^{\mu\nu} d\Omega_\nu, \tag{1.23}$$

where tensor $\widetilde{\mathcal{P}}^{\mu\nu}$ is taken in the wave zone and integration goes over the closed hypersurface with the element $d\Omega_\nu$ (see Fig. 1.2)

As in the rest system the world line of the particle transforms into a straight line parallel to the time axis, then $d\Omega_\nu$ is a spacelike four-vector satisfying the condition

$$v_\mu d\Omega^\mu = 0.$$

Hence it follows that

$$d\Omega^\mu = e^\mu c d\tau \tilde{r}^2 d\Omega, \tag{1.24}$$

where $d\Omega$ is an element of a solid angle, and e^μ is the unit spacelike four-vector

$$e^\mu = -c\frac{n^\mu}{n_\rho v^\rho} - \frac{1}{c}v^\mu, \quad v_\mu e^\mu = 0, e_\mu e^\mu = 1. \tag{1.25}$$

In the rest system e^μ takes the form $e_0^\mu = (0, \boldsymbol{n}_0)$.

Four-vector e^μ is related to the components n^μ by the formula

$$n^\mu = -\frac{1}{c}n_\rho v^\rho\left(e^\mu + \frac{1}{c}v^\mu\right). \tag{1.26}$$

Eq. (1.23) with account of Eq. (1.24) can be rewritten in the form

$$\frac{d\widetilde{\mathcal{P}}^\mu}{d\tau} = \tilde{r}^2 \oint d\Omega \widetilde{\mathcal{P}}^\mu{}_\nu e^\nu.$$

It is possible to show that in the wave zone

$$\widetilde{\mathcal{P}}^{\mu\nu} = \frac{\overset{\circ}{\widetilde{A}}_\lambda \overset{\circ}{\widetilde{A}}{}^\lambda}{(n_\rho v^\rho)^2} n^\mu n^\nu$$

according to Eqs (1.20)—(1.22) and also Eqs (1.1) and (1.18). As $n_\mu e^\mu = -n_\mu v^\mu / c$, then

$$\frac{d\widetilde{\mathcal{P}}^\mu}{d\tau} = -\frac{\tilde{r}^2}{4\pi c}\oint d\Omega \frac{\overset{\circ}{\widetilde{A}}_\lambda \overset{\circ}{\widetilde{A}}{}^\lambda}{n_\rho v^\rho} n^\mu. \tag{1.27}$$

According to Eq. (1.19) we have

$$\overset{\circ}{\widetilde{A}}_\lambda \overset{\circ}{\widetilde{A}}{}^\lambda = \frac{e^2}{\tilde{r}^2}\frac{w_\rho w^\rho - (e_\rho w^\rho)^2}{(n_\rho v^\rho)^2}.$$

And, therefore, taking into account Eq. (1.26) and further integration over the angles we get

$$\frac{d\widetilde{\mathcal{P}}^\mu}{d\tau} = \frac{e^2}{4\pi c^3}\oint d\Omega \frac{w_\rho w^\rho - (e_\rho w^\rho)^2}{(n_\rho v^\rho)^2} v^\mu. \tag{1.28}$$

Integrating over the angles we obtain

$$\frac{d\widetilde{\mathcal{P}}^\mu}{d\tau} = \frac{2}{3}\frac{e^2}{c^5} w_\rho w^\rho v^\mu. \tag{1.29}$$

During integration the formulas[18]

$$\begin{aligned} \frac{c^2}{4\pi}\oint d\Omega \frac{1}{(n_\rho v^\rho)^2} &= 1, \quad \frac{c^2}{4\pi}\oint d\Omega \frac{e^\mu}{(n_\rho v^\rho)^2} = 0, \\ \frac{c^2}{4\pi}\oint d\Omega \frac{e^\mu e^\nu}{(n_\rho v^\rho)^2} &= \frac{1}{3}\left(g^{\mu\nu} + \frac{1}{c^2}v^\mu v^\nu\right) \end{aligned} \tag{1.30}$$

were used.

If we put $\mu = 0$ in Eq. (1.29), it is possible to obtain the total power of instantaneous radiation

$$W = \frac{d\mathcal{E}}{dt} = \frac{2}{3}\frac{e^2}{c^3}\, w_\mu w^\mu .$$

Hence it is obvious that this value is invariant. In other notation

$$W = \frac{2}{3}\frac{e^2}{c^3}\,\gamma^6\left(a^2 - [\boldsymbol{\beta}\boldsymbol{a}]^2\right). \tag{1.31}$$

When $\boldsymbol{\beta} \perp \boldsymbol{a}$ we have synchrotron radiation power

$$W_{\text{SR}} = \frac{2}{3}\frac{e^2}{c^3}\, a^2\gamma^4. \tag{1.32}$$

This formula was obtained by A.A. Liénard [6] as far back as 1898. In the nonrelativistic approximation $\gamma \to 1$ it comes to the Larmor formula[19].

1.3. Angular Distribution of Instantaneous Radiation Power

In a general case, the angular distribution of instantaneous radiation power can be obtained by Eq. (1.27) at $\mu = 0$

$$dW = -\frac{e^2}{4\pi\gamma}\, d\Omega\left[\frac{w_\alpha w^\alpha}{(n_\rho v^\rho)^3} - c^2\frac{n_\alpha w^\alpha}{(n_\rho v^\rho)^5}\right]. \tag{1.33}$$

The angular distribution in terms of fields can be obtained after a substitution of $\overset{\circ}{\widetilde{A}}_\lambda \overset{\circ}{\widetilde{A}}{}^\lambda$ from Eq. (1.21) into Eq. (1.27)[e]

$$dW = \frac{e}{4\pi}\,\tilde{r}^2\, d\Omega \widetilde{E}^2[1 - (\boldsymbol{n}\boldsymbol{\beta})].$$

If $\widetilde{E}^2$ is substituted here from Eq. (1.7) then the angular distribution of radiation will take the form

$$\frac{dW}{d\Omega} = \frac{e^2}{4\pi c^3}\,\frac{a^2[1-(\boldsymbol{n}\boldsymbol{\beta})]^2 - (\boldsymbol{n}\boldsymbol{a})^2(1-\beta^2) + 2(\boldsymbol{\beta}\boldsymbol{a})(\boldsymbol{n}\boldsymbol{a})[1-(\boldsymbol{n}\boldsymbol{\beta})]}{[1-(\boldsymbol{n}\boldsymbol{\beta})]^5}. \tag{1.35}$$

Of course, Eq. (1.33) gives the same result.

[e] We note that the reference application of the Poynting vector $\boldsymbol{S} = c/4\pi[\boldsymbol{E}\boldsymbol{H}]$ for calculation of the angular distribution does not give the power but the radiation intensity

$$dI = (\boldsymbol{S}d\boldsymbol{\Omega}) = \frac{c}{4\pi}\, d\Omega \widetilde{E}^2, \tag{1.34}$$

which manifests itself here in absence of factor of $1 - (\boldsymbol{n}\boldsymbol{\beta})$. The radiation intensity is a noninvariant value and depends on the reference frame. Failure to respond in this problem can lead to serious misunderstandings (see discussion in Refs[28–31] and also Ref.[32])

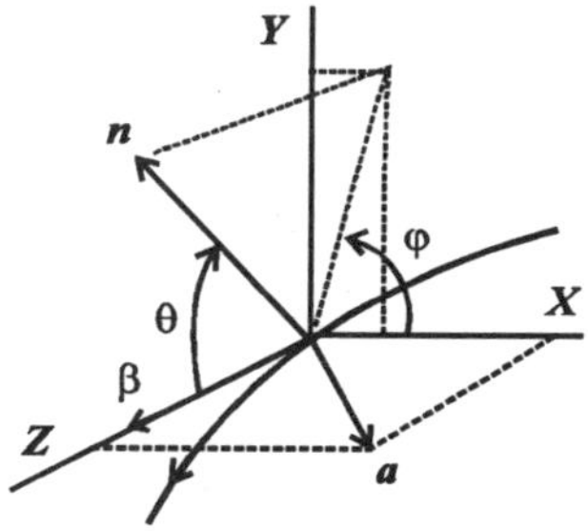

Fig. 1.3. Coordinate system for analysis of the angular distribution of the total power of instantaneous radiation. Plane ZX coincides with the plane of velocity and acceleration vectors. The components of the vectors $\boldsymbol{n}$, $\boldsymbol{\beta}$, and $\boldsymbol{a}$ are equal to $\boldsymbol{n} = (\sin\theta\cos\varphi, \sin\theta\sin\varphi, \cos\theta)$, $\boldsymbol{\beta} = \beta(0,0,1)$, $\boldsymbol{a} = a(\sin\alpha, 0, \cos\alpha)$.

Space anisotropy of angular distribution essentially depends on the value of $1 - (\boldsymbol{n\beta})$ in the denominator of this expression. In the ultrarelativistic case, when the angle ψ included between the vectors $\boldsymbol{n}$ and $\boldsymbol{\beta}$ is very small

$$1 - (\boldsymbol{n\beta}) \approx \frac{1}{2}\left(\frac{1}{\gamma^2} + \psi^2\right)$$

this value is close to zero and the angular distribution is highly extended along the direction of the particle velocity (projector effect).

For the analysis of the angular distribution it is more convenient to use a coordinate system shown in Fig. 1.3. In the general case of arbitrary orientation of the velocity and acceleration vectors, the angular distribution of the total power of instantaneous radiation has the form[26]

$$\frac{dW}{d\Omega} = \frac{e^2a^2}{4\pi c^3}\rho(\theta,\varphi;\alpha),$$

$$\rho(\theta,\varphi;\alpha) = \frac{\sin^2\varphi\sin^2\alpha}{(1-\beta\cos\theta)^3} + \frac{[(\beta-\cos\theta)\cos\varphi\sin\alpha + \sin\theta\cos\alpha]^2}{(1-\beta\cos\theta)^5}. \tag{1.36}$$

If ρ is interpreted as a radius-vector, then it will describe a surface in the spherical coordinate system which can be called an indicatrix of radiation. It follows from Eq. (1.36) that even in the general case, this surface is symmetrical with respect to $\boldsymbol{\beta a}$ plane and nonsymmetrical with respect to $\boldsymbol{H\beta}$ plane (at the expense of α). All singularities of radiation are in the $\boldsymbol{\beta a}$ plane. There are always two such directions exactly in this plane where the radiation is absent completely. According to Eq. (1.36), these directions are determined by the condition[1,10]

$$\cos\theta_{\min} = \beta\sin^2\alpha \pm \cos\alpha\sqrt{1-\beta^2\sin^2\alpha}.$$

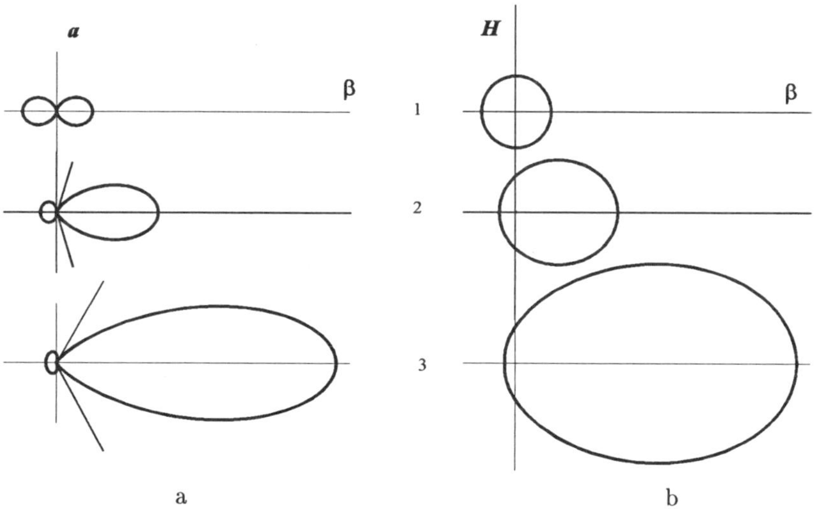

Fig. 1.4. $\boldsymbol{\beta a}$ (a) and $\boldsymbol{H\beta}$ (b) plane sections of the indicatrix of SR at $\beta = 0.01$ (1), $\beta = 0.3$ (2), and $\beta = 0.5$ (3).

We note that in these directions the wave zone tends to infinity.

Let us consider some particular cases. SR is of the greatest interest. In this case, the velocity of the particle is orthogonal to acceleration ($\alpha = \pi/2$). The indicatrix of SR has the form

$$\rho(\theta, \varphi) = \frac{(1 - \beta\cos\theta)^2 - (1 - \beta^2)\sin^2\theta\cos^2\varphi}{(1 - \beta\cos\theta)^5}.$$

One can see that this function is symmetrical with respect to both $\boldsymbol{\beta a}$ and $\boldsymbol{H\beta}$ planes (see Fig. 1.4). We obtain sufficiently complete information on the indicatrix by investigation of its sections in these planes

$$\rho(\theta, \varphi = 0, \pi) = \frac{(\beta - \cos\theta)^2}{(1 - \beta\cos\theta)^5}, \quad \rho(\theta, \varphi = \pm\frac{\pi}{2}) = \frac{1}{(1 - \beta\cos\theta)^5}.$$

The angle $\theta = 0$ for SR corresponds to the principal maximum of the radiation power. The directions where the radiation is absent are determined by the condition of $\cos\theta_{\min} = \beta$. In the nonrelativistic case, they almost coincide with the forward or opposite direction of acceleration ($\theta_{\min} \approx \pi/2$). The angle $\theta_{\min}$ decreases with increasing velocity and at $0.63 \leq \beta < 1$ the secondary maximums appear. It is of

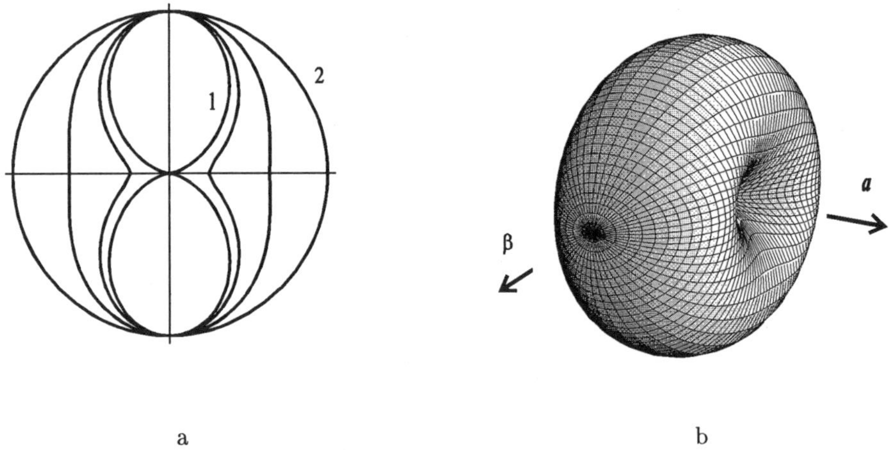

Fig. 1.5. Evolution of $\boldsymbol{aH}$ plane section of SR indicatrix as a function of β from $\beta = 0.01$ (1) to $\beta = 1$ (2) and the space indicatrix of SR at $\beta = 0.01$.

interest to estimate the relation "forward-backward"

$$\frac{\rho(\theta = 0, \varphi)}{\rho(\theta = \pi, \varphi)} = \left(\frac{1+\beta}{1-\beta}\right)^3 \Bigg|_{\beta \to 1} \approx 2^6 \gamma^6$$

and the relation of the principal maximum to the secondary one

$$\frac{\rho(\theta = 0, \varphi)}{\rho(\theta_{\max'}, \varphi = 0)} = \frac{5^5}{216} \beta^2 (1+\beta)^3 \,|_{\beta \to 1} \approx 231.48.$$

These sections for various quantities of β are shown in Fig. 1.4a,b.

The $\boldsymbol{aH}$ plane section is given by the curves

$$\rho(\theta = \pi/2, \varphi) = \sin^2 \varphi + \beta^2 \cos^2 \varphi.$$

The evolution of this section as a function of β is shown in Fig. 1.5a.

Space indicatrixes for various quantities of β are shown in Fig. 1.5b and Fig. 1.6.

Now let us consider another extreme case when the velocity and acceleration are parallel. The corresponding indicatrix has the form

$$\rho(\theta, \varphi) = \frac{\sin^2 \theta}{(1 - \beta \cos \theta)^5}.$$

This radiation possesses azimuthal symmetry in the plane orthogonal to velocity. Thereat in the direction of $\beta = 0$ the radiation is absent at all. It is interesting that

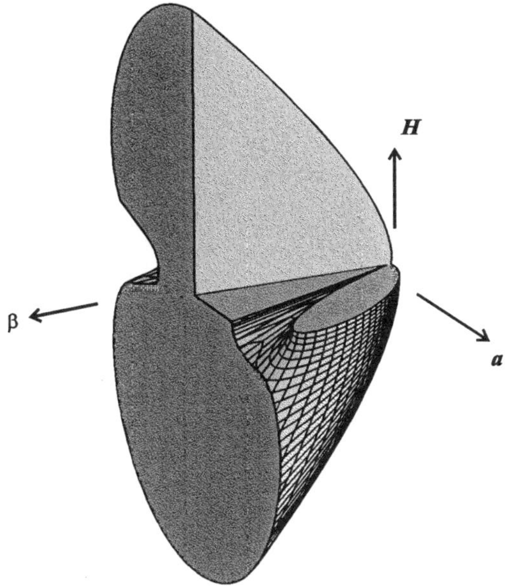

Fig. 1.6. Space indicatrix of SR for $\beta = 0.9$.

at $\theta = \pi/2$ the cross section does not depend on the velocity. The evolution of the longitudinal section as the velocity increases is shown in Fig. 1.7a. The angles related to the radiation maximum are determined by the formula

$$\cos\theta_{\max} = \frac{5\beta}{1+\sqrt{1+15\beta^2}} \,.$$

Space distribution of this radiation is shown in Fig. 1.8. The evolution of the maximum of radiation is shown in Fig. 1.7b.

In the general case, in presence of both longitudinal and orthogonal acceleration the evolution of a ZX plane section of the angular distribution is shown in Fig. 1.9 and corresponding indicatrix is shown in Fig. 1.10.

Of course, further integration over the angles leads to the total power of instantaneous radiation which coincides with Eq. (1.31).

In conclusion of this section we note that the angular distribution of the radiation power can be obtained in another way by means of Lorentz transformations[1,32]. Thereat it is necessary to bear in mind that $d\widetilde{\mathcal{P}}^\mu$ is a four-vector which, according to Eq. (1.27), satisfies the condition

$$n_\mu d\widetilde{\mathcal{P}}^\mu = 0,$$

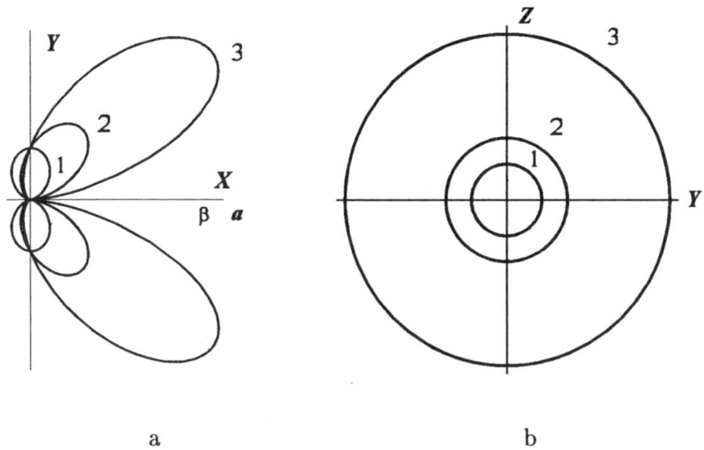

Fig. 1.7. Evolution of the longitudinal section of the indicatrix as the velocity increases when $\boldsymbol{\beta} \parallel \boldsymbol{a}$ (a); the evolution of the maximum of radiation in this case (b); $\beta = 0.01$ (1), $\beta = 0.3$ (2), $\beta = 0.5$ (3).

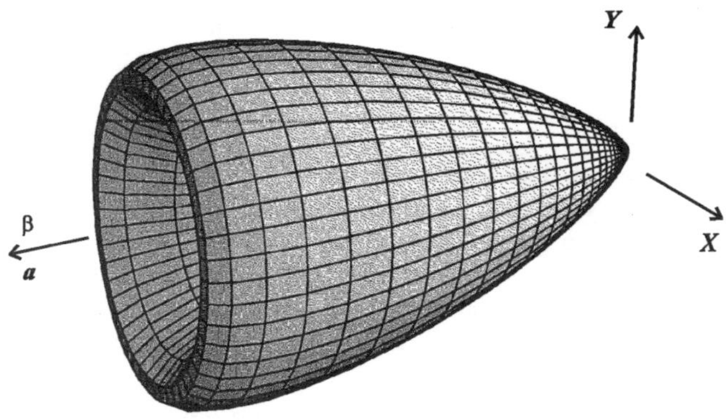

Fig. 1.8. Space distribution of radiation when $\boldsymbol{\beta} \parallel \boldsymbol{a}$.

Fig. 1.9. A section of the indicatrix of the instantaneous radiation of an arbitraty moving charge by ZX plane containing the vectors $\boldsymbol{\beta}$ and $\boldsymbol{a}$ (see the next Fig.).

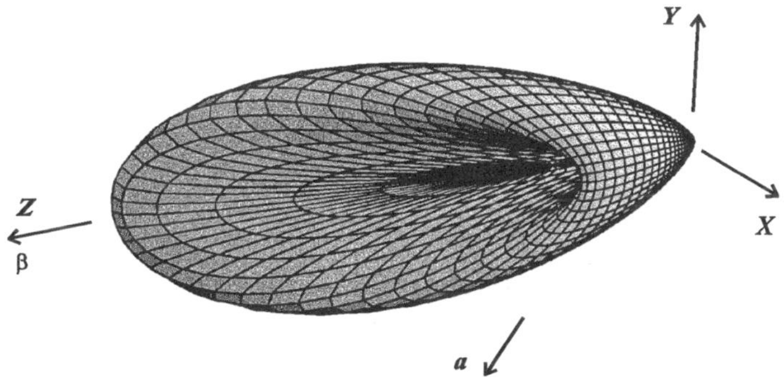

Fig. 1.10. Space indicatrix of radiation when $\alpha = \pi/4$.

as $n_\mu n^\mu = 0$. Hence it follows that the radiation energy is related to the momentum by the simple formula

$$d\mathcal{E} = c d\widetilde{\mathcal{P}}.$$

Now it is not difficult to write the Lorentz transformations for $d\mathcal{E}$. Supposing that the charge rests at the origin of a primed coordinate system (see Fig. 1.3) we have

$$d\mathcal{E}' = \frac{1-\beta\cos\theta}{\sqrt{1-\beta^2}} d\mathcal{E}.$$

Taking into account that $dt/d\tau = \sqrt{1-\beta^2}$ we obtain

$$dW = \frac{1-\beta^2}{1-\beta\cos\theta} dW'.$$

In the system of rest the angular distribution of radiation is determined by Eq. (1.35) where we it should be put $\beta = 0$. Then we obtain

$$dW' = \frac{e^2}{4\pi c^3}[a'^2 - (\boldsymbol{n}'\boldsymbol{a}')^2] d\Omega'.$$

The expression in the square brackets is an invariant functional

$$I_a(\theta, \varphi) = a'^2 - (\boldsymbol{n}'\boldsymbol{a}')^2 = w_\mu w^\mu - (e_\mu w^\mu)^2. \tag{1.37}$$

Here we have taken into a consideration that

$$w^\mu\,|_{\beta=0} = (0, \boldsymbol{a}'), \quad e^\mu\,|_{\beta=0} = (0, \boldsymbol{n}').$$

One can prove that the direct calculation of the invariant I_a on the basis of Eqs (1.3), (1.6), and (1.26) gives the same result.

The solid angle is transformed by the rule[1]

$$d\Omega' = \frac{d\Omega}{\gamma^2(1-\beta\cos\theta)^2}.$$

As $(1-\beta\cos\theta) = -n_\rho v^\rho/\gamma$ we finally have

$$dW = -\frac{e^2}{4\pi\gamma}\frac{d\Omega}{(n_\rho v^\rho)^2}\left[w_\alpha w^\alpha - c^2\frac{(n_\alpha w^\alpha)^2}{(n_\rho v^\rho)^2}\right],$$

which coincides with Eq. (1.33).

1.4. Linearly Polarized SR and Its Angular Distribution

As well as the angular distribution of instantaneous radiation power, the linearly polarized radiation and its angular distribution can be analyzed in the most general case without regard to a particular trajectory which follows from the solution of equations of motion.

To describe the radiation component with linear polarization we have to split the vector of the electric field in the wave zone into mutually orthogonal components

$$\widetilde{\boldsymbol{E}} = \sum_s \widetilde{E}_s \boldsymbol{n}_s, \quad \widetilde{E}_s = (\boldsymbol{n}_s \widetilde{\boldsymbol{E}}) \tag{1.38}$$

with the two arbitrary unit vectors $\boldsymbol{n}_s$ being orthogonal to each other as well as the direction of propagation of the plane electromagnetic wave. The definition of unit vectors depends on the choice of the coordinate system. The spherical coordinate system, where the vectors $\boldsymbol{n}_s$ coincide with the unit vectors $\boldsymbol{n}_\theta$ and $\boldsymbol{n}_\varphi$, is the most convenient. Then the unit vectors $\boldsymbol{n}_\theta$ and $\boldsymbol{n}_\varphi$ may be chosen as

$$\boldsymbol{n}_\varphi = [\boldsymbol{n}\boldsymbol{n}_\theta], \quad \boldsymbol{n}_\theta = -[\boldsymbol{n}\boldsymbol{n}_\varphi], \quad \boldsymbol{n} = [\boldsymbol{n}_\theta \boldsymbol{n}_\varphi],$$

$$\boldsymbol{n}_\varphi = \frac{[\boldsymbol{k}\boldsymbol{n}]}{\sqrt{1-(\boldsymbol{k}\boldsymbol{n})^2}}, \quad \boldsymbol{n}_\theta = \frac{\boldsymbol{n}(\boldsymbol{k}\boldsymbol{n}) - \boldsymbol{k}}{\sqrt{1-(\boldsymbol{k}\boldsymbol{n})^2}},$$

$$(\boldsymbol{n}\boldsymbol{n}_\theta) = (\boldsymbol{n}\boldsymbol{n}_\varphi) = (\boldsymbol{n}_\theta \boldsymbol{n}_\varphi) = 0,$$

where the unit vector $\boldsymbol{k}$ coincides with the orientation of Z-axis. The direction of the unit vector $\boldsymbol{k}$ and, at the same time, the direction of Z-axis[f] can be connected with one of the physically distinguished directions. For example, there is a magnetic field direction, a velocity direction or an acceleration one[34].

Further, we shall limit ourselves to consideration of SR linear polarization. In this case, there are three possible types of the linear polarization of radiation shown in Fig. 1.4.

The most wide-spread polarization type is the first, when Z-axis of the spherical coordinate system coincides with the magnetic field direction (for $e < 0$). In this case, the vector of φ-component of the electromagnetic field oscillates in the orbit plane and this component is sometimes called σ-component of linear polarization, while θ-component is referred to as a π-component. In the second polarization type the coordinate system coincides with that used for the analysis of the angular distribution of the total power of instantaneous radiation. The third polarization type when Z-axis is directed along the acceleration of particle is less known but, as we shall see, none the less interesting.

According to Eq. (1.38), the linear polarization components of the electric field $\widetilde{E}_s = (\boldsymbol{n}_s \widetilde{\boldsymbol{E}})$ have the form $(s = \theta, \varphi)$

$$\widetilde{E}_s = -\frac{e}{\tilde{r}} \frac{1}{(n_\rho v^\rho)^2} \left[(\boldsymbol{w}\boldsymbol{n}_s) - (\boldsymbol{v}\boldsymbol{n}_s) \frac{n_\alpha w^\alpha}{n_\rho v^\rho} \right]. \tag{1.39}$$

[f]Of course, the unit vector $\boldsymbol{k}$ can not be connected with Z-axis orientation, however in this case the description of the linear polarization of radiation becomes more complicate and is not so unified.

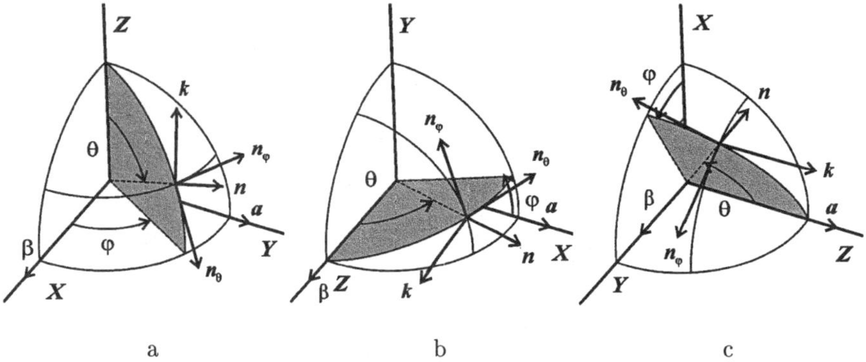

Fig. 1.11. There are three types of SR linear polarization: a) $\boldsymbol{k} \parallel [\boldsymbol{\beta a}]$, $\boldsymbol{\beta} = \beta(1,0,0)$, $\boldsymbol{a} = a(0,1,0)$, $(\boldsymbol{n\beta}) = \beta \sin\theta \cos\varphi$, $(\boldsymbol{na}) = a \sin\theta \sin\varphi$; b) $\boldsymbol{k} \parallel \boldsymbol{\beta}$, $\boldsymbol{\beta} = \beta(0,0,1)$, $\boldsymbol{a} = a(1,0,0)$, $(\boldsymbol{n\beta}) = \beta\cos\theta$, $(\boldsymbol{na}) = a\sin\theta\cos\varphi$; c) $\boldsymbol{k} \parallel \boldsymbol{a}$, $\boldsymbol{\beta} = \beta(0,1,0)$, $\boldsymbol{a} = a(0,0,1)$, $(\boldsymbol{n\beta}) = \beta\sin\theta\sin\varphi$, $(\boldsymbol{na}) = a\cos\theta$. In all cases $\boldsymbol{n} = (\sin\theta\cos\varphi, \sin\theta\sin\varphi, \cos\theta)$, $\boldsymbol{n}_\varphi = (-\sin\varphi, \cos\varphi, 0)$, $\boldsymbol{n}_\theta = (\cos\theta\cos\varphi, \cos\theta\sin\varphi, -\sin\theta)$.

In the case of SR under consideration

$$w^\mu = \gamma^2(0, \boldsymbol{a})$$

and for the above shown polarization types we have (here $e > 0$)

$$\begin{aligned} E_\theta^H &= -\frac{ea}{\tilde{r}c^2}\frac{\cos\theta\sin\varphi}{(1-\beta\sin\theta\cos\varphi)^3}, \quad E_\varphi^H = \frac{ea}{\tilde{r}c^2}\frac{\beta\sin\theta-\cos\varphi}{(1-\beta\sin\theta\cos\varphi)^3}; \\ E_\theta^\beta &= -\frac{ea}{\tilde{r}c^2}\frac{(\cos\theta-\beta)\cos\varphi}{(1-\beta\cos\theta)^3}, \quad E_\varphi^\beta = \frac{ea}{\tilde{r}c^2}\frac{\sin\varphi}{(1-\beta\cos\theta)^2}; \\ E_\theta^a &= \frac{ea}{\tilde{r}c^2}\frac{\sin\theta-\beta\sin\varphi}{(1-\beta\sin\theta\sin\varphi)^3}, \quad E_\varphi^a = -\frac{ea}{\tilde{r}c^2}\frac{\beta\cos\theta\cos\varphi}{(1-\beta\sin\theta\sin\varphi)^3}. \end{aligned} \tag{1.40}$$

The angular distribution of the radiation with the polarization components of Eq. (1.40) has the form ($i = H, \beta, a$)

$$\frac{dW_s}{d\Omega} = \frac{e^2a^2}{4\pi c^2}\rho_s^i(\theta,\varphi).$$

Polarization indicatrixes are described by the functions $\rho_s^i(\theta,\varphi)$. In the considered cases, we have

$$\rho_\theta^H(\theta,\varphi) = \frac{\cos^2\theta\sin^2\varphi}{(1-\beta\sin\theta\cos\varphi)^5}, \quad \rho_\varphi^H(\theta,\varphi) = \frac{(\beta\sin\theta-\cos\varphi)^2}{(1-\beta\sin\theta\cos\varphi)^5};$$

$$\rho_\theta^\beta(\theta,\varphi) = \frac{(\cos\theta-\beta)^2\cos^2\varphi}{(1-\beta\cos\theta)^5}, \quad \rho_\varphi^\beta(\theta,\varphi) = \frac{\sin^2\varphi}{(1-\beta\cos\theta)^3};$$
$$\rho_\theta^a(\theta,\varphi) = \frac{(\sin\theta-\beta\sin\varphi)^2}{(1-\beta\sin\theta\sin\varphi)^5}, \quad \rho_\varphi^a(\theta,\varphi) = \frac{\beta^2\cos^2\theta\cos^2\varphi}{(1-\beta\sin\theta\sin\varphi)^5}. \tag{1.41}$$

Some of these indicatrixes are shown in Fig. 1.12, Fig. 1.4.
Further integration over the angles yields the total polarization of radiation

$$W_\theta^H = \frac{2-\beta^2}{8}W_{\rm SR}, \quad W_\varphi^H = \frac{6+\beta^2}{8}W_{\rm SR};$$
$$W_\theta^\beta = \frac{1}{4}W_{\rm SR}, \quad W_\varphi^\beta = \frac{3}{4}W_{\rm SR}; \tag{1.42}$$
$$W_\theta^a = \frac{8-2\beta^2+\beta^4}{8}W_{\rm SR}, \quad W_\varphi^a = \frac{\beta^2(2-\beta^2)}{8}W_{\rm SR}.$$

The important polarization feature is the polarization degree

$$\Delta = \left|\frac{W_\theta - W_\varphi}{W_\theta + W_\varphi}\right|.$$

In the latter case, this value is a maximum and equal to

$$\Delta^a = \frac{4-2\beta^2+\beta^4}{4}.$$

In the nonrelativistic case, it tends to a limit equal to 1. It is easy to understand that E_φ^a-component is absent at all at $\beta = 0$ according to Eq. (1.40). With increasing β the quantity Δ^a decreases monotonously and in the ultrarelativistic case, takes on a value equal to 0.75. On the contrary, the maximum value

$$\Delta^H = \frac{2+\beta^2}{4}$$

is attained in the ultrarelativistic case and coincides with the minimum value of Δ^a. Δ^H takes on the minimum quantity at $\beta = 0$. It coincides with constant value Δ^β which does not depend on the velocity and is equal to 0.5. Thus we always have

$$0.5 = \Delta^\beta \le \Delta^H < \Delta^a \le 1.$$

The polarization properties of radiation can be also described by means of parameters introduced by G.G. Stokes[35] (see also Refs[1,36]) which represent physically measurable properties and characterize completely the polarization of a plane electromagnetic wave. The discussion on SR polarization in notation of the Stokes parameters can be found in Ref.[37] (see also Refs[31,38]). Amongst the variety of recent works of special interest is the first measurement of the Stokes parameters of SR[39].

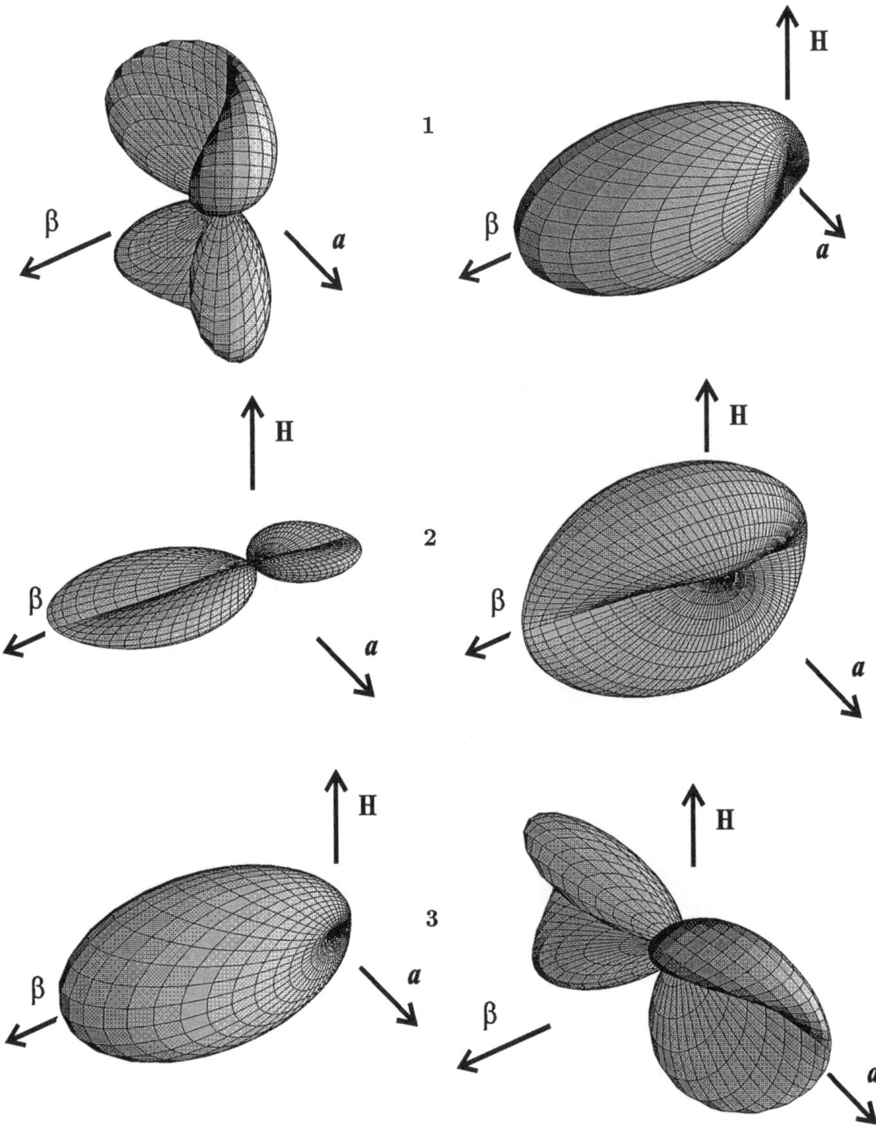

Fig. 1.12. Polarization indicatrixes of SR in the nonrelativistic case: 1,2,3 correspond to $\rho_{\theta}^{H,\beta,a}$ (left) and $\rho_{\varphi}^{H,\beta,a}$ (right). In the latter case, almost all radiation is linearly polarized in the plane orthogonal to the plane of the orbit and π-component coincides with the indicatrix of the total power of SR. For this reason, the scales of σ- and π-components are different: $\beta = 0.4$ for π-component and $\beta = 0.01$ for σ-component.

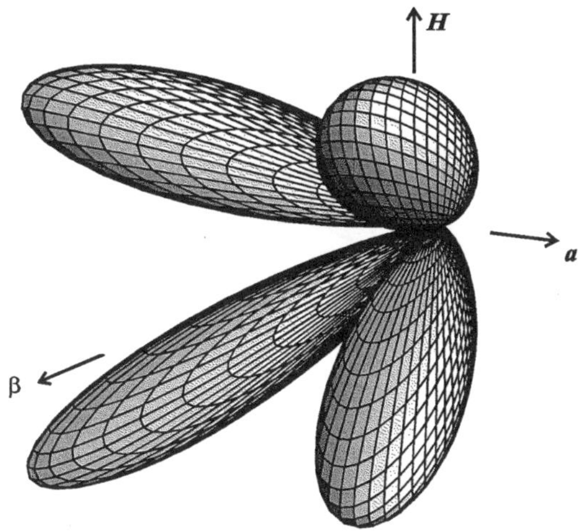

1

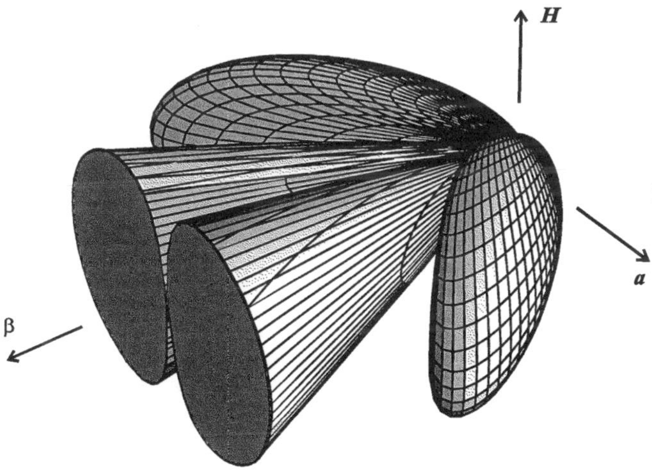

2

Fig. 1.13. Relativistic transformation of the π-components of SR shown in Fig. 1.12 (*1,2*) for $\beta = 0.9$. As the electron moves in a circle in a uniform magnetic field, then, in the first case, only two maxima of radiation disposed symmetrically with respect to the orbit plane containing the vectors $\boldsymbol{\beta}$ and $\boldsymbol{a}$ will be observed after averaging over the period of motion (or over the azimuthal angle φ) [see also Fig. 1.16].

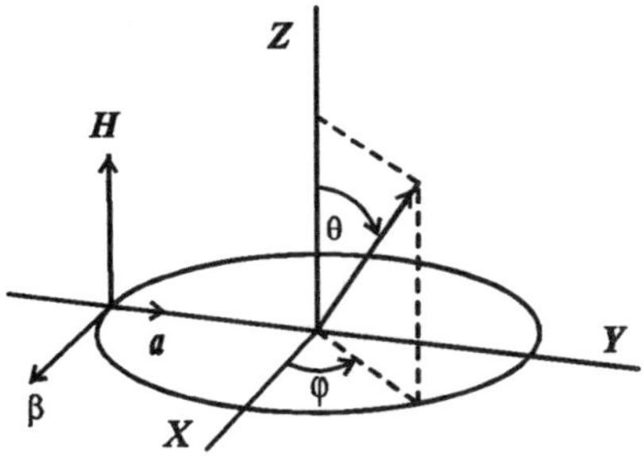

Fig. 1.14. System of coordinates in the problem of the electron SR.

1.5. Electron Radiation in a Uniform Magnetic Field

Here we shall consider the electron radiation in a uniform magnetic field on the basis of the exact solutions of equations of motion. With this end in view we shall use a coordinate system shown in Fig. 1.14. Without restriction of generality we shall consider that the electron at initial time is at the point with the coordinates $\boldsymbol{r} = (0, -\rho, 0)$. In this case the initial conditions are adapted to choice of coordinate system of type Fig. 1.4a.

The necessary solutions of the equations of motion have the form ($e = -e_0 < 0$)

$$\begin{aligned} r^\mu(t) &= (ct, \rho \sin \omega t, -\rho \cos \omega t, c\beta_{\|} t), \\ v^\mu(t) &= c\gamma(1, \beta_\perp \cos \omega t, \beta_\perp \sin \omega t, \beta_{\|}), \\ w^\mu(t) &= c\gamma^2(0, -\omega\beta_\perp \sin \omega t, \omega\beta_\perp \cos \omega t, 0), \end{aligned} \tag{1.43}$$

where

$$\rho = \frac{c\beta}{\omega}, \quad \omega = \frac{e_0 H}{m_0 c\gamma}, \quad \beta_{\|} = \left(\boldsymbol{\beta}\frac{\boldsymbol{H}}{H}\right), \quad \beta_\perp = \sqrt{\beta^2 - \beta_{\|}^2}.$$

Taking account of

$$n^\rho = (1, \boldsymbol{n}) = (1, \sin\theta\cos\varphi, \sin\theta\sin\varphi, \cos\theta)$$

we have

$$\begin{aligned} n_\rho v^\rho &= -c\gamma(1 - \beta_{\|}\cos\theta)\left[1 - \frac{\beta_\perp \sin\theta}{1 - \beta_{\|}\cos\theta}\cos(\omega t - \varphi)\right], \\ n_\rho w^\rho &= -c\gamma^2\beta_\perp\omega\sin\theta\sin(\omega t - \varphi). \end{aligned}$$

Substituting these expressions into Eq. (1.31) we obtain the angular distribution of radiation

$$\frac{dW}{d\Omega} = \frac{C}{A^3}\left[\frac{1}{(1 - a\cos x)^3} - \frac{c^2\sin^2\theta\sin^2 x}{A^2(1 - a\cos x)^5}\right],$$

where $x = \omega t - \varphi$,

$$A = c\gamma(1 - \beta_{\|}\cos\theta), \quad a = \frac{\beta_{\perp}\cos\theta}{1 - \beta_{\|}\cos\theta}, \quad C = \frac{1}{4\pi}e_0^2c^2\gamma^3\beta_{\perp}^2\omega^2.$$

At $\beta_{\perp} = 0$, that corresponds to uniform and rectilinear motion along Z-axis, the radiation is absent.

Further we shall carry out averaging of the angular distribution over the period T of revolution of an electron in an orbit

$$\overline{\frac{dW}{d\Omega}} = \frac{1}{T}\int_0^T \frac{dW}{d\Omega}dt.$$

Integration is carried out by means of the expressions

$$\int_0^{2\pi} \frac{dx}{(1 - a\cos x)^3} = \pi\frac{2 + a^2}{(1 - a^2)^{5/2}}, \quad \int_0^{2\pi} \frac{\sin^2 x dx}{(1 - a\cos x)^5} = \frac{\pi}{4}\frac{4 + a^2}{(1 - a^2)^{7/2}}.$$

After averaging over the period the dependence on the azimuthal angle φ disappears, therefore, integration over φ is reduced to a simple multiplication by 2π. Finally, we obtain

$$\overline{W} = \pi\frac{C}{A^3}\left[\frac{2 + a^2}{(1 - a^2)^{5/2}} - \frac{c^2\sin^2\theta(4 + a^2)}{4A^2(1 - a^2)^{7/2}}\right]\sin\theta d\theta. \tag{1.44}$$

Integration over θ is carried out by means of Lorentz transformations for angles[1] (the primed system moves along Z-axis at the velocity $c\beta_{\|}$)

$$\cos\theta = \frac{\cos\theta' + \beta_{\|}}{1 + \beta_{\|}\cos\theta'}, \quad \sin\theta = \frac{\sqrt{1 - \beta_{\|}^2}\sin\theta'}{1 + \beta_{\|}\cos\theta'},$$

$$\sin\theta d\theta = \frac{1 - \beta_{\|}^2}{(1 + \beta_{\|}\cos\theta')^2}\sin\theta' d\theta'.$$

Finally, we obtain

$$\overline{W} = \frac{2}{3}\frac{e_0^4H^2}{m_0^2c^3}\frac{\beta^2 - \beta_{\|}^2}{1 - \beta^2}. \tag{1.45}$$

Thus, over period the average of electron radiation power in a uniform magnetic field coincides with instantaneous radiation power. It appears that a similar assertion (the theorem about the average power of radiation) holds for SR when $\beta_{\|} = 0$, too.

One can carry out the above calculations in another way determining, first, the polarization components of the electric field of radiation.

Let us show this with the example of SR. In the chosen coordinate system we have

$$\begin{aligned} n_\rho v^\rho &= -c\gamma[1-\beta\sin\theta\cos(\omega t-\varphi)], \quad n_\rho w^\rho = -\gamma^2 a\sin\theta\sin(\omega t-\varphi), \\ (\boldsymbol{n_\theta v}) &= c\gamma\beta\cos\theta\cos(\omega t-\varphi), \quad (\boldsymbol{n_\varphi v}) = c\gamma\beta\sin(\omega t-\varphi), \\ (\boldsymbol{n_\theta w}) &= -\gamma^2 a\cos\theta\sin(\omega t-\varphi), \quad (\boldsymbol{n_\varphi w}) = \gamma^2 a\cos(\omega t-\varphi). \end{aligned}$$

The initial position of an electron in our coordinate system (see Fig. 1.14) is phased so that at $t=0$ these relations coincide with the corresponding relations in the system of a-type in Sec. 1.4 (see Fig. 1.4a).

Now it is not difficult to find the polarization components of an electric field. According to Eq. (1.39), we obtain

$$\widetilde{E}_\theta = -\frac{e_0 a}{\widetilde{r}c^2}\frac{\cos\theta\sin(\omega t-\varphi)}{1-\beta\sin\theta\cos(\omega t-\varphi)}, \quad \widetilde{E}_\varphi = -\frac{e_0 a}{\widetilde{r}c^2}\frac{\beta\sin\theta-\cos(\omega t-\varphi)}{1-\beta\sin\theta\cos(\omega t-\varphi)}. \tag{1.46}$$

At $t=0$ we have Eq. (1.40) for $E^H_{\theta,\varphi}$ where it should be replaced by $e\to -e_0$. Averaging over the period T should be carried out for the values

$$\begin{aligned} \overline{[1-(\boldsymbol{n\beta})]\widetilde{E}^2_\theta} &= \frac{e_0^2a^2}{\widetilde{r}^2c^3}\frac{\cos^2\theta(4+\beta^2\sin^2\theta)}{8(1-\beta^2\sin^2\theta)^{7/2}}, \\ \overline{[1-(\boldsymbol{n\beta})]\widetilde{E}^2_\varphi} &= \frac{e_0^2a^2}{\widetilde{r}^2c^3}\frac{4+3\beta^2\sin^2\theta}{8(1-\beta^2\sin^2\theta)^{5/2}}. \end{aligned}$$

After that we find the components of the radiation power

$$\begin{aligned} \overline{W}_s &= \frac{cr^2}{4\pi}\oint d\Omega\overline{[1-(\boldsymbol{n\beta})]E^2_s}, \\ \overline{W}_\theta &= \frac{e_0^2a^2}{16c^3}\int_0^\pi \frac{\cos^2\theta(4+\beta^2\sin^2\theta)}{(1-\beta^2\sin^2\theta)^{7/2}}\sin\theta d\theta, \\ \overline{W}_\varphi &= \frac{e_0^2a^2}{16c^3}\int_0^\pi \frac{4+3\beta^2\sin^2\theta}{(1-\beta^2\sin^2\theta)^{5/2}}\sin\theta d\theta. \end{aligned} \tag{1.47}$$

The total radiation can be represented in the form

$$\overline{W} = \overline{W}_\theta + \overline{W}_\varphi = \frac{e_0^2a^2}{4c^3}\int_0^\pi\left[\frac{2+\beta^2\sin^2\theta}{(1-\beta^2\sin^2\theta)^{5/2}} - \frac{\sin^2\theta(4+\beta^2\sin^2\theta)}{4\gamma^2(1-\beta^2\sin^2\theta)^{7/2}}\right]\sin\theta d\theta. \tag{1.48}$$

It is easy to see that this expression coincides with Eq. (1.44) if we put there

$$\beta=\beta_\perp, \quad A=c\gamma, \quad a=\beta\sin\theta, \quad C=e_0^2a^2\gamma^3/4\pi.$$

Thus, integration over φ and averaging over the period of motion give the same angular distributions of SR with respect to θ. Further integration over θ in Eqs (1.47) and (1.48) gives the known results of Eq. (1.42)

$$\begin{aligned} \overline{W}_\theta &= \frac{2-\beta^2}{8} W_{\text{SR}}, \quad \overline{W}_\varphi = \frac{6+\beta^2}{8} W_{\text{SR}}, \\ \overline{W} &= \overline{W}_\theta + \overline{W}_\varphi = W_{\text{SR}}. \end{aligned} \tag{1.49}$$

We have in the nonrelativistic case ($\beta \ll 1$)

$$\overline{W}_\theta = \frac{1}{4} W_{\text{SR}}, \quad \overline{W}_\varphi = \frac{3}{4} W_{\text{SR}}, \tag{1.50}$$

and in the ultrarelativistic one ($\beta \to 1$)

$$\overline{W}_\theta = \frac{1}{8} W_{\text{SR}}, \quad \overline{W}_\varphi = \frac{7}{8} W_{\text{SR}}. \tag{1.51}$$

The obtained results for electron radiation in the homogeneous magnetic field have the most general character including the radiation during the spiral motion at a high velocity (helical undulator). We also note that the above theorems on the average power and angular distribution of SR hold true for nonhomogeneous magnetic fields for the motion in the plane orthogonal to magnetic field direction, too[40,41].

1.6. Spectral-Angular Distribution and Circular Polarization of SR

The spectral properties of radiation are determined by the law of motion of a radiating particle. Here we shall consider the radiation with a completely formed spectrum. The problem of formation of the relativistic charged particle spectrum is discussed in detail in Refs[42–44] (see also Chapter 4).

1.6.1. General Information

The energy quantity received by the observer within the solid angle $d\Omega$ is determined by Eq. (1.34)

$$\frac{d}{d\Omega}\left(\frac{\partial \tilde{\mathcal{E}}}{\partial \tilde{t}}\right) = \frac{\tilde{r}^2 c}{4\pi} \widetilde{E}^2. \tag{1.52}$$

To determine the spectrum of radiation it is necessary to integrate this expression over $\tilde{t}$ within the limits of total observation time and use Parseval's theorem[1]:

$$\int_{-\infty}^{\infty} \widetilde{E}^2(\tilde{t}) d\tilde{t} = \frac{1}{\pi} \int_0^{\infty} \left|\widetilde{\boldsymbol{E}}_{\tilde{\omega}}\right|^2 d\tilde{\omega}, \quad \widetilde{\boldsymbol{E}}_{\tilde{\omega}} = \int_{-\infty}^{\infty} \widetilde{\boldsymbol{E}}(\tilde{t}) e^{i\tilde{\omega}\tilde{t}} d\tilde{t} = \widetilde{\boldsymbol{E}}^*_{-\tilde{\omega}},$$

where the complex conjugation is denoted by $*$.

Integrating over $\tilde{t}$ we obtain the spectral-angular distribution of the energy emitted in the polarization components ($s = \theta, \varphi$)

$$\frac{d^2\tilde{\mathcal{E}}^s}{d\tilde{\omega}d\Omega} = \frac{\tilde{r}^2 c}{4\pi^2}\left|\widetilde{\boldsymbol{E}}^s_{\tilde{\omega}}\right|^2. \tag{1.53}$$

Dividing by the period $T = c\beta/2\pi\rho$ and applying a theorem on the average power of SR we find the spectral-angular distribution of SR power

$$\overline{\frac{d^2W^s_{\mathrm{SR}}}{d\tilde{\omega}d\Omega}} = \frac{\tilde{r}^2 c^2\beta}{8\pi^3\rho}\left|\widetilde{\boldsymbol{E}}^s_{\tilde{\omega}}\right|^2. \tag{1.54}$$

Similar arguments can be applied for periodical motion. Integrating Eq. (1.52) over $\tilde{t}$ within the limits of one period we obtain

$$\frac{d\tilde{\mathcal{E}}}{d\Omega} = \frac{\tilde{r}^2 c}{4\pi}\int\limits_0^T \tilde{E}^2(\tilde{t})d\tilde{t}.$$

Hence it follows

$$\overline{\frac{dW_{\mathrm{SR}}}{d\Omega}} = \frac{\tilde{r}c}{4\pi}\overline{\widetilde{\boldsymbol{E}}^2(\tilde{t})}. \tag{1.55}$$

The radiation fields $\widetilde{\boldsymbol{E}}(\tilde{t})$ and $\widetilde{\boldsymbol{H}}(\tilde{t})$ are composed of plane monochromatic waves with frequencies divisible to the frequency of the periodical motion $\tilde{\omega} = n\omega$

$$\widetilde{\boldsymbol{E}}(\tilde{t}) = \sum_{n=1}^{\infty}\boldsymbol{e}_n(\tilde{t}), \quad \widetilde{\boldsymbol{H}}(\tilde{t}) = [\boldsymbol{n}\widetilde{\boldsymbol{E}}(\tilde{t})] = \sum_{n=1}^{\infty}\boldsymbol{h}_n(\tilde{t}).$$

As in the case with the total field, the field of the plane monochromatic wave can be expanded into the components of linear polarization, too

$$\boldsymbol{e}_n(\tilde{t}) = e_{n\varphi}(\tilde{t})\boldsymbol{n}_\varphi + e_{n\theta}(\tilde{t})\boldsymbol{n}_\theta,$$

$$\boldsymbol{h}_n(\tilde{t}) = e_{n\theta}(\tilde{t})\boldsymbol{n}_\varphi - e_{n\varphi}(\tilde{t})\boldsymbol{n}_\theta.$$

Any plane monochromatic wave in a general case has elliptic polarization. The right-polarized elliptic oscillations of the vector $\boldsymbol{e}_n(\tilde{t})$ at the observation point are shown in Fig. 1.15. The resulting motion of the vector $\boldsymbol{e}_n(\tilde{t})$ can be conditioned by either two mutually opposite circular oscillations with amplitudes $a_n^{(\pm)}/\sqrt{2}$ (the reason of the appearance of the factor $1/\sqrt{2}$ will be explained further)

$$e^{(+)}_{n\varphi}(\tilde{t}) = \frac{a_n^{(+)}}{\sqrt{2}}\cos n\omega\tilde{t}, \quad e^{(+)}_{n\theta}(\tilde{t}) = -\frac{a_n^{(+)}}{\sqrt{2}}\sin n\omega\tilde{t},$$

$$e^{(-)}_{n\varphi}(\tilde{t}) = \frac{a_n^{(-)}}{\sqrt{2}}\cos n\omega\tilde{t}, \quad e^{(-)}_{n\theta}(\tilde{t}) = \frac{a_n^{(-)}}{\sqrt{2}}\sin n\omega\tilde{t},$$

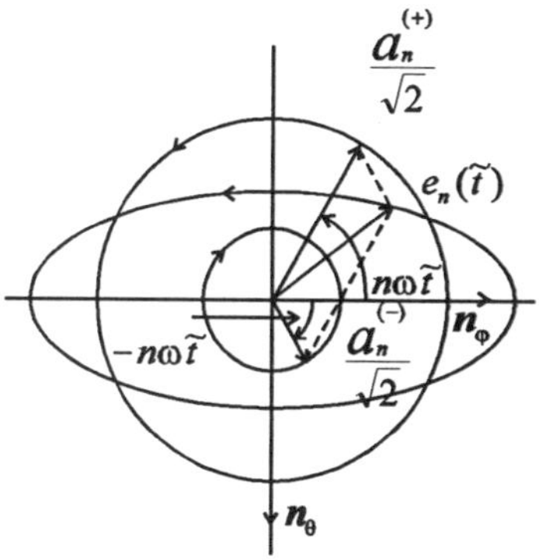

Fig. 1.15. Elliptic polarization of the n-th harmonic of a plane monochromatic wave propagating from the plane of the figure to the reader. The unit vectors of linear polarization $\boldsymbol{n}_{\theta,\varphi}$ are adapted for the coordinate system of the type of Fig. 1.4a. The elliptic axes coincide in direction with $\boldsymbol{n}_\theta$ and $\boldsymbol{n}_\varphi$. The initial moment $\tilde{t}$ is chosen so that $e_n(0) = e_{n,\max}$. The index $(+)$ corresponds to the radiation with right-circular polarization and $(-)$ to that with left-circular polarization.

or by linearly polarized oscillations

$$e_{n\varphi}(\tilde{t}) = \frac{a_n^{(+)} + a_n^{(-)}}{\sqrt{2}} \cos n\omega\tilde{t}, \quad e_{n\theta}(\tilde{t}) = -\frac{a_n^{(+)} - a_n^{(-)}}{\sqrt{2}} \sin n\omega\tilde{t}.$$

We denote the amplitudes of linearly polarized oscillations by

$$b_{n\varphi} = \frac{a_n^{(+)} + a_n^{(-)}}{\sqrt{2}}, \quad b_{n\theta} = -\frac{a_n^{(+)} - a_n^{(-)}}{\sqrt{2}}.$$

The reverse relations have the form

$$a_n^{(+)} = \frac{b_{n\varphi} + b_{n\theta}}{\sqrt{2}}, \quad a_n^{(-)} = \frac{b_{n\varphi} - b_{n\theta}}{\sqrt{2}}. \tag{1.56}$$

The degree of ellipticity of oscillations is given by the relation

$$\delta_n = \frac{a_n^{(+)} - a_n^{(-)}}{a_n^{(+)} + a_n^{(-)}} = \frac{b_{n\theta}}{b_{n\varphi}}.$$

If $\delta_n > 0$, then the resulting oscillation will be right-polarized. If $\delta_n < 0$ it will be left-polarized and at $\delta_n = 0$ it will be linearly polarized.

The square of the vector $\boldsymbol{e}_n(\tilde{t})$ averaged over the oscillation period is connected with the amplitudes $a_n^{(\pm)}$ and $b_{n\theta,\varphi}$ by the simple relation

$$\overline{e_n^2(\tilde{t})} = \frac{1}{2}\left[\left(a_n^{(+)}\right)^2 + \left(a_n^{(-)}\right)^2\right] = \frac{1}{2}\left(b_{n\varphi}^2 + b_{n\theta}^2\right).$$

Now it is clear why the factor $1/\sqrt{2}$ was introduced into the amplitudes $a_n^{(\pm)}$. Thus we have

$$\widetilde{\boldsymbol{E}}(\tilde{t}) = \sum_{n=1}^{\infty} \left(b_{n\varphi} \cos n\omega\tilde{t}\boldsymbol{n}_\varphi - b_{n\theta} \sin n\omega\tilde{t}\boldsymbol{n}_\theta \right).$$

It is easy to find on this basis the Fourier-components of the electric field of radiation. According to definition[1]

$$\widetilde{\boldsymbol{E}}_n = \frac{1}{T}\int_0^T \widetilde{\boldsymbol{E}}(\tilde{t}) e^{in\omega\tilde{t}} d\tilde{t}$$

we have

$$\widetilde{\boldsymbol{E}}_n = \frac{1}{2}\left(b_{n\varphi}\boldsymbol{n}_\varphi - ib_{n\theta}\boldsymbol{n}_\theta\right), \quad n > 0.$$

Hence it follows that

$$\left|\widetilde{\boldsymbol{E}}_n\right|^2 = \frac{1}{4}\left(b^2_{n\varphi} + b^2_{n\theta}\right)$$

and

$$\overline{E^2(\tilde{t})} = 2\sum_{n=1}^{\infty}\left|\widetilde{\boldsymbol{E}}_n\right|^2 = \frac{1}{2}\left(b^2_{n\varphi} + b^2_{n\theta}\right).$$

Coming back to Eq. (1.55) we obtain the angular distribution of radiation of each harmonic taken separately

$$\frac{\overline{dW_{n\mathrm{SR}}}}{d\Omega} = \frac{\tilde{r}^2 c}{8\pi}\left(b^2_{n\varphi} + b^2_{n\theta}\right) = \frac{\tilde{r}^2 c}{8\pi}\left[\left(a_n^{(+)}\right)^2 + \left(a_n^{(-)}\right)^2\right]. \tag{1.57}$$

1.6.2. Fourier-Components of Radiation Fields

Now we pass from these general arguments to more definite ones by application of the tensor of the electromagnetic radiation field $\widetilde{H}^{\mu\nu}$. With this end in view we rewrite the expression for tensor $\widetilde{H}^{\mu\nu}$ in the wave zone given by Eq. (1.20) in the form

$$\widetilde{H}^{\mu\nu}((\tilde{t})) = -\frac{1}{c}\frac{\partial}{\partial\tilde{t}}\, n^{[\mu}\widetilde{A}^{\nu]} = -\frac{e}{\tilde{r}c}\frac{\partial}{\partial\tilde{t}}\frac{n^{[\mu}\beta^{\nu]}}{1-(\boldsymbol{n}\boldsymbol{\beta})},$$

where $\beta^\nu = (1, \boldsymbol{\beta})$.

Further application of this formula arouses difficulties because the objects of differentiation with respect to $\tilde{t}$ depend on the trajectory time t. This difficulty may be avoided by means of δ-function[9]

$$\delta(t-t^*) = [1-(\boldsymbol{n\beta})]\,\delta\left(t+\frac{1}{c}\tilde{r}(t)-\tilde{t}\right).$$

Thus we obtain

$$\widetilde{H}^{\mu\nu}(\tilde{t}) = -\frac{e}{\tilde{r}c}\frac{\partial}{\partial\tilde{t}}\int_{-\infty}^{\infty} n^{[\mu}\beta^{\nu]}\delta\left(t-\tilde{t}+\frac{\tilde{r}}{c}\right)dt.$$

Using the standard representation of δ-function

$$\delta\left(t-\tilde{t}+\frac{\tilde{r}}{c}\right) = \frac{1}{2\pi}\int_{-\infty}^{\infty} e^{i\tilde{\omega}\left(t-\tilde{t}+\frac{\tilde{r}}{c}\right)}d\tilde{\omega},$$

where $\tilde{r} = R-(\boldsymbol{nr})$ and, therefore,

$$\delta\left(t-\tilde{t}+\frac{\tilde{r}}{c}\right) \approx \frac{1}{2\pi}\int_{-\infty}^{\infty} e^{i\tilde{\omega}\left[t-\frac{1}{c}(\boldsymbol{nr})\right]-i\tilde{\omega}\tilde{t}+i\frac{\tilde{\omega}}{c}R}d\tilde{\omega}$$

with an accuracy to the phase factor $\exp(i\tilde{\omega}R/c)$ which is not essential in further calculations we find

$$\widetilde{H}^{\mu\nu} = \frac{1}{2\pi}\int_{-\infty}^{\infty}\widetilde{H}^{\mu\nu}_{\tilde{\omega}}e^{-i\tilde{\omega}\tilde{t}}d\tilde{\omega},$$

where

$$\widetilde{H}^{\mu\nu}_{\tilde{\omega}} = \frac{ie}{\tilde{r}c}\tilde{\omega}\int_{-\infty}^{\infty} n^{[\mu}\beta^{\nu]}e^{i\tilde{\omega}\left[t-\frac{1}{c}(\boldsymbol{nr})\right]}dt.$$

Hence it is easy to find

$$\widetilde{\boldsymbol{E}}_{\tilde{\omega}} = \frac{ie\tilde{\omega}}{\tilde{r}c}\int_{-\infty}^{\infty}(\boldsymbol{\beta}-\boldsymbol{n})e^{i\tilde{\omega}\left[t-\frac{1}{c}(\boldsymbol{nr})\right]}dt, \quad \widetilde{\boldsymbol{H}}_{\tilde{\omega}} = [\boldsymbol{n}\widetilde{\boldsymbol{E}}_{\tilde{\omega}}]. \tag{1.58}$$

The unit vector $\boldsymbol{n}$ outside the exponent can be omitted as the contribution of longitudinal component to the total radiation is equal to zero

$$(\boldsymbol{n}\widetilde{\boldsymbol{E}}_{\tilde{\omega}}) \sim \tilde{\omega}\delta(\tilde{\omega}) \to 0.$$

---

[9]This formula can be obtained by means of the known property of δ-function having a complex argument[4,45]

$$\delta[f(t)] = \sum_{\alpha}\frac{\delta(t-t_\alpha)}{|\partial f/\partial t_\alpha|},$$

where $f(t) = t+\frac{1}{c}\tilde{r}(t)-\tilde{t}$, t_α are the roots of the equation $\tilde{r}(t) = c(\tilde{t}-t) = 0$. The latter requirement is an expression of causality condition according to which $t = t_\alpha$ is an emission moment and $\tilde{t}$ is a moment of radiation fixation. In the considering theory ($\beta < 1$) there is only one root $t_\alpha = t^*$ that regards to one-to-one correspondence between the emission time t^* and radiation fixation time $\tilde{t}$. Another situation appears in the medium where $\beta > 1$[46,47].

By analogy, the Eq. (1.58) can be obtained directly from the Liénard-Wiechert potentials. In this case

$$\widetilde{A}^{\mu} = \frac{1}{2\pi}\int\limits_{-\infty}^{\infty} \widetilde{A}^{\mu}_{\widetilde{\omega}} e^{-i\widetilde{\omega}\tilde{t}} d\widetilde{\omega}$$

and

$$\widetilde{A}^{\mu}_{\widetilde{\omega}} = \frac{e}{\tilde{r}}\widetilde{\omega}\int\limits_{-\infty}^{\infty} \beta^{\mu} e^{i\widetilde{\omega}\left[t-\frac{1}{c}(\boldsymbol{nr})\right]} dt.$$

Further we shall use the relations which follow from Eq. (1.58)

$$\widetilde{\boldsymbol{E}}_{\widetilde{\omega}} = -i\frac{\omega}{c}[\boldsymbol{n}[\boldsymbol{n}\widetilde{\boldsymbol{A}}_{\widetilde{\omega}}]], \quad \widetilde{\boldsymbol{H}}_{\widetilde{\omega}} = i\frac{\omega}{c}[\boldsymbol{n}\widetilde{\boldsymbol{A}}_{\widetilde{\omega}}].$$

The components $\widetilde{\boldsymbol{E}}_{\widetilde{\omega}}$ and $\widetilde{\boldsymbol{H}}_{\widetilde{\omega}}$ obtained in this way will coincide with Eq. (1.58) if the above notation about zero contribution of the longitudinal component of $\widetilde{\boldsymbol{E}}_{\widetilde{\omega}}$ to radiation is taken into account. Automatically, the absence of the longitudinal component of the electric field in the plane wave is ensured by Coulomb's gauge of potentials[1]. Then

$$\widetilde{\boldsymbol{E}} = -\frac{1}{c}\dot{\widetilde{\boldsymbol{A}}}, \quad \widetilde{\boldsymbol{E}}_{\widetilde{\omega}} = i\frac{\widetilde{\omega}}{c}\widetilde{\boldsymbol{A}}_{\widetilde{\omega}}$$

and the fields of Eq. (1.58) are obtained immediately without a longitudinal component of the field.

The analogous arguments for periodical fields give the following result

$$\widetilde{H}^{\mu\nu}(\tilde{t}) = \sum_{n=-\infty}^{\infty} \widetilde{H}^{\mu\nu}_{n} e^{-in\omega\tilde{t}}, \quad \widetilde{H}^{\mu\nu}_{n} = \frac{ie\omega^2 n}{2\pi\tilde{r}c}\int\limits_{0}^{T} n^{[\mu}\beta^{\nu]} e^{in\omega\left[t-\frac{1}{c}(\boldsymbol{nr})\right]} dt.$$

Hence we find the Fourier-components of the radiation field

$$\widetilde{\boldsymbol{E}}_{n} = \frac{ie\omega n}{\tilde{r}c}\frac{1}{T}\int\limits_{0}^{T} \boldsymbol{\beta} e^{in\omega\left[t-\frac{1}{c}(\boldsymbol{nr})\right]} dt.$$

1.6.3. Schott's Formula

Now we consider the polarization of Fourier components of the electric field of SR. In the chosen coordinate system of a-type (see Fig. 1.4a) we have

$$(\boldsymbol{nr}) = \rho \sin\theta \sin(\omega t - \varphi),$$

$$(\boldsymbol{\beta n_{\theta}}) = \beta\cos\theta\cos(\omega t - \varphi), \quad (\boldsymbol{\beta n_{\varphi}}) = \beta\sin(\omega t - \varphi).$$

Further we shall use the integral representation of Bessel functions[47]

$$\begin{aligned} J_n(nx) &= \frac{1}{2\pi}\int_{-\pi}^{\pi} e^{in(\Phi - x\sin\Phi)}d\Phi, \\ iJ_n'(nx) &= \frac{1}{2\pi}\int_{-\pi}^{\pi} \sin\Phi e^{in(\Phi - x\sin\Phi)}d\Phi, \\ \frac{1}{x}J_n(nx) &= \frac{1}{2\pi}\int_{-\pi}^{\pi} \cos\Phi e^{in(\Phi - x\sin\Phi)}d\Phi. \end{aligned} \tag{1.59}$$

By means of Eq. (1.59) we easily find for SR ($x = \beta\sin\theta, \Phi = \omega\ t - \varphi$)

$$\begin{aligned} \widetilde{E}_{n\varphi} &= \frac{e_0\beta^2}{\tilde{r}\rho} nJ_n'(n\beta\sin\theta)e^{in\varphi}, \\ \widetilde{E}_{n\theta} &= -i\frac{e_0\beta}{\tilde{r}\rho}\cot\theta nJ_n(n\beta\sin\theta)e^{in\varphi}. \end{aligned} \tag{1.60}$$

It is possible now to write the Fourier series for polarization components. In the real form

$$\begin{aligned} \widetilde{E}_{\varphi}(\tilde{t}) &= \frac{e_0\beta^2}{\tilde{r}\rho}\sum_{n=-\infty}^{\infty} nJ_n'(n\beta\sin\theta)\cos n(\omega\tilde{t} - \alpha), \\ \widetilde{E}_{\theta}(\tilde{t}) &= -\frac{e_0\beta}{\tilde{r}\rho}\cot\theta\sum_{n=-\infty}^{\infty} nJ_n(n\beta\sin\theta)\sin n(\omega\tilde{t} - \alpha), \end{aligned} \tag{1.61}$$

where $\alpha = \omega R/c + \varphi$.

Using the sign rules for the Bessel functions

$$J_n(nx) = J_{-n}(-nx), \quad J_n'(nx) = -J_{-n}'(-nx),$$

we obtain

$$\widetilde{E}_{\varphi}(\tilde{t}) = \sum_{n=1}^{\infty} b_{n\varphi}\cos n(\omega\tilde{t} - \alpha), \quad \widetilde{E}_{\theta}(\tilde{t}) = -\sum_{n=1}^{\infty} b_{n\theta}\sin n(\omega\tilde{t} - \alpha), \tag{1.62}$$

where

$$b_{n\varphi} = 2\frac{e_0\beta^2}{\tilde{r}\rho} nJ_n'(n\beta\sin\theta), \quad b_{n\theta} = 2\frac{e_0\beta}{\tilde{r}\rho} n\cot\theta J_n(n\beta\sin\theta). \tag{1.63}$$

Substituting the latter expression into Eq. (1.57) we obtain the spectral-angular distribution of SR

$$\overline{\frac{dW_{n\mathrm{SR}}}{d\Omega}} = \frac{e_0^2 c\beta^2}{2\pi\rho^2} n^2\left[\beta^2 J_n'^2(n\beta\sin\theta) + \cot^2\theta J_n^2(n\beta\sin\theta)\right]. \tag{1.64}$$

The first term in the square brackets corresponds to the σ-component and the second one to the π-component of the linear polarization. Eq. (1.64) was derived in 1907 by G.A. Schott[49].[h] It proves to be correct for arbitrary velocities.

A reservation should be made, nevertheless, that in the superstrong magnetic fields at

$$\gamma > \gamma_c = \frac{H^*}{H}, \quad H^* = \frac{m_0^2 c^3}{e\hbar},$$

where $H^* = 4.41 \cdot 10^{13}$ Oe is Schwinger's magnetic field, it is impossible to apply either Schott's formula or the known expression of the total power of SR because the quantum effects occur and the classical radiation theory loses sense (for more detail see Chapter 7).

As $b_{n\theta,\varphi}$ is known, using Eqs (1.56) and (1.63) we find

$$a_n^{(\pm)} = \sqrt{2}\,\frac{e_0\beta}{\tilde{r}\rho} n\left[\beta J_n'(n\beta\sin\theta) \pm \cot\theta J_n(n\beta\sin\theta)\right].$$

The spectral-angular distribution of the circular polarization of radiation has the form

$$\frac{\overline{dW_{\rm SR}^{(\pm)}}}{d\Omega} = \frac{1}{2}\,\frac{e_0^2 c\beta^2}{2\pi\rho^2}\, n^2\left[\beta J_n'(n\beta\sin\theta) \pm \cot\theta J_n(n\beta\sin\theta)\right]^2. \tag{1.65}$$

It is of some interest to generalize Schott's formula to the case of electron motion along a spiral in the uniform magnetic field. This problem was solved in Ref.[30] (see also Ref.[32]). Here we adduce the final expression

$$\frac{\overline{dW_n}}{d\Omega} = \frac{e_0^2 c\beta_\perp^2}{2\pi\rho^2}\, n^2\, \frac{\beta_\perp^2 J_n'^2(x) + (\cos\theta - \beta_{||})^2/\sin^2\theta J_n^2(x)}{(1-\beta_{||}\cos\theta)^3}, \tag{1.66}$$

where

$$x = \beta_\perp \frac{n\sin\theta}{1-\beta_{||}\cos\theta}.$$

But we note that this formula can be obtained in a simple way by means of the Lorentz transformations. Actually, we assume that the plane motion takes place in the primed coordinate system which moves along the direction of the magnetic field at the velocity of $c\beta_{||}$. Then the Lorentz transformations for the angular distribution of radiation have the form

$$\frac{\overline{dW_n}}{d\Omega} = \frac{1}{\gamma_{||}^4(1-\beta_{||}\cos\theta)^3}\,\frac{\overline{dW_n'}}{d\Omega'},$$

where $\overline{dW_n'}/d\Omega'$ corresponds to Schott's formula where all transformed values are denoted by primes. To replace them, we use the transformations

$$\beta_\perp' = \gamma_{||}\beta_\perp, \quad \cot\theta' = \gamma_{||}\frac{\cos\theta - \beta_{||}}{\sin\theta}, \quad \beta_\perp'\sin\theta' = x.$$

[h] Schott tried to explain the stability of rotational electron states in atoms by means of this formula. Later this formula fell into oblivion, and about 40 years later it was employed quite unexpectedly in the theory of synchrotron radiation.

After that we come to Eq. (1.66).

Before analyzing Eqs (1.64) and (1.65) we consider the spectral composition of radiation.

1.6.4. Spectral Composition

The spectral distribution of radiation can be obtained by integration of Schott's formula over the angles (really over θ). This problem is solved by means of integrals

$$\int_0^\pi J_n^2(n\beta\sin\theta)\sin\theta d\theta = \frac{1}{n\beta}\int_0^{2n\beta} J_{2n}(x)dx, \quad \int_0^\pi J_n^2(n\beta\sin\theta)\frac{d\theta}{\sin\theta} = 2\int_0^{2n\beta} J_{2n}(x)dx,$$

$$\int_0^\pi J_n'^2(n\beta\sin\theta)\sin\theta d\theta = \frac{2}{n\beta}J_{2n}'(2n\beta) + \frac{1}{n\beta}\int_0^{2n\beta} J_{2n}(x)dx - \frac{2}{\beta^2}\int_0^{2n\beta} J_{2n}(x)\frac{dx}{x}.$$

These formulas can be easily verified by the termwise integration of the Bessel functions series. Integrating in Eq. (1.64) over $d\Omega = 2\pi\sin\theta\ d\theta$ we obtain

$$\overline{W_n} = \overline{(W_n)_\theta} + \overline{(W_n)_\varphi},$$

$$\overline{(W_n)_\theta} = \frac{e_0^2 c\beta^2}{\rho^2} n \left[2n\int_0^{2n\beta} J_{2n}(x)\frac{dx}{x} - \frac{1}{\beta}\int_0^{2n\beta} J_{2n}(x)dx\right],$$

$$\overline{(W_n)_\varphi} = \frac{e_0^2 c\beta^2}{\rho^2} n \left[2\beta J_{2n}'(2n\beta) + \beta\int_0^{2n\beta} J_{2n}(x)dx - 2n\int_0^{2n\beta} J_{2n}(x)\frac{dx}{x}\right],$$

$$\overline{W_n} = \frac{e_0^2 c\beta^2}{\rho^2} n \left[2\beta^2 J_{2n}'(2n\beta) - (1-\beta^2)\int_0^{2n\beta} J_{2n}(x)dx\right].$$

The total radiation power is obtained by summation over the spectrum. For this purpose it is necessary to use the formulas

$$\sum_{n=1}^\infty 2nJ_{2n}'(2n\beta) = \frac{\beta}{(1-\beta^2)^2}, \quad \sum_{n=1}^\infty n\int_0^{2n\beta} J_{2n}(x)dx = \frac{1}{3}\frac{\beta^3}{(1-\beta^2)^3},$$

which can be reduced from the well known sums for Bessel functions[48].

After the calculation of the spectral sums we obtain the known results

$$\sum_{n=1}^\infty \overline{W_n} = \sum_{n=1}^\infty \left[\overline{(W_n)_\theta} + \overline{(W_n)_\varphi}\right] = \overline{W}_\theta + \overline{W}_\varphi = \overline{W} = W_{\text{SR}},$$

where $\overline{W}_\theta, \overline{W}_\varphi$, and $\overline{W}$ are given by Eqs (1.47) and (1.44).

Summation over the spectrum can be also carried out directly in Schott's formula even before the integration over the angles. Here the sums must be used[50]

$$\sum_{n=1}^{\infty} J_n^2(nb) = \frac{1}{16}\frac{b^2(4+b^2)}{(1-b^2)^{7/2}}, \quad \sum_{n=1}^{\infty} n^2\left[J_n'(nb)\right]^2 = \frac{4+3b^2}{16(1-b^2)^{5/2}}. \tag{1.67}$$

Then we obtain the angular distributions of SR given by Eqs (1.47) and (1.44).

In the general case for an arbitrary $\boldsymbol{\beta}$, the analysis of the spectral distribution of SR is not a simple problem. However, in the nonrelativistic case $\beta \ll 1$, it is not difficult to do this. In this case, it is possible to use an approximative formula for the Bessel functions

$$J_{2n}(2n\beta) \approx \frac{(n\beta)^{2n}}{(2n)!}. \tag{1.68}$$

Applying this formula we obtain ($a = c^2\beta^2/\rho$)

$$\begin{gathered} \overline{W_n}\,|_{\beta\approx 0} \approx \frac{3(n+1)}{(2n+1)!}n^{2n+1}\beta^{2n+2}W_1, \quad W_1 = \frac{2}{3}\frac{e_0^2a^2}{c^3}, \\ \overline{W_{n\theta}} = \frac{1}{2(n+1)}W_n, \quad \overline{W_{n\varphi}} = \frac{2n+1}{2(n+1)}W_n. \end{gathered} \tag{1.69}$$

Hence it follows that in the nonrelativistic approximation the maximum of radiation falls on the first harmonic $n = 1$ which corresponds to the dipole radiation. The radiation power of the rest of the harmonics decreases quickly with increasing n at the expense of the factor of β^{2n+2}.

Now it is possible to analyze the polarization of nonrelativistic radiation. First of all, we note that, according to Eq. (1.69), the total linear polarization at $n = 1$ coincides with known values of Eq. (1.50). The spectral-angular distribution can be obtained by means of the approximated formulas

$$J_1(\beta\sin\theta) \approx \frac{1}{2}\beta\sin\theta, \quad J_1'(\beta\sin\theta) \approx \frac{1}{2}$$

applying these we easily find

$$b_{1\varphi} = \frac{e_0\beta^2}{\tilde{r}\rho}, \quad b_{1\theta} = \frac{e_0\beta^2}{\tilde{r}\rho}\cos\theta.$$

The degree of ellipticity of oscillations of the first harmonic is $\delta_1 = \cos\theta$. Hence it follows that at $\theta = 0$ (along the magnetic field direction) the electron radiation possesses only the right-hand circular polarization and at $\theta = \pi$ (opposite the magnetic field direction) only the left-hand one. At $\theta = \pi/2$ the radiation is completely polarized in the orbit plane as only σ-component is not equal to zero.

The electric fields of the first harmonics $\widetilde{E}_\varphi(\tilde{t})$ and $\widetilde{E}_\theta(\tilde{t})$ in the nonrelativistic approximation have the form

$$\widetilde{E}_\varphi(\tilde{t}) \approx -\frac{e_0a}{\tilde{r}c^2}\cos\theta\sin(\omega\tilde{t}-\alpha), \quad \widetilde{E}_\theta(\tilde{t}) \approx \frac{e_0a}{\tilde{r}c^2}\cos(\omega\tilde{t}-\alpha)$$

and coincide with the total fields in Eq. (1.46) if we put there $\beta \approx 0$. It is interesting that they rotate in the same direction as an electron.

In the most interesting ultrarelativistic case when $\beta \to 1$ the analysis of spectral-angular distribution is not so trivial and it should be considered separately.

1.6.5. The Helical Undulator

Let us consider an important case of ultrarelativistic electron motion along the magnetic field direction when the velocity of orthogonal motion is not high and satisfies the condition

$$\beta_\perp \ll \sqrt{1-\beta_\parallel^2}.$$

In this case, the radiation is concentrated near the direction $\theta = 0$ and is emitted at the basic frequency transformed by the longitudinal Doppler effect

$$\tilde{\omega} = \frac{\omega}{1-\beta_\parallel \cos\theta}, \quad \omega = \frac{eH}{m_0 c\gamma}.$$

The argument of Bessel functions in the generalized Schott formula of Eq. (1.66) is much less than unity and it is possible to take

$$J_1(x) \approx \frac{1}{2}x = \frac{\beta_\perp \sin\theta}{2(1-\beta_\parallel\cos\theta)}, \quad J_1'(x) \approx \frac{1}{2}.$$

If the new variable $\chi = \gamma_\parallel \theta$ ($\gamma_\parallel = 1/\sqrt{1-\beta_\parallel^2} \approx \gamma$) is introduced, it is possible to write down the approximative relations

$$1-\beta_\parallel\cos\theta \approx \frac{1}{2\gamma_\parallel^2}(1+\chi^2), \quad \cos\theta - \beta_\parallel \approx \frac{1}{2\gamma_\parallel^2}(1-\chi^2), \quad \sin\theta d\theta \approx \frac{1}{\gamma_\parallel^2}\chi d\chi.$$

Substituting all these into Eq. (1.66) we obtain

$$\overline{dW_1} = 4\,\frac{e_0^2 c\beta_\perp^4}{\rho^2}\,\gamma_\parallel^4\,\frac{1+\chi^4}{(1+\chi^2)^5}\,\chi d\chi.$$

Integrating over χ by means of the integrals[48]

$$\int_{-\infty}^{\infty} \frac{\chi^{m-1}d\chi}{(1+\chi^2)^n} = \frac{\Gamma\left(\frac{m}{2}\right)\Gamma\left(n-\frac{m}{2}\right)}{\Gamma(n)}, \tag{1.70}$$

we find the total power of radiation

$$\overline{W_1} = \frac{2}{3}\,\frac{e_0^2 c\beta_\perp^4}{\rho^2}\,\gamma_\parallel^4. \tag{1.71}$$

We see that there are two opposite tendencies. In spite of the great increase in radiation power at $\beta_\parallel \to 1$ and $\beta_\perp \neq 0$ (factor of γ^4) it completely vanishes when $\beta_\perp = 0$. An electron moving uniformly and rectilinearly (without regard for an intrinsic magnetic moment, see Chapter 4), does not emit radiation in a magnetic field. The polarization of radiation has some interesting properties. It follows from the generalized Schott formula of Eq. (1.66) that a complete linear polarization is observed at the angle

$$\cos\theta_c = \beta_{||}. \tag{1.72}$$

In the case of SR, the radiation is completely polarized in the orbit plane (σ-component) because $\beta_{||} = 0$ and $\theta_c = \pi/2$. With the appearance of the longitudinal velocity the angle θ_c starts to decrease and in the ultrarelativistic case being considered it is small $\theta_c = \sqrt{1-\beta_{||}^2}$, $(\chi = 1)$. As the solid angle within which the radiation is concentrated is determined by the same value of $\theta = \theta_c$, almost all the radiation has the right-hand circular polarization. This property of a helical undulator makes it an unique source of the radiation with circular polarization.

Simple calculations show that the angular distribution of circularly polarized radiation has the form

$$\frac{\overline{dW^{(\pm)}}}{d\chi} = \frac{e_0^2 c\beta_\perp^4}{\rho^2}\gamma_{||}^4 \frac{1}{(1+\chi^2)^3}\left(\frac{1-\chi^2}{1+\chi^2} \pm 1\right)^2.$$

From here all of the above conclusions follow.

The spectral distribution of radiation can be found by means of δ-function

$$\delta\left(\tilde{\omega} - \frac{\omega}{1-\beta_{||}\cos\theta}\right) \approx \delta\left(\tilde{\omega} - \frac{\omega_{\max}}{1+\chi^2}\right),$$

where $\omega_{\max} = 2\gamma_{||}\omega$. Next, it is convenient to introduce the variables $\xi = 1+\chi^2$ and $y = \tilde{\omega}/\omega_{\max}$. Using the rule of multiplication by unit

$$d\tilde{\omega}\delta\left(\tilde{\omega} - \frac{\omega}{1-\beta_{||}\cos\theta}\right) \approx \xi\delta\left(\xi - \frac{1}{y}\right)\frac{y}{dy},$$

the spectral-angular distribution of radiation can be represented in the form

$$\frac{\overline{d^2W_1}}{d\xi dy} = 2\,\frac{e_0^2 c\beta_\perp^4}{\rho^2}\gamma_{||}^4 \frac{2-2\xi+\xi^2}{y\xi^4}\,\delta\left(\xi - \frac{1}{y}\right).$$

Integrating over the angular variable ξ we obtain the spectral distribution of radiation

$$\frac{\overline{dW_1}}{dy} = 2\,\frac{e_0^2 c\beta_\perp^4}{\rho^2}\gamma_{||}^4\,(1-2y+2y^2)y. \tag{1.73}$$

Further integration over y leads again to Eq. (1.71).

Analogous calculations carried out for the second harmonic give the result[51]

$$\frac{\overline{dW_2}}{dy} = 4\,\frac{e_0^2 c\beta_\perp^6}{\rho^2}\gamma_{||}^6\left(1 - \frac{1}{2}y\right)\left(1 - y + \frac{1}{2}y^2\right)y^2.$$

As compared with the first harmonic, the second one is considerably suppressed by the factor of $\beta_\perp^2\gamma_\|^2 \ll 1$.

Thus, the considered system gives a highly directed radiation of high frequency and can be used as a source of undulator radiation (for more detail see Chapter 5).

Similarly, it is possible to consider the devices with a very high velocity of orthogonal translational motion. For example, in the constant and uniform mutually orthogonal electric and magnetic fields $\boldsymbol{E} = -[\overline{\boldsymbol{\beta}}\boldsymbol{H}]$ the periodical motion at the constant drift velocity $\overline{\boldsymbol{\beta}}$ in the direction of $(\overline{\boldsymbol{\beta}}\boldsymbol{H}) = 0$ is realized. The radiation in the fields of this type was considered in Refs[27,52]. Drift motion is possible in nonuniform magnetic fields, too. The radiation accompanying the above motion can possess highly unusual characteristics along with the typical SR properties[40].

1.7. SR of the Ultrarelativistic Electron

We have already discussed some properties of ultrarelativistic SR. These are high directivity along the velocity and strong linear polarization. To get a deeper insight into the polarizational and spectral properties of SR in the ultrarelativistic limit $\beta \to 1$ it is necessary to work out a special mathematical procedure.

1.7.1. Radiation of the Ultrarelativistic Charge as SR

Let us analyze the expression for instantaneous radiation power of Eq. (1.31) of an arbitrarily moving charge as a function of the acceleration type. With this end in view we represent the terms with acceleration in Eq. (1.31) in the following way

$$\boldsymbol{a} = \boldsymbol{a}_\perp + \boldsymbol{a}_\|, \quad a_\| = \frac{(\boldsymbol{\beta}\boldsymbol{a})}{\beta}, \quad a^2 - [\boldsymbol{\beta}\boldsymbol{a}]^2 = \frac{1}{\gamma^2}a_\perp^2 + a_\|^2.$$

After that the total power of instantaneous radiation takes the form

$$W = \frac{2}{3}\frac{e^2}{c^3}\,\gamma^4(a_\perp^2 + \gamma^2 a_\|^2). \tag{1.74}$$

One would think that as the velocity increases so that $\gamma \gg 1$, the second term related to the acceleration longitudinal with respect to the velocity dominates here. However, it is not valid. The point is that the acceleration is completely determined by the external electromagnetic field[1]

$$\boldsymbol{a} = \frac{e}{m_0\gamma}\left(\boldsymbol{E} - \boldsymbol{\beta}(\boldsymbol{\beta}\boldsymbol{E}) + [\boldsymbol{\beta}\boldsymbol{H}]\right).$$

Hence it follows that

$$a_\| = \frac{e}{m_0\gamma^3\beta}(\boldsymbol{\beta}\boldsymbol{E}), \quad a_\perp = \frac{e}{m_0\gamma\beta}\,|[\boldsymbol{\beta}\boldsymbol{E}] + [\boldsymbol{\beta}[\boldsymbol{\beta}\boldsymbol{H}]]| \tag{1.75}$$

and, therefore,

$$\frac{a_{\parallel}^2}{a_{\perp}^2} = \frac{1}{\gamma^4}\frac{(\boldsymbol{\beta E})^2}{([\boldsymbol{\beta E}] + [\boldsymbol{\beta}[\boldsymbol{\beta H}]])^2}.$$

With the exception of a special case when $\boldsymbol{H} = 0$ and $\boldsymbol{\beta} \parallel \boldsymbol{E}$ this relation at $\beta \to 1$ falls off as γ^{-4}. Therefore, at high velocities the main contribution to the instantaneous radiation power will be given by the term with orthogonal acceleration which corresponds to SR. Substituting Eq. (1.75) into Eq. (1.74) we obtain

$$W = \frac{2}{3}\frac{e^2}{c^3}\gamma^4 a_{\perp}^2 \left\{1 + \frac{1}{\gamma^2}\frac{(\boldsymbol{\beta E})^2}{([\boldsymbol{\beta E}] + [\boldsymbol{\beta}[\boldsymbol{\beta H}]])^2}\right\}. \tag{1.76}$$

Now it is evident that

$$W\,|_{\beta\to 1} \to W_{\text{SR}}.$$

Further we shall see that the other radiation characteristics in the ultrarelativistic case coincide with the synchrotron radiation characteristics, too.

Beginning with M. Born[20] and W. Pauli[21] (see also Ref.[22]) the exotic case when acceleration is parallel to velocity has been widely discussed in literature[23–25]. It is not our task to take part in this discussion. Although we still have some unsolved questions connected with more profound understanding of electromagnetic radiation.

If the acceleration in Eq. (1.76) is completely excepted by the use of Eq. (1.75), we come to the instantaneous radiation power expressed in terms of the electromagnetic fields

$$W = \frac{2}{3}\frac{e^4\gamma^2}{m_0^2c^3}\left\{E^2\left[1 - \left(\boldsymbol{\beta}\frac{\boldsymbol{E}}{E}\right)^2\right] + H^2\left[\beta^2 - \left(\boldsymbol{\beta}\frac{\boldsymbol{H}}{H}\right)^2\right] - 2(\boldsymbol{\beta}[\boldsymbol{EH}])\right\}.$$

In the particular cases of $\boldsymbol{E} = 0$ we have the SR power [see also Eq. (1.45)]

$$W_H = \frac{2}{3}\frac{e^4H^2}{m_0^2c^3}\frac{\beta^2 - \beta_{\parallel}^2}{1 - \beta^2}, \quad \beta_{\parallel} = \left(\boldsymbol{\beta}\frac{\boldsymbol{H}}{H}\right), \tag{1.77}$$

and at $\boldsymbol{H} = 0$ we obtain the radiation power in the electric field

$$W_E = \frac{2}{3}\frac{e^4E^2}{m_0^2c^3}\frac{1 - \beta_{\parallel}^2}{1 - \beta^2}, \quad \beta_{\parallel} = \left(\boldsymbol{\beta}\frac{\boldsymbol{E}}{E}\right). \tag{1.78}$$

When $\boldsymbol{H} = [\boldsymbol{nE}]$ and $(\boldsymbol{EH}) = 0$ we have the instantaneous radiation power in the plane electromagnetic wave

$$W_{\sim} = \frac{2}{3}\frac{e^4E^2}{m_0^2c^3}\frac{(1 - \beta_{\perp})^2}{1 - \beta^2}, \quad \beta_{\perp} = \frac{|[\boldsymbol{EH}]|}{EH}. \tag{1.79}$$

All these values are invariant. The absolute value of velocity in the magnetic field is constant. And if the intensity of the magnetic field along the particle trajectory does not vary, the instantaneous radiation power will be a constant value in time. In a general case, the latter statement does is not valid. In more detail these questions have been considered in Refs[26,27].

1.7.2. Angular Distribution and Linear Polarization of Radiation

As the specral-angular distribution of SR (Schott's formula) does not depend on the azimuthal angle φ, we shall assume without a loss of generality that $\varphi = 0$. Then the angles θ and $\psi = \pi/2 - \theta$ will be in the plane XZ (see Fig. 1.19). Further we introduce the new variable

$$\chi = \frac{\cos\theta}{\sqrt{1-\beta^2}} = \gamma \sin\psi. \tag{1.80}$$

In the ultrarelativistic case, the angles ψ are small and in this approximation

$$1 - \beta^2 \sin\theta^2 \approx \frac{1}{\gamma^2}(1+\chi^2) = \left(\frac{1}{\gamma^2} + \psi^2\right) = \eta^2. \tag{1.81}$$

Let us consider the angular distributions of linearly polarized SR. According to Eqs (1.47), (1.48), (1.80), and (1.81), the angular distributions of σ- and π-components have the form

$$\frac{\overline{dW}}{d\chi} = X(\chi) W_{\text{SR}}, \quad \frac{\overline{dW_\varphi}}{d\chi} = X_\varphi(\chi) W_{\text{SR}}, \quad \frac{\overline{dW_\theta}}{d\chi} = X_\theta(\chi) W_{\text{SR}},$$

$$X(\chi) = \frac{3}{32}\,\frac{7+12\chi^2}{(1+\chi^2)^{7/2}}, \tag{1.82}$$

$$X_\varphi(\chi) = \frac{21}{32}\,\frac{1}{(1+\chi^2)^{5/2}}, \quad X_\theta(\chi) = \frac{15}{32}\,\frac{\chi^2}{(1+\chi^2)^{7/2}},$$

$$X = X_\varphi + X_\theta, \quad \int_{-\infty}^{\infty} X(\chi) d\chi = 1.$$

In Fig. 1.16 the graphs of the functions X, X_φ, and X_θ are represented. The total radiation and σ-component have an equal-in-magnitude central maximum at $\chi = 0$ since in this case, the π-component tends to zero. The maximum values of π-component fall on $\chi = \pm\sqrt{0.4}$. The relation of maximums is

$$\frac{X(\chi=0)}{X_\theta(\chi=\pm\sqrt{0.4})} = \frac{1}{2}\,\frac{7^{9/2}}{5^{7/2}} \approx 11.33.$$

Of course, if integration over χ is carried out in Eq. (1.82), we obtain the known results of Eq. (1.49). The integration is carried out by means of integrals of the type given by Eq. (1.70).

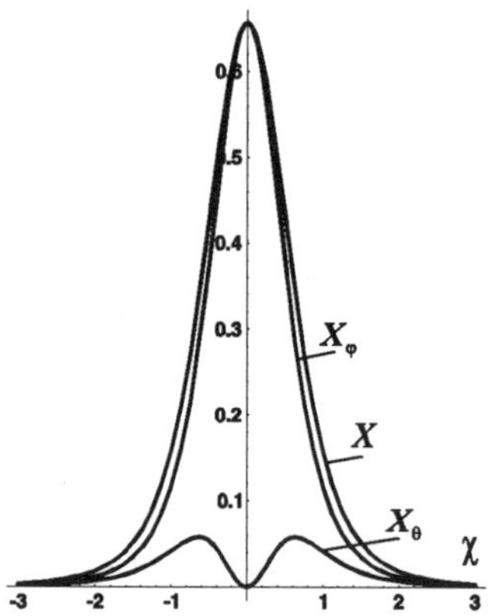

Fig. 1.16. The linear polarization as a function of the emission angle. The functions X_φ and X_θ describe the σ- and π-components of the linear polarization, respectively, and $X = X_\varphi + X_\theta$ is the normalized function of the total angular distribution of SR.

It is interesting that for the first time the formulas of the linear polarization of SR in the ultrarelativistic case were obtained by A.A. Sokolov and I.M. Ternov in the framework of the quantum theory[53] and after that in the classical one[54].

These derivations were experimentally verified by F.A. Korolev and collaborators[55,56] on the FIAN synchrotron at the energies from about 250 to 680 MeV (see also Refs[57,58]). They investigated the polarization and the angular distribution of radiation intensity. The σ- and π-components of the radiation have been found experimentally in good agreement with the theory. The experiment confirmed the main theoretical derivations which agreed well with the theory.

1.7.3. Spectral Composition and the Circular Polarization of Radiation

Let us turn to the consideration of the spectral-angular composition of radiation. In the case of an ultrarelativistic electron, a direct investigation of Schott's exact formula of Eq. (1.64) leads to some mathematical difficulties. These are connected with the fact that the maximum of the emission corresponds to the highest harmonics $\tilde{\omega}$ which are in the argument and in the index of the Bessel functions. These difficulties are easily overcome if asymptotic representation of Bessel functions of a higher order are used in the analysis of SR[59,60,61,62]

$$J_n(\nu) \approx \frac{1}{\pi\sqrt{3}}\left(1-\frac{\nu^2}{n^2}\right)^{1/2} K_{1/3}\left[\frac{n}{3}\left(1-\frac{\nu^2}{n^2}\right)^{3/2}\right], \qquad (1.83)$$

$$J'_n(\nu) \approx \frac{1}{\pi\sqrt{3}}\left(1-\frac{\nu^2}{n^2}\right)K_{2/3}\left[\frac{n}{3}\left(1-\frac{\nu^2}{n^2}\right)^{3/2}\right],$$

where $K_{1/3}$ and $K_{2/3}$ are the modified Bessel functions.

Taking account this approximation and others in Schott's formula at $\beta \to 1$ and $n \gg 1$, it is necessary to carry out the following transformations

$$\omega \approx \frac{c}{\rho}, \quad dn = \frac{d\tilde{\omega}}{\omega} = \frac{\rho}{c}d\tilde{\omega}, \quad d\Omega = \frac{2\pi}{\gamma}d\chi,$$

$$J_n(n\beta\sin\theta) = \frac{1}{\pi\sqrt{3}}\,\frac{1}{\gamma}\,(1+\chi^2)^{1/2}K_{1/3}(x),$$
$$J'_n(n\beta\sin\theta) = \frac{1}{\pi\sqrt{3}}\,\frac{1}{\gamma^2}\,(1+\chi^2)K_{2/3}(x),$$

where

$$x = \frac{1}{3}\frac{\tilde{\omega}\rho}{c\gamma^3}(1+\chi^2)^{3/2} = \frac{1}{3}\frac{\tilde{\omega}}{c}\rho\eta^3.$$

Then instead of Eq. (1.64) we shall have[i]

$$\overline{\frac{d^2W_{\rm SR}}{d\tilde{\omega}d\Omega}} = \frac{e_0^2\tilde{\omega}^2\rho}{6\pi^3c^2\gamma^4}\left[(1+\chi^2)^2K^2_{2/3}(x) + \chi^2(1+\chi^2)K^2_{1/3}(x)\right]. \tag{1.85}$$

As before, the first term in the square brackets corresponds to the σ-component and the second one to the π-component of the linear polarization of radiation.

In order to obtain the spectrum, it is necessary to integrate the Eq. (1.85) over the angles (actually over χ, see the above definition of $d\Omega$). For this purpose we need the integrals (see, for example, Ref.[62])

$$\int_{-\infty}^{\infty}(1+\chi^2)^2K^2_{2/3}(x)d\chi = \frac{\pi}{\sqrt{3}y}\left[\int_y^{\infty}K_{5/3}(x)dx + K_{2/3}(y)\right], \tag{1.86}$$
$$\int_{-\infty}^{\infty}\chi^2(1+\chi^2)K^2_{1/3}(x)d\chi = \frac{\pi}{\sqrt{3}y}\left[\int_y^{\infty}K_{5/3}(x)dx - K_{2/3}(y)\right].$$

---

[i]The ultrarelativistic approximation of Schott's formula was obtained by V.V. Vladimirskii[63] in terms of the Airy functions (see also Ref.[64])

$$Ai(z) = \frac{1}{\pi}\sqrt{\frac{z}{3}}K_{1/3}(x), \quad Ai'(z) = -\frac{1}{\pi}\frac{z}{\sqrt{3}}K_{2/3}(x), \quad x = \frac{2}{3}z^{3/2}.$$

His result can be represented in the form

$$\overline{\frac{d^2W}{d\tilde{\omega}d\Omega}} = \frac{e^2}{2\pi\rho}\,\frac{\tilde{\omega}^2}{\omega_V^2}\left[\frac{\omega^2}{\omega_V^2}Ai'^2(z) + \psi^2Ai^2(z)\right], \quad \omega_V = \left(\frac{1}{2}\tilde{\omega}\omega^2\right)^{1/3}, \tag{1.84}$$

where $\omega \approx c/\rho$. The modern symbols for Airy functions are used here and a trivial transformation of variables is performed. As a result of the exact integration of Eq. (1.84) over the angular variables by means of the tables of Airy functions composed by V.A. Fock[59], the spectral distribution of SR power was represented, for the first time, as a well-known curve Y (see Fig. 1.17).

The variable

$$y = \frac{2}{3}\frac{\tilde{\omega}\rho}{c\gamma^3}$$

is more convenient for the analysis of the spectral composition of radiation. Taking into account that

$$\tilde{\omega}^2 d\tilde{\omega} = \frac{27}{16}\frac{c^2\gamma^9}{\rho^3}\, y^2 dy$$

we find the asymptotic ($\beta \to 1$) formula

$$\frac{\overline{dW}}{dy} = Y(y) W_{\rm SR}^{\rm as}, \tag{1.87}$$

where

$$W_{\rm SR}^{\rm as} = \frac{2}{3}\frac{e_0^2 c}{\rho^2}\gamma^4, \tag{1.88}$$

$$Y(y) = \frac{9\sqrt{3}}{8\pi} y \int_y^\infty K_{5/3}(x)dx, \qquad \int_0^\infty Y(y)dy = 1.$$

The function $Y(y)$ determines the SR spectrum and as it is normalized on unit, after the integration over the spectrum we obtain

$$\overline{W} = W_{\rm SR}^{\rm as}.$$

For the first time, these results were obtained by D.D. Ivanenko and A.A. Sokolov[65]. It is not difficult to find the spectral composition of the linear polarization of radiation as well

$$\begin{aligned} Y_\varphi(y) &= \frac{9\sqrt{3}}{16\pi} y \left[\int_y^\infty K_{5/3}(x)dx + K_{2/3}(y)\right], \\ Y_\theta(y) &= \frac{9\sqrt{3}}{16\pi} y \left[\int_y^\infty K_{5/3}(x)dx - K_{2/3}(y)\right]. \end{aligned}$$

Further integration over y can be carried out by means of the integrals

$$\int_0^\infty y \int_y^\infty K_{5/3}(x)dxdy = \frac{8\pi}{9\sqrt{3}}, \qquad \int_0^\infty y K_{2/3}(y)dy = \frac{2\pi}{3\sqrt{3}}$$

after that we obtain Eq. (1.49) again.

The plots of the functions Y, Y_φ and Y_θ are represented in Fig.1.17.

According to Eq. (1.65), the circular polarization of radiation is defined by the expression

$$\frac{\overline{dW_{\rm SR}^{(\pm)}}}{d\tilde{\omega}} = \frac{1}{2}\frac{e_0^2\tilde{\omega}^2\rho}{3\pi^3 c^2\gamma^5} \int_{-\infty}^{\infty} \left[(1+\chi^2)K_{2/3}(x) \pm \chi(1+\chi^2)K_{1/3}(x)\right]^2 d\chi.$$

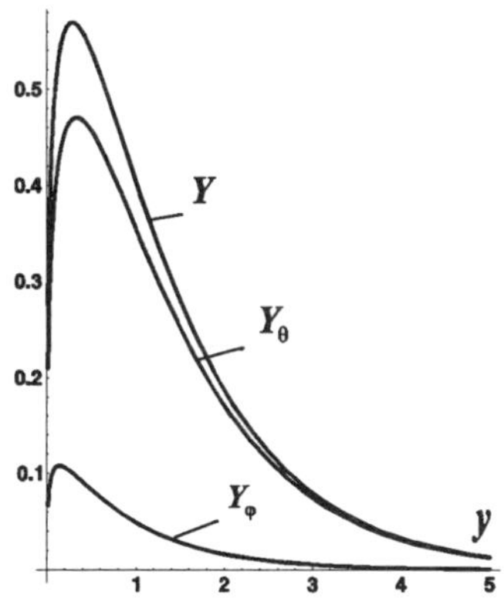

Fig. 1.17. The linear polarization as a function of the harmonic number. The functions Y_φ and Y_θ describe the σ- and π-components of the spectral distribution, respectively, and $Y = Y_\varphi + Y_\theta$ is the normalized function of the total spectral distribution of SR.

During integration over χ one more integral in the mixed term appears. However, it tends to zero as the integrand is an odd function. Therefore, after the integration over the spectrum we obtain

$$\overline{W^{(+)}} = \overline{W^{(-)}} = \frac{1}{2}\overline{W}.$$

Hence it follows that the degree of the circular polarization of radiation tends to zero

$$\Delta^{(\pm)} = \frac{\overline{W^{(+)}} - \overline{W^{(-)}}}{\overline{W^{(+)}} + \overline{W^{(-)}}} = 0.$$

In order to explain the qualitative picture of the spectral distribution, it is sufficient to investigate the behaviour of the function $Y(y)$ in the limiting cases of $y \ll 1$ and $y \gg 1$.

Noticing that for small y[64]

$$K_\mu(y) \approx \frac{1}{2}\Gamma(\mu)\left(\frac{y}{2}\right)^{-\mu}$$

and using the l'Hospital rule one can find that

$$y\int_y^\infty K_{5/3}(x)dx = 2^{2/3}\Gamma\left(\frac{2}{3}\right)y^{1/3}.$$

Hence, it is easy to obtain at $y \ll 1$

$$Y(y) \approx \frac{9\sqrt{3}}{2^{7/3}\pi}\,\Gamma\left(\frac{2}{3}\right) y^{1/3},$$

i.e. the radiation power increases in proportion to the cube root of the radiation frequency $\widetilde{\omega} = y\omega^*$, where

$$\omega^* = \frac{3}{2}\,\frac{c}{\rho}\,\gamma^3 \tag{1.89}$$

is the critical radiation frequency.

In another limiting case of $y \gg 1$ taking into account the asymptotic behaviour of the function[64]

$$K_\mu(y) \approx \sqrt{\frac{\pi}{2y}}\, e^{-y} \tag{1.90}$$

we find

$$y\int_y^\infty K_{5/3}(x)dx \approx \sqrt{\frac{\pi}{2y}}\, e^{-y} y^{1/2},$$

and

$$Y(y) \approx \frac{9}{8}\sqrt{\frac{3}{2\pi}}\, e^{-y} y^{1/2}.$$

Thus, for the frequencies higher than the critical one $\omega \gg \omega^*$ the radiation power falls off exponentially with an increase in the harmonic number. Hence, it is possible to conclude that the main part of the radiation is emitted in frequencies close to the critical one.

This was inferred for the first time by L.A. Arzimowich and I.Ya. Pomeranchuk in 1945[66]. Later on we shall prove (see Sec. 1.7) that this and the others features of SR can be deduced in a simple qualitative way.

As we see from Eq. (1.89) for very high frequencies of the rotating electron the critical radiation frequency (in the vicinity of which the maximum of radiation occurs) may shift into the visible (and also into the ultraviolet and X-ray) part of the spectrum, i.e. the electron becomes a source of visible radiation ("luminous" electron). This radiation has been visually detected in 1947 in Pollock's laboratory (USA) during the acceleration of electrons in a synchrotron[67]. In full agreement with the conclusions from the classical theory the synchrotron radiation was observed in the forward direction with respect to the moving charge as a reddish spot at 30 MeV and as a bluish-white one at 70 MeV of such brightness that it was visible by daylight. The validity of the classical theory is completely confirmed by further investigations (see Ref.[68]) of the spectral composition of SR.

If the spectral-angular distribution of radiation in Eq. (1.66) is integrated over the angles, we obtain the above Eqs (1.82) for the angular distribution of radiation, as it should be expected. During integration the formula should be used[48]

$$\int_0^\infty K_\rho(x)K_\nu(x)x^{\mu-1}dx =$$

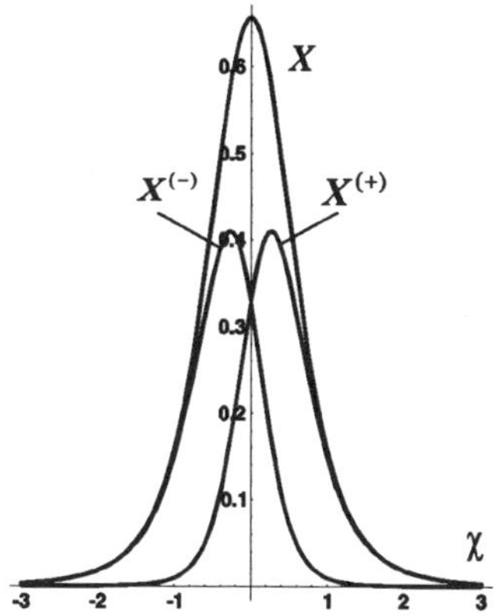

Fig. 1.18. The circular polarization as a function of the emission angle. The function $X^{(\pm)}$ describes the right $(+)$ and the left $(-)$ circular polarizations of the radiation, respectively, and $X = X^{(+)} + X^{(-)}$ is the normalized function of the total angular distribution of SR.

$$= \frac{2^{\nu-3}}{\Gamma(\mu)}\Gamma\left(\frac{\mu+\nu+\rho}{2}\right)\Gamma\left(\frac{\mu+\nu-\rho}{2}\right)\Gamma\left(\frac{\mu-\nu+\rho}{2}\right)\Gamma\left(\frac{\mu-\nu-\rho}{2}\right),$$

where $\mu > \nu + \rho$. Applying this we find the necessary integrals

$$\int_0^\infty x^2 K_{1/3}^2(x)dx = \frac{5\pi^2}{2^4 3^2}, \quad \int_0^\infty x^2 K_{2/3}^2(x)dx = \frac{7\pi^2}{2^4 3^2}.$$

To obtain the angular distribution of the circular polarization of radiation it is necessary to know one more integral

$$\int_0^\infty x^2 K_{1/3}(x) K_{2/3}(x)dx = \frac{2\pi}{9\sqrt{3}}.$$

Then by analogy we obtain

$$X^{(\pm)}(\chi) = \frac{3}{32}\left(\frac{7+12\chi^2}{2(1+\chi^2)^{7/2}} \pm \frac{32}{\pi\sqrt{3}}\frac{\chi}{(1+\chi^2)^3}\right).$$

The diagrams of the components of the circular polarization are given in Fig. 1.18 showing the functions $X^{(\pm)}(\chi)$ to have maxima at $\chi \approx \pm 0.34$, respectively.

The sum of the components of circular polarization must be equal to the sum of the linearly polarized components, i.e.

$$X = X^{(+)} + X^{(-)} = X_\varphi + X_\theta.$$

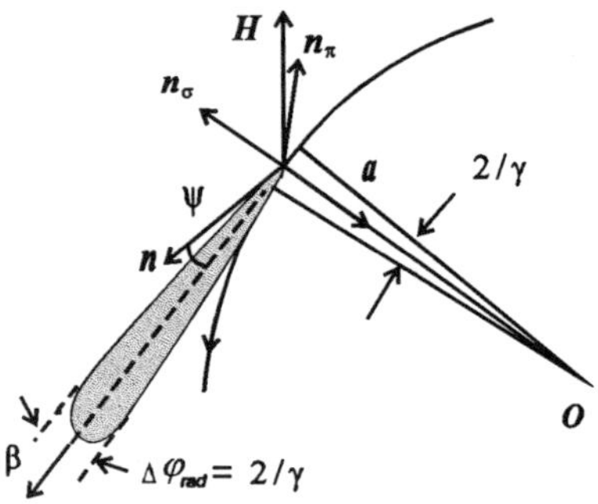

Fig. 1.19. The comoving frame for the ultrarelativistic case. Here $\boldsymbol{n}_\pi = -\boldsymbol{n}_\theta = \boldsymbol{n}_\psi$, $\boldsymbol{n}_\sigma = -\boldsymbol{n}_\varphi$.

The theoretical study of the linear and circular polarization was found in good qualitative and quantitative agreement with the experiment[55,56] (for more detail see Ref.[58]).

1.7.4. Schwinger's Approximation

As it was shown by J. Schwinger[61], all the above properties of ultrarelativistic SR can be obtained by a more simple and graceful method (see also Ref.[2]).

Schwinger's method is based on the fact that ultrarelativistic SR is almost completely concentrated not far from the orbit plane and practically all the radiation emitted from a tiny section of the circular arc $\Delta l_{\rm rad}$ in the direction of the electron velocity (see Fig. 1.19) is seen by the observer to be located in the plane of $\boldsymbol{n}$ and $\boldsymbol{\beta}$. The length of the radiation arc $\Delta l_{\rm rad} = \rho\Delta\varphi$ can be determined within small limits if $\Delta\varphi$ is taken as an angle between two directions where the radiation is absent completely (see Sec. 1.6). In the rest frame these directions are orthogonal to the velocity and coincide with the direction of acceleration (see Fig. 1.19), i.e. $\varphi' = \pi/2, 3\pi/2$. In the laboratory frame, according to the Lorentz transformation for angles, we have

$$\sin\varphi = \frac{\sqrt{1-\beta^2}\sin\varphi'}{1+\beta\cos\varphi'} = \frac{1}{\gamma}.$$

Hence it follows that $\Delta\varphi_{\rm rad} = 2/\gamma$. Thus, the length of the arc, where the radiation in this direction is formed (the radiation formation length, see also Ref.[42]), is

$$\Delta l_{\rm rad} = \rho\Delta\varphi_{\rm rad} = \frac{2\rho}{\gamma}. \tag{1.91}$$

At the observation plane a splash of electromagnetic fields coming from $\Delta l_{\rm rad}$ will be fixed. It is obvious that the radiation power of moving source will be maximal for the electromagnetic waves with the phases differences $\Delta\alpha$ and the radiation formation length $\Delta l_{\rm rad}$ if

$$\Delta\alpha = \widetilde{\omega}_{\max}\left[\frac{1}{u}\Delta l_{\rm rad} + \frac{1}{c}\left(\boldsymbol{n}\Delta\boldsymbol{l}_{\rm rad}\right)\right] = \frac{1}{u}\widetilde{\omega}_{\max}(1-\beta\cos\psi)\Delta l_{\rm rad} \approx \pi,$$

since their amplitudes in this case are adding up. Thus, the radiation formation length is equal to

$$\Delta l_{\rm rad} \approx \frac{\pi u}{\widetilde{\omega}_{\max}}\frac{1}{1-\beta\cos\psi}.$$

In the ultrarelativistic case, it is possible to take on $u = c$ and for the radiation "forward" ($\psi = 0$) we obtain

$$\Delta l_{\rm rad} \approx \frac{2\pi c}{\widetilde{\omega}_{\max}}\gamma^2 = \lambda_{\max}\gamma^2. \tag{1.92}$$

Comparing Eqs (1.91) and (1.92) we determine the value $\widetilde{\omega}_{\max}$ which is

$$\widetilde{\omega}_{\max} = \pi\frac{c}{\rho}\gamma^3.$$

This value is almost an exact expression of Eq. (1.89) which was obtained in the systematic radiation theory.

The sense of Schwinger's method is that all variables are expanded into series in terms of the small parameters of γ^{-1}, ct/ρ, and ψ. It is possible to assume without restriction of generality that ψ is the angle between the vector $\boldsymbol{n}$ and the orbital plane (the projection of $\boldsymbol{n}$ onto the orbital plane is parallel to $\boldsymbol{\beta}$). After that the necessary variables take on the form ($e < 0$)

$$\boldsymbol{\beta} = \left(1 - \frac{1}{2\gamma^2} - \frac{\psi^2}{2} - \frac{1}{2}\frac{c^2t^2}{\rho^2}\right)\boldsymbol{n} + \frac{ct}{\rho}\boldsymbol{n}_\sigma - \psi\boldsymbol{n}_\pi,$$

$$\boldsymbol{n} = \left(1 - \frac{\psi^2}{2}, 0, \psi\right), \quad \boldsymbol{n}_\sigma = (0,-1,0), \quad \boldsymbol{n}_\pi = (-\psi, 0, 1 - \frac{\psi^2}{2}),$$

which is correct to the second order of smallness. The unit vectors of polarization are adapted to the coordinate system of the a-type in Fig. 1.4. The phase of the electromagnetic wave of Eq. (1.58)

$$\widetilde{\boldsymbol{E}}_\omega = -i\frac{e_0\widetilde{\omega}}{\widetilde{r}c}\int_{-\infty}^{\infty}\boldsymbol{\beta}e^{i\widetilde{\omega}\left[t-\frac{1}{c}(\boldsymbol{nr})\right]}dt$$

can be represented in the form

$$\widetilde{\omega}\left[t - \frac{1}{c}(\boldsymbol{nr})\right] = \frac{3}{2}x\left(\xi - \frac{1}{3}\xi^3\right),$$

where

$$\xi = \frac{ct}{\rho}\eta^{-1}, \quad x = \frac{1}{3}\frac{\widetilde{\omega}\rho}{c}\eta^3, \quad \eta = \frac{1}{\gamma^2} + \psi^2.$$

Using Airy's integrals[64]

$$\frac{2}{\sqrt{3}} K_{1/3}(x) = \int_{-\infty}^{\infty} e^{i\frac{3}{2}x\left(\xi+\frac{1}{3}\xi^3\right)} d\xi, \quad \frac{2i}{\sqrt{3}} K_{2/3}(x) = \int_{-\infty}^{\infty} e^{i\frac{3}{2}x\left(\xi+\frac{1}{3}\xi^3\right)} \xi d\xi,$$

we obtain

$$\widetilde{E}^{\sigma}_{\widetilde{\omega}} = -\frac{2}{\sqrt{3}} \frac{e\widetilde{\omega}\rho}{\widetilde{r}c^2} K_{2/3}(x), \quad \widetilde{E}^{\pi}_{\widetilde{\omega}} = i\frac{2}{\sqrt{3}} \frac{e\widetilde{\omega}\rho}{\widetilde{r}c^2} \psi\left(\frac{1}{\gamma^2} + \psi^2\right) K_{1/3}(x).$$

After substitution of these expressions into Eq. (1.53) we obtain the already known spectral-angular distribution of SR (see Eq. (1.64)).

1.7.5. Limits of Validity of the Classical Theory

It is known that in the ultrarelativistic case, an electron is operated by various factors which in the conventional classical radiation theory are not considered at all. These are the force of radiative damping and the quantum and spin effects. It is known that the first signs of SR were discovered by J.P. Blewett[69] due to a radiative decrease in the orbital radius of electrons in the 100 MeV betatron. Excellent reviews of the problem of radiative damping can be found in Refs[13,25,70–72] (see also Refs[73–76]). The equation of motion of an electron under the SR-damping was considered in Refs[78–83]. A detailed discussion of the general questions connected with the radiative damping force is beyond the framework of our task. We shall discuss here only some questions related to the radiative damping force and the validity of the classical radiation theory in the presence of the considerable action of the radiative damping force and quantum effects.

First of all, it is very important to note that the condition of smallness of the radiative damping force in the nonrelativistic approximation (in the rest frame) and that in the ultrarelativistic case differ from each other very much. Really, in the rest frame the equation of motion of the charge in a constant magnetic field $\boldsymbol{H}$ with account of the radiative damping force has the form (see, for example, Ref.[1])

$$m_0\frac{d\boldsymbol{u}}{dt} = e\left(\boldsymbol{E}_0 + [\boldsymbol{\beta}\boldsymbol{H}_0]\right) + \frac{2}{3}\frac{e^2}{c^3}\frac{d^2\boldsymbol{u}}{dt^2} = \boldsymbol{f}_0^{\mathrm{Lor}} + \boldsymbol{f}_0^{\mathrm{rad}}, \tag{1.93}$$

where

$$\boldsymbol{E}_0 = \gamma[\boldsymbol{\beta}\boldsymbol{H}], \quad \boldsymbol{H}_0 = \gamma\boldsymbol{H}$$

are fields in the rest frame. Assuming that the radiative damping force is a small one can obtain

$$\boldsymbol{f}_0^{\mathrm{rad}} = \frac{2}{3}\frac{e^3}{m_0^2c^4}[\boldsymbol{E}_0\boldsymbol{H}_0] = -\frac{2}{3}\frac{e^4}{m_0^2c^4}\gamma^2H^2\boldsymbol{\beta}. \tag{1.94}$$

The radiative damping force is directed opposite the velocity, as it should be expected. According to Eqs (1.93) and (1.94), the criterion of smallness of the radiative damping force in the rest system has the form

$$\frac{f_0^{\text{rad}}}{f_0^{\text{Lor}}} \approx \frac{2}{3} \frac{e^4}{m_0^2 c^4} \gamma H \ll 1$$

or

$$\alpha\gamma \frac{H}{H^*} \ll 1, \tag{1.95}$$

where $\alpha = e^2/\hbar c$ is the fine structure constant, $H^* = m_0^2 c^3/e\hbar$ is Schwinger's magnetic field.

As it is known[1], this criterion is at the same time a condition of validity of the classical theory.

The same criterion derived from the explicit form of the ultrarelativistic force of the radiative damping in the laboratory frame has the form

$$\frac{f^{\text{rad}}}{f^{\text{Lor}}} \approx \frac{2}{3} \frac{e^2}{m_0^2 c^4} \gamma^2 H \ll 1$$

or

$$\alpha\gamma^2 \frac{H}{H^*} \ll 1. \tag{1.96}$$

It is clear that if Eq. (1.96) is valid, then Eq. (1.95) will be valid too. However, the reverse statement may not hold, especially at $\gamma \gg 1$. That means that the radiative damping force in the ultrarelativistic case can be the main force acting on a charge. Thereat the validity of classical electrodynamics is not cancelled. This interesting situation can be easily understood if one takes into account that the condition of Eq. (1.96) as well as Eq. (1.95) is covariant but not invariant. On the contrary, the total power of SR is a covariant value. Therefore, if radiation corrections of SR are small in one inertial frame (in the rest frame), they will be small in any other coordinate system.

The same can be proved in another way, too. We have seen that the ultrarelativistic radiation received by an observer is formed within the small section of the circular arc $\Delta l_{\text{rad}} \approx \rho/\gamma$ in the time interval

$$\Delta t_{\text{rad}} = \frac{\Delta l_{\text{rad}}}{c} = \frac{1}{\gamma\omega} = \frac{m_0 c}{eH}.$$

The loss of energy due to radiation in this time interval will be small if

$$W\Delta t \ll m_0 c^2 \gamma.$$

Substituting here the power W of SR as a function of the magnetic field intensity [see Eq. (1.45) with $\beta_{\|} = 0$] we obtain the same criterion of Eq. (1.95). Classical electrodynamics is valid in spite of the great radiation loss while this criterion is valid.

Quantum effects in radiation begin to appear in the case when the energy of radiated photons becomes comparable with the electron energy

$$\hbar\tilde{\omega} \approx m_0 c^2 \gamma.$$

Substituting the frequency $\tilde{\omega}_{\rm max}$, which corresponds to the maximum of SR spectrum one obtains the smallness condition for the corrections to the classical radiation theory[j]

$$\gamma \frac{H}{H^*} \ll 1. \tag{1.97}$$

Hence it follows that due to the quantum effects the classical theory of SR can be invalid in weaker fields than those which follow from the condition of Eq. (1.95).

The range of fields where the classical theory is valid but the radiative damping force is stronger than the Lorentz force, is determined by the inequality

$$\frac{1}{\alpha\gamma^2} H^* < H \ll \frac{1}{\gamma} H^*.$$

The quantum and spin effects in the theory of SR will be considered in Chapter 4 in more detail.

1.8. Coherence Properties of SR

All the above properties of SR were assigned to the radiation of a single electron. In real physical devices a great number of electrons variously distributed in space are simultaneously emitting. Electrons can either uniformly fill the entire orbit (betatron) or be clustered in separate bunches of $10^{12} - 10^{13}$ particles (synchrotron). Thereat a situation can arise when due to interference of electromagnetic waves the total power of the radiation of electrons will not be equal to the sum of the radiation power emitted by each electron. Then they speak about the appearance of coherent radiation.[k]

[j] Unlike Eq. (1.96) this criterion is invariant because the parameter

$$\chi = \gamma \frac{H}{H^*} = \frac{\hbar \gamma^2}{m_0 c \rho} = \frac{\hbar}{m_0 c^3} \sqrt{w_\mu w^\mu} = \frac{2}{3}\xi = inv.$$

[k] G.A. Schott was the first to consider the problem of coherent radiation[49]. He tried to solve the problem of the atomic spectra on the basis of H. Nagaoka's atomic model. The model presumed that a large number of electrons move simultaneously in a closed orbit. Studying the radiation of a cluster of electrons uniformly distributed along the circle, Schott came to the conclusion that the classical model of an atom is inconsistent, since the conclusions of the theory contradicted the observed data.

Let us consider the problem of radiation coherence in more detail. According to Eq. (1.61), the electric radiation fields of the i-th electron are defined by the expressions [compare with Eq. (1.62)]

$$\widetilde{E}_{\varphi i}(\tilde{t}) = \sum_{n=0}^{\infty} b_{n\varphi} \cos n(\omega\tilde{t} - \alpha_i), \quad \widetilde{E}_{\theta i}(\tilde{t}) = -\sum_{n=0}^{\infty} b_{n\theta} \sin n(\omega\tilde{t} - \alpha_i),$$

where $\alpha_i = \omega R/c + \varphi$ is the initial phase of the i-th electron, $i = 1, 2, 3, \ldots, Z$.

Averaging the quadratic combinations of radiation fields in the spectral distribution of the radiation power of Eq. (1.54) is carried out by means of the relations

$$\frac{1}{T}\int_0^T \sin n(\omega\tilde{t} - \alpha_i) \cos n'(\omega\tilde{t} - \alpha_j) d\tilde{t} = 0,$$

$$\frac{1}{T}\int_0^T \sin n(\omega\tilde{t} - \alpha_i) \sin n'(\omega\tilde{t} - \alpha_j) d\tilde{t} =$$

$$= \frac{1}{T}\int_0^T \cos n(\omega\tilde{t} - \alpha_i) \cos n'(\omega\tilde{t} - \alpha_j) d\tilde{t} = \frac{1}{2}\delta_{nn'} \cos n(\alpha_i - \alpha_j).$$

After that the total radiation power of the n-th harmonic is represented in the form

$$W_{nZ} = S_Z W_n,$$

where W_n is the radiation power of a single electron,

$$S_Z = Z + \sum_{i,j=1 (i\neq j)}^{Z} \cos n(\alpha_i - \alpha_j)$$

is the coherence factor. If the electrons are spread chaotically, then $S_Z = Z$ and the radiation is incoherent. If electrons are distributed uniformly, then

$$S_Z = Z \sum_{i=1}^{Z} \cos n \frac{2\pi}{Z} i.$$

Further we shall use the sum

$$\sum_{i=1}^{Z} \cos n \frac{2\pi}{Z} i = (-1)^n \frac{\sin \pi Z}{\tan \frac{\pi n}{Z}}.$$

Hence it follows that if ratio $s = n/Z$ is not integral number, $S_Z = 0$, i.e. only those harmonics will be emitted for which s is an integer. It is possible to say that s is the integral number which numbers the nonzero harmonics of the coherent radiation.

In this case, $S_Z = Z^2$. The total power of coherent radiation is defined by the expression

$$W_Z = \sum_{s=1}^{\infty} W_s, \tag{1.98}$$

$$W_s = \frac{e^2 c\beta Z^3}{\rho^2} s \left[2\beta^2 J'_{2sZ}(2sZ_\beta) - (1-\beta^2) \int_0^{2sZ_\beta} J_{2sZ}(x)dx \right].$$

In the nonrelativistic case at given Z, the peak of coherent radiation falls on $s = 1$. Using the nonrelativistic approximation of the Bessel functions of Eq. (1.68) we have

$$W_{s=1} = \frac{2e^2 c\beta^2}{\rho^2} Z^3 (\beta Z)^{2Z} \frac{Z+1}{(2Z+1)!}. \tag{1.99}$$

This formula shows the radiation power of the first harmonic as a function of the concentration of electrons uniformly distributed in a circle. The maximum of radiation coincides with the radiation power of a single electron ($Z = 1$). The radiation emitted by two and more electrons possesses the property of mutual extinguishing. For example, two electrons disposed symmetrically in a circle will extinguish the main tone ($n = 1$) and emit the second harmonic ($n = 2, s = 1$) which is approximately β^2 times less than the first harmonic power. In the same way, three electrons will extinguish not only the first harmonic but also the second harmonic, as in this case of $s = 1$ at $Z = 3$, thereat the third harmonic power (the first allowed) is approximately β^4 times less than the first one. Finally, this property leads to the fact that the continuous current ($Z \gg 1$) does not emit at all.

Eq. (1.99) was obtained by J. Thomson (see Ref.[49]).

In the ultrarelativistic case, instead of Eq. (1.99) other approximations should be used, namely Eqs (1.83), (1.86) where it should be replaced by

$$y \to \frac{2}{3} sZ\gamma^{-3}, \quad dy = \frac{2}{3} Z\gamma^{-3}.$$

Then we obtain (cf. Eq. (1.87))

$$W_s^{\text{coh}} = W_{\text{SR}}^{as} \frac{\sqrt{3}}{2\pi} \frac{sZ^4}{\gamma^6} \int_y^{\infty} K_{5/3}(x)dx,$$

where W_{SR}^{as} is the asymptotical radiation power of a single electron given by Eq. (1.88). The radiation frequency, in which the maximum of radiation lies, is defined by the relation

$$\widetilde{\omega}_{\max} = s_{\max} Z\omega \approx \omega^* = \frac{2}{3}\omega\gamma^3.$$

Hence it follows that the allowed harmonic, in which the maximum of coherent radiation lies, is

$$s_{\max} \approx \frac{2}{3}\frac{\gamma^3}{Z}.$$

It shows that the spectrum structure depends on the electron concentration very much. It is possible to distinguish three cases: $Z \ll \gamma^3$, $Z \approx \gamma^3$, and $Z \gg \gamma^3$. In the first case, $s_{\max} \gg 1$ but a situation when $s_{\max} < s$ is possible, too. In general, at small electron concentration the regularities of the radiation spectrum of a single electron are retained although the number of the allowed s-harmonics becomes smaller. With an increase in the electron concentration in a circle, the number of the allowed s-harmonics still decreases. Moreover, the first of them tends to $s_{\max}$ and at $Z \approx \gamma^3$ (the second case) it is located near the maximum of SR spectral distribution of a single electron. The low-frequency spectral range completely disappears in this case. Finally, in the third case of $Z \gg \gamma^3$ the first allowed harmonic corresponds to the far higher-frequency range, moreover it is strongly suppressed as this is a range where $y \gg 1$ (see Eq. (1.90)). At a very high concentration of electrons the radiation completely disappears, as it should be expected. As in the nonrelativistic case, the continuous current does not emit. It follows from these arguments that the coherent SR of the electrons uniformly distributed in a circle appears most markedly in the long-wave range at $Z \leq \gamma^3$. Really, the SR is formed in a small section of a circular arc $\Delta\varphi \approx 1/\gamma$ (see also Chapter 4) and, therefore, the number of the electrons participating in the observed radiation decreases: $Z \to Z/2\pi\gamma$. However, the uniform distribution of the electrons in a circle is too strong idealization.[l]

In this connection, it is of interest to consider SR of the electrons bunch. In this case, the coherence factor can be presented in the form[84]

$$S_Z = Z + (Z^2 - Z)f_s = Z(1 - f_s) + Z^2 f_s.$$

The radiation is fully coherent for $f_s = 1$ and is incoherent for $f_s = 0$. In the assumption that the distribution of the electrons in a bunch is symmetrical about the center (for example, at zero azimuth), the function f_s is given by

$$f_s = \left(\int d\alpha \, w(\alpha) \cos s\alpha_Z \right)^2,$$

where $\alpha_Z = \alpha Z$ and $w(\alpha)d\alpha$ is the probability that the electrons are in the orbit in the angular range from α to $\alpha + d\alpha$. In particular, if the electrons are uniformly distributed within the angular interval $\widehat{\alpha}$, then

$$w(\alpha) = \frac{1}{\widehat{\alpha}}, \quad -\frac{\widehat{\alpha}}{2} \leq \alpha \leq \frac{\widehat{\alpha}}{2}$$

[l]The case in which electrons uniformly fill the orbit was discussed in connection with possible losses of the energy of particles in a betatron. Concerns were voiced that radiation from the electrons would be suppressed in a betatron owing to its coherence. However, the analysis of the problem carried out by L.A. Arzimovich and I.Pomeranchuk[66], and also the decisive experiment made by J.P. Blewett[69], showed that the process is not coherent, at least at the peak of the radiation. It was also noted that the fluctuations of density prevent particles from filling the orbit regularly. In contrast to the case of a betatron, electrons do not fill the entire orbit in a synchrotron (or in a storage ring) but move in separate bunches.

and

$$f_s = \left(\frac{\sin s\widehat{\alpha}_z/2}{s\widehat{\alpha}_z/2}\right)^2, \quad \widehat{\alpha}_z = \widehat{\alpha}Z.$$

If the distribution follows the Gauss law, then

$$w(\alpha) = \frac{1}{\widehat{\alpha}\sqrt{\pi}} e^{-\alpha^2/\widehat{\alpha}^2}$$

and correspondingly

$$f_s = e^{-s^2\widehat{\alpha}_z^2/2}.$$

We shall consider coherent radiation in the long-wave range where it is most essential. In this case, the total power of coherent radiation has the asymptotics

$$W_z = \int_0^\infty W_s^{\rm coh} f_s ds,$$

where $W_s^{\rm coh}$ is defined by the asymptotic ($y \ll 1$)

$$W_s^{\rm coh} = W_{\rm SR}^{\rm as} \frac{3^{7/6}}{2\pi} \Gamma\left(\frac{2}{3}\right) s^{1/3} \frac{Z^{10/3}}{\gamma^4} = \frac{e^2 c}{\rho^2} \frac{3^{1/6}\Gamma\left(\frac{2}{3}\right)}{\pi} (sZ)^{1/3}.$$

During integration with respect to all s-harmonics we shall use the formulas from Ref.[85]

$$\begin{aligned}
\int_0^\infty \frac{\sin^2 ax}{x^b} dx &= \frac{\pi}{4} \frac{(2a)^{b-1}}{\Gamma(b)\sin\left(\frac{b-1}{2}\pi\right)}, \\
\int_0^\infty x^b e^{-ax^2} dx &= \frac{\Gamma\left(\frac{b+1}{2}\right)}{2a^{\frac{b+1}{2}}}.
\end{aligned}$$

Then we obtain

$$\begin{aligned}
W_Z^{\rm unif} &= W_{\rm SR}^{\rm as} \frac{3^{5/3}}{2} \frac{Z^2}{\gamma^4} \widehat{\alpha}^{-4/3} = Z^2 \frac{e^2 c}{\rho^2} 3^{2/3} \widehat{\alpha}^{-4/3}, \\
W_Z^{\rm Gauss} &= W_{\rm SR}^{\rm as} \frac{3^{7/6}}{2^{4/3}\pi} \Gamma\left(\frac{2}{3}\right)^2 \frac{Z^2}{\gamma^4} \widehat{\alpha}^{-4/3} = Z^2 \frac{e^2 c}{\rho^2} \frac{3^{1/6}}{2^{1/3}} \Gamma\left(\frac{2}{3}\right)^2 \widehat{\alpha}^{-4/3},
\end{aligned}$$

whence it follows that the coherent radiation power in the long-wave range is independent of the energy of an electron and the shape of a bunch. This is the case when electrons move in bunches of the size comparable with the wave length of the radiation. In this case, the radiation can increase drastically $\sim Z^2$. The spectrum of coherent SR in the far infrared was observed experimentally in Japan[86,87].

Thus, the SR is incoherent at least in the high-frequency range where the total power of radiation is proportional to the number of electrons. This is the reason why experimentalists are satisfied with the theory of SR of a single electron.

Other aspects of SR coherence were discussed in Refs[72,88–92].

More complete information about the modern status of experimental investigations of SR can be found in Refs[58,93–95].

Bibliography

1. L.D. Landau and E.M. Lifshitz, *The Classical Theory of Fields* (Pergamon Press, Oxford, 1980).
2. J.D. Jackson, *Classical Electrodynamics* (John Wiley, New York, 1962).
3. W.K.H. Panofsky and M. Phillips (Addison-Wesley, Reading, Mass., 1961).
4. A.A. Sokolov, I.M. Ternov, V.Ch. Zhukovskiĭ, A.V. Borisov, *Quantum Electrodynamics* (Mir, Moscow, 1988).
5. V.A. Bordovitsyn and V.S. Gushchina, *Izv. Vuz. Fiz.* **36** 2 (1993) 113 [*Sov. Phys. J.* **36** (1993) 148].
6. A.A. Liénard, *L'Eclairage Electr.* **16** (1898) 5, 53, 106.
7. E. Wiechert, *Archives Neerland* **5** (1900), 549; *Ann. d. Phys.* **4** (1901) 667.
8. M. Abraham, *Ann. d. Phys.* **14** (1904) 236.
9. V.G. Bagrov, *Optika i Spektrosk.* **18** (1965) 541 [in Russian].
10. V.G. Bagrov, V.A. Bordovitsyn, G.F. Kopytov, *Izv. Vuz. Fiz.* **15** 3 (1972) 30 [in Russian].
11. A. Schild, *J. Math. Analysis & Appl.* **1** (1960) 127.
12. F. Rohrlich, *Nuovo Cim.* **21** (1961) 811.
13. F. Rohrlich, *Classical Charged Particles* (Addison-Wesley, New York, 1965).
14. C. Teitelboim, *Phys. Rev.*, **D1** (1970) 1572; *Phys. Rev.* **D3** (1971) 297.
15. S.R. de Groot and L.G. Suttorp, *Foundation of Electrodynamics* (North-Holland, Amsterdam, 1972).
16. E. Marx, *Intern. J. of Theoret. Phys.* **14** (1975) 55.
17. B.P. Kosyakov, *Usp. Fiz. Nauk* **162** (1992) 161 [in Russian].
18. V.A. Bordovitsyn, G.K. Razina, N.N. Bysov, *Izv. Vuz. Fiz.* **23** 10 (1980) 33 [in Russian].
19. J. Larmor, *Phil. Mag.* **44** (1897) 503.
20. M. Born, *Ann. d. Phys.* **30** (1901) 1.
21. W. Pauli, in *Encykl. Math. Wiss.* *5*, Teil 2 (1921) 539; *Theory of Relativity* (Pergamon Press, London, 1958).
22. S.R. Milner, *Phil. Mag.* **41** (1821) 405.
23. T. Fulton and F. Rohrlich, *Ann. of Phys.* **9** (1960) 499.
24. V.L. Ginzburg, *Usp. Fiz. Nauk* **98** (1969) 569 [in Russian].
25. V.L. Ginzburg, *Theoretical Physics and Astrophysics* (Pergamon, Oxford, 1969).
26. V.G. Bagrov and Yu.A. Markin, *Izv. Vuz. Fiz.* **10** 5 (1967) 37 [in Russian].
27. V.A. Bordovitsyn, *Izv. Vuz. Fiz.* **14** 1 (1971) 131 [in Russian].
28. E.G. Mychelkin, *Astron. Zhurn.* **45** (1968) 408 [in Russian].
29. V.L. Ginzburg, N. Sazonov, S.I. Syrovatskii, *Usp. Fiz. Nauk* **94** (1968) 63 [in

Russian].
30. A.A. Sokolov, I.M. Ternov, V.G. Bagrov et al. *Izv. Vuz. Fiz.* **11** 5 (1968) 13 [in Russian]; *Zs. Phys.* **211** (1968) 1.
31. G. Bekefi, *Radiation Processes in Plasma* Russian ed. (Moskwa, Mir, 1971).
32. A.A. Sokolov and I.M. Ternov, *Radiation from Relativistic Electrons* (ATP, New York, 1986).
33. I.M. Ternov, V.A. Bordovitsyn, *Izv. Vuz. Fiz.* **25** 2 (1982) 120 [in Russian].
34. V.G. Bagrov, V.A. Bordovitsyn, G.F. Kopytov, V.Ya. Epp, *Izv. Vuz. Fiz.* **17** 1 (1974) 46 [in Russian].
35. G.G. Stokes, *Trans. Cambridge Phil. Soc.* **9** (1852) 399.
36. M. Born and E. Wolf, *Principles of Optics* (Pergamon, Oxford, 1968).
37. W.H. McMaster, *Rev. Mod. Phys.* **33** (1961) 8.
38. A.G. Pacholczyk, *Radio Astrophysics* (W.H. Frecman, San Francisko, 1970).
39. B.A. Grishanin, A.V. Titov et al. *Sb. Stateĭ Fiz. Inst. Akad. Nauk (FIAN)* **7** (1984) 37 [in Russian].
40. V.G. Bagrov, V.A. Bordovitsyn, A.F. Zhurawlev, *Vestn. Moskow. Univ., III* 3 (1968) 107 [in Russian].
41. I.M. Ternov, V.G. Bagrov, V.A. Bordovitsyn, *Vestn. Moskow. Univ., III* 2 (1972) 248 [in Russian].
42. B.M. Bolotovskii, *Trudy Fiz. Inst. Akad. Nauk (FIAN)* **140** (1982) 95 [in Russian].
43. A.G. Bonch-Osmolovskii and M.M. Podgoretzkii, *Zhurn. Tekhn. Fiz.* **65** (1985) 1500 [in Russian].
44. I.I. Abbasov, B.M. Bolotovskii, V.A. Davydov, *Usp. Fiz. Nauk* **149** (1986) 709 [in Russian].
45. M.J. Lighthill, *Introduction to Fourier Analysis and Generalized Functions* (Cambridge Univ. Press, 1956).
46. B.M. Bolotovskiĭ and V.P. Bykov, *Usp. Fiz. Nauk* **160** (1990) 141 [in Russian].
47. B.M. Bolotovskiĭ and S.N. Stolyarov, *Usp. Fiz. Nauk* **162** (1992) 178 [in Russian].
48. I.S. Gradsteyn and I.M. Ryzhik, *Tables of Integrals, Series, and Products* (Academic, New York, 1965).
49. G.A. Schott, *Phil. Mag.* **13** (1907) 189; *Ann. d. Phys.* **24** (1907) 635; *Electromagnetic Radiation* (Cambr. Univ. Press, 1912).
50. G. Watson, *A Treatise on the Theory of Bessel Fuctions* (Cambr. Univ. Press, 1922).
51. A.A. Sokolov, D.V. Galtzov, V.Ch. Zhukovskii, *Zhurn. Tekhn. Fiz.* **43** (1973) 682 [in Russian].
52. V.G. Bagrov and V.A. Bordovitsyn, *Izv. Politekhn. Inst.* **170** (1969) 72 [in Russian].
53. A.A. Sokolov and I.M. Ternov, *Zhurn. Eksp. Teor. Fiz.* **31** (1956) 473 [in Russian].
54. E. Le Roux, *Ann d'Astrophys.* **24** (1961) 71.
55. F.A. Korolev et al. *Dokl. Akad. Nauk SSSR* **110** (1956) 542 [in Russian].
56. F.A. Korolev, O.F. Kulikov, A.S. Yarov, *Zhurn. Eksp. Teor. Fiz.* **43** (1962) 1653 [in Russian], see also O.F. Kulikov *Trudy Fiz. Inst. Akad. Nauk (FIAN)* **80** (1965)

3 [in Russian].
57. A.A. Sokolov and I.M. Ternov, *Synchrotron Radiation* (Berlin, Akademie-Verlag; New York, Pergamon Press, 1968).
58. I.M. Ternov, V.V. Mikhailin, V.R. Khalilov, *Synchrotron Radiation and its applications* (Chur, Switzerland, Harwood, 1985).
59. V. Fock, *Dokl. Akad. Nauk SSSR* **1** (1934) 97 [in Russian].
60. D. Ivanenko and A.A. Sokolov, *Klassicheskaya Teoriya Polya* (Gostekhizdat, Moskwa-Leningrad, 1949) [in Russian].
61. J. Schwinger, *Phys. Rev.* **75** (1949) 1912 (an erratum list for this paper: J. Schwinger. *Phys. Rev.* **D7** (1973) 1696).
62. K.C. Westfold, *Astrophys. J.* **130** (1959) 241.
63. V.V. Vladimirskii, *Zhurn. Eksp. Teor. Fiz.* **18** (1948) 332 [in Russian].
64. M. Abramovitz and I.A. Stegun, *Handbook of Mathematical Functions* (N.Y., Dover, 1970).
65. D. Ivanenko and A.A. Sokolov, *Dokl. Akad. Nauk SSSR* **59** (1948) 1551 [in Russian].
66. L.A. Arzimovich and I. Pomeranchuk. *Zhurn. Eksp. Teor. Fiz.* **16** (1946) 379 [*J. Phys. USSR* **91** (1945) 267].
67. H.C. Pollok et al. *Phys. Rev.* **71** (1947) 829; **74** (1948) 52, see also *Am. J. Phys.* **51** (1983) 278.
68. P. Joos, *Phys. Rev. Lett.* **4** (1960) 558.
69. J.P. Blewett, *Phys. Rev.* **69** (1946) 87.
70. T. Erber, *Fortschr. Physik* **9** (1961) 343.
71. C.S. Chen, *Phys. Rev.* **D6** (1972) 2736.
72. N.P. Klepikov, *Usp. Fiz. Nauk* **146** (1985) 317 [in Russian].
73. D.V. Galtsov, V.A. Losev, *Yad. Fiz.* **17** (1979) 109; *Vestn. Moskow. Univ., III* 1 (1975) 37 [in Russian].
74. W.B. Bonnor, *Proc. Roy. Soc.* **A337** (1974) 591.
75. P. Paskual and L. Mas, *Phys. Rev.* **D9** (1974) 2482.
76. R.W. Nelson and I. Wasserman, *Astrophys. J.* **371** (1991) 265.
77. G.N. Plass, *Rev. Mod. Phys.* **33** (1972) 2736.
78. G.E. Gernet, *Dokl. Akad. Nauk SSSR* **168** (1966) 63 [in Russian].
79. N.D. Sen Gupta, *Phys. Letters* **A32** (1970) 103; *Int. J. Theor. Phys.* **4** (1971) 389; *Phys. Rev.* **D5** (1972) 1546.
80. A.A. Sokolov and M.M. Kolesnikova, *Vestn. Moskow. Univ., III* 2 (1971) 198 [in Russian].
81. A.G. Bonch-Osmolovskii and M.I. Podgoretzkii, *Zhurn. Tekhn. Fiz.* **49** (1979) 449 [in Russian].
82. J.C. Herera, *Phys. Rev.* **D21** (1980) 384.
83. R. Lieu, *J. Phys.* **A20** (1987) 2405.
84. L.J. Schiff, *Rev. Sci. Instr.* **17** (1946) 6.
85. H.B. Dwight, *Tables of Integrals and other Mathematical Data* (Macmillan, New York, 1961).

86. K. Ishi,..., Y. Kondo, T. Nakazato et al. *Phys. Rev.* **A43** (1991) 5597.
87. Y. Shibata,..., Y. Kondo, S. Urazawa et al. *Phys. Rev.* **A44** (1991) 3445.
88. P. Goldreich and D.A. Keely. *Astrophys. J.* **170** (1971) 463.
89. N.P. Klepikov and A.K. Jashchenko, *Vestn. Moskow. Univ., III* 5 (1985) 15 [in Russian].
90. N.P. Klepikov and I.M. Ternov, *Nucl. Instrum. and Meth.* **A308** (1991) 113.
91. I.M. Ternov, *Usp. Fiz. Nauk* **165** (1995) 429.
92. N.P. Klepikov, I.M. Ternov, V.Ya. Epp, *Nucl. Instrum. and Meth.* **A282** (1989) 413.
93. *Synchrotron Radiation. Techniques and Application.* Ed. by C. Kunz, with contribution by K. Colding, W. Gudat, E.E. Koch et al. (Springer, Berlin, 1979).
94. H. Winick, S. Doniach, eds. *Synchrotron Radiation Research* (Plenum, New York, 1980).
95. H. Winick, ed. *Synchrotron Radiation Sources: A Primer. In Series on Synchrotron Radiation Techniques and Applications.* (World Scientific, 1995; Plenum, New York, 1980).

Chapter 2

Spin Properties of Relativistic Particles

V.A. Bordovitsyn, V.S. Gushchina
Physics Faculty, Tomsk State University
634050, Tomsk, GSP-14, Russia[a]

2.1. Spin Kinematics in the Classical and Quantum Theories

Systematic investigations of the spin properties of relativistic particles started from the development of group theory methods in the quantum theory, first of all, due to the works of E. Wigner, V. Bargmann[1-3] and Yu. Shirokov[4] (see also Refs[5,7,8]). As a result of these investigations, it was found that the notion of spin and its kinematic properties are inherent in the structure of the irreducible representation of the Poincaré group, which corresponds to one-particle system and is characterized by one (positive) sign of energy and two invariants (mass and spin). But let us begin with the classical theory of spin.

The angular momentum of a point spinless particle in relativistic classical mechanics is defined by the tensor of the orbital angular momentum

$$L^{\mu\nu} = r^\mu P^\nu - r^\nu P^\mu = r^{[\mu}P^{\nu]}.$$

The components $L^{\mu\nu}$ form the matrix

$$L^{\mu\nu} = \begin{pmatrix} 0 & -K_x & -K_y & -K_z \\ K_x & 0 & L_z & -L_y \\ K_y & -L_z & 0 & L_x \\ K_z & L_y & -L_x & 0 \end{pmatrix} = (\boldsymbol{K}, \boldsymbol{L}), \tag{2.1}$$

where

$$\boldsymbol{L} = [\boldsymbol{r}\boldsymbol{P}] = m_0\gamma[\boldsymbol{r}\boldsymbol{u}], \quad \boldsymbol{K} = m_0\gamma(\boldsymbol{r} - \boldsymbol{u}t).$$

[a]E-mail: bord@urania.tomsk.su

In a general case, the tensor $L^{\mu\nu}$ is timelike as in the rest system

$$L_0^{\mu\nu} = (\boldsymbol{K}_0, \boldsymbol{0}).$$

On the contrary, spin, or the intrinsic angular momentum, is characterized by the spacelike tensor $S^{\mu\nu} = (\boldsymbol{T}, \boldsymbol{S})$ of a similar structure, which has only space components in the rest system

$$S_0^{\mu\nu} = (\boldsymbol{0}, \boldsymbol{S}_0).$$

Thus, if $M^{\mu\nu}$ is a total angular momentum, i.e.

$$M^{\mu\nu} = L^{\mu\nu} + S^{\mu\nu}, \tag{2.2}$$

then it is possible to define the spin tensor $S^{\mu\nu}$ as a spacelike part of $M^{\mu\nu}$

$$S^{\mu\nu} = \overrightarrow{M^{\mu\nu}} = G^{\mu\alpha}G^{\nu\beta}M_{\alpha\beta}, \tag{2.3}$$

where

$$G^{\mu\alpha} = g^{\rho\alpha} + \frac{1}{c^2}v^{\mu}v^{\alpha}.$$

According to Eq. (2.3), tensor $S^{\mu\nu}$ satisfies the Frenkel condition[11]

$$S^{\mu\nu}v_{\nu} = 0, \tag{2.4}$$

which leads to

$$\boldsymbol{T} = [\boldsymbol{\beta}\boldsymbol{S}]. \tag{2.5}$$

On account of the asymmetry and condition of Eq. (2.5) the tensor $S^{\mu\nu}$ has three independent components, two of them are associated with rotary motion. That is why it is possible to impose an additional condition on the tensor $S^{\mu\nu}$ fixing the numerical value of the invariant

$$\frac{1}{2}S_{\mu\nu}S^{\mu\nu} = \boldsymbol{S}^2 - \boldsymbol{T}^2 = \frac{\hbar^2}{4}\zeta^2.$$

Here on the right-hand side of the expression we have separated the dimensional factor of $\hbar/2$. Number ζ characterizes the value of spin in the rest system and requires a more accurate definition. Forestalling (see also footnote b) we note that in the one-particle spin theory one can takes $\zeta = 1$.

In Eq. (2.3) the tensor of the total angular momentum can be separated

$$M^{\mu\nu} = -\frac{1}{c^2}M^{[\mu\rho}v^{\nu]}v_{\rho} + S^{\mu\nu}.$$

The first term of this expression coincides formally and really with the orbital angular momentum

$$L^{\mu\nu} = R^{[\mu}P^{\nu]}, \tag{2.6}$$

if we put

$$R^\mu = -\frac{M^{\mu\rho} v_\rho}{m_0 c^2}.$$

The four-dimensional vector R^μ does not coincide with the ordinary coordinate r^μ and is spacelike:

$$v_\mu R^\mu = 0.$$

In this connection

$$R^0 = (\boldsymbol{\beta R}).$$

Taking into account Eq. (2.3) and the Frenkel condition given by Eq. (2.4), we also have

$$R^\mu = -\frac{L^{\mu\rho} v_\rho}{m_0 c^2}. \tag{2.7}$$

Since the four-dimensional vector R^μ describes the orbital motion of the particle, it is natural to speak of its components as coordinates of the mass center.

To describe spin, besides the tensor $S^{\mu\nu}$, one can also use the four-dimensional spacelike vector $s^\alpha = (s^0, \boldsymbol{s})$ so that

$$S^{\mu\nu} = \frac{1}{2}\, \varepsilon^{\mu\nu\alpha\beta} s_\alpha v_\beta. \tag{2.8}$$

In this case, the Frenkel condition given by Eq. (2.4) is automatically valid. The relation inverse to Eq. (2.8) has the form

$$s^\sigma = \frac{1}{2c}\, \varepsilon^{\mu\nu\rho\sigma} S_{\mu\nu} v_\rho. \tag{2.9}$$

It is easy to see that s^α satisfies Tamm's condition[12]

$$v_\alpha s^\alpha = 0 \tag{2.10}$$

and, therefore,

$$s^0 = (\boldsymbol{\beta S}).$$

Thus, we have three independent components s^α. If we require, however, that

$$s_\alpha s^\alpha = \frac{1}{2}\, S_{\mu\nu} S^{\mu\nu} = \frac{\hbar^2}{4}\, \zeta^2, \tag{2.11}$$

then only two independent spin components will be associated with the rotary degrees of freedom. It can be shown that Eq. (2.11) is the well-known Lubański-Pauli invariant[13] in the Poincaré group (see Ref.[3])

$$I_{\rm LP} = \frac{1}{2}\, M_{\mu\nu} M^{\mu\nu} + \frac{1}{m_0^2 c^2}\, M_{\mu\rho} M^{\rho\nu} P^\mu P_\nu. \tag{2.12}$$

By means of the Lorentz transformations one can find the relation of $S^{\mu\nu}$ and s^{α} to the spin vector in the rest system $\boldsymbol{\zeta}$. For convenience and simplicity we put

$$s^{\alpha} = \frac{\hbar}{2}\,\pi^{\alpha}, \quad S^{\mu\nu} = \frac{\hbar}{2}\,\Pi^{\mu\nu}, \tag{2.13}$$

and taking into account Eq. (2.11), we obtain

$$\pi_{\alpha}\pi^{\alpha} = \frac{1}{2}\,\Pi_{\mu\nu}\Pi^{\mu\nu} = \zeta^2. \tag{2.14}$$

If in the rest system $\Pi_0^{\mu\nu} = (\mathbf{0}, \boldsymbol{\zeta})$ and $\pi_0^{\alpha} = (0, \boldsymbol{\zeta})$ then

$$\begin{aligned} \pi^{\alpha} &= \left((\boldsymbol{b\zeta}),\ \boldsymbol{\zeta} + \frac{\boldsymbol{b}(\boldsymbol{b\zeta})}{\gamma+1}\right) = (\pi^0, \boldsymbol{\pi}), \\ \Pi^{\mu\nu} &= \left([\boldsymbol{b\zeta}],\ \gamma\boldsymbol{\zeta} - \frac{\boldsymbol{b}(\boldsymbol{b\zeta})}{\gamma+1}\right) = (\boldsymbol{\Phi}, \boldsymbol{\Pi}), \\ \boldsymbol{\zeta} &= \frac{1}{\gamma}\boldsymbol{\Pi} + \frac{\boldsymbol{b}(\boldsymbol{b\Pi})}{\gamma(\gamma+1)} = \boldsymbol{s} - \frac{\boldsymbol{b}(\boldsymbol{bs})}{\gamma(\gamma+1)}, \end{aligned} \tag{2.15}$$

where $\boldsymbol{b} = \boldsymbol{v}/c = \gamma\boldsymbol{\beta}$. From Eqs (2.8), (2.9), and (2.13) the mutual dependence of the spin components follows

$$\pi^{\alpha} = \left((\boldsymbol{b\Pi}),\ \frac{1}{\gamma}\boldsymbol{\Pi} + \boldsymbol{b}(\boldsymbol{b\Pi})\right), \quad \Pi^{\mu\nu} = \left([\boldsymbol{b\pi}],\ \gamma\boldsymbol{\pi} - \frac{1}{\gamma}\boldsymbol{b}(\boldsymbol{b\pi})\right), \tag{2.16}$$

It is known that the orbital angular momentum $L^{\mu\nu} = M^{\mu\nu}$ of the spinless particle is retained due to the invariance of the Lagrange function with respect to the space-time rotations generated by the tensor $L^{\mu\nu}$. In this case, variation of the Lagrange function must be equal to zero

$$\delta\mathcal{L} = \frac{1}{2}\,\omega_{\mu\nu}\,\frac{d}{d\tau}\,M^{\mu\nu} = 0, \tag{2.17}$$

where $\omega^{\mu\nu}$ is the asymmetrical rotation tensor. If rotations take place on the plane orthogonal to the four-dimensional momentum, then

$$\omega_{\mu\nu}P^{\nu} = 0, \tag{2.18}$$

and taking into account Eq. (2.4) for the particle with a spin, we have

$$\omega_{\mu\nu}M^{\mu\nu} = \omega_{\mu\nu}S^{\mu\nu}. \tag{2.19}$$

Rotations of the type in Eq. (2.18) can also be described by the four-dimensional vector

$$\omega^{\sigma} = \frac{1}{2m_0c}\,\varepsilon^{\mu\nu\rho\sigma}\omega_{\mu\nu}P_{\rho}, \quad \omega^{\mu\nu} = \frac{1}{m_0c}\,\varepsilon^{\mu\nu\alpha\beta}\omega_{\alpha}P_{\beta}, \tag{2.20}$$

thereat

$$\omega_\sigma S^\sigma = \frac{1}{2}\,\omega_{\mu\nu}S^{\mu\nu}, \quad \omega_\sigma P^\sigma = 0.$$

Thus,

$$\delta\mathcal{L} = \frac{1}{2}\,\omega_{\mu\nu}\,\frac{dS^{\mu\nu}}{d\tau} = \omega_\alpha\,\frac{ds^\alpha}{d\tau} = 0. \tag{2.21}$$

Hence it follows that the invariance of the Lagrange function under the rotations generated by the tensor or vector of spin on the plane orthogonal to the four-dimensional momentum leads to the corresponding spin conservation law. Such rotations belong to the so-called little Lorentz group.

Analogous consideration can be also carried out in the quantum theory on the basis of the Dirac equations

$$(m_0 c - i\gamma_\mu \widehat{P}^\mu)\Psi = 0, \tag{2.22}$$

where

$$\gamma^\mu = (\gamma^0, \boldsymbol{\gamma}) = i\rho_3(I, \boldsymbol{\alpha})$$

are the Dirac-matrices, $\widehat{P}^\mu = -i\hbar\partial^\mu$ is the momentum operator of a free particle.

In the quantum theory the vector- and tensor-spin operators as well as the coordinates of the mass center operator take a definite form. Let us show how it can be done.

It is known that the retaining value in the Dirac theory is only a sum of projections of the orbital and spin angular momentum even for the free particle (see, for example, Ref.[14]). This also follows from the variational principle of least action, which in the quantum theory leads to an expression[35] similar to Eq. (2.17):

$$\delta S = \frac{1}{2}\,\omega_{\mu\nu}\oint \mathcal{M}^{\mu\nu\rho}d\Omega_\rho = \frac{1}{2}\,\omega_{\mu\nu}\int \partial_\rho\mathcal{M}^{\mu\nu\rho}d\Upsilon = 0, \tag{2.23}$$

where the surface integral is taken over the closed surface Ω^ρ in four-dimensional space, which volume element is $d\Upsilon$,

$$\mathcal{M}^{\mu\nu\rho} = -\frac{i}{2}\,\overline{\Psi}\gamma^\rho\widehat{M}^{\mu\nu}\Psi \; + \; h.c.,$$

$M^{\mu\nu}$ is the operator of the total angular momentum

$$\widehat{M}^{\mu\nu} = r^\mu\widehat{P}^\nu - r^\nu\widehat{P}^\mu + \frac{\hbar}{2}\sigma^{\mu\nu}, \tag{2.24}$$

and *h.c.* are the Hermitian conjugate values.

Further steps in our procedure are quite obvious. Choosing the hypersurface in Eq. (2.23) orthogonal to the time axis for arbitrary space-time rotations with respect to the fixed direction of momentum

$$\widehat{P}^\mu\Psi = P^\mu\Psi \tag{2.25}$$

we obtain the conservation law for the total angular momentum

$$M^{\mu\nu} = \int \mathcal{M}^{\mu\nu 0} dV == \int \Psi^{+} \widehat{M}^{\mu\nu} \Psi dV.$$

Similarly, if we consider the specialized rotations of Eq. (2.20) from the little Lorentz group taking into account Eq. (2.19) and the condition given by Eq. (2.25), we shall have spin conservation laws for tensor- and vector-spin operators

$$\widehat{S}^{\mu\nu} = \frac{1}{m_0 c}\, \varepsilon^{\mu\nu\alpha\beta} \widehat{s}_\alpha \widehat{P}_\beta, \quad \widehat{s}^\sigma = \frac{1}{2m_0 c}\, \varepsilon^{\mu\nu\rho\sigma} \widehat{S}_{\mu\nu} \widehat{P}_\sigma, \tag{2.26}$$

thereat, as in the classical theory

$$\widehat{S}^{\mu\nu} = \overrightarrow{\widehat{M}^{\mu\nu}}\,. \tag{2.27}$$

It is very interesting that Eqs (2.8), (2.9), and (2.26) were obtained almost simultaneously and independently both in the classical[15–17] and quantum[4,7,18–20] theories. But in the quantum theory unlike the classical one $\widehat{M}^{\mu\nu}$ is the concrete operator of Eq. (2.24) and, therefore, $\widehat{S}^{\mu\nu}$ and $\widehat{s}^\alpha$ according to Eqs (2.24) and (2.27), are also quite definite operators

$$\widehat{S}^{\mu\nu} = \frac{\hbar}{2}\left[\sigma^{\mu\nu} - \frac{1}{m_0 c}\left(\gamma^\mu \widehat{P}^\nu - \gamma^\nu \widehat{P}^\mu\right)\right], \quad \widehat{s}^\alpha = \frac{\hbar}{2}\left(\gamma^\alpha + \frac{i}{m_0 c}\widehat{P}^\alpha\right)\gamma^5. \tag{2.28}$$

The operator $\widehat{S}^{\mu\nu}$ was introduced by J. Hilgevoord and S.A. Wouthuysen[21] (see also Ref.[22]) and $\widehat{s}^\alpha$ was introduced by V. Bargmann and E.P. Wigner[3] (see also Refs[22–24]).

The components $\widehat{S}^{\mu\nu}$ and $\widehat{s}^\alpha$ have the form

$$\begin{aligned}
\widehat{S}^{\mu\nu} &= \frac{\hbar}{2}\left(\rho_3 \frac{[\boldsymbol{\sigma}\boldsymbol{P}]}{m_0 c},\ \boldsymbol{\sigma} + \rho_2 \frac{[\boldsymbol{\sigma}\boldsymbol{P}]}{m_0 c}\right), \\
\widehat{s}^\alpha &= \frac{\hbar}{2}\left(\frac{(\boldsymbol{\sigma}\boldsymbol{P})}{m_0 c},\ \rho_3 \boldsymbol{\sigma} + \rho_1 \frac{\boldsymbol{P}}{m_0 c}\right).
\end{aligned} \tag{2.29}$$

Some of them had been used by V. Fock[25], M.H.L. Price[26] and A. Sokolov[27] (see also Ref.[28]) long before the covariant definition of these operators was given.

In accordance with Eq. (2.29), the operational Lubański-Pauli invariant given by Eq. (2.12) is equal to[b]

$$\widehat{I}_{\mathrm{LP}} = \hat{s}_\alpha \hat{s}^\alpha = \frac{1}{2}\, \widehat{S}_{\mu\nu} \widehat{S}^{\mu\nu} = \frac{\hbar^2}{4}\, \boldsymbol{\sigma}^2 = \frac{3}{4}\, \hbar^2. \tag{2.30}$$

[b]One would think that according to Eqs (2.30) and (2.11), $\zeta = \sqrt{3}$. However, in connection with noncommutativity of the Pauli-matrices, only one spin component can be observed at a single measurement and then, as in the classical theory, we shall have $\zeta = 1$. But if the question is about the variation of the spin absolute value, then in fact $\zeta = \sqrt{3}$. The problem of the possibility of observation of spin is one of the principal questions of quantum mechanics[30,31]. We shall return to this question repeatedly [see Eqs (7.3), (2.46) and Eqs (2.52), (2.53)].

As in the quantum theory (see, for example, Ref.[29])

$$\widehat{I}_{\rm LP} = \hbar^2 s(s+1).$$

Hence Eqs (2.30) correspond to the half-spin particle $s = 1/2$.

Now let us consider the operator of the coordinate of the mass center $\widehat{R}^\mu$. From operational generalization of Eq. (2.3) and definitions of the operators $\widehat{M}^{\mu\nu}$ in Eq. (2.27) and $\widehat{S}^{\mu\nu}$ in Eq. (2.28) one can find an operator of the orbital angular momentum

$$\widehat{L}^{\mu\nu} = \widehat{M}^{\mu\nu} - \widehat{S}^{\mu\nu} = \widehat{R}^\mu \widehat{P}^\nu - \widehat{R}^\nu \widehat{P}^\mu, \tag{2.31}$$

where

$$\widehat{R}^\mu = \overrightarrow{r^\mu} - \frac{\hbar}{2m_0^2c^2}\,\sigma^{\mu\nu}\widehat{P}_\nu \tag{2.32}$$

is space-like operator of the coordinate of the mass center. In the rest system $\widehat{R}^\mu$ coincides with the Fock[25]-Price[26] coordinate operator

$$\widehat{R}_0^\mu = \left(0,\ \boldsymbol{r} + \frac{i\hbar}{2m_0c}\rho_3\boldsymbol{\alpha}\right).$$

Taking into account that for the solutions of the Dirac equations[c]

$$\sigma^{\mu\nu}\widehat{P}_\nu \div -m_0c\gamma^\mu + i\widehat{P}^\mu,$$

we have also

$$\widehat{R}^\mu_{\rm FBC} \div r^\mu + \frac{\hbar}{2m_0c}\gamma^\mu.$$

This operator is more known in the literature as the Feynman[32]-Bunge[33]-Corben[34] operator, although, it was found for the first time in the work of V. Fock[25]. We note that $\widehat{R}^\mu_{\rm FBC}$ unlike $\widehat{R}^\mu$ does not satisfy the relativistically invariant commutative relations.

The set of the relativistically invariant commutative relations for the operators $\widehat{P}^\mu$, $\widehat{M}^{\mu\nu}$, $\widehat{L}^{\mu\nu}$, $\widehat{S}^{\mu\nu}$, $\widehat{s}^\alpha$ and $\widehat{R}^\mu$ has the following form[35]

$$\begin{aligned}
\left[\widehat{P}^\mu, \widehat{L}^{\alpha\beta}\right] &= i\hbar(\widehat{P}^\alpha g^{\mu\beta} - \widehat{P}^\beta g^{\mu\alpha}) = \left[\widehat{P}^\mu, \widehat{M}^{\alpha\beta}\right], \quad \left[\widehat{P}^\mu, \widehat{S}^{\alpha\beta}\right] = 0,\\
\left[\widehat{L}^{\mu\nu}, \widehat{L}^{\alpha\beta}\right] &= i\hbar(\widehat{L}^{\mu\alpha}\widehat{G}^{\nu\beta} - \widehat{L}^{\mu\beta}\widehat{G}^{\nu\alpha} - \widehat{L}^{\nu\alpha}\widehat{G}^{\mu\beta} + \widehat{L}^{\nu\beta}\widehat{G}^{\mu\alpha}) +\\
&+ \frac{i\hbar}{m_0^2c^2}(\widehat{S}^{\mu\alpha}\widehat{P}^\nu\widehat{P}^\beta - \widehat{S}^{\mu\beta}\widehat{P}^\nu\widehat{P}^\alpha - \widehat{S}^{\nu\alpha}\widehat{P}^\mu\widehat{P}^\beta + \widehat{S}^{\nu\beta}\widehat{P}^\mu\widehat{P}^\alpha), \qquad (2.33)\\
\left[\widehat{S}^{\mu\nu}, \widehat{S}^{\alpha\beta}\right] &= i\hbar(\widehat{S}^{\mu\alpha}\widehat{G}^{\nu\beta} - \widehat{S}^{\mu\beta}\widehat{G}^{\nu\alpha} - \widehat{S}^{\nu\alpha}\widehat{G}^{\mu\beta} + \widehat{S}^{\nu\beta}\widehat{G}^{\mu\alpha}),\\
\left[\widehat{L}^{\mu\nu}, \widehat{S}^{\alpha\beta}\right] &= -\frac{i\hbar}{m_0^2c^2}(\widehat{M}^{\mu\alpha}\widehat{P}^\nu\widehat{P}^\beta - \widehat{M}^{\mu\beta}\widehat{P}^\nu\widehat{P}^\alpha - \widehat{M}^{\nu\alpha}\widehat{P}^\mu\widehat{P}^\beta + \widehat{M}^{\nu\beta}\widehat{P}^\mu\widehat{P}^\alpha),\\
\left[\widehat{R}^\mu\widehat{R}^\nu\right] &= \frac{i\hbar}{m_0^2c^2}\widehat{M}^{\mu\nu}, \quad \left[\widehat{S}^\mu\widehat{S}^\nu\right] = i\hbar\widehat{S}^{\mu\nu}, \quad \left[\widehat{R}^\mu, \widehat{S}^\nu\right] = \frac{i\hbar}{m_0^2c^2}\widehat{S}^\mu\widehat{P}^\nu,\\
&\left[\widehat{P}^\mu\widehat{R}^\nu\right] = -i\hbar\widehat{G}^{\mu\nu}, \quad \left[\widehat{P}^\mu\widehat{S}^\nu\right] = 0.
\end{aligned}$$

[c]This rule means that realizing operator must be extreme to the right.

These formulas follow from the well-known (see, for example, Ref.[8]) set of commutative relations, which express the relativistic invariance of the quantum theory in the nonreducible representation of the Poincaré group

$$[P^\mu P^\nu] = 0, \quad [P^\mu, M^{\alpha\beta}] = i\hbar(P^\alpha g^{\mu\beta} - P^\beta g^{\mu\alpha}),$$

$$[M^{\mu\nu}, M^{\alpha\beta}] = i\hbar(M^{\mu\alpha}g^{\nu\beta} - M^{\mu\beta}g^{\nu\alpha} - M^{\nu\alpha}g^{\mu\beta} + M^{\nu\beta}g^{\mu\alpha}),$$

but all commutators of Eq. (2.33) can be also verified by direct application of the above operators.

2.2. Physical Interpretation of the Spin Operators

The operators $\widehat{S}^{\mu\nu}$ and $\widehat{s}^\alpha$ have directly to do with Pauli's spin operator $\frac{\hbar}{2}\boldsymbol{\sigma}'$. It is easy to show by the example of the free particle for which the solution of the Dirac equations given by Eq. (2.22) has the form

$$\Psi(r^\mu) = \psi_\varepsilon \exp\left(i\frac{\varepsilon}{\hbar}r_\mu P^\mu\right).$$

The amplitude equation for a particle with the positive energy sign ($\varepsilon = 1$) has the form

$$(1 - i\gamma_\mu b^\mu)\psi = 0, \tag{2.34}$$

where $b^\mu = P^\mu/m_0c$, $b_\mu b^\mu = -1$. Solution of the set of the algebraic equations (2.34) is represented in the form

$$\psi = \sqrt{\frac{\gamma+1}{2}}\begin{pmatrix} I \\ \frac{(\boldsymbol{\sigma}'\boldsymbol{b})}{\gamma+1} \end{pmatrix}\varphi \tag{2.35}$$

with the normalization condition $\overline{\psi}\psi = 1$, where $\overline{\psi} = \psi^+\rho_3$.

The two-component amplitudes φ are determined by the condition of spin orientation in the rest system

$$(\boldsymbol{\sigma}'\boldsymbol{\zeta})\varphi = \zeta\varphi, \tag{2.36}$$

from which it follows that $\zeta = \pm 1$ and

$$\varphi = \frac{1}{\sqrt{2}}\begin{pmatrix} \zeta & \sqrt{1+\zeta\cos\theta} \\ & \sqrt{1-\zeta\cos\theta} & e^{i\varphi} \end{pmatrix}.$$

The angles θ and φ indicate the direction of the unit spin vector in the rest system

$$\boldsymbol{\zeta} = \langle\boldsymbol{\sigma}'\rangle_0 = \overset{+}{\varphi}\,\boldsymbol{\sigma}'\varphi = \zeta(\sin\theta\cos\varphi, \sin\theta\sin\varphi, \cos\theta). \tag{2.37}$$

Let us draw attention to the fact that

$$\langle\boldsymbol{\sigma}'\rangle_0^2 = \zeta^2 = 1, \tag{2.38}$$

while $\boldsymbol{\sigma}'^2 = 3$.

Let us find the amplitude densities of bilinear covariant forms of the Dirac matrices

$$\Omega = -i\gamma^\mu, \quad \sigma^{\mu\nu} = \frac{\gamma^\mu\gamma^\nu - \gamma^\nu\gamma^\mu}{2i}, \quad \gamma^\mu\gamma^5, \quad \gamma^5 \tag{2.39}$$

or

$$\begin{aligned}
-i\gamma^\mu &= \frac{i}{3!}\varepsilon^{\mu\nu\rho\sigma}\gamma_\nu\gamma_\rho\gamma_\sigma\gamma^5, \\
\sigma^{\mu\nu} &= \frac{i}{2}\varepsilon^{\mu\nu\rho\sigma}\gamma_\rho\gamma_\sigma\gamma^5, \\
\gamma^\mu\gamma^5 &= \frac{1}{3!}\varepsilon^{\mu\nu\rho\sigma}\gamma_\nu\gamma_\rho\gamma_\sigma, \\
\gamma^5 &= -\frac{1}{4!}\varepsilon^{\mu\nu\rho\sigma}\gamma_\mu\gamma_\nu\gamma_\rho\gamma_\sigma.
\end{aligned}$$

In the metrics we use

$$\gamma^\mu = i\rho_3(I, \boldsymbol{\alpha}), \quad \sigma^{\mu\nu} = (-i\boldsymbol{\alpha}, \boldsymbol{\sigma}), \quad \gamma^\mu\gamma^5 = (i\rho_2, \rho_3\boldsymbol{\sigma}), \quad \gamma^5 = -i\rho_1. \tag{2.40}$$

One can show that at the amplitudes of Eq. (2.35) in the notation of Eq. (2.36) the following relations can be obtained

$$\begin{aligned}
\langle -i\gamma^\mu \rangle &= b^\mu, \\
\langle \sigma^{\mu\nu} \rangle &= \left([\boldsymbol{b\zeta}],\ \gamma\boldsymbol{\zeta} - \frac{\boldsymbol{b}(\boldsymbol{b\zeta})}{\gamma + 1} \right) = \Pi^{\mu\nu}, \\
\langle \gamma^\alpha\gamma^5 \rangle &= \left((\boldsymbol{b\zeta}),\ \boldsymbol{\zeta} + \frac{\boldsymbol{b}(\boldsymbol{b\zeta})}{\gamma + 1} \right) = \pi^\alpha, \\
\langle \gamma^5 \rangle &= 0,
\end{aligned} \tag{2.41}$$

where $\langle(\ldots)\rangle = \overline{\psi}(\ldots)\psi$. Here we have taken account of the Lorentz transformations for the spin given by Eq. (2.15).

In the dimensionless form the spin operators $\widehat{S}^{\mu\nu}$ and $\hat{s}^\alpha$ [see Eq. (2.26)] for the free particle have the form

$$\widehat{\Pi}^{\mu\nu} = \sigma^{\mu\nu} - (\gamma^\mu b^\nu - \gamma^\nu b^\mu), \quad \hat{\pi}^\alpha = (\gamma^\alpha + ib^\alpha)\gamma^5, \tag{2.42}$$

thereat

$$\hat{\pi}_\alpha\hat{\pi}^\alpha = \frac{1}{2}\,\widehat{\Pi}_{\mu\nu}\widehat{\Pi}^{\mu\nu} = 3. \tag{2.43}$$

It is found that

$$\langle \sigma^{\mu\nu} \rangle = \langle \widehat{\Pi}^{\mu\nu} \rangle, \quad \langle \gamma^\alpha\gamma^5 \rangle = \langle \hat{\pi}^\alpha \rangle. \tag{2.44}$$

With the amplitude factors in the system of rest

$$\overline{\psi}_0 = (\varphi^+, 0), \quad \psi_0 = \begin{pmatrix} \varphi \\ 0 \end{pmatrix}$$

the relations given by Eq. (2.41) take the form

$$\begin{aligned}
\langle -i\gamma^\mu\rangle_0 &= (1,\mathbf{0}),\\
\langle \sigma^{\mu\nu}\rangle_0 &= \langle \widehat{\Pi}^{\mu\nu}\rangle_0 = (\mathbf{0},\boldsymbol{\zeta}) = \Pi_0^{\mu\nu},\\
\langle \gamma^\alpha\gamma^5\rangle_0 &= \langle \hat{\pi}^\alpha\rangle_0 = (0,\boldsymbol{\zeta}) = \pi_0^\alpha,\\
\langle \gamma^5\rangle_0 &= 0.
\end{aligned} \tag{2.45}$$

Hence it follows that unlike Eq. (2.43)

$$\langle \hat{\pi}_\alpha\rangle\langle \hat{\pi}^\alpha\rangle = \frac{1}{2}\langle \widehat{\Pi}_{\mu\nu}\rangle\langle \widehat{\Pi}^{\mu\nu}\rangle = \boldsymbol{\zeta}^2 = 1. \tag{2.46}$$

That corresponds to the classical value of the Lubanśki-Pauli invariant with $\zeta = 1$ (see Eqs (2.37) and (2.38)).

Thus, the amplitude densities of the bilinear covariant forms get an obvious classical interpretation (cf. Eq. (2.14)).

It follows from the above relations that the classical values of b^μ, $\Pi^{\mu\nu}$, and π^α can be used to project the corresponding operators onto the proper values in the quantum theory

$$\begin{aligned}
(i\gamma_\mu b^\mu - \varepsilon)\psi_\varepsilon(\zeta) &= 0,\\
\left(\frac{1}{2}\sigma_{\mu\nu}\Pi^{\mu\nu} - \zeta\right)\psi_\varepsilon(\zeta) &= 0,\\
(\gamma_\alpha\gamma^5\pi^\alpha - \zeta)\psi_\varepsilon(\zeta) &= 0.
\end{aligned}$$

Spin equations at $\varepsilon = 1$ can be verified by the direct substitution of Eq. (2.35). If the Lorentz transformations given by Eq. (2.15) are used, the above equations take quite a simple form

$$[(\boldsymbol{\sigma}'\boldsymbol{\zeta}) - \zeta]\varphi = 0,$$

which coincides with Eq. (2.36).

It is possible to introduce the projecting operators

$$\begin{aligned}
\widehat{\Lambda}_\varepsilon &= \frac{1 + \varepsilon i\gamma_\mu b^\mu}{2},\\
\widehat{Z}_\zeta &= \frac{1 + \zeta\gamma_\mu\gamma^5\pi^\mu}{2} = \frac{1 + \frac{1}{2}\zeta\sigma_{\mu\nu}\Pi^{\mu\nu}}{2}.
\end{aligned} \tag{2.47}$$

Their properties are as follows

$$\widehat{\Lambda}_{\varepsilon'}\widehat{\Lambda}_\varepsilon = \frac{1+\varepsilon\varepsilon'}{2}\widehat{\Lambda}_\varepsilon,$$

i.e.

$$\widehat{\Lambda}_\pm^2 = \widehat{\Lambda}_\pm, \quad \widehat{\Lambda}_\pm\widehat{\Lambda}_\mp = 0, \quad \widehat{\Lambda}_+ + \widehat{\Lambda}_- = 1 \tag{2.48}$$

and analogous for $\widehat{Z}_\zeta$.

For a fixed energy sign ($\varepsilon = 1$) the action of the projecting operators is regulated by the relations

$$\widehat{\Lambda}_+\psi = \psi, \quad \widehat{\Lambda}_-\psi = 0, \quad \widehat{Z}_\zeta\psi = \psi, \quad \widehat{Z}_{-\zeta}\psi = 0,$$

where $\psi = \psi_1(\zeta)$.

Earlier we have shown that the spin operators in the laboratory system satisfy the rather complicated set of the commutative relations [see Eq. (2.33)]. A question arises whether it is possible to obtain such operators in the laboratory system which would retain the simple commutative relations for the Pauli matrices

$$[\sigma_i', \sigma_j'] = 2i\varepsilon_{ijk}\sigma_k' \tag{2.49}$$

if we know the spin operators in the rest system.

In principle, this problem can be solved by means of the unitary Foldy-Wouthuysen (FW) transformation[36]. Let us recollect some properties of this transformation. Under the action of FW transformation the Hamiltonian of the Dirac equation in our notation

$$\widehat{\gamma} = \rho_3 + (\boldsymbol{\alpha b}) \div \gamma$$

takes the diagonal form

$$\widehat{U}\widehat{\gamma}\widehat{U}^{-1} = \rho_3\gamma,$$

where

$$\widehat{U} = \sqrt{\frac{\gamma+1}{2}}\left(1 + \frac{\rho_3(\boldsymbol{\alpha b})}{\gamma+1}\right), \quad \widehat{U}^+ = \widehat{U}^{-1}, \quad \overleftarrow{\widehat{U}^{-1}} \div \overrightarrow{\widehat{U}}\,.$$

At the same time, solutions of the Dirac equation (2.35) actually transform into the rest frame

$$\overline{\psi}_0 = \overline{\psi}\,\frac{\widehat{U}}{\sqrt{\gamma}}, \quad \frac{\widehat{U}}{\sqrt{\gamma}}\,\psi = \psi_0.$$

According to the inverse transformation

$$\overset{\widehat{U}}{\Omega} = \widehat{U}^{-1}\Omega\widehat{U}$$

the Dirac bilinear covariant form of Eq. (2.39) goes over from the rest system to the laboratory system according to the rule

$$\langle\Omega\rangle_0 = \frac{1}{\gamma}\langle\widehat{U}^2\,\overset{\widehat{U}}{\Omega}\rangle \div \frac{1}{\gamma}\langle\rho_3\,\overset{\widehat{U}}{\Omega}\rangle. \tag{2.50}$$

As a point of departure for the construction of laboratory spin operators with the Pauli matrices properties according to Eq. (2.40), the spin matrices $\boldsymbol{\sigma}$ should be taken from $\sigma^{\mu\nu}$ and $\rho_3\boldsymbol{\sigma}$ from $\gamma^\mu\gamma^5$. In the first case, we shall obtain

$$\overset{\widehat{U}}{\boldsymbol{\sigma}} = \frac{1}{\gamma}\left(\boldsymbol{\sigma} - i\rho_3[\boldsymbol{\alpha b}] + \frac{\boldsymbol{b}(\boldsymbol{\sigma b})}{\gamma+1}\right) = \overset{\widehat{U}}{\widehat{\boldsymbol{\Sigma}}}_{\boldsymbol{\sigma}}\,.$$

For the solutions of the Dirac equations this operator takes the form

$$\overset{\widehat{U}}{\widehat{\boldsymbol{\Sigma}}}_{\boldsymbol{\sigma}} \div \boldsymbol{\sigma} + (1-\rho_3)\frac{i[\boldsymbol{\alpha b}]}{\gamma+1} = \widehat{\boldsymbol{\Sigma}}.$$

There is another form of this operator[8] (see also Ref.[39])

$$\overset{\widehat{U}}{\widehat{\boldsymbol{\Sigma}}}_{\boldsymbol{\sigma}} \div \boldsymbol{\sigma} - \frac{i\rho_3[\boldsymbol{\alpha b}]}{\gamma} - \frac{[\boldsymbol{b}[\boldsymbol{\sigma b}]]}{\gamma(\gamma+1)} = \widehat{\boldsymbol{\Sigma}}_{\mathbf{FW}}.$$

In the second case

$$\overset{\widehat{U}}{\widehat{\boldsymbol{\Sigma}}}_{\rho_3\boldsymbol{\sigma}} = \rho_3\boldsymbol{\sigma} + \frac{1}{\gamma}\rho_1\boldsymbol{b} - \rho_3\frac{(\boldsymbol{\sigma b})\boldsymbol{b}}{\gamma+1} \div \rho_3\boldsymbol{\sigma} + (1-\rho_3)\frac{(\boldsymbol{\sigma b})\boldsymbol{b}}{\gamma^2-1}.$$

This spin operator was considered in the Refs[37,38].
It is possible to show that

$$\overset{\widehat{U}}{\widehat{\boldsymbol{\Sigma}}}_{\boldsymbol{\sigma}} \div \overset{\widehat{U}}{\widehat{\boldsymbol{\Sigma}}}_{\rho_3\boldsymbol{\sigma}} \div \widehat{\boldsymbol{\Sigma}}_{\mathbf{FW}} \div \widehat{\boldsymbol{\Sigma}}.$$

It is interesting that the relation

$$\langle\rho_3\boldsymbol{\Sigma}\rangle = \gamma\langle\boldsymbol{\Sigma}\rangle = \gamma\boldsymbol{\zeta}.$$

is true for all these operators and, therefore, for the vector

$$\boldsymbol{\zeta} = \langle\boldsymbol{\sigma}\rangle_0 = \langle\rho_3\boldsymbol{\sigma}\rangle_0$$

the above mentioned chain of the equalities [see Eq. (2.50)] closes on itself.

The spin operators $\widehat{\boldsymbol{\Sigma}}$ can be also obtained by means of the direct Lorentz transformation

$$\begin{aligned} \overset{L}{\Omega} &= \widehat{L}^{-1}\Omega\widehat{L}, \\ \widehat{L} &= \sqrt{\frac{\gamma+1}{2}}\left(1+\frac{(\boldsymbol{\alpha b})}{\gamma+1}\right), \quad \widehat{L}^{+} = \widehat{L} \neq \widehat{L}^{-1}, \\ \overline{\psi} &= \overline{\psi}_0\widehat{L}^{-1}, \quad \widehat{L}\psi_0 = \psi. \end{aligned} \tag{2.51}$$

Analogous to Eq. (2.50) the transformation has the form

$$\langle\widehat{\Omega}\rangle_0 = \langle L^{-1}\widehat{\Omega}L\rangle = \langle\overset{L}{\widehat{\Omega}}\rangle.$$

Connection of this transformation with FW transformation can be determined by means of the relations

$$\frac{\widehat{U}}{\sqrt{\gamma}} \div \widehat{L}^{-1} \div \frac{1+\rho_3}{\sqrt{2(\gamma+1)}}.$$

Spin operators obtained by means of the Lorentz transformation of Eq. (2.51) have the form

$$\overset{L}{\widehat{\boldsymbol{\Sigma}}}_{\boldsymbol{\sigma}} = \gamma\boldsymbol{\sigma} + i[\boldsymbol{\alpha b}] - \frac{(\boldsymbol{b}(\boldsymbol{\sigma b}))}{\gamma+1} \div \overset{\widehat{U}}{\widehat{\boldsymbol{\Sigma}}}_{\boldsymbol{\sigma}},$$

$$\overset{L}{\widehat{\boldsymbol{\Sigma}}}_{\rho_3\boldsymbol{\sigma}} = \rho_3\left(\boldsymbol{\sigma} + \frac{\boldsymbol{b}(\boldsymbol{\sigma b})}{\gamma+1}\right) - i\rho_2\boldsymbol{b} \div \overset{\widehat{U}}{\widehat{\boldsymbol{\Sigma}}}_{\rho_3\boldsymbol{\sigma}}.$$

Thus, all the spin operators

$$\widehat{\boldsymbol{\Sigma}} \div \overset{\widehat{U}}{\widehat{\boldsymbol{\Sigma}}}_{\boldsymbol{\sigma}} \div \overset{\widehat{U}}{\widehat{\boldsymbol{\Sigma}}}_{\rho_3\boldsymbol{\sigma}} \div \overset{L}{\widehat{\boldsymbol{\Sigma}}}_{\boldsymbol{\sigma}} \div \overset{L}{\widehat{\boldsymbol{\Sigma}}}_{\rho_3\boldsymbol{\sigma}} \div \widehat{\boldsymbol{\Sigma}}_{\mathrm{FW}}$$

acting in the laboratory system possess the properties of the Pauli-matrices

$$\widehat{\Sigma}_i\widehat{\Sigma}_j = I\delta_{ij} + i\varepsilon_{ijk}\widehat{\Sigma}_{jk}, \quad \widehat{\boldsymbol{\Sigma}}^2 = 3 \tag{2.52}$$

and have the same mean value as in the system of rest, as it should be expected

$$\langle\boldsymbol{\sigma}\rangle_0 = \langle\widehat{\boldsymbol{\Sigma}}\rangle = \boldsymbol{\zeta}, \quad \left(\langle\widehat{\boldsymbol{\Sigma}}\rangle\langle\widehat{\boldsymbol{\Sigma}}\rangle\right) = 1. \tag{2.53}$$

It is possible to construct a spacelike tensor and a vector of spin on the basis of the operator $\boldsymbol{\Sigma}$. According to Eq. (2.45), we have

$$\widehat{\Sigma}^{\mu\nu} = \left(\mathbf{0}, \widehat{\boldsymbol{\Sigma}}\right), \quad \widehat{\Sigma}^{\alpha} = \left(0, \widehat{\boldsymbol{\Sigma}}\right).$$

Finally, if these operators are subjected to the Lorentz transformation, we come to the known spin operators in the laboratory system (see Eq. (2.42))

$$\widehat{\Pi}^{\mu\nu} = \alpha^{\mu\rho}\alpha^{\nu\sigma}\widehat{\Sigma}_{\rho\sigma}, \quad \widehat{\pi}^{\mu} = \alpha^{\mu\nu}\widehat{\Sigma}_{\nu},$$

where $\alpha^{\mu\nu}$ are the coefficients of the Lorentz transformations (2.15).

But now the proper values are also varied. In particular, if we require that

$$\widehat{\Pi}_z\psi = \zeta\lambda_\perp\psi, \quad \widehat{\pi}_z\psi = \zeta\lambda_\parallel\psi,$$

we obtain

$$\lambda_\perp = \sqrt{1+b_\perp^2}, \quad \lambda_\parallel = \sqrt{1+b_\parallel^2}$$

correspondingly.

It is possible to show that the components of the operators $\widehat{\Pi}^{\mu\nu} = ([\boldsymbol{\beta\Pi}], \boldsymbol{\Pi})$ and $\widehat{\pi}^{\mu} = ((\boldsymbol{\beta\pi}), \boldsymbol{\pi})$ are linked to each other by the similar relations [see Eq. (2.16)] as in the classical theory. Hence it follows that the classical vectors $\boldsymbol{\zeta}$, $\boldsymbol{\Pi}$, and $\boldsymbol{\pi}$ can be used to determine of the spin projections adequate for all operators

$$\left(\boldsymbol{\zeta}\widehat{\boldsymbol{\Sigma}}\right) = \frac{1}{\gamma}\left(\boldsymbol{\pi}\widehat{\boldsymbol{\Pi}}\right) = \frac{1}{\gamma}\left(\widehat{\boldsymbol{\pi}}\boldsymbol{\Pi}\right).$$

For the free particle all the components of $\widehat{\Pi}^{\mu\nu}$ and $\widehat{\pi}^{\mu}$ commutate with Hamiltonian $\widehat{\gamma}$ and can be used to determine spin states. In the presence of the external field some of them are no longer the motion integrals. But it is a separate question that we shall consider later.

2.3. Classical Spin Equations of Motion

According to the Uhlenbeck and Goudsmit hypothesis[9] the intrinsic magnetic and mechanical moments of the particle are proportional to each other

$$\boldsymbol{\mu} = \frac{e}{m_0 c}\boldsymbol{m} = \mu_0 \boldsymbol{\zeta}. \tag{2.54}$$

Here $\boldsymbol{\mu}$ is the magnetic moment of a particle, $\boldsymbol{m}$ is the mechanical one, $\mu_0 = e\hbar/2m_0c$ is the Bohr magneton, $\boldsymbol{\zeta}$ is the unit vector of spin. In the most modern representation one should take $\mu = g\mu_0 s$, where s is the spin number (for electron $s = 1/2$), g is the Lande factor, which, in a general case, is a complex function of the electron energy state and the magnetic field intensity[40]. In further consideration, we limit ourselves by the Schrödinger value[41] of $g = 2(1+\alpha/2\pi)$, where $\alpha = e^2/\hbar c$ is the fine structure constant. Thus, the electron magnetic moment is the sum of the Bohr magneton μ_0 and the anomalous magnetic moment μ_a

$$\mu = \mu_0 + \mu_a = \mu_0\left(1 + \frac{\alpha}{2\pi}\right) = \frac{g-2}{2}\mu_0.$$

The simplest way to determine equations of motion is to do it in the rest system. According to the Poisson formula, the unit vector motion obeys the equation

$$\frac{d\boldsymbol{\zeta}}{d\tau} = [\boldsymbol{\Omega}_0 \boldsymbol{\zeta}].$$

The frequency $\boldsymbol{\Omega}_0$ can be found from the equation of the magnetic moment precession in classical electrodynamics[42]

$$\frac{d\boldsymbol{m}}{d\tau} = [\boldsymbol{\mu}\boldsymbol{H}].$$

Hence, in accordance with Eq. (2.54), we find that

$$\boldsymbol{\Omega}_0 = -\frac{ge\boldsymbol{H}_0}{2m_0c}$$

and, therefore,

$$\frac{d\boldsymbol{\zeta}}{d\tau} = \frac{ge}{2m_0c}[\boldsymbol{\zeta}\boldsymbol{H}_0]. \tag{2.55}$$

Let us represent the covariant generalization of this equation in the form

$$\frac{d\Pi^{\mu\nu}}{d\tau} = \frac{ge}{2m_0c}H^{[\mu}{}_{\rho}\Pi^{\rho\nu]} + \frac{1}{c^2}v^{[\mu}W^{\nu]}. \tag{2.56}$$

An additional term on the right-hand side of the equation containing W^ν is applied in order to keep in time the Frenkel condition given by Eq. (2.4), i.e.

$$\Pi^{\mu\nu} w_\nu + \frac{d\Pi^{\mu\nu}}{d\tau} v_\nu = 0.$$

Hence, it is easy to find that

$$\Pi^{\mu\nu} w_\nu + \left(g^{\mu\nu} + \frac{1}{c^2} v^\mu v^\nu \right) W_\nu = 0. \tag{2.57}$$

The second term is a spacelike part

$$\overrightarrow{W^\mu} = G^{\mu\rho} W_\rho = -\Pi^{\mu\nu} w_\nu. \tag{2.58}$$

The timelike part $\widetilde{W}^\mu$ satisfies the relation

$$\widetilde{W}^\mu + \frac{1}{c^2} v^\mu v_\rho \widetilde{W}^\rho = 0$$

and does not give any contribution to Eq. (2.56).

It remains to determine the acceleration w^μ of the particle with a spin. Here it means the following. The balance of forces acting to the charged particle with a spin in the rest system is determined by the expression

$$\boldsymbol{f}_0 = e\boldsymbol{E}_0 + \mu \boldsymbol{\nabla}(\boldsymbol{\zeta} \boldsymbol{H}_0). \tag{2.59}$$

The second term on the right-hand side of the expression is the Stern and Gerlach force, which in its turn is conditioned by the potential energy of the magnetic moment in the magnetic field

$$U = -(\boldsymbol{\mu} \boldsymbol{H}_0).$$

However, in Eq. (2.59) unlike Eq. (2.55), the terms on the right-hand side have a different order of the Plank constant $\hbar$. If we restrict ourselves to the same approximation in $\hbar$ as in Eq. (2.56), then w^μ will contain only the Lorentz force[d]

$$w^\mu = \frac{e}{m_0 c} H^{\mu\nu} v_\nu. \tag{2.60}$$

With these reserves and with due regard for Eq. (2.58), Eq. (2.56) takes on the form

$$\frac{d\Pi^{\mu\nu}}{d\tau} = \frac{eg}{2m_0 c} H^{[\mu}{}_\rho \Pi^{\rho\nu]} + \frac{g-2}{2m_0 c^3} v^{[\mu} \Pi^{\nu\rho]} H_{\rho\sigma} v^\sigma. \tag{2.61}$$

It is possible to verify that obeying the Frenkel condition given by Eq. (2.4) and the obvious relation

$$\Pi_{\mu\rho} H^\rho{}_\nu \Pi^{\nu\mu} = 0$$

[d]The proof of this assertion will be given in the quasi-classical spin theory and in the more strict quantum theory (see further).

the equality

$$\Pi_{\mu\nu}\frac{d\Pi^{\mu\nu}}{d\tau} = 0$$

is valid and, therefore, the value

$$\frac{1}{2}\,\Pi_{\mu\nu}\Pi^{\mu\nu} = \zeta^2$$

is constant and equal to unit, as it should be expected.

Eq. (2.61) in the above approximation can be found in the works of many authors (see Refs[43–45] and others).

Analogous arguments allow obtaining a spin equation in the vector form

$$\frac{d\pi^\alpha}{d\tau} = \frac{eg}{2m_0c}\,H^{\alpha\beta}\pi_\beta + \frac{e}{m_0c^3}\,\frac{g-2}{2}\,v^\alpha v_\beta H^{\beta\gamma}\pi_\gamma. \tag{2.62}$$

Moreover, this equation can be obtained from Eq. (2.61) directly by means of Eq. (2.16). The spin equation (2.62) is known as the Bargmann-Michel-Telegdi (BMT) equation[43].

Let us pay attention to an interesting peculiarity of spin equations. They can be simplified considerably if some effective external field is introduced

$$H^{\mu\nu}_{\text{eff}} = \overrightarrow{H^{\mu\nu}} + \frac{2m_0}{egc}\,v^{[\mu}w^{\nu]}, \tag{2.63}$$

where $\overrightarrow{H^{\mu\nu}}$ is the spacelike part of the tensor $H^{\mu\nu}$, which is determined on the analogy of Eq. (2.3) by the relation

$$\overrightarrow{H^{\mu\nu}} = H^{\mu\nu} + \frac{1}{c^2}\,v^{[\mu}v_\rho H^{\rho\nu]}. \tag{2.64}$$

Then, instead of Eqs (2.61) and (2.62), we shall obtain

$$\frac{d\Pi^{\mu\nu}}{d\tau} = \frac{eg}{2m_0c}\,H^{[\mu\rho}_{\text{eff}}\Pi_\rho{}^{\nu]}, \tag{2.65}$$

$$\frac{d\pi^\alpha}{d\tau} = \frac{eg}{2m_0c}\,H^{\alpha\beta}_{\text{eff}}\pi_\beta. \tag{2.66}$$

We note that

$$\frac{dv^\mu}{d\tau} = \frac{eg}{2m_0c}\,H^{\mu\nu}_{\text{eff}}v_\nu \tag{2.67}$$

but unlike Eqs (2.65) and (2.66) this equation is a simple consequence of the definition $H^{\mu\nu}_{\text{eff}}$ in Eq. (2.63).

It is possible to write down the spacelike parts of Eqs (2.65)—(2.67) in a similar form

$$\frac{\overrightarrow{d\Pi^{\mu\nu}}}{d\tau} = \frac{eg}{2m_0c}\,\overrightarrow{H^{[\mu\rho}}\,\Pi_\rho^{\ \nu]} \neq \frac{d\Pi^{\mu\nu}}{d\tau},$$

$$\frac{\overrightarrow{d\pi^\alpha}}{d\tau} = \frac{eg}{2m_0c}\,\overrightarrow{H^{\alpha\beta}}\,\pi_\beta \neq \frac{d\pi^\alpha}{d\tau}.$$

However

$$\frac{\overrightarrow{dv^\mu}}{d\tau} = \frac{eg}{2m_0c} H^{\mu\nu}_{\text{eff}}\, v_\nu = \frac{dv^\mu}{d\tau}.$$

Thus, the spin equations of motion include the timelike terms having a purely relativistic kinematic origin. Additional spin precession referred to as the Thomas precession[10] is generated by the Lorentz transformations at the particle motion along the curved trajectory.

Let us demonstrate this assertion by the example of the BMT-equation

$$\left(\frac{d\pi^\alpha}{d\tau}\right)_{\text{Th}} = \frac{d\pi^\alpha}{d\tau} - \frac{\overrightarrow{d\pi^\alpha}}{d\tau} = \frac{1}{c^2}\, v^{[\alpha} w^{\beta]} \pi_\beta.$$

Coming to the spin vector $\boldsymbol{\zeta}$ in the system of rest in accordance with Eq. (2.15) we obtain the equation

$$\left(\frac{d\boldsymbol{\zeta}}{d\tau}\right)_{\text{Th}} = \frac{1}{c}\,\frac{\gamma^3}{\gamma+1}\,[[\boldsymbol{\beta a}]\boldsymbol{\zeta}] = \frac{1}{\gamma}\,[\boldsymbol{\Omega}_{Th}\boldsymbol{\zeta}],$$

where $\boldsymbol{a}$ is the Newton acceleration,

$$\boldsymbol{\Omega}_{\text{Th}} = -\frac{1}{c}\,\frac{\gamma^2}{\gamma+1}\,[\boldsymbol{\beta a}],$$

is the angular velocity of the Thomas precession.

In a general case, the BMT-equation in terms of $\boldsymbol{\zeta}$ is represented in the form

$$\begin{aligned}\frac{d\boldsymbol{\zeta}}{d\tau} &= \frac{eg}{2m_0c}\,[\boldsymbol{\zeta H}] - \frac{e(g-2)}{2m_0c}\,\frac{\gamma^2}{\gamma+1}\,[\boldsymbol{\zeta}[\boldsymbol{\beta}[\boldsymbol{\beta H}]]] \\ &- \frac{e}{m_0c}\,\gamma\left(\frac{g}{2} - \frac{\gamma}{\gamma+1}\right)[\boldsymbol{\zeta}[\boldsymbol{\beta E}]],\end{aligned} \tag{2.68}$$

or

$$\frac{d\boldsymbol{\zeta}}{d\tau} = \frac{1}{\gamma}\,[\boldsymbol{\Omega\zeta}], \tag{2.69}$$

where vector

$$\boldsymbol{\Omega} = -\frac{e}{m_0c}\left\{\frac{g}{2}\,\frac{\boldsymbol{H}}{\gamma} - \left(\frac{g}{2} - \frac{\gamma}{\gamma+1}\right)[\boldsymbol{\beta E}] - \frac{g-2}{2}\,\frac{\gamma}{\gamma+1}\,[\boldsymbol{\beta}[\boldsymbol{\beta H}]]\right\}$$

is a sum of the angular velocities of the Larmor- and Thomas spin precession[44,45]

$$\boldsymbol{\Omega} = \boldsymbol{\Omega}_{\rm L} + \boldsymbol{\Omega}_{\rm Th},$$

$$\boldsymbol{\Omega}_{\rm L} = -\frac{eg}{2m_0c}\left(\frac{1}{\gamma}\,\boldsymbol{H} - [\boldsymbol{\beta}\boldsymbol{E}] - \frac{\gamma}{\gamma+1}\,[\boldsymbol{\beta}[\boldsymbol{\beta}\boldsymbol{H}]]\right),$$

$$\boldsymbol{\Omega}_{\rm Th} = -\frac{e}{m_0c}\,\frac{\gamma}{\gamma+1}\,([\boldsymbol{\beta}\boldsymbol{E}] + [\boldsymbol{\beta}[\boldsymbol{\beta}\boldsymbol{H}]])\,.$$

Using the Lorentz transformations to the rest system for fields in Eq. (2.68)

$$\boldsymbol{H} = \gamma\left(\boldsymbol{H}_0 + [\boldsymbol{\beta}\boldsymbol{E}_0] - \frac{\gamma}{\gamma+1}\,\boldsymbol{\beta}(\boldsymbol{\beta}\boldsymbol{H}_0)\right),$$

$$\boldsymbol{E} = \gamma\left(\boldsymbol{E}_0 - [\boldsymbol{\beta}\boldsymbol{H}_0] - \frac{\gamma}{\gamma+1}\,\boldsymbol{\beta}(\boldsymbol{\beta}\boldsymbol{E}_0)\right),$$

we obtain a very simple result

$$\frac{d\boldsymbol{\zeta}}{d\tau} = \frac{eg}{2m_0c}\,[\boldsymbol{\zeta}\boldsymbol{H}_0] + \frac{e}{m_0c}\,\frac{\gamma}{\gamma+1}\,[\boldsymbol{\zeta}[\boldsymbol{\beta}\boldsymbol{E}_0]]. \tag{2.70}$$

Naturally, at $\beta = 0$ this equation coincides with Eq. (2.55).

Abundant literature is devoted to the Thomas precession[10–12] (see, for example, Refs[44–61] and many others). We have adduced here only some of the most general results related to this phenomenon.

2.4. Quasi-Classical Theory of Spin

The quasi-classical approximation of the Dirac equations were considered by many authors[39,63–75] beginning with Pauli[62]. The theory developed below[75] is built on the special proper time representation of the Dirac equations. In this case, all the relations of the quasi-classical theory including the BMT-equation are immediately obtained in a covariant form if $\hbar \to 0$. Evidently, the quasi-classical conditions must be satisfied. In accordance to them, the wave function is to be localized in the vicinity of the particle path and if $\hbar \to 0$ the density of momentum operators must be equal to their instantaneous classical value

$$\langle \widehat{P}^{\mu}\rangle_0 = P^{\mu} = m_0 v^{\mu} = m_0 c b^{\mu}.$$

Initially this method had been introduced by Fock[76] for the squared Dirac equations since then it has been used by many authors[75,77–82].

Following Fock's method we introduce

$$\Psi = \int \Phi d\tau,$$

where integration is carried out along a closed path on the complex plane or between the constant limits. Then the proper time Dirac equation can be represented in the form

$$i\,\frac{\hbar}{c}\,\frac{\partial\Phi}{\partial\tau} = \left(m_0c - i\gamma_\mu \widehat{P}^\mu\right)\Phi \tag{2.71}$$

provided

$$\int\limits_C \frac{\partial\Phi}{\partial\tau}d\tau = \Phi\mid_c = 0.$$

Eq. (2.71) is referred to as the Dirac τ-equation. It has an ordinary quantum mechanics form

$$i\,\frac{\hbar}{c}\,\frac{\partial\Phi}{\partial\tau} = \widehat{H}\Phi$$

with the Hamiltonian

$$\widehat{H} = m_0c - i\gamma_\mu \widehat{P}^\mu. \tag{2.72}$$

We represent the wave function of the Dirac τ-equation in the ordinary quasi-classical form

$$\overline{\Phi} = \overline{F}e^{-i\frac{S}{\hbar}}, \quad \Phi = e^{i\frac{S}{\hbar}}F, \tag{2.73}$$

where $S(\tau)$ is the proper time function of action satisfying Fock's relativistic quasi-classical equation[76]

$$\frac{\partial S}{\partial\tau} + \frac{1}{2m_0c}\left[\left(\partial^\mu S - \frac{e}{c}A^\mu\right)\left(\partial_\mu S - \frac{e}{c}A_\mu\right) + m_0^2c^2\right] = 0. \tag{2.74}$$

Assuming here that

$$\partial^\mu S - \frac{e}{c}\,A^\mu = m_0v^\mu \tag{2.75}$$

we impose the condition

$$\frac{\partial S}{\partial\tau} = 0 \tag{2.76}$$

which ensures that τ will be a proper time and S will be a relativistic function of action.

As usually, let us represent the spinor F in the form of series expansion in Planck's constant

$$F = f_0 + \hbar f_1 + \hbar^2 f_2 + \ldots. \tag{2.77}$$

Then, taking into account Eq. (2.75) in a zero approximation in $\hbar$ we have

$$\overline{\Phi}\,\overleftarrow{\widehat{P}^\mu} = -m_0v^\mu\overline{\Phi}, \quad \overrightarrow{\widehat{P}^\mu}\,\Phi = m_0v^\mu\Phi,$$

$$\partial_\mu\overline{\Phi}\Phi = 0.$$

Since

$$i\hbar\frac{\partial\Phi}{\partial\tau} = e^{i\frac{S}{\hbar}}\left(i\hbar\frac{\partial F}{\partial\tau} - \frac{\partial S}{\partial\tau}F\right),$$

so taking into account Eqs (2.76) and (2.77), we obtain

$$i\hbar\left.\frac{\partial\Phi}{\partial\tau}\right|_{\hbar\to 0} = 0.$$

Then, in the same approximation for the solutions of the Dirac τ-equation the following equalities will be valid

$$\begin{aligned} &\langle 1 - i\gamma_\mu b^\mu\rangle \div 0, & \\ &\langle (1 + i\gamma_\mu b^\mu)\,\gamma^\rho\rangle \div 0, \quad & \langle \gamma^\rho\,(1 - i\gamma_\mu b^\mu)\rangle \div 0, \\ &\langle (1 + i\gamma_\mu b^\mu)\,\sigma^{\alpha\beta}\rangle \div 0, \quad & \langle \sigma^{\alpha\beta}\,(1 - i\gamma_\mu b^\mu)\rangle \div 0, \\ &\langle (1 + i\gamma_\mu b^\mu)\,\gamma^\rho\gamma^5\rangle \div 0, \quad & \langle \gamma^\rho\gamma^5\,(1 - i\gamma_\mu b^\mu)\rangle \div 0, \end{aligned} \tag{2.78}$$

here the notation $\langle(\ldots)\rangle = \overline{f}(\ldots)f$ has been used for the normalized spinors

$$\overline{f} = \frac{f_0}{\sqrt{\overline{f}_0 f_0}}, \quad f = \frac{f_0}{\sqrt{\overline{f}_0 f_0}}.$$

Using the properties of the γ-matrices given by Eq. (2.39)

$$\begin{aligned} \gamma^\mu\gamma^\nu + \gamma^\nu\gamma^\mu &= 2g^{\mu\nu}, \\ \sigma^{\mu\nu}\gamma^\alpha + \gamma^\alpha\sigma^{\mu\nu} &= 2i\varepsilon^{\mu\nu\alpha\beta}\gamma_\beta\gamma^5, \\ \sigma^{\mu\nu}\gamma^\alpha - \gamma^\alpha\sigma^{\mu\nu} &= 2i\left(g^{\mu\alpha}\gamma^\nu - g^{\nu\alpha}\gamma^\mu\right), \\ \gamma^5\sigma^{\mu\nu} &= \frac{1}{2}\varepsilon^{\mu\nu\alpha\beta}\sigma_{\alpha\beta} = \frac{i}{2}\varepsilon^{\mu\nu\alpha\beta}\gamma_\alpha\gamma_\beta\gamma^5 = \sigma^{\mu\nu}\gamma^5, \end{aligned} \tag{2.79}$$

Eqs (2.78) can be reduced to the form

$$\begin{aligned} \langle -i\gamma^\mu\rangle &= b^\mu, \\ \langle\sigma^{\mu\nu}\rangle &= \varepsilon^{\mu\nu\alpha\beta}\langle\gamma_\alpha\gamma^5\rangle b_\beta, \\ \langle\gamma^\mu\gamma^5\rangle &= \langle\sigma^{\mu\nu}\gamma^5\rangle b_\nu, \\ \langle\gamma^5\rangle &= 0 \end{aligned}$$

Moreover,

$$\begin{aligned} b_\mu\langle -i\gamma^\mu\rangle &= -1, \\ b_\mu\langle\gamma^\mu\gamma^5\rangle &= b_\mu\langle\sigma^{\mu\nu}\rangle = 0. \end{aligned}$$

Hence it follows that in a zero approximation in $\hbar$ for the spin operators $\widehat{\Pi}^{\mu\nu}$ and $\widehat{\pi}^{\mu\nu}$ (see Eq. (2.42)) the relations are the same as for a free particle

$$\langle\widehat{\Pi}^{\mu\nu}\rangle = \langle\sigma^{\mu\nu}\rangle, \quad \langle\widehat{\pi}^\alpha\rangle = \langle\gamma^5\gamma^\alpha\rangle,$$

which with the γ-matrix properties given by Eq. (2.79) take on the form of the purely classical Eqs (2.8) and (2.9) in the dimensionless form

$$\begin{aligned} \langle\widehat{\Pi}^{\mu\nu}\rangle &= \varepsilon^{\mu\nu\alpha\beta}\langle\widehat{\pi}_\alpha\rangle b_\beta, \\ \langle\widehat{\pi}^\alpha\rangle &= \frac{1}{2}\varepsilon^{\alpha\beta\mu\nu}\langle\widehat{\Pi}_{\mu\nu}\rangle b_\beta, \end{aligned}$$

with an additional Frenkel's condition and Tamm's condition [see Eqs (2.4) and (2.10)]

$$b_\mu\langle\widehat{\Pi}^{\mu\nu}\rangle = 0, \quad b_\mu\langle\widehat{\pi}^\mu\rangle = 0.$$

Taking account of the anomalous magnetic moment $\mu_a = (g-2)\mu/2$ the Dirac τ-equation (2.71) comes to the form

$$i\,\frac{\hbar}{c}\,\frac{\partial\Phi}{\partial\tau} = \left(m_0 c - i\gamma_\mu\widehat{P}^\mu - \frac{\mu_a}{c}\sigma_{\alpha\beta}H^{\alpha\beta}\right)\Phi = 0.$$

Substituting the quasi-classical wave function given by Eq. (2.73) into the latter and equating the terms of the same order in $\hbar$ to zero we obtain a set of consistent approaches for determining the spinors f_0 , f_1,...

$$\begin{aligned} (1 - i\gamma_\mu b^\mu)f_0 &= 0, \\ (1 - i\gamma_\mu b^\mu)f_1 &= \frac{1}{m_0 c}\left(\frac{i}{c}\frac{\partial}{\partial\tau} + \gamma^\mu\partial_\mu + \frac{1}{2}\frac{\mu_a}{c\hbar}\sigma_{\alpha\beta}H^{\alpha\beta}\right)f_0, \\ (1 - i\gamma_\mu b^\mu)f_2 &= \frac{1}{m_0 c}\left(\frac{i}{c}\frac{\partial}{\partial\tau} + \gamma^\mu\partial_\mu + \frac{1}{2}\frac{\mu_a}{c\hbar}\sigma_{\alpha\beta}H^{\alpha\beta}\right)f_1 \\ \cdots\cdots &\cdots \cdots\cdots\cdots\cdots \end{aligned}$$

It follows from the first equation that

$$f_0 = i\gamma^\mu b_\mu f_0. \tag{2.80}$$

Further we see that if all the following equations are multiplied by $1 + i\gamma_\mu b^\mu$, then their left parts will be equal to zero and after some transformations the second equation takes the form

$$\frac{1}{c}\,\frac{df_0}{d\tau} + \frac{1}{2}\left(\partial_\mu b^\mu\right)f_0 = i\,\frac{\mu}{2\hbar c}\,\sigma_{\alpha\beta}H^{\alpha\beta}f_0 - i\,\frac{\mu_a}{c\hbar}\,H^{\mu\nu}\gamma_\mu b_\nu f_0. \tag{2.81}$$

Here we have used Eq. (2.80) and the representation of the derivative $d/d\tau$ in the form

$$\frac{1}{c}\,\frac{d}{d\tau} = \frac{1}{c}\,\frac{\partial}{\partial\tau} + \frac{\gamma^\mu\gamma^\nu + \gamma^\nu\gamma^\mu}{2}\,b_\mu\partial_\nu.$$

Introducing the normalized spinor f instead of f_0 in Eq. (2.81) and taking into consideration that

$$\frac{1}{c}\,\frac{df_0}{d\tau} + \frac{1}{2}\,(\partial_\mu b^\mu)f_0 == \sqrt{\overline{f}_0 f_0}\,\frac{df}{d\tau}$$

we find

$$\frac{1}{c}\,\frac{df}{d\tau} = i\,\frac{\mu}{2\hbar c}\,\sigma_{\alpha\beta}H^{\alpha\beta}f - i\,\frac{\mu_a}{c\hbar}\,H^{\mu\nu}\gamma_\mu b_\nu f. \tag{2.82}$$

If the anomalous magnetic moment is neglected, this equation coincides with Fock's equation in Ref.[76] (see also Ref.[64]).

In the same way, from the conjugate Dirac τ-equation one can find

$$\frac{1}{c}\,\frac{d\overline{f}}{d\tau} = i\,\frac{\mu}{2\hbar c}\,\overline{f}\sigma_{\alpha\beta}H^{\alpha\beta} - i\,\frac{\mu_a}{c\hbar}\,\overline{f}H^{\mu\nu}\gamma_\mu b_\nu. \tag{2.83}$$

These derivatives can be used for deducing the charge and spin equations in the quasi-classical approximation.

At first, one should point out that according to Eqs (2.82) and (2.83) $d(\overline{f}f)/d\tau = 0$. We have already used this property when we put $\overline{f}f = 1$. Further, it may be shown that if $\hbar \to 0$, the equation of motion of a charged particle determined by the bilinear form $\langle -i\gamma^\mu\rangle = b^\mu$ has the ordinary Lorentz form

$$m_0\,\frac{db^\mu}{d\tau} = \frac{e}{c}\,H^{\mu\nu}b_\nu.$$

This result confirms that in a zero approximation in $\hbar$ the center of gravity of the quasi-classical wave packet moves along the classical path.

The spin equations for $\langle\widehat{\Pi}^{\mu\nu}\rangle$ and $\langle\widehat{\pi}^\alpha\rangle$ coincide with the BMT-equations (2.62) and (2.61)

$$\begin{aligned}\frac{d\langle\widehat{\Pi}^{\mu\nu}\rangle}{d\tau} &= \frac{eg}{2m_0c}\,H^{[\mu}{}_{\alpha}\langle\widehat{\Pi}^{\alpha\nu]}\rangle,\\ \frac{d\langle\widehat{\pi}^\alpha\rangle}{d\tau} &= \frac{eg}{2m_0c}\,H^{\mu\nu}\langle\widehat{\pi}_\nu\rangle + \frac{e\,(g-2)}{2m_0c}\,b^\mu H^{\alpha\beta}b_\alpha\langle\widehat{\pi}^\beta\rangle.\end{aligned}$$

To obtain these relations we have used the relation

$$\langle\sigma^{\alpha\beta}\gamma^5\rangle = b^{[\alpha}\langle\pi^{\beta]}\rangle$$

and the identity

$$\varepsilon^{\mu\nu\alpha\beta}\langle\pi_\alpha\rangle H_{\beta\gamma}b^\gamma = b^{[\mu}\langle\Pi^{\beta\nu]}\rangle H_{\beta\gamma}b^\gamma.$$

All the equations can be deduced on the basis of the squared Fock's τ-equation (see Ref.[77]).

Thus, we have shown that in the quasi-classical limit $\hbar \to 0$ the quasi-classical theory of spin is completely realized.

2.5. Spin in the Quantum Theory with the Definite Evenness of Operators

Earlier in the framework of the nonreducible representation of the Poincaré group for a free Dirac particle with a positive energy sign we have obtained spin operators in the real form and determined their classical origin. In the presence of the external field the quasiclassical spin equations of motion also coincide with their classical BMT-analogue at $\hbar \to 0$. However, it is impossible to ignore the negative energy sign in the Dirac theory, and it leads to various paradoxes which are beyond the classical theory framework. For example, if in the classical theory both of the energy signs are taken into account, then the current density

$$j^{\mu}_{\mathrm{D}} = -iec\overline{\Psi}\gamma^{\mu}\Psi \div \frac{e}{m}\left(\overline{\Psi}\widehat{P}^{\mu}\Psi + i\overline{\Psi}\sigma^{\mu\nu}\widehat{P}_{\nu}\Psi\right)$$

will be a sum of the ordinary (orbital) current density and spin addition which get excited on account of interference of the energy states different in sign (see, for example, Ref.[83]). As a result, the separation of the orbital and spin motions become impossible even for a free particle. Probably, in this connection an erroneous opinion came up that the classical spin theory does not exist (see Refs[29,84,85], and also H. Corben's commentary in Ref.[61]).

The interference of energy-conjugate states in the Dirac theory leads to the appearance of the Zitterbewegung effect, i.e. quick-oscillating electron motion with the frequency $\omega_{\mathrm{Zb}} = 2m_0c^2/\hbar \approx 1.6 \cdot 10^{21}\ sec^{-1}$ and the amplitude of the Compton wavelength of the order of $\bar{\lambda}_{\mathrm{C}} = \hbar/m_0c \approx 3.9 \cdot\ 10^{-11}\ cm$. Only in the nonrelativistic case, when the negative energy solutions of the Dirac equation are well suppressed, Zitterbewegung disappears and the orbital and spin angular momentum are separated. In this case, the Dirac equation becomes two-component and in the weak uniform magnetic field takes the form[83]

$$i\hbar\,\frac{\partial\varphi}{\partial t} = \left[\frac{\hat{p}^2}{2m_0} - \frac{e}{2m_0c}((\widehat{\boldsymbol{L}} + 2\hat{\boldsymbol{s}})\boldsymbol{H})\right]\varphi,$$

where $\widehat{\boldsymbol{L}} = [\boldsymbol{r}\widehat{\boldsymbol{P}}]$ is the operator of the orbital angular momentum, $\hat{\boldsymbol{s}} = \frac{\hbar}{2}\boldsymbol{\sigma}'$ is the nonrelativistic spin operator with proper values $\pm\hbar/2$. We note that the term which describes the interaction of the magnetic field and the intrinsic magnetic moment gives the correct quantity of the electron g-factor equal to 2.

Thus, the correct description of the spin properties of particles in the Dirac theory is related to the problem of the relativistically invariant separation of Zitterbewegung and localization of the wave packet near the classical trajectory. However, the length l of this region has to be much greater than the value of the Compton wavelength of order of $\bar{\lambda}_{\mathrm{C}}$. At the same time, the external field has to vary sufficiently smoothly at the distances of the order of the wave packet width l. Thus, the conditions

$$l > \bar{\lambda}_{\mathrm{C}} = \frac{\hbar}{m_0c}, \quad l\left|\frac{\partial H}{\partial \boldsymbol{r}}\right| \ll H. \tag{2.84}$$

must be carried out.

Moreover, in the quasiclassical theory the particle momentum should vary little at the distance of the order of $\hbar/|\boldsymbol{P}|$ and $\hbar/l$. According to Eq. (2.25), this condition is satisfied best of all when momentum projection is fixed.

In principle, the conditions given by Eq. (2.84) can always be satisfied in the weakly nonuniform magnetic field since the amplitude of the Zitterbewegung is, as a rule, much less than the radius ρ of the trajectory curvature

$$\frac{\hbar}{m_0 c} \ll \rho = \frac{m_0 c^2 \gamma}{eH}.$$

Hence it follows that the condition (cf. Eq. (1.97))

$$H \ll \frac{1}{\gamma} H^*, \tag{2.85}$$

where $H^* = m_0^2 c^3 / e\hbar = 4.41 \cdot 10^{13}\ Oe$ is Schwinger's magnetic field, must be always valid.

It is clear that Eq. (2.85), which is related to the stability of the electron-positron vacuum,[e] will be valid even for the ultrarelativistic electrons with $\gamma \approx 10^5$.

It is important to note that the condition (2.85) is within the limits of the validity of the classical theory. Actually, the fields, where the classical electrodynamics laws are still valid, are determined by the inequality (cf. Eq. (1.95))

$$H \ll \frac{m_0^2 c^3}{e} = \frac{H^*}{\alpha} \approx 6 \cdot 10^{15}\ Oe \tag{2.86}$$

while, according to Eq. (2.85) at $\gamma \approx 10^5$

$$H \ll \frac{H^*}{\gamma} \approx 4 \cdot 10^8\ Oe. \tag{2.87}$$

Thus, if Eq. (2.86) is valid, then Eq. (2.87) will be valid too.

Separation of the energy states different in sign can be carried out by means of the FW transformation. Thereat the Dirac equation comes to the form

$$i\hbar \frac{\partial \Psi_{\rm FW}}{\partial t} = m_0 c^2 \rho_3 \gamma \Psi_{\rm FW}. \tag{2.88}$$

Hence it follows that the upper components of the wave function are related to the positive energy sign particle, while the lower components to the negative one. Separation of the wave function in sign of energy

$$\Psi_{\rm FW} = \Psi_{\rm FW}^{(+)} + \Psi_{\rm FW}^{(-)}$$

---

[e] The simplest way to obtain the Schwinger's magnetic field is to use the equality

$$\hbar\omega_0 = \hbar \frac{eH^*}{m_0 c} = m_0 c^2.$$

As the magnetic field does not do work, the electromagnetic vacuum turns out to be stable at any quantities of H including $H > H^*$. On the contrary, the electric field $E^* = m_0^2 c^3 / e\hbar$ does work, which leads to vacuum instability. As the virtual electron-positron pair can exist during the time $\delta\tau = \hbar/\Delta\ E_0 = \hbar/m_0 c^2$ and go away to the distance of $c\delta\tau \approx \hbar/m_0 c = \bar{\lambda}_C$ then the particles can acquire the energy $m_0 c^2$ and the birth of the pair will take place at $eE^* \bar{\lambda}_C = m_0 c^2$. From here the quantity of E^* follows.

is carried out by means of the transformation

$$\Psi_{\mathrm{FW}}^{(+)} = \frac{1}{2}(1+\rho_3)\Psi_{\mathrm{FW}}, \quad \Psi_{\mathrm{FW}}^{(-)} = \frac{1}{2}(1-\rho_3)\Psi_{\mathrm{FW}}$$

and then Eq. (2.88) is devided into two two-component equations with a different sign of energy

$$i\hbar\,\frac{\partial \Psi_{\mathrm{FW}}^{(+)}}{\partial t} = m_0 c^2 \gamma \Psi_{\mathrm{FW}}^{(+)}, \quad i\hbar\,\frac{\partial \Psi_{\mathrm{FW}}^{(-)}}{\partial t} = -m_0 c^2 \gamma \Psi_{\mathrm{FW}}^{(-)}.$$

Earlier we saw (see Sec. 2.2) that in this case, a more profound understanding of the correspondence principle in the spin theory is attained. However, the FW transformation is unattractive because it is not covariant and, therefore, in the presence of the external field the calculation is considerably complicated.[f]

It is possible to do it in another way[86] if, instead of the transformation to the two-component form of the Dirac equation with the two-component wave function distinctive in the sign of energy (it is not covariant) one would come to the two charge-conjugate Dirac equations ($e = \pm e_0,\ e_0 > 0$), each of that is related to the wave function with the positive sign of energy (it is covariant!).

This problem is solved on the basis of the Feynman and Gell-Mann transformation[87] for the Dirac equation with the fixed charge sign (for definiteness we assume that $e > 0$). The conjugate Dirac equations have the form

$$\overline{\Psi}(m_0 c - i\gamma_\mu \overleftarrow{\widehat{P}^\mu}) = 0, \quad (m_0 c - i\gamma_\mu \overrightarrow{\widehat{P}^\mu})\Psi = 0, \tag{2.89}$$

where

$$\overleftarrow{\widehat{P}^\mu} = -i\hbar \overleftarrow{\partial^\mu} + \frac{e}{c}A^\mu, \quad \overrightarrow{\widehat{P}^\mu} = -i\hbar \overrightarrow{\partial^\mu} - \frac{e}{c}A^\mu,$$

the arrows indicate the direction of the derivative action.

According to the Feynman and Gell-Mann transformation, we put

$$\overset{\vee}{\overline{\Psi}} = \overline{\Psi}\,\frac{1}{2}\,(1 - i\gamma^5), \quad \frac{1}{2}\,(1 + i\gamma^5)\Psi = \overset{\vee}{\Psi}\,.$$

The reverse relations for the solutions of the Dirac equations (2.89) have the form

$$\overline{\Psi} \div \overset{\vee}{\overline{\Psi}}\,\frac{1}{m_0 c}\,(m_0 c - i\gamma_\mu \overleftarrow{\widehat{P}}{}^\mu), \quad \frac{1}{m_0 c}\,(m_0 c + i\gamma_\mu \overrightarrow{\widehat{P}}{}^\mu)\,\overset{\vee}{\Psi} \div \Psi.$$

[f]From the mathematical point of view the realization of the FW transformation is related to the transition to the so-called even operator form as the action of even operators to the wave function does not mix up the upper and lower components. The algorithm of this operation as expansion in the Compton wavelength power series is well known (see, for example, Ref.[83]). However, the unwieldy mathematical procedure does not allow to completely analyze Zitterbewegung. Usually, it is limited by the first Darvin correction.

Hence it follows that the wave functions $\overset{\vee}{\Psi}$ and $\overset{\vee}{\overline{\Psi}}$ satisfy the squared Dirac equations which were obtained by V. Fock[76]

$$(m_0^2c^2+ \overrightarrow{\widehat{P}^{\mu}}\overrightarrow{\widehat{P}^{\mu}} - m_0\mu_0\sigma_{\alpha\beta}H^{\alpha\beta})\, \overset{\vee}{\Psi}= 0, \tag{2.90}$$

$$\overset{\vee}{\overline{\Psi}}\, (m_0^2c^2+ \overleftarrow{\widehat{P}^{\mu}}\overleftarrow{\widehat{P}^{\mu}} - m_0\mu_0\sigma_{\alpha\beta}H^{\alpha\beta}) = 0.$$

After that the Dirac equations (2.89) are decomposed into two pairs of crosswise charge-conjugate equations

$$\overline{\Psi}_+(m_0c + i\gamma_\mu \overleftarrow{\widehat{P}^{\mu}}) = 0. \quad (m_0c - i\gamma_\mu \overrightarrow{\widehat{P}^{\mu}})\Psi_+ = 0,$$

$$\overline{\Psi}_-(m_0c - i\gamma_\mu \overleftarrow{\widehat{P}^{\mu}}) = 0. \quad (m_0c + i\gamma_\mu \overrightarrow{\widehat{P}^{\mu}})\Psi_- = 0,$$

where $\Psi_+ = \Psi$ $(e > 0)$ while $\Psi_- = i\gamma^5\Psi$ $(e < 0)$. As we have decided to restrict ourselves to the case of $e > 0$, then Ψ_- should be excepted from consideration.

That problem can be solved by the use of the projection operators given by Eq. (2.47) which in a more general entry have the form

$$\overleftarrow{\widehat{\Lambda}}_\pm= \frac{1}{2m_0c}\,(m_0c \mp i\gamma_\mu \overleftarrow{\widehat{P}^{\mu}}), \quad \frac{1}{2m_0c}\,(m_0c \pm i\gamma_\mu \overrightarrow{\widehat{P}^{\mu}}) =\overrightarrow{\widehat{\Lambda}}_\pm .$$

Consequently, we have

$$\left.\begin{matrix}\overline{\Psi}_+\\ 0\end{matrix}\right\} = \overline{\Psi}\, \overleftarrow{\widehat{\Lambda}}_\pm, \quad \overrightarrow{\widehat{\Lambda}}_\pm\, \Psi = \left\{\begin{matrix}\Psi_+\\ 0\end{matrix}\right. .$$

In a general case $(e = \pm e_0,\ e_0 > 0)$, the projection operators separate the charge conjugate states automatically

$$\begin{aligned}\overline{\Psi}_\pm &= \overline{\Psi}\, \overleftarrow{\widehat{\Lambda}}_\pm, & \overrightarrow{\widehat{\Lambda}}_\pm\, \Psi_\pm &= \Psi_\pm,\\ 0 &= \overline{\Psi}_\mp\, \overleftarrow{\widehat{\Lambda}}_\pm, & \overrightarrow{\widehat{\Lambda}}_\pm\, \Psi_\mp &= 0.\end{aligned} \tag{2.91}$$

The properties of the projection operators also allow constructing operators which will act only in the invariantly separated spaces of the charge conjugate functions Ψ_+ or Ψ_-. Really, any operator acting on $\Psi_\pm$ can be represented in the form

$$\widehat{Q} = (\widehat{\Lambda}_\pm + \widehat{\Lambda}_\mp)\widehat{Q}(\widehat{\Lambda}_\pm + \widehat{\Lambda}_\mp) \div \widehat{\Lambda}_\pm\widehat{Q}\widehat{\Lambda}_\pm + \widehat{\Lambda}_\mp\widehat{Q}\widehat{\Lambda}_\pm = \overline{\widehat{Q}}_\pm + \tilde{\widehat{Q}}_\pm,$$

where

$$\overline{\widehat{Q}}_\pm = \widehat{\Lambda}_\pm\widehat{Q}\widehat{\Lambda}_\pm, \quad \tilde{\widehat{Q}}_\pm = \widehat{\Lambda}_\mp\widehat{Q}\widehat{\Lambda}_\pm \tag{2.92}$$

are even and odd operators, respectively. The connection of these operators with the traditional scheme of even and odd operators (see, for example, Ref.[88]) can be determined by means of the sign operators

$$\widehat{\varepsilon} = \frac{i\gamma_\mu \widehat{P}^\mu}{m_0 c}, \tag{2.93}$$

so that

$$\varepsilon\Psi_\pm = \pm\Psi_\pm, \quad \varepsilon^2\Psi_\pm = \pm\varepsilon\Psi_\pm = \Psi_\pm.$$

The sign operator $\widehat{\varepsilon}$ is connected with the projection operator $\widehat{\Lambda}_\pm$ by the obvious relation

$$\widehat{\Lambda}_\pm = \frac{1}{2}\,(1 \pm \widehat{\varepsilon}). \tag{2.94}$$

According to Eqs (2.93) and (2.94), for $\overline{\widehat{Q}}_\pm$ and $\widetilde{\widehat{Q}}$ we shall have

$$\overline{\widehat{Q}}_\pm = \frac{1}{2}\,(\widehat{Q} + \widehat{\varepsilon}\widehat{Q}\widehat{\varepsilon}), \quad \widetilde{\widehat{Q}}_\pm = \frac{1}{2}\,(\widehat{Q} - \widehat{\varepsilon}\widehat{Q}\widehat{\varepsilon}),$$

that corresponds to the names of these operators cited above.

It follows from Eq. (2.92) that the even operators retain the proper values of operators on the basis of which they have been constructed, while the odd operators put them to zero. For this reason, the even operators will be further of the greatest interest.

Let us find the even operators on the basis of the matrices of the Dirac bilinear forms given by Eq. (2.39). According to Eq. (2.92), we have

$$\overline{\widehat{Q}} \div \widehat{Q} + \frac{1}{2m_0 c}\,[i\gamma_\mu \widehat{P}^\mu, \widehat{Q}].$$

This formula is more convenient for practical calculations. Here substituting $Q = \Omega$ from Eq. (2.39) we obtain

$$\begin{aligned} -\overline{i\gamma^\mu} &= -i(\gamma^\mu + \sigma^{\mu\nu}\hat{b}_\nu) \div \hat{b}^\mu, \\ \overline{\sigma^{\mu\nu}} &= \sigma^{\mu\nu} - (\gamma^\mu\hat{b}^\nu - \gamma^\nu\hat{b}^\mu) = \widehat{\Pi}^{\mu\nu}, \\ \overline{\gamma^\mu\gamma^5} &= (\gamma^\mu + i\hat{b}^\mu)\gamma^5 = \widehat{\pi}^\alpha, \\ \overline{\gamma^5} &= 0. \end{aligned} \tag{2.95}$$

Here $\hat{b}^\mu = \widehat{P}^\mu/m_0 c$ is the dimensionless momentum operator, $\widehat{\Pi}^{\mu\nu}$ and $\widehat{\pi}^\alpha$ coincide with the spin operators given by Eq. (2.42) obtained earlier in another way.

The coordinate operator r^μ written down in the even form coincides with Fock's operator of the coordinate of the mass center

$$\overline{r^\mu} = r^\mu + \frac{\hbar}{2m_0 c}\,\gamma^\mu. \tag{2.96}$$

As compared with the ordinary Dirac operators, the even operators more adequately conform with the corresponding classical values. For example, the current density constructed on the basis of $-i\overline{\gamma^\mu}$ according to Eq. (2.95), has the standard quantum-mechanical form

$$\overline{j^\mu} = -iec\,\overline{\Psi}\,\overline{\gamma^\mu}\,\Psi \doteqdot \frac{e}{m_0}\,\overline{\Psi}\widehat{P}^\mu\Psi$$

and the above paradox of the orbital and spin currents lay-up does not come up.

The kinematical relations between even operators also have the well-known "classical" form. Really, in the framework of τ-formalism (see Sec. 2.4) Heisenberg's equations of the operators motion have the form

$$\frac{dQ}{d\tau} = \frac{i\hbar}{c}\,[\widehat{H}, Q], \tag{2.97}$$

where $\widehat{H}$ is the Hamiltonian of the Dirac τ-equation (2.72). With respect to r^μ, according to Eq. (2.97), we obtain the well-known relation of the Dirac theory

$$\frac{d\hat{r}^\mu}{d\tau} = -ic\gamma^\mu = \hat{v}^\mu.$$

However, the wide-spread interpretation of $\hat{v}^\mu$ as a "velocity" operator of the particle is at least unsatisfactory because $\hat{v}_\mu\hat{v}^\mu = -4c^2$, while it must be $-c^2$ (see also Refs[15,88]).

Things are quite different with the even operator $\overline{r^\mu}$ given by Eq. (2.96). In this case, according to Eq. (2.97) we have

$$\frac{d\overline{r^\mu}}{d\tau} = -ic\overline{\gamma^\mu} \doteqdot \widehat{V}^\mu, \tag{2.98}$$

where $\widehat{V}^\mu = \widehat{P}^\mu/m_0$ and in the absence of the external fields $\widehat{V}_\mu\widehat{V}^\mu \doteqdot -c^2$.

The external field greatly changes both kinematics and dynamics of the Dirac particles. In order to research these phenomena, it is necessary to elaborate a special mathematical procedure of the quantum theory with definite evenness of the operators.

Further, we shall take into account both the charge signs $e = \pm e_0$, $e_0 > 0$. Then, instead of $(dQ/d\tau)$ in Eq. (2.97), it should be written

$$\left(\frac{d\widehat{Q}}{d\tau}\right)_\pm = \frac{ic}{\hbar}\,[\widehat{H}_\pm, \widehat{Q}], \tag{2.99}$$

where

$$\widehat{H}_\pm = m_0c \mp i\gamma_\mu\widehat{P}^\mu.$$

By analogy, the equation of motion for even operators is written as

$$\left(\frac{d\overline{\widehat{Q}}}{d\tau}\right)_\pm = \frac{ic}{\hbar}[\widehat{H}_\pm, \overline{\widehat{Q}}_\pm] = \frac{d\overline{\widehat{Q}}_\pm}{d\tau},$$

where $\overline{\widehat{Q}}_\pm$ is determined according to Eq. (2.92).

Taking into account that $\widehat{H}_\pm\widehat{\Lambda}_\pm = \widehat{\Lambda}_\pm\widehat{H}_\pm$, the latter equation can be written in the form

$$\overline{\widehat{Q}}_\pm = \widehat{Q} + \frac{i\hbar}{2m_0c^2}\left(\frac{d\widehat{Q}}{d\tau}\right)_\pm . \tag{2.100}$$

This expression is convenient for iterations. According to Eq. (2.100), any operator can be expressed as an expansion in even operators

$$\widehat{Q} = \overline{Q}_\pm \; + \; \sum_{k=1}^{n}\left(\frac{1}{2m_0c}\right)^k \underbrace{[H_\pm[H_\pm,...[H_\pm,\overline{\widehat{Q}}_\pm]...]]}_{k} +$$
$$+ \; \left(\frac{1}{2m_0c}\right)^{n+1} \underbrace{[H_\pm[H_\pm,...[H_\pm,\widehat{Q}]...]]}_{n+1},$$

or

$$\widehat{Q} = \overline{\widehat{Q}}_\pm + \sum_{k=1}^{n}\left(\frac{i\hbar}{2m_0c^2}\right)^k \frac{d^k\overline{\widehat{Q}}_\pm}{d\tau^k} + \left(\frac{1}{2m_0c}\right)^{n+1}\frac{d^{n+1}\widehat{Q}}{d\tau^{n+1}}.$$

Developing this scheme further the whole of the quantum theory can be constructed in terms of even operators if the last term containing the usual operator $\widehat{Q}$ is neglected. Obviously, this can always be done, since the expansion parameter $\hbar/2m_0c^2 = 1/\omega_{\mathrm{Zb}} \approx 6\cdot 10^{-22}$ *sec* is very small.

Resuming the procedure of adopting evenness, according to Eq. (2.100), doubly even operators may be written in the form

$$\overline{\overline{\widehat{Q}}} = \overline{\widehat{Q}}_\pm + \frac{i\hbar}{2m_0c^2}\frac{d\overline{\widehat{Q}}_\pm}{d\tau} = \widehat{Q} + \frac{i\hbar}{m_0c^2}\frac{d\widehat{Q}}{d\tau} - \frac{\hbar^2}{4m_0^2c^4}\frac{d^2\widehat{Q}}{d\tau^2}$$

and so on.

The equation of motion of even operators can be obtained by differentiating Eq. (2.100) with respect to τ. Then[89,90]

$$\frac{d\widehat{Q}}{d\tau} + \frac{i\hbar}{2m_0c^2}\frac{d^2\widehat{Q}}{d\tau^2} \div [\widehat{G}_\pm, \widehat{Q}], \tag{2.101}$$

where

$$\widehat{G} = \frac{1}{2m_0c}\left(\widehat{P}_\mu\widehat{P}^\mu + m_0^2c^2 - \mu_0 m_0\sigma_{\alpha\beta}H^{\alpha\beta}\right) \tag{2.102}$$

is the Hamiltonian of the Dirac equations put in the squared form (see Fock's equation (2.90)). Since this operator does not depend on the sign, the symbol $\pm$ can be omitted in the equation.

Eq. (2.101) is evidently valid for operators of any degree of evenness. Note also that

$$\widehat{G} = \overline{\widehat{G}} = \overline{\overline{\widehat{G}}} = \ldots = inv.$$

Thus, the quantum theory with definite evenness of operators is invariant with respect to the charge in the degree of operator evenness and is a closed theory.

More convenient notation may be introduced for the derivatives with respect to the proper time

$$\frac{D}{D\tau} = \frac{d}{d\tau} + \frac{i\hbar}{2m_0c^2}\frac{d^2}{d\tau^2}.$$

Then Eq. (2.101) takes the more simple form

$$\frac{D\widehat{Q}}{D\tau} \div \frac{ic}{\hbar}\,[\widehat{G},\widehat{Q}]. \tag{2.103}$$

Double use of the generalized derivative gives the equation

$$\frac{D^2\widehat{Q}}{D\tau^2} \div -\frac{c^2}{\hbar^2}\,[\widehat{G}[\widehat{G},\widehat{Q}]].$$

Let us consider the application of the obtained formulas for $e > 0$ to the operators given by Eq. (2.95) taking into account the anomalous magnetic moment of the electron when

$$\widehat{H} = m_0c - i\gamma_\mu\widehat{P}^\mu - \frac{\mu_a}{2c}\,\sigma_{\alpha\beta}H^{\alpha\beta}$$

and

$$\widehat{G} = \left(\widehat{P}_\alpha\widehat{P}^\alpha + m_0^2c^2 - \frac{g}{2}m_0\mu_0\sigma_{\alpha\beta}H^{\alpha\beta} + \frac{2\mu_a}{c}H_{\mu\nu}\gamma^\mu P^\nu\right)\Big/2m_0c.$$

It is possible to show[91] that thereat the same equations as above are obtained if

$$\frac{\mu_a}{2m_0c^2}\,|\langle\sigma_{\alpha\beta}H^{\alpha\beta}\rangle| \ll 1.$$

Substituting $\widehat{b}^\mu$, $\widehat{\Pi}^{\mu\nu}$ and $\widehat{\pi}^\mu$ into Eq. (2.103) we obtain ($e > 0$, terms proportional to $\hbar\mu_a$ are omitted)

$$\frac{i\hbar}{2m_0c^2}\frac{d^2\widehat{P}^\mu}{d\tau^2} + \frac{d\widehat{P}^\mu}{d\tau} = \frac{e}{m_0c}H^{\mu\nu}\widehat{P}_\nu + \frac{ge\hbar}{8m_0c}\sigma_{\alpha\beta}\partial^\mu H^{\alpha\beta}, \tag{2.104}$$

$$\begin{aligned}\frac{i\hbar}{2m_0c^2}\frac{d^2\widehat{\Pi}^{\mu\nu}}{d\tau^2} + \frac{d\widehat{\Pi}^{\mu\nu}}{d\tau} &= \frac{ge}{2m_0c}H^{[\mu\alpha}\widehat{\Pi}_\alpha{}^{\nu]} + \frac{g-2}{2}\frac{e}{m_0^2c^3}\widehat{P}^{[\mu}\widehat{\Pi}^{\nu]\alpha}H_{\alpha\beta}\widehat{P}^\beta + \\ &+ \frac{ge\hbar}{4m_0c}i\gamma_\alpha(\partial^\alpha H^{\mu\nu} - \gamma^5\partial^\alpha E^{\mu\nu}),\end{aligned} \tag{2.105}$$

$$\begin{aligned}\frac{i\hbar}{2m_0c^2}\frac{d^2\widehat{\pi}^\mu}{d\tau^2} + \frac{d\widehat{\pi}^\mu}{d\tau} &= \frac{ge}{2m_0c}H^{\mu\alpha}\widehat{\pi}_\alpha + \\ &+ \frac{g-2}{2}\frac{e}{m_0^2c^3}H^{\alpha\beta}\widehat{P}_\alpha\widehat{\pi}_\beta\widehat{P}^\mu + \frac{ge\hbar}{8m_0c}i\sigma_{\alpha\beta}\partial^\mu E^{\alpha\beta}.\end{aligned} \tag{2.106}$$

If the conditions

$$\frac{g}{8}\,|\langle\sigma_{\alpha\beta}\partial^\mu H^{\alpha\beta}\rangle| \ll |\langle H^{\mu\nu}\widehat{P}_\nu\rangle| \quad etc.$$

are satisfied, the evenness of all the operators of Eqs (2.104)—(2.106) is retained. In this approximation the commutativity of the operators can be ignored, and then we obtain semiclassical equations of motion of the operators $\widehat{P}^{\mu}$, $\widehat{\Pi}^{\mu\nu}$ and $\widehat{\pi}^{\mu}$. The following substitution $\widehat{Q} \to \langle \widehat{Q} \rangle = Q$ corresponds to the transition to "classical" equations, including the terms proportional to Plank's constant $\hbar$. These terms obviously describe Zitterbewegung, too. If we ignore the terms containing $\hbar$, we obtain the purely classical Lorentz equation for the charge and the BMT equation for the spin.

2.6. Frenkel-Corben Classical Spin Equations

L.H. Thomas[10] was the first to apply a special relativity theory to the classical spin theory after Uhlenbeck and Goudsmit[9] had set up their hypothesis. However, the basic laws of the relativistic classical spin theory were advanced by J. Frenkel[11]. The Frenkel theory was constructed on the basis of his tensor of magnetic moments $\mu^{\alpha\beta}$, which is proportional to the spin tensor[g]

$$\mu^{\alpha\beta} = \mu_0 \Pi^{\alpha\beta},$$

where $\mu_0 = e\hbar/2m_0c$ is the Bohr magneton.

After the discovery of the anomalous magnetic moment[41] the Frenkel equations were generalized by H.C. Corben[17,34] (FC-equations). In the particular case of constant and uniform fields, the spin FC-equations transform into the BMT-equations.

The first relativistic quantum spin theory was built up by W. Pauli[94]. The two-component Schrödinger equation was taken as a basis of this theory. P.A.M. Dirac formulated the relativistic quantum spin theory[95] which later was generalized in order to include the electron anomalous magnetic moment[96].

For a long time the classical and quantum spin theories were developed independently. In the Refs[63–66] the quasiclassical origin of the BMT-equations and their connection with the Dirac equation were determined. After that the classical and quantum spin theories began to develop intersecting and mutually enriching each other.

Thus, history has confirmed that Bohr's remarks on the importance of the classical hypothesis of the rotating electron within the framework of the correspondence principle of classical and quantum mechanics concerning G.E. Uhlenbeck's and S. Goudsmit's paper were correct.

In spite of the swift development of the quantum spin theory[1–8], the classical spin theory in connection with its conveniences of interpretation of the polarization effects of relativistic particles and determination of the fundamental constants[105–107] is no less urgent and continues to develop even now.

Using various methods we have shown that the BMT-equation is an approximative expression of the correspondence principle in the spin theory. Here we shall determine

[g]Somewhat later the classical theory on the basis of four-dimensional spin vector s^{α} was built by I. Tamm[12]. In spite of I. Tamm's criticism of Frenkel's theory it is possible to show that both the theories are equal[92,93].

that the spin FC-equations and the BMT-equations, at the same time, allow an exact quantum analogue, too.

The spin and force FC-equations can be obtained if the Stern and Gerlach force of Eq. (2.59) acting on the charge is taken into account in the equation of charge motion in the rest system. Then instead of Eq. (2.60), we obtain for acceleration w^μ

$$w^\mu = \frac{e}{m_0 c}\, H^{\mu\nu} v_\nu + \frac{\mu}{2m_0}\, \Pi_{\rho\sigma} D^\mu H^{\rho\sigma} \tag{2.107}$$

where $\mu = g\mu_0/2$.

To satisfy the condition $v_\mu w^\mu = 0$, generalization of the derivative was carried out here

$$\partial^\mu \to D^\mu = \partial^\mu + \frac{1}{c^2}\, v^\mu \partial_\rho v^\rho .$$

Then the spacelike part in the spin equation (2.56) takes the form

$$\overrightarrow{W^\mu} = \frac{g-2}{2}\,\frac{e}{m_0 c}\, \Pi^{[\mu\rho} H_\rho{}^{\nu]} - \frac{\mu}{2}\, \Pi^{\mu\nu} \Pi_{\rho\sigma} D_\nu H^{\rho\sigma} .$$

After that instead of Eq. (2.58) we shall obtain the equation[h]

$$\frac{d\Pi^{\mu\nu}}{d\tau} = \frac{ge}{2m_0 c}\, H^{[\mu}_{\ \ \rho}\Pi^{\rho\nu]} + \frac{g-2}{2m_0 c^3}\, v^{[\mu}\Pi^{\nu]\alpha} H_{\alpha\beta} v^\beta - \frac{\mu}{2c^2}\, v^{[\mu}\Pi^{\nu]\alpha}\Pi_{\rho\sigma} D_\alpha H^{\rho\sigma} . \tag{2.109}$$

We note that in the terms $H^{\mu\nu}_{\text{eff}}$ with w^μ from Eq. (2.107) this equation does not differ from Eq. (2.61) which was obtained earlier without taking account of the force given by Eq. (2.59).

After separation of the acceleration w^μ Eq. (2.109), in accordance with Eq. (2.107), is easily transformed into the equations obtained by Frenkel[11]. In the dimensional form we have

$$\frac{dS^{\mu\nu}}{d\tau} = \frac{ge}{2m_0 c^3}\, H^{[\mu\rho} S_\rho{}^{\nu]} + \frac{1}{c^2}\, v^{[\mu} S^{\nu]\rho} a_\rho , \tag{2.110}$$

where

$$a^\rho = \frac{eg}{2m_0 c}\, H^{\rho\sigma} v_\sigma - w^\rho . \tag{2.111}$$

Of course, in the Dirac equation it was $g = 2$ as at that time the anomalous magnetic moment was unknown yet.

In the uniform magnetic field with account of Eq. (2.107) we have

$$a^\rho = \frac{g-2}{2m_0 c}\, H^{\rho\sigma} v_\sigma$$

---

[h]It is possible to obtain the vector version of this equation[108–110]

$$\frac{d\pi^\alpha}{d\tau} = \frac{ge}{2m_0 c}\, H^{\alpha\beta}\pi_\beta + \frac{g-2}{2m_0 c^3}\, v^\alpha v_\beta H^{\beta\gamma}\pi_\gamma + \frac{\mu}{2c^2}\, v^\alpha \pi^\beta \partial_\beta E^{\gamma\delta}\pi_\gamma v_\delta . \tag{2.108}$$

In constant and uniform fields this equation coincides with the BMT-equation. Equivalency of Eqs (2.109) and (2.108) is shown in the Refs[92,93].

and the Frenkel equations go over into the tensor version of the BMT-equation (2.61).

According to Corben[17] the spin equation (2.110) is represented in the form

$$\frac{dS^{\mu\nu}}{d\tau} + v^{[\mu}P^{\nu]} = \frac{ge}{2m_0c}\, H^{[\mu\rho}S_\rho{}^{\nu]}, \tag{2.112}$$

where

$$P^\mu = -\frac{1}{c^2}\, v^\mu v_\rho P^\rho - \frac{1}{c^2}\, S^{\mu\nu}a_\nu. \tag{2.113}$$

It is possible to show that the Corben equation (2.112) is a consequence of the balance of force moments in the magnetic field

$$\frac{d}{d\tau}\left([\boldsymbol{rP}] + \boldsymbol{S}\right) = [\boldsymbol{rF}] + [\boldsymbol{\mu H}].$$

Really, assuming that $\boldsymbol{F} = d\boldsymbol{P}/d\tau$ we obtain from here

$$\frac{d\boldsymbol{S}}{d\tau} + [\boldsymbol{vP}] = [\boldsymbol{\mu H}]. \tag{2.114}$$

For $H^{\mu\nu} = (-\boldsymbol{E}, \boldsymbol{H})$ and $S^{\mu\nu} = (\boldsymbol{Q}, \boldsymbol{S})$ in Eq. (2.112) we have

$$H^{[\mu\rho}S_\rho{}^{\nu]} = (-[\boldsymbol{SE}] - [\boldsymbol{HQ}],\ [\boldsymbol{SH}] + [\boldsymbol{QE}]).$$

At $\boldsymbol{E} = 0$ the purely spacelike components of the Corben equation transform into Eq. (2.114), which proves the above assertion. At the same time it follows from Eq. (2.114) that the orbital momentum $\boldsymbol{P}$ and the instantaneous particle velocity $\boldsymbol{v}$ are not parallel to each other. However, we note that at $H^{\mu\nu} \to H^{\mu\nu}_{\text{eff}}$ in accordance with Eq. (2.63), a^ρ vanishes, the momentum P^μ becomes timelike and we come to Eq. (2.65) again.

If we put

$$v_\mu P^\mu = -m_0c^2, \tag{2.115}$$

then the orbital momentum will coincide with the usual definition of $P^\mu = m_0 w^\mu$ exactly. In a more general case, the effective mass of the particle with a spin is introduced[i]

$$M = m_0 - \frac{\mu}{2c^2}\, H_{\alpha\beta}\Pi^{\alpha\beta} \tag{2.116}$$

and then, instead of Eq. (2.115), we have

$$v_\mu P^\mu = -Mc^2.$$

In this case, according to Eq. (2.113) we obtain

$$P^\mu = Mv^\mu - \frac{1}{c^2}\, S^{\mu\nu}a_\nu = Mv^\mu + \overrightarrow{P^\mu}, \tag{2.117}$$

[i] In the classical theories the effective mass of the particles appears in many Refs[111–122]. It was substantiated in the quantum theory Ref.[123]

where $\overrightarrow{P^\mu}$ is the spacelike part of P^μ.

In these notations the force equations for P^μ obtained by Frenkel[11] and Corben[17] have the form

$$\frac{d}{d\tau}\left(Mv^\mu + \overrightarrow{P^\mu}\right) = \frac{e}{c}\, H^{\mu\nu} v_\nu + \frac{\mu}{2}\,\Pi_{\rho\sigma}\partial^\mu H^{\rho\sigma}. \tag{2.118}$$

If one assumes again that the spin interacts with the external electromagnetic field by means of H^{eff} then since

$$\Pi_{\rho\sigma} H^{\rho\sigma}_{\text{eff}} = \Pi_{\rho\sigma} H^{\rho\sigma}$$

instead of Eq. (2.118) we obtain

$$\frac{d}{d\tau}\left(Mv^\mu\right) = \frac{e}{c}\, H^{\mu\nu} v_\nu + \frac{\mu}{2}\,\Pi_{\rho\sigma}\partial^\mu H^{\rho\sigma}.$$

It is easy to see that hence it follows

$$M\,\frac{dv^\mu}{d\tau} = \frac{e}{c}\, H^{\mu\nu} v_\nu + \frac{\mu}{2}\,\Pi_{\rho\sigma} D^\mu H^{\rho\sigma}. \tag{2.119}$$

At the replacement of $M \to m_0$ we have the acceleration coinciding with w^μ in Eq. (2.107). We note that Eq. (2.119) allows a simple Lagrange formulation with the Lagrange function[j]

$$\mathcal{L} = \frac{1}{2}\, M v_\rho v^\rho + \frac{e}{c}\, v_\rho A^\rho + \frac{\mu}{4}\, H_{\alpha\beta}\Pi^{\alpha\beta}.$$

The spin-like addition to mass is very small and in the best case for $H \approx 10^5\, Oe$, $\gamma \approx 10^4$ we have (cf. Eqs (1.97) and (2.85))

$$\frac{\Delta m}{m_0} = \frac{1}{2}\,\frac{H}{H_*}\,\gamma \approx 10^{-4} \div 10^{-5}.$$

Now we shall prove that all the obtained relations have an exact quantum analogue. With this view to start with, we pay attention to the fact that the square

[j] For the first time the Lagrangian form for the classical spin theory was introduced by J. Frenkel[11]. To take into account the additional conditions of $v_\mu v^\mu = -c^2$ and $v_\mu \Pi^{\mu\nu} = 0$ in the principle of the least action

$$\delta \int \mathcal{L} d\tau = 0$$

J. Frenkel used the method of the Lagrange multipliers λ and a_α:

$$\lambda v_\alpha \delta v^\alpha = 0, \quad a_\alpha \delta\left(\Pi^{\alpha\beta} v_\beta\right) = 0.$$

One of the Lagrange multipliers $\lambda = M$ is included into the above Lagrangian $\mathcal{L}$ (in linear approximation with respect to μ). As a result, the first condition of $v_\mu v^\mu = -c^2$ is satisfied automatically by virtue of the equations of motion. From the Lagrangian introduced by A.O. Barut[121] $\mathcal{L}$ is formally distinguished by the constant value $m_0c^2/2$.

Hamiltonian G given by Eq. (2.102) of the Dirac-Pauli equation in notations of even operators has the "classical" form[123] to an accuracy of the terms $\sim \hbar^2(g-2)$

$$\widehat{G} = \frac{1}{2m_0 c}\left(\widehat{P}_\mu \widehat{P}^\mu + \overset{\vee}{\overline{M}}{}^2 c^2\right),$$

where

$$\overset{\vee}{\overline{M}} = m_0 - \frac{\mu}{2c^2}\, H_{\alpha\beta}\widehat{\Pi}^{\alpha\beta}. \tag{2.120}$$

There is an exact quantum analogue of the effective mass of the particle with a spin [see Eq. (2.116)]. The initial operator of the zero-order of evenness has the form

$$\overset{\vee}{M} = m_0 - \frac{\mu}{2c^2}\, H_{\alpha\beta}\sigma^{\alpha\beta}. \tag{2.121}$$

In the operational form the force FC-equation can be obtained if the proper time Hamiltonian $\widehat{H}$ is represented in the modified form as compared with Eq. (2.72)

$$\overset{\vee}{H} = \overset{\vee}{M} c + \frac{1}{c}\, \overset{\vee}{v}_\mu \,\widehat{P}^\mu, \tag{2.122}$$

where $\overset{\vee}{M}$ corresponds to Eq. (2.121) and the velocity operator $\overset{\vee}{v}{}^\mu$ differs from the previous one $\widehat{v}^\mu$ [see Eq. (2.98)] by factor of 1/2 (see also Ref.[118])

$$\overset{\vee}{v}{}^\mu = -\frac{1}{2}\, ic\gamma^\mu.$$

Using the Hamiltonian $\overset{\vee}{H}$ according to Eq. (2.97), we find the acceleration operator

$$\frac{d\overset{\vee}{v}{}^\mu}{d\tau} = \frac{e}{m_0 c}\, H^{\mu\nu} v_\nu - \frac{c^2}{2\hbar}\, \sigma^{\mu\nu}\widehat{P}_\nu.$$

Introduction of $\overset{\vee}{v}{}^\mu$ and $\overset{\vee}{w}{}^\mu$ operators is justified by the "classical" relations[k]

$$\overset{\vee}{v}_\mu \overset{\vee}{v}{}^\mu = -c^2, \quad (\overset{\vee}{v}_\mu, \overset{\vee}{w}{}^\mu) = 0.$$

If the spin operator is used in its standard form

$$\overset{\vee}{S}{}^{\mu\nu} = \frac{\hbar}{2}\, \sigma^{\mu\nu},$$

then the classical Frenkel condition will be valid, too

$$(\overset{\vee}{v}_\mu, \overset{\vee}{S}{}^\mu) = 0.$$

---

[k]Here we have introduced the notation $(\overset{\vee}{a}, \overset{\vee}{b}) = (\overset{\vee}{a}\overset{\vee}{b} + \overset{\vee}{b}\overset{\vee}{a})/2$.

In accordance with identical redefining of the momentum

$$\check{P}^{\mu} = \frac{4}{c^2}\,(\check{v}^{\mu}, \check{v}_{\alpha}\check{P}^{\alpha})$$

for the solutions of the proper time equation with the Hamiltonian $\check{H}$ given by Eq. (2.122) we obtain

$$\check{P}^{\mu} \div (\widehat{M}, \check{v}^{\mu}) - \frac{1}{c^2}\,(\check{S}^{\mu\rho}, \check{a}_{\rho}), \tag{2.123}$$

where

$$\check{a}^{\rho} = \frac{eg}{2m_0c}\,H^{\rho\sigma}\,\check{v}_{\sigma} - \check{w}^{\rho}.$$

These operators are the exact quantum analogues of the corresponding classical values [see Eqs (2.113) and (2.111)].

The force equation for $\check{P}^{\mu}$ which is obtained according to the general procedure [see Eq. (2.97)] with the Hamiltonian $\check{H}$ of Eq. (2.122), also has the form of FC-equation (2.118)

$$\frac{d}{d\tau}\left[\left((\check{M}, \check{v}^{\mu}) - \frac{1}{c^2}\,(\check{S}^{\mu\nu}, \check{a}_{\nu})\right)\right] = \frac{e}{c}\,H^{\mu\nu}\,\check{v}_{\nu} + \frac{\mu}{2}\,\check{S}_{\alpha\beta}\partial^{\mu}H^{\alpha\beta}.$$

Similarly the spin equation in the Corben form [see Eq. (2.112)] is obtained

$$\frac{d\check{S}^{\mu\nu}}{d\tau} = \frac{ge}{2m_0c}\,H^{[\mu\rho}\,\check{S}_{\rho}{}^{\nu]} - \check{v}^{[\mu}\,\check{P}^{\nu]}.$$

Here substituting $\check{P}^{\mu}$ from Eq. (2.123) we obtain the equation

$$\frac{d\check{S}^{\mu\nu}}{d\tau} = \frac{ge}{2m_0c}\,H^{[\mu\rho}\,\check{S}_{\rho}{}^{\nu]} + \frac{1}{c^2}\left(\left(\check{v}^{[\mu}, \check{S}^{\nu]\rho}\right), a_{\rho}\right),$$

which is the exact analogue of the classical Frenkel equation (2.110).

Of course, the standpoint presented here is not indisputable. However, we think, it may be worth being taken into consideration.

2.7. Some Exact Solutions of the Spin Equations of Motion

The spin properties of electrons are of great importance in the precise measurements of the fundamental physical values[124,125].

Study of the spin precession on the basis of the classical BMT equation has enabled measuring the anomalous magnetic moment of the electron and other light particles[97,98,104] with a great accuracy. High accuracy of the measurement of masses of the elementary particles via electron-positron collisions, especially for ψ and Υ groups, was attained in the experiments on the superprecise measurement of electron energy in the storage rings on the basis of correspondence of the spin precession frequency and the electron revolution frequency in the orthogonal magnetic field of a storage ring[105,106]. Experimental measurements of the spin precession frequency of the particles in a storage ring have been carried out by observation of the polarization degree of beam under a high-frequency electromagnetic field, as the spin resonances have a great influence on beam polarization[100–102].

Polarizational analysis on the basis of the intense sources of polarized particles is applied more and more extensively in the atomic physics, solid-state physics, and physics of surface[125]. The discussion of these experiments is not our task. Here we give only some exact solutions of the BMT-equation which are necessary for further consideration.

2.7.1. Magnetic Top

In the most pure form the spin precession and the intrinsic magnetic moment associated with spin manifest themselves in the example of the electrically neutral particle (magnetic top) moving in a constant and uniform electromagnetic field. It is possible, in this case, to obtain an exact solution for the arbitrary configuration of field intensities[126].

For an electrically neutral top with a magnetic moment $\mu > 0$ we can write

$$\frac{ge}{2m_0c} = \frac{g}{\hbar}\,\mu_0 = \frac{\mu}{\hbar s}.$$

The following notation will be also useful

$$h^{\mu\nu} = \frac{ge}{2m_0c}\,H^{\mu\nu}, \quad q^{\mu\nu} = \frac{ge}{2m_0c}\,\overrightarrow{H^{\mu\nu}},$$

where $\overrightarrow{H^{\mu\nu}}$ is the spacelike part of the external electromagnetic field tensor given by Eq. (2.64).

In this notation the vector BMT-equation (2.62) for $w^\nu = 0$ and $H^{\mu\nu}_{\text{eff}} = \overrightarrow{H^{\mu\nu}}$ acquires a simple form

$$\frac{d\pi^\mu}{d\tau} = q^{\mu\nu}\pi_\nu, \tag{2.124}$$

where

$$q^{\mu\nu} = h^{\mu\nu} + \frac{1}{c^2}\,v^{[\mu}v_\rho h^{\rho\nu]}, \quad q^{\mu\nu}v_\nu = 0.$$

The components of the tensor $q^{\mu\nu} = (-\boldsymbol{d}, \boldsymbol{q})$ are equal to

$$\boldsymbol{d} = \gamma^2([\boldsymbol{\beta}[\boldsymbol{\beta}\boldsymbol{e}]] - [\boldsymbol{\beta}\boldsymbol{h}]), \quad \boldsymbol{q} = \gamma^2(\boldsymbol{h} - \boldsymbol{\beta}(\boldsymbol{\beta}\boldsymbol{h}) - [\boldsymbol{\beta}\boldsymbol{e}]),$$

where $\boldsymbol{\beta} = \boldsymbol{u}/c$, $\boldsymbol{u}$ is the velocity of the magnetic top.

Of major importance are the invariants of the tensor $q^{\mu\nu}$

$$\frac{1}{2}\,q_{\mu\nu}q^{\mu\nu} = q^2 - d^2 = \lambda^2, \quad \frac{1}{4}\,q_{\mu\nu}\tilde{q}^{\mu\nu} = -(\boldsymbol{q}\boldsymbol{d}) = 0,$$

where

$$\tilde{q}^{\mu\nu} = \frac{1}{2}\varepsilon^{\mu\nu\alpha\beta}q_{\alpha\beta},$$

and

$$\lambda = \gamma\sqrt{(\boldsymbol{h} - [\boldsymbol{\beta e}])^2 - (\boldsymbol{\beta h})^2}.$$

Eq. (2.124) represents a system of ordinary differential equations with constant coefficients. The general solution has the form[126]

$$\pi^{\mu} = \left[g^{\mu\nu} + \frac{1}{\lambda}q^{\mu\nu}\sin\lambda\tau + \frac{1}{\lambda^2}q^{\mu\rho}q_{\rho}{}^{\nu}(1-\cos\lambda\tau)\right]\pi_{\nu 1}, \tag{2.125}$$

where $\pi_{\nu 1}$ denotes the initial spin components, which satisfy the condition of $v_{\mu}\pi_1^{\mu} = 0$, so that $\pi_1^0 = (\boldsymbol{\beta\pi}_1)$. These relations, in turn, follow from the spacelike property of the spin vector $v_{\mu}\pi^{\mu} = 0$.

We write the solution of the tensor equation (2.61)

$$\frac{d\Pi^{\mu\nu}}{d\tau} = q^{[\mu\rho}\Pi_{\rho}{}^{\nu]}$$

in the form

$$\Pi^{\mu\nu} = \Pi_1^{\mu\nu} + \frac{1}{\lambda}q^{[\mu\alpha}\Pi_{\alpha 1}{}^{\nu]}\sin\lambda\tau - \left(2\Pi_1^{\mu\nu} + \frac{1}{\lambda^2}q^{[\mu\alpha}q_{\alpha\beta}\Pi_1^{\beta\nu]}\right)(1-\cos\lambda\tau), \tag{2.126}$$

where $v_{\mu}\Pi_1^{\mu\nu} = 0$, so that $\boldsymbol{\Phi}_1 = [\boldsymbol{\beta\Pi}_1]$. The latter result is also a consequence of the spacelike property of the spin tensor $v_{\mu}\Pi^{\mu\nu} = 0$.

According to Eqs (2.8), (2.9), and (2.13) each of the solutions given by Eqs (2.125) and (2.126) can be transformed one-to-one into the other. The solutions can also be checked by means of Eqs (2.16), which are deduced from Eqs (2.8) and (2.9).

Eqs (2.125) and (2.126) give the spin precession in the form

$$\omega = \frac{\lambda}{\gamma} = \sqrt{(\boldsymbol{h} - [\boldsymbol{\beta e}])^2 - (\boldsymbol{\beta h})^2}.$$

Let us consider special cases.

1. First we examine the uniform rectilinear motion of a magnetic top in a uniform magnetic field $\boldsymbol{H} = (0, 0, H)$. We assume that the translational motion of the top takes place in the XZ plane $\boldsymbol{\beta} = \beta(\sin\alpha, 0, \cos\alpha)$.

Using the invariant

$$\pi_{\mu}\pi^{\mu} = \pi_{\mu 1}\pi^{\mu 1} = \zeta^2 \tag{2.127}$$

we can subject the initial values of $\boldsymbol{\pi}_1$ to the auxiliary condition

$$\pi_{1z} = \frac{\beta^2\sin\alpha\cos\alpha}{1-\beta^2\cos^2\alpha}\,\pi_{1x}.$$

Then

$$\pi_{\mu 1}\pi^{\mu 1} = \frac{1}{\gamma^2(1-\beta^2\cos^2\alpha)}\,\pi_{1x}^2 + \pi_{1y}^2 = \zeta^2.$$

Setting

$$\pi_{1x} = -\zeta\gamma\sqrt{1-\beta^2\cos^2\alpha}\ \sin\varphi_1, \quad \pi_{1y} = \zeta\cos\varphi_1$$

we obtain ($\mu < 0$)

$$\begin{aligned}
\pi_x &= -\zeta\gamma\sqrt{1-\beta^2\cos^2\alpha}\ \sin(\lambda\tau+\varphi_1), \\
\pi_y &= \zeta\cos(\lambda\tau+\varphi_1), \\
\pi_z &= -\zeta\,\frac{\gamma\beta^2\sin\alpha\cos\alpha}{\sqrt{1-\beta^2\cos^2\alpha}}\ \sin(\lambda\tau+\varphi_1).
\end{aligned} \tag{2.128}$$

Similarly from Eq. (2.126) we have

$$\begin{aligned}
\Pi_x &= -\frac{1}{\gamma\sqrt{1-\beta^2\cos^2\alpha}}\,\Pi_{1y}\sin\lambda\tau + \Pi_{1x}\cos\lambda\tau, \\
\Pi_y &= \gamma\sqrt{1-\beta^2\cos^2\alpha}\,\Pi_{1x}\sin\lambda\tau + \Pi_{1y}\cos\lambda\tau, \quad \Pi_z = \Pi_{1z}.
\end{aligned} \tag{2.129}$$

On the basis of the invariant

$$\frac{1}{2}\,\Pi_{\mu\nu}\Pi^{\mu\nu} = \frac{1}{2}\,\Pi_{\mu\nu 1}\Pi_1^{\mu\nu} = \Pi_1^2 - \Phi_1^2 = \frac{1}{\gamma^2}\,\Pi_1^2 + (\boldsymbol{\beta}\boldsymbol{\Pi}_1)^2 = \zeta^2 \tag{2.130}$$

we formulate the auxiliary condition $\Pi_{1z} = 0$, so that also $\Pi_z = 0$.

Setting ($\zeta = \zeta_\perp$)

$$\Pi_{1x} = -\frac{\zeta_\perp}{\sqrt{1-\beta^2\cos^2\alpha}}\ \sin\varphi_1, \quad \Pi_{1y} = \zeta_\perp\gamma\cos\varphi_1 \tag{2.131}$$

we obtain

$$\begin{aligned}
\Pi_x &= -\frac{\zeta_\perp}{\sqrt{1-\beta^2\cos^2\alpha}}\ \sin(\lambda\tau+\varphi_1), \\
\Pi_y &= \zeta_\perp\gamma\cos(\lambda\tau+\varphi_1), \quad \Pi_z = 0.
\end{aligned} \tag{2.132}$$

The spin precession frequency is

$$\omega = \frac{\lambda}{\gamma} = h\sqrt{1-\beta^2\cos^2\alpha}. \tag{2.133}$$

For an electrically neutral particle with a negative magnetic moment (neutron) we have $h = |\mu|H/\hbar s$.

It is interesting that the spin precession has azimuthal symmetry and the frequency $\omega = \lambda/\gamma$ for $\alpha = 0$ (motion in the direction of the magnetic field), whereas for $\alpha = \pi/2$ (motion across the magnetic field) the amplitude of the components (π_x, Π_y) increases by the factor of γ, and the precession frequency $\omega = h$ is totally independent of the velocity.

2. We now consider the motion of a magnetic top (charged or electrically neutral) in the constant and orthogonal electromagnetic fields $E < H$:

$$\boldsymbol{H} = (0, 0, H), \quad \boldsymbol{E} = (0, E, 0), \quad E = H\sin\eta, \quad \boldsymbol{\beta} = (\sin\eta, 0, 0),$$

where η is a certain constant angle. We know that in this case (the Wien filter), a charge can move along the X axis uniformly and rectilinearly.

Here, making use of the fact that $\lambda = h\cos\eta$, and assuming also that

$$\pi_{1x} = \zeta\,\frac{\sin\varphi_1}{\cos\eta}, \quad \pi_{1y} = \zeta\cos\varphi_1, \quad \pi_{1z} = 0,$$

we obtain ($\mu > 0$)

$$\pi_x = \frac{\zeta}{\cos\eta}\,\sin(\lambda\tau - \varphi_1), \quad \pi_{1y} = \zeta\cos(\lambda\tau - \varphi_1), \quad \pi_z = 0.$$

If $\eta = 0$, this solution coincides with Eq. (2.128), where it must be assumed that $\alpha = \pi/2$.

Similarly, for the tensor modification with the auxiliary condition

$$\Pi_{1x} = -\zeta_\perp\sin\varphi_1, \quad \Pi_{1y} = \frac{\zeta_\perp}{\cos\eta}\,\cos\varphi_1, \quad \Pi_{1z} = 0$$

we have

$$\Pi_x = -\zeta_\perp\sin(\lambda\tau - \varphi_1), \quad \Pi_y = \frac{\zeta_\perp}{\cos\eta}\,\cos(\lambda\tau - \varphi_1), \quad \Pi_z = 0.$$

When $\eta = 0$ this solution is obtained from Eq. (2.132) for $\alpha = \pi/2$.

The spin precession frequency is

$$\omega = \frac{\lambda}{\gamma} = \frac{h}{\gamma}\cos\eta = \frac{h}{\gamma^2}.$$

Considering the motion of a magnetic top in the plane of the constant and orthogonal electromagnetic fields

$$\boldsymbol{H} = (0, 0, Q\cos\eta), \quad \boldsymbol{E} = (0, Q\sin\eta, 0), \quad \boldsymbol{\beta} = \beta(0, \sin\alpha, \cos\alpha),$$

where $Q = \sqrt{E^2 + H^2}$, $\tan\eta = E/H$, in the same way, as in the previous case, and using the invariant given by Eq. (2.127) we determine the auxiliary condition in the form

$$\pi_{1z} = \frac{\beta\cos\alpha(-\pi_{1x}\sin\eta + \pi_{1y}\beta\cos\eta\sin\alpha)}{\cos\eta(1 - \beta^2\cos^2\alpha)}$$

and set the initial components

$$\begin{aligned}
\pi_{1x} &= \zeta\gamma \frac{q\cos\eta}{\lambda} \sqrt{1-\beta^2\cos^2\alpha}\cos\varphi_1, \\
\pi_{1y} &= \zeta\gamma\sqrt{1-\beta^2\cos^2\alpha}\ \sin\varphi_1, \\
\pi_{1z} &= \frac{\zeta\gamma\beta\cos\alpha}{\sqrt{1-\beta^2\cos^2\alpha}}\left\{\beta\sin\alpha\sin\varphi_1 - \frac{q}{\lambda}\sin\eta\cos\varphi_1\right\},
\end{aligned}$$

where $q = (2\mu Q)/\hbar$.

In this case, the components of the spin vector are as follows

$$\begin{aligned}
\pi_x &= \zeta\gamma \frac{q\cos\eta}{\lambda}\sqrt{1-\beta^2\cos^2\alpha}\ \cos(\lambda\tau - \varphi_1), \\
\pi_y &= -\zeta\gamma\sqrt{1-\beta^2\cos^2\alpha}\ \sin(\lambda\tau - \varphi_1), \\
\pi_z &= \zeta\gamma \frac{\beta\cos\alpha}{\sqrt{1-\beta^2\cos^2\alpha}}\left\{\frac{q}{\lambda}\sin\eta\cos(\lambda\tau-\varphi_1) - \beta\sin\alpha\sin(\lambda\tau-\varphi_1)\right\}.
\end{aligned}$$

In the similar way, it is possible to obtain the spin tensor components from Eq. (2.126). However, to simplify calculations one can use the result for components of the spin vector and Eq. (2.16). Thus, we have

$$\begin{aligned}
\Pi_x &= \zeta\gamma^2 \frac{q\cos\eta}{\lambda}\sqrt{1-\beta^2\cos^2\alpha}\ \cos(\lambda\tau-\varphi_1), \qquad (2.134)\\
\Pi_y &= \frac{\zeta}{\sqrt{1-\beta^2\cos^2\alpha}}\left\{\frac{q}{\lambda}\gamma^2\beta^3\sin\eta\sin\alpha\cos^2\alpha\cos(\lambda\tau-\varphi_1) - \sin(\lambda\tau-\varphi_1)\right\}, \\
\Pi_z &= -\zeta\gamma^2\frac{q\sin\eta}{\lambda}\ \beta\cos\alpha\sqrt{1-\beta^2\cos^2\alpha}\ \cos(\lambda\tau-\varphi_1).
\end{aligned}$$

In this case, the spin precession frequency takes on the form

$$\omega = \frac{\lambda}{\gamma} = q\sqrt{\cos^2\eta(1-\beta^2\cos^2\alpha) + \beta^2\sin^2\eta\cos^2\alpha}.$$

Now we consider the uniform motion of a magnetic top orthogonal to the plane of the fields

$$\boldsymbol{H} = (0,0,Q\cos\eta), \quad \boldsymbol{E} = (0,Q\sin\eta,0), \quad \boldsymbol{\beta} = (\beta,0,0).$$

Using Eq. (2.127) we set the auxiliary condition $\pi_{1z} = 0$ which involves $\pi_z = 0$. Further assuming

$$\pi_{1x} = \gamma\zeta\cos\varphi_1, \quad \pi_{1y} = \zeta\sin\varphi_1,$$

we get

$$\pi_x = \gamma\zeta\cos(\lambda\tau-\varphi_1), \quad \pi_y = -\zeta\sin(\lambda\tau-\varphi_1).$$

Hence it follows that

$$\Pi_x = \zeta\cos(\lambda\tau-\varphi_1), \quad \Pi_y = -\gamma\zeta\sin(\lambda\tau-\varphi_1). \qquad (2.135)$$

The obtained results depend on the parameter η only through the spin precession frequency

$$\omega = \frac{\lambda}{\gamma} = q(\cos\eta - \beta\sin\eta).$$

3. It is also interesting to consider the motion of a magnetic top in constant and parallel electromagnetic fields

$$\boldsymbol{H} = (0, 0, Q\cos\eta), \quad \boldsymbol{E} = (0, 0, Q\sin\eta), \quad \boldsymbol{\beta} = \beta(\sin\alpha, 0, \cos\alpha).$$

Here we shall act as in the previous cases. Determining the auxiliary condition

$$\pi_{1z} = \frac{\beta\sin\alpha(\pi_{1x}\beta\cos\alpha\cos\eta - \pi_{1y}\sin\eta)}{cos\eta(1-\beta^2\cos^2\alpha)},$$

and setting

$$\begin{aligned}
\pi_{1x} &= \gamma\zeta\sqrt{1-\beta^2\cos^2\alpha}\ \cos\varphi_1, \\
\pi_{1y} &= \gamma\zeta\frac{q}{\lambda}\cos\eta\sqrt{1-\beta^2\cos^2\alpha}\ \sin\varphi_1, \\
\pi_{1z} &= \frac{\gamma\zeta\beta\sin\alpha}{\sqrt{1-\beta^2\cos^2\alpha}}\left\{\beta\cos\alpha\cos\varphi_1 - \frac{q}{\lambda}\sin\eta\sin\varphi_1\right\}.
\end{aligned}$$

we obtain the following expression for the spin vector components

$$\begin{aligned}
\pi_x &= \gamma\zeta\sqrt{1-\beta^2\cos^2\alpha}\ \cos(\lambda\tau - \varphi_1), \\
\pi_y &= -\gamma\zeta\frac{q}{\lambda}\cos\eta\sqrt{1-\beta^2\cos^2\alpha}\ \sin(\lambda\tau - \varphi_1), \\
\pi_z &= \frac{\gamma\zeta\beta\sin\alpha}{\sqrt{1-\beta^2\cos^2\alpha}}\left\{\beta\cos\alpha\cos(\lambda\tau - \varphi_1) + \frac{q}{\lambda}\sin\eta\sin(\lambda\tau - \varphi_1)\right\}
\end{aligned}$$

and for the spin tensor components respectively

$$\begin{aligned}
\Pi_x &= \frac{\zeta}{\sqrt{1-\beta^2\cos^2\alpha}}\left\{\cos(\lambda\tau - \varphi_1) - \frac{q}{\lambda}\gamma^2\beta^3\sin^2\alpha\cos\alpha\sin\eta\sin(\lambda\tau - \varphi_1)\right\}, \\
\Pi_y &= -\gamma^2\zeta\frac{q}{\lambda}\cos\eta\sqrt{1-\beta^2\cos^2\alpha}\ \sin(\lambda\tau - \varphi_1), \qquad (2.136) \\
\Pi_z &= \gamma^2\zeta\frac{q}{\lambda}\beta\sin\alpha\sin\eta\sqrt{1-\beta^2\cos^2\alpha}\ \sin(\lambda\tau - \varphi_1).
\end{aligned}$$

The spin precession frequency has the form

$$\omega = \frac{\lambda}{\gamma} = q\sqrt{\cos^2\eta(1-\beta^2\cos^2\alpha) + \beta^2\sin^2\eta\sin^2\alpha}.$$

4. We also consider the spin precession in the rest frame of a magnetic top. The BMT-equation (2.62) has now the form ($\mu < 0$)

$$\frac{d\boldsymbol{\zeta}}{d\tau} = -\frac{|\mu|}{\hbar s}\,[\boldsymbol{\zeta}\boldsymbol{H}_0], \qquad (2.137)$$

where $\boldsymbol{H}_0$ is the magnetic field in the rest frame of the top.

The solution of Eq. (2.137) with the same parameters as in the first case has the form

$$\begin{aligned}\zeta_x &= -\zeta\,\frac{1-\frac{\gamma-1}{\gamma}\cos^2\alpha}{\sqrt{1-\beta^2\cos^2\alpha}}\,\sin(\omega t+\varphi_1), \quad \zeta_y=\zeta\cos(\omega t+\varphi_1),\\ \zeta_z &= -\zeta\,\frac{\frac{\gamma-1}{\gamma}\sin\alpha\cos\alpha}{\sqrt{1-\beta^2\cos^2\alpha}}\,\sin(\omega t+\varphi_1).\end{aligned}$$

The precession frequency is

$$\omega=\frac{|\mu|H}{\hbar s}\sqrt{1-\beta^2\cos^2\alpha}.$$

2.7.2. Electron in a Constant and Uniform Magnetic Field

In this case, the vector BMT-equation (2.62) has the form

$$\frac{d\boldsymbol{\pi}}{d\tau}=\frac{eg}{2m_0c}\,[\boldsymbol{\pi}\boldsymbol{H}]+\frac{e(g-2)}{2m_0c^3}\,\boldsymbol{v}(\boldsymbol{v}[\boldsymbol{\pi}\boldsymbol{H}]). \tag{2.138}$$

The charge velocity included here $\boldsymbol{v}=\gamma c\boldsymbol{\beta}$ is determined by the relations ($e>0$, $H \parallel Z$)

$$v_x=v_\perp\sin\omega t, \quad v_y=v_\perp\cos\omega t, \quad v_z=v_\parallel \tag{2.139}$$

with the frequency

$$\omega=\frac{eH}{m_0c\gamma}.$$

We represent the solution of Eq. (2.138) in the form[127]

$$\begin{aligned}\pi_x &= \zeta_\perp(\cos\omega t\cos\omega_a t-q\sin\omega t\sin\omega_a t),\\ \pi_y &= -\zeta_\perp(\sin\omega t\cos\omega_a t+q\cos\omega t\sin\omega_a t),\\ \pi_z &= (\zeta_\parallel-\gamma\beta_\perp\beta_\parallel\zeta_\perp\sin\omega t)/\sqrt{1-\beta_\parallel^2},\end{aligned} \tag{2.140}$$

where

$$q=\sqrt{1+\gamma^2\beta_\perp^2}=\gamma\sqrt{1-\beta_\parallel^2}, \quad \omega_a=\frac{g-2}{2}\,q\omega.$$

In connection with this

$$(\boldsymbol{\beta\pi})=(\zeta_\parallel\beta_\parallel-\gamma\beta_\perp\zeta_\perp\sin\omega_a t)\,\Big/\sqrt{1-\beta_\parallel^2}$$

and (cf. Eq. (2.14))

$$\pi_\mu\pi^\mu=\zeta_\perp^2+\zeta_\parallel^2=\zeta^2.$$

We note that

$$(\boldsymbol{\beta}\boldsymbol{\pi})\big|_{g=2} = \frac{\beta_{\|}\zeta_{\|}}{\sqrt{1-\beta_{\|}^2}} = const.$$

The tensor BMT-equation (2.61)

$$\frac{d\boldsymbol{\Pi}}{d\tau} = \frac{eg}{2m_0c}\left[\boldsymbol{\Pi}\boldsymbol{H}\right] + \frac{e(g-2)}{2m_0c^3}\left(\boldsymbol{v}\boldsymbol{\Pi})[\boldsymbol{v}\boldsymbol{H}]\right)$$

has the solution

$$\begin{aligned}
\Pi_x &= \zeta_\perp\left(\gamma\cos\omega t\cos\omega_a t - \frac{\sin\omega t\sin\omega_a t}{\sqrt{1-\beta_{\|}^2}}\right) - \frac{\gamma\beta_\perp\beta_{\|}}{\sqrt{1-\beta_{\|}^2}}\zeta_{\|}\sin\omega t,\\
\Pi_y &= -\zeta_\perp\left(\gamma\sin\omega t\cos\omega_a t + \frac{\cos\omega t\sin\omega_a t}{\sqrt{1-\beta_{\|}^2}}\right) - \frac{\gamma\beta_\perp\beta_{\|}}{\sqrt{1-\beta_{\|}^2}}\zeta_{\|}\cos\omega t, \qquad (2.141)\\
\Pi_z &= \gamma\sqrt{1-\beta_{\|}^2}\zeta_{\|},
\end{aligned}$$

thereat (cf. Eq. (2.14))

$$\frac{1}{2}\,\Pi_{\mu\nu}\Pi^{\mu\nu} = \zeta_\perp^2 + \zeta_{\|}^2 = \zeta^2,$$

and

$$(\boldsymbol{\beta}\boldsymbol{\pi}) = \gamma(\boldsymbol{\beta}\boldsymbol{\Pi}).$$

By means of Eq. (2.16) one can verify that the solutions given by Eqs (2.140) and (2.141) mutually transform from one into the other. The conditions of spin invariance [see Eqs (2.127) and (2.130)] are also valid.

Of more interest are the solutions of the BMT-equation for the spin vector in the rest system given by Eq. (2.68)

$$\frac{d\boldsymbol{\zeta}}{d\tau} = [\boldsymbol{\Omega}\boldsymbol{\zeta}],$$

where

$$\boldsymbol{\Omega} = -\frac{e}{m_0c}\left(\frac{g}{2}\frac{\boldsymbol{H}}{\gamma} - \frac{g-2}{2}\,\frac{\gamma}{\gamma+1}\left[\boldsymbol{\beta}[\boldsymbol{\beta}\boldsymbol{H}]\right]\right). \qquad (2.142)$$

As we have seen in Sec. 2.3, this frequency is made up of the Larmor and Thomas precession frequencies

$$\begin{aligned}
\boldsymbol{\Omega}_{\mathrm{L}} &= -\frac{eg}{2m_0c}\left(\frac{1}{\gamma}\,\boldsymbol{H} - \frac{\gamma}{\gamma+1}\left[\boldsymbol{\beta}[\boldsymbol{\beta}\boldsymbol{H}]\right]\right),\\
\boldsymbol{\Omega}_{\mathrm{Th}} &= -\frac{e}{m_0c}\,\frac{\gamma}{\gamma+1}\left[\boldsymbol{\beta}[\boldsymbol{\beta}\boldsymbol{H}]\right],
\end{aligned}$$

i.e.

$$\boldsymbol{\Omega} = \boldsymbol{\Omega}_{\mathrm{L}} + \boldsymbol{\Omega}_{\mathrm{Th}}.$$

All frequencies $\boldsymbol{\Omega}$, $\boldsymbol{\Omega}_{\rm L}$, and $\boldsymbol{\Omega}_{\rm Th}$ can be calculated on the basis of the given charge motion in a uniform magnetic field (see Eq. (2.139)). As a result, we obtain

$$\begin{aligned}
\boldsymbol{\Omega}_{\rm L} &= \frac{g}{2}\omega\frac{\gamma^2}{\gamma+1}\left\{\beta_\perp\beta_\parallel(\sin\omega t,\cos\omega t), -\left(\frac{1}{\gamma}+1-\beta_\parallel^2\right)\right\},\\
\boldsymbol{\Omega}_{\rm Th} &= -\omega\frac{\gamma^2}{\gamma+1}\left\{\beta_\perp\beta_\parallel(\sin\omega t,\cos\omega t), -\beta_\perp^2\right\},\\
\boldsymbol{\Omega} &= \omega\frac{\gamma^2}{\gamma+1}\left\{\frac{g-2}{2}\beta_\perp\beta_\parallel(\sin\omega t,\cos\omega t), \beta_\perp^2-\frac{g}{2}\left(\frac{1}{\gamma}+1-\beta_\parallel^2\right)\right\}
\end{aligned}$$

and the absolute values

$$\Omega_{\rm L} = \frac{g}{2}q\omega, \quad \Omega_{\rm Th} = \omega(\gamma-1),$$

$$\Omega = \omega\gamma\sqrt{\frac{g^2}{4}(1-\beta_\parallel^2) - \frac{(g-1)\gamma+1}{\gamma+1}\beta_\perp^2}.$$

We note that in the nonrelativistic case, the frequency $\Omega_{\rm Th}$ takes the known form

$$\Omega_{\rm Th} \approx \frac{1}{2}\beta^2\omega$$

and the total spin precession frequency Ω at $g=2$ coincides with that of the charge orbital revolution

$$\Omega|_{g=2} = \omega.$$

A solution of Eq. (2.137) has the form

$$\begin{aligned}
\zeta_x &= \zeta_\perp\left(\cos\omega t\cos\omega_a t - \delta_\parallel\sin\omega t\sin\omega_a t\right) - \frac{\gamma}{\gamma+1}\frac{\beta_\perp\beta_\parallel}{\sqrt{1-\beta_\parallel^2}}\zeta_\parallel\sin\omega t,\\
\zeta_y &= -\zeta_\perp\left(\sin\omega t\cos\omega_a t + \delta_\parallel\cos\omega t\sin\omega_a t\right) - \frac{\gamma}{\gamma+1}\frac{\beta_\perp\beta_\parallel}{\sqrt{1-\beta_\parallel^2}}\zeta_\parallel\cos\omega t,\\
\zeta_z &= \delta_\parallel\zeta_\parallel - \delta_\perp\zeta_\perp\sin\omega_a t, \qquad (2.143)
\end{aligned}$$

where

$$\delta_\perp = \frac{\gamma}{\gamma+1}\frac{\beta_\perp\beta_\parallel}{\sqrt{1-\beta_\parallel^2}}, \quad \delta_\parallel = \frac{\gamma(1-\beta_\parallel^2)+1}{(\gamma+1)\sqrt{1-\beta_\parallel^2}},$$

$$\delta_\perp^2 + \delta_\parallel^2 = 1.$$

One can verify that Eq. (2.143) is related to Eqs (2.140) and (2.141) by the Lorentz transformation given by Eq. (2.15).

Let us consider some particular cases.

For a purely orthogonal motion we obtain

$$\beta_\perp = \beta, \quad \beta_\parallel = 0, \quad \delta_\perp = 0, \quad \delta_\parallel = 1, \quad q = \gamma$$

and

$$\zeta_x = \zeta_\perp \cos\Omega_\| t, \quad \zeta_y = -\zeta_\perp \sin\Omega_\| t, \quad \zeta_z = \zeta_\|, \tag{2.144}$$

where

$$\Omega_\| = \omega + \omega_a = \omega\left(\frac{g-2}{2}\gamma + 1\right).$$

The latter formula for $\Omega_\|$ was a basis of the precise measurements of the μ^+ meson anomalous magnetic moment[98]. It was also used for precision calibration of the energy of electron beams in the storage rings by means of the resonance depolarization method in the experiments on high-frequency determination of masses of elementary particles and resonances ($\varphi, K^0, K^\pm, \omega, \psi, \Upsilon$) produced via electron-positron collisions[105,106].

In the case of the purely longitudinal motion, we have

$$\beta_\perp = 0, \quad \beta_\| = \beta, \quad \delta_\perp = 0, \quad \delta_\| = 1, \quad q = 1$$

and the same Eq. (2.144), where

$$\Omega_\| = \frac{g}{2}\,\omega.$$

2.7.3. Spin-Resonance Solutions

Spin resonances strongly affect electron beam polarization. Therefore, the research of spin resonances is of great importance in the production of relativistic polarized beams, conservation and transformation of polarization[100–102,106,124]. The BMT-equation gives a simple and reliable method of spin resonances research.

Let us consider first a simple model of the relativistic magnetic spin-resonance[128]. With this in view, we define an external field of the magnetic resonance as usual in the form

$$\boldsymbol{H}_{\rm res} = h(\boldsymbol{i}\cos\tilde{\omega}t + \boldsymbol{j}\sin\tilde{\omega}t) + \boldsymbol{k}H,$$

where $\boldsymbol{i}$, $\boldsymbol{j}$, $\boldsymbol{k}$ are the unit vectors of the Cartesian coordinate system, H is the uniform magnetic field, and h is the amplitude of the rotating magnetic field.

To describe spin cinematics we use the BMT-equation for the spin-vector given in the rest frame of the particle ($e = -e_0 < 0, g = 2$)

$$\frac{d\boldsymbol{\zeta}}{dt} = [\boldsymbol{\Omega}\boldsymbol{\zeta}], \tag{2.145}$$

where

$$\boldsymbol{\Omega} = \frac{e_0}{m_0 c\gamma}\,\boldsymbol{H}_{\rm res}.$$

In a coordinate system rotating with the angular velocity $\tilde{\omega}$ Eq. (2.145) has the form

$$\frac{d\boldsymbol{\zeta}'}{dt} = [\boldsymbol{\Omega}'\boldsymbol{\zeta}'], \tag{2.146}$$

where

$$\boldsymbol{\Omega}' = (\omega_h, 0, \delta), \quad \delta = \omega - \widetilde{\omega}, \quad \omega = \frac{eH}{m_0 c\gamma}, \quad \omega_h = \frac{eh}{m_0 c\gamma}.$$

Further, it will be more convenient to write down Eq. (2.146) in a matrix form

$$\frac{d\boldsymbol{\zeta}'}{d\tau} = \begin{vmatrix} 0 & -\delta & 0 \\ \delta & 0 & -\omega_h \\ 0 & \omega_h & 0 \end{vmatrix} \boldsymbol{\zeta}', \quad \boldsymbol{\zeta}' = \begin{vmatrix} \cos\widetilde{\omega}t & \sin\widetilde{\omega}t & 0 \\ -\sin\widetilde{\omega}t & \cos\widetilde{\omega}t & 0 \\ 0 & 0 & 1 \end{vmatrix} \boldsymbol{\zeta}. \tag{2.147}$$

Eq. (2.147) formally coincides with Bloch's equation for the magnetization vector in the case when the relaxation terms are negligibly small[129,130]. A solution of this equation is found by means of the composition of three consecutive rotations

$$\boldsymbol{\zeta}' = \begin{vmatrix} \cos\xi & 0 & -\sin\xi \\ 0 & 1 & 0 \\ \sin\xi & 0 & \cos\xi \end{vmatrix} \begin{vmatrix} 1 & 0 & 0 \\ 0 & \cos\Omega' & -\sin\Omega' \\ 0 & \sin\Omega' & \cos\Omega' \end{vmatrix} \begin{vmatrix} \cos\xi & 0 & \sin\xi \\ 0 & 1 & 0 \\ -\sin\xi & 0 & \cos\xi \end{vmatrix} \boldsymbol{\zeta}_1', \tag{2.148}$$

where

$$\sin\xi = \frac{\delta}{\Omega'}, \quad \cos\xi = \frac{\omega_h}{\Omega'}, \quad \omega' = \sqrt{\delta^2 + \omega_h^2},$$

$\boldsymbol{\zeta}_1'$ is the spin vector at the time $t = 0$. It is easy to find from Eq. (2.148) all projections of $\boldsymbol{\zeta}'$

$$\begin{aligned} \zeta_x' &= \frac{\omega_h^2 + \delta^2 \cos\Omega' t}{\omega'^2}\zeta_{1x}' - \frac{\delta}{\Omega'^2}\sin\Omega' t\, \zeta_{1y}' + \frac{\omega_h\delta}{\Omega'^2}(1 - \cos\Omega' t)\zeta_{1z}', \\ \zeta_y' &= \frac{\delta}{\Omega'^2}\sin\Omega' t\, \zeta_{1x}' + \cos\Omega' t\, \zeta_{1y}' - \frac{\omega_h}{\Omega'^2}\sin\Omega' t\, \zeta_{1z}', \\ \zeta_z' &= \frac{\omega_h\delta}{\Omega'^2}(1 - \cos\Omega' t)\zeta_{1x}' + \frac{\omega_h}{\Omega'^2}\sin\Omega' t\, \zeta_{1y}' + \frac{\delta^2 + \omega_h^2\cos\Omega' t}{\omega'^2}\zeta_{1z}'. \end{aligned}$$

It is not difficult to verify that $\boldsymbol{\zeta}'^2 = \boldsymbol{\zeta}_1'^2$. In particular, when $\boldsymbol{\zeta}_1' = (0, 0, 1)$ we have a relativistic generalization of Rabi's formula

$$\zeta_z' = 1 - \frac{2\omega_h^2}{\omega_h^2 + \delta^2}\sin^2\frac{\Omega' t}{2}.$$

In quantum mechanics this formula is obtained by a different method[130].

In the case of the resonance, $\delta = 0$ and the vector $\boldsymbol{\zeta}'$ rotates (under the same initial conditions) in the plane $X'Z'$ which contains H

$$\boldsymbol{\zeta}' = (0, -\sin\omega_h t, \cos\omega_h t).$$

In the other extreme case, when $\omega_h \ll \delta$ ($\alpha \cong \pm\pi/2$) we have, according to Eq. (2.148),

$$\zeta_x' = \cos\delta t\zeta_{1x}' \mp \sin\delta t\zeta_{1y}', \quad \zeta_y' = \pm\sin\delta\zeta_{1x}' t + \cos\delta t\zeta_{1y}', \quad \zeta_z' = \zeta_{1z}'.$$

In a general case

$$(\boldsymbol{\zeta}'\boldsymbol{\Omega}') = \omega_h \zeta'_{1x} + \delta \zeta'_{1z}.$$

If the frequency defect δ changes adiabatically slowly, then the angle between the vectors $\boldsymbol{\zeta}'$ and $\boldsymbol{\Omega}'$ in practice remains fixed, and the vector $\boldsymbol{\zeta}'$ smoothly follows Ω' precessing around this vector.

We note that the model of relativistic magnetic resonance considered here describes very well the passage resonance of polarized electrons in cyclic accelerators[100,124].

Now let us consider the behavior of the spin of a relativistic electron moving in the focusing azimuthally symmetrical magnetic field

$$H_\rho = -q\frac{z}{\rho}\left(\frac{R}{\rho}\right)^q H_R, \quad H_\varphi = 0, \quad H_z = \left(1 - \frac{q^2}{2}\frac{z^2}{\rho^2}\right)\left(\frac{R}{\rho}\right)^q H_R,$$

where q is the parameter characterizing the radial nonuniformity of the magnetic field, which in the case of stable equilibrium lies within the limits of $0 < q < 1$ (weak focusing). In the plane of the equilibrium orbit

$$z \to 0, \quad \rho \to R = \frac{m_0 c^2 \beta\gamma}{eH_R}$$

and

$$H_\rho, H_\varphi \to 0, \quad H_z \to H_R.$$

The classical theory of electron motion in the fields of this type was considered, for example, in Ref.[131], and the quantum theory in Ref.[132] Solution of the classical equation of charge motion ($e = -e_0 < 0$) in a linear approximation in the amplitudes of betatron oscillations has the form[133]

$$\begin{aligned} \rho &= R[1 + a_\rho \sin(\omega_\rho t + \alpha_\rho)], \\ z &= Ra_z \sin(\omega_z t + \alpha_z), \\ \varphi &= \omega t + \frac{a_\rho}{\sqrt{1-q}} \cos(\omega_\rho t + \alpha_\rho), \end{aligned} \tag{2.149}$$

where

$$\omega_\rho = \sqrt{1-q}\,\omega, \quad \omega_z = \sqrt{q}\,\omega, \quad \omega = eH/m_0 c\gamma$$

and $a_\rho = A_\rho/R$, $a_z = A_z/R$ are the relative amplitudes of betatron oscillations.

We shall consider the spin precession for the motion given by Eq. (2.149) on the basis of BMT-equation (2.145) with the vector of spin precession frequency given by Eq. (2.142). In the cylindrical coordinate system taking into account Eq. (2.149) we have

$$\begin{aligned} \Omega_\rho &= -(\omega + \omega_a) q a_z \sin(\omega_z t + \alpha_z), \\ \Omega_\varphi &= -\omega_a \sqrt{q} a_z \cos(\omega_z t + \alpha_z), \\ \Omega_z &= (\omega + \omega_a)[1 - q a_\rho \sin(\omega_\rho t + \alpha_\rho)], \end{aligned}$$

where

$$\omega_a = \frac{g-2}{2}\frac{eH}{m_0 c}.$$

We shall find a solution of the BMT-equation (2.145) by means of the consecutive approximation method. In a zero-order approximation, i.e. in the absence of betatron oscillations, we have a simple solution

$$\zeta_\rho^0 = \zeta_\perp \cos(\omega_a t + \alpha_a), \quad \zeta_\varphi^0 = \zeta_\perp \sin(\omega_a t + \alpha_a), \quad \zeta_z = \zeta_\parallel,$$

which in the Cartesian coordinate system coincides with that obtained earlier for a uniform magnetic field ($e = -e_0 < 0$, $\beta_\parallel = 0$).

In a general case, the solution of Eq. (2.145) has the form

$$\begin{aligned}
\zeta_\rho &= \zeta_\perp \cos(\omega_a t + \alpha_a) - \\
&- \zeta_\perp \frac{(1-q)\omega + q\omega_a}{\omega_\rho} a_\rho \sin(\omega_a t + \alpha_a)\cos(\omega_\rho t + \alpha_\rho) - \\
&- \zeta_\parallel (\omega - \omega_a + k\omega_z)\frac{q\omega_a}{\omega_z^2 - \omega_a^2} a_z \sin(\omega_z t + \alpha_z), \\
\zeta_\varphi &= \zeta_\perp \sin(\omega_a t + \alpha_a) + \\
&+ \zeta_\perp \frac{(1-q)\omega + q\omega_a}{\omega_\rho} a_\rho \cos(\omega_a t + \alpha_a)\cos(\omega_\rho t + \alpha_\rho) + \\
&+ \zeta_\parallel [(\omega - \omega_a)\omega_z + k\omega_a^2]\frac{q}{\omega_z^2 - \omega_a^2} a_z \cos(\omega_z t + \alpha_z), \\
\zeta_z &= \zeta_\parallel - \\
&- \zeta_\perp \frac{q}{2}\left\{[\omega - (1+k)\omega_a]\frac{\sin[(\omega_z + \omega_a)t + \alpha_z + \alpha_a]}{\omega_z + \omega_a} - \right. \\
&- \left. [\omega - (1+k)\omega_a]\frac{\sin[(\omega_z - \omega_a)t + \alpha_z - \alpha_a]}{\omega_z - \omega_a}\right\} a_z,
\end{aligned} \tag{2.150}$$

where

$$k = \frac{1}{\sqrt{q}}\frac{\gamma}{\gamma+1}.$$

Let us turn to the resonance character of these expressions: all three components have a gap at $\omega_z = \omega_a$. We note that in spite of that, the absolute value $|\boldsymbol{\zeta}|$ remains constant in quantity.

Similarly, one can obtain solutions for $\Pi^{\mu\nu}$ and π^α of the BMT-equations. For example, z-component of the solution of the tensor equation has the form (see also Ref.[134])

$$\begin{aligned}
\Pi_z &= \gamma\left\{\zeta_\parallel + \zeta_\perp \frac{q}{2}\eta\left[\frac{\sin[(\omega_z + \omega_a)t + \alpha_z - \alpha_a]}{\sqrt{q} + \eta} - \right.\right. \\
&- \left.\left. \frac{\sin[(\omega_z - \omega_a)t + \alpha_z + \alpha_a]}{\sqrt{q} - \eta}\right] a_z\right\},
\end{aligned} \tag{2.151}$$

where

$$\eta = \frac{\omega_a}{\omega_z} = \frac{g-2}{2}\,\gamma.$$

Eqs (2.150) are in complete agreement with the quantum theory. Putting in Eq. (2.151) $\zeta_{\|} = \cos\Phi$, $\zeta_{\perp} = \sin\Phi$ we obtain Eq. (35) of Ref.[132], in which the quantum theory of spin resonances was considered for the first time.

2.7.4. Classical Interpretation of Spin-Flip Transitions

Quantum spin-flip transitions are of great importance in the SR theories. For better understanding of the nature of this phenomenon, it is necessary to except the effects connected with the electric charge radiation from observation. This fact explains the suggested choice of the spin-flip radiation model in the form of radiation of the electric neutral Dirac-Pauli particle moving in the uniform magnetic field. It is known[135] that in this case, the total radiation in the quantum theory is conditioned by spin-flip transitions. The idea is that spin-flip radiation is represented as a nonstationary process connected with spin precession. From this point of view, we shall show how to construct a solution of the classical equation of spin precession in the BMT theory having the exact solution of the Dirac-Pauli equation. Thus, one will find the connection of the quantum spin-flip transitions with classical spin precession[136].

According to the uncertainty principle, for the spin-flip transition with the characteristic frequency of SR we obtain

$$\omega_{\max} \approx \omega_0\gamma^3 = \frac{2|\mu|H}{\hbar}\,\gamma^2, \quad \omega_0 = \frac{eH}{m_0 c\gamma} \approx \frac{c}{\rho},$$

where μ is the magnetic moment, ρ is the radius of curvature of the relativistic electron trajectory in the uniform magnetic field H. The ratio of the transition time $\Delta\tilde{t}$ and the time of the radiation forming Δt on the circular arc $dl \approx \gamma\rho$ is equal

$$\Delta\tilde{t}/\Delta t \approx 2\pi\gamma^{-4} \ll 1.$$

That means that in the ultrarelativistic case, spin-flip transitions are practically non-inertial, that is why the electron motion can be considered for the time $\Delta\tilde{t}$ as uniform and rectilinear.

Let us consider the electric neutral Dirac-Pauli particle, which moves in the uniform magnetic field $\boldsymbol{H} = (0,0,H)$. As is known, the wave function of the above particle has the form

$$\Psi(\boldsymbol{r},t) = L^{-3/2}\begin{pmatrix} c_1 \\ c_2 e^{i\varphi} \\ c_3 \\ c_4 e^{-i\varphi}\end{pmatrix} \exp(ib_\mu\xi^\mu) = \psi_\zeta(\boldsymbol{r})\exp\left(-i\frac{m_0c^2}{\hbar}\gamma_\zeta t\right),$$

where $b^\mu = \gamma_\zeta(1, \boldsymbol{\beta})$, $\xi^\mu = (m_0c/\hbar)r^\mu$ are the dimensionless energy-momentum and the coordinates of the particle,

$$\gamma_\zeta = \gamma\left(1 + \zeta S\sqrt{1+\gamma^2\beta_\perp^2}\right), \quad \zeta = \pm 1,$$

and moreover, $S = |\mu|H/m_0c^2$, $\gamma = 1/\sqrt{1-\beta^2}$.

The spin coefficients c_i $(i = 1, 2, 3, 4)$ are determined from the condition that the stationary wave function ψ_ζ is an eigenfunction of the spin operator

$$\widehat{\Pi}_z = \rho_2[\boldsymbol{\sigma}\widehat{\boldsymbol{b}}]_z + \sigma_z,$$

which is a motion integral and describes the spin projection on the magnetic field direction with the quantum number ζ.

If the particle moves in the plane XZ $(\varphi = 0)$, then for $\boldsymbol{\beta} = \beta(\sin\alpha, 0, \cos\alpha)$ the spin coefficients take on the form

$$\begin{pmatrix} c_1 \\ c_2 \\ c_3 \\ c_4 \end{pmatrix} = \begin{pmatrix} \frac{\zeta}{2}\sqrt{\frac{1}{2}\left(1+\frac{\zeta}{q}\right)}\left(\sqrt{1+\beta_z}+\zeta\sqrt{1-\beta_z}\right) \\ -\frac{1}{2}\sqrt{\frac{1}{2}\left(1-\frac{\zeta}{q}\right)}\left(\sqrt{1+\beta_z}-\zeta\sqrt{1-\beta_z}\right) \\ \frac{\zeta}{2}\sqrt{\frac{1}{2}\left(1+\frac{\zeta}{q}\right)}\left(\sqrt{1+\beta_z}-\zeta\sqrt{1-\beta_z}\right) \\ \frac{1}{2}\sqrt{\frac{1}{2}\left(1-\frac{\zeta}{q}\right)}\left(\sqrt{1+\beta_z}-\zeta\sqrt{1-\beta_z}\right) \end{pmatrix} = |\zeta\rangle,$$

where $q = \gamma\sqrt{1-\beta^2\cos^2\alpha}$. We note that the spin coefficients written here in the first approximation on $\hbar$ differ from zero (this notation concerns β_z, cf. Ref.[135]).

In the quantum theory the spin precession is described by the nonstationary wave function which is a superposition of the spin states $\zeta = \pm 1$

$$\widetilde{\Psi}(\boldsymbol{r}, t) = A\psi_1(\boldsymbol{r})\exp\left\{-i\frac{m_0c^2}{\hbar}\gamma_1 t\right\} + B\psi_{-1}(\boldsymbol{r})\exp\left\{-i\frac{m_0c^2}{\hbar}\gamma_{-1}t\right\}.$$

The constant factors A and B are determined from the initial condition

$$(\widehat{\boldsymbol{\Pi}}\boldsymbol{n})\widetilde{\Psi}(\boldsymbol{r}, 0) = \lambda\widetilde{\Psi}(\boldsymbol{r}, 0),$$

where $\boldsymbol{n} = (\sin\lambda\cos\nu, \sin\lambda\sin\nu, \cos\lambda)$. For example, in the case of the initial spin orientation along the X-axis, we shall have

$$A = -\frac{\zeta_1}{\sqrt{2}}\sqrt{1 - \frac{\zeta_1\gamma\beta^2\sin\alpha\cos\alpha}{\sqrt{1+\gamma^2\beta^4\sin^2\alpha\cos^2\alpha}}},$$

$$B = \frac{1}{\sqrt{2}}\sqrt{1 + \frac{\zeta_1\gamma\beta^2\sin\alpha\cos\alpha}{\sqrt{1+\gamma^2\beta^4\sin^2\alpha\cos^2\alpha}}},$$

$$\lambda = \zeta_1 \sqrt{1 + \gamma^2 \beta^2 \cos^2 \alpha},$$

along the Y-axis

$$A = \frac{i}{\sqrt{2}}, \quad B = \frac{\zeta_1}{\sqrt{2}}, \quad \lambda = \zeta_1 \gamma,$$

along the Z-axis

$$A = (1, 0), \quad B = (0, 1), \quad \lambda = \zeta \, \frac{\cos^2 \alpha + \gamma^2 \sin^2 \alpha}{\gamma \sqrt{1 - \beta^2 \cos^2 \alpha}}.$$

We note that in the second case, these relations have the simplest form, as the spin tensor component $\widehat{\Pi}_y$ is really not subjected to the Lorentz transformations: according to Eq. (2.15), we have $\Pi_y = \gamma \zeta_y$. Value $\zeta_1 = \pm 1$ characterizes the direction of the initial spin orientation. In the latter case, $\zeta_1 = \zeta$.

To calculate the mean value of all components of $\widehat{\Pi}^{\mu\nu}$ tensor it is also necessary to use the relations

$$\langle \zeta | \widehat{\Pi}_x | \zeta \rangle = -\zeta \frac{\gamma \beta^2 \sin \alpha \cos \alpha}{\sqrt{1 - \beta^2 \cos^2 \alpha}}, \qquad \langle -\zeta | \widehat{\Pi}_x | \zeta \rangle = -\frac{1}{\sqrt{1 - \beta^2 \cos^2 \alpha}},$$
$$\langle \zeta | \widehat{\Pi}_y | \zeta \rangle = 0, \qquad \langle -\zeta | \widehat{\Pi}_y | \zeta \rangle = -i \zeta \gamma,$$
$$\langle \zeta | \widehat{\Pi}_z | \zeta \rangle = \zeta \gamma \sqrt{1 - \beta^2 \cos^2 \alpha}, \qquad \langle -\zeta | \widehat{\Pi}_z | \zeta \rangle = 0.$$

The mean values of the components $\overline{\langle \widehat{\boldsymbol{\Pi}} \rangle_t}$ will essentially differ from one another in dependence on the initial spin orientation. In the most common case when $\overline{\langle \widehat{\boldsymbol{\Pi}} \rangle_0} = \overline{\langle \widehat{\Pi}_y \rangle_t}$ we have

$$\overline{\langle \widehat{\Pi}_x \rangle_t} = -\frac{\zeta_1}{\sqrt{1 - \beta^2 \cos^2 \alpha}} \sin \omega t,$$

$$\overline{\langle \widehat{\Pi}_y \rangle_t} = \zeta_1 \gamma \cos \omega t, \quad \overline{\langle \widehat{\Pi}_z \rangle_t} = 0.$$

As a matter of fact these expressions are not different from the solution of the tensor BMT-equation of spin precession given by Eq. (2.132) for $\mu < 0$, $\zeta_\perp = \zeta_1$, $\zeta_\parallel = 0$.

The frequency of spin precession determined by the condition

$$\omega = \frac{m_0 c^2}{\hbar} \left(\gamma_\zeta - \gamma_{-\zeta} \right) = \zeta \, \frac{2|\mu| H}{\hbar} \, \sqrt{1 - \beta^2 \cos^2 \alpha}$$

coincides with Eq. (2.133).

For the initial spin orientation along the X axis all calculations omitted here become more complex although nothing changes in principle.

Interesting results are obtained at the initial spin orientation along the Z axis

$$\overline{\langle \widehat{\Pi}_x \rangle_t} = -\zeta \, \frac{\gamma \beta^2 \sin \alpha \cos \alpha}{\sqrt{1 - \beta^2 \cos^2 \alpha}},$$

$$\overline{\langle\widehat{\Pi}_y\rangle_t} = 0, \quad \overline{\langle\widehat{\Pi}_z\rangle_t} = \zeta\gamma\sqrt{1-\beta^2\cos^2\alpha},$$

that is all the components are constant.

Calculation of the spin invariant is of great interest

$$I = \frac{1}{2}\,\overline{\langle\widehat{\Pi}_{\mu\nu}\rangle_t}\;\overline{\langle\widehat{\Pi}^{\mu\nu}\rangle_t} = \frac{1}{\gamma^2}\,\overline{\langle\widehat{\boldsymbol{\Pi}}\rangle_t}^2 - \overline{\langle(\boldsymbol{\beta}\widehat{\Pi})\rangle_t}^2.$$

Simple but unwieldy calculations show that in any case $I = 1$, which corresponds to the classical expression of this invariant (see Eq. (2.46)).

This method also allows to determine the precession of the spin projection on the direction of motion $\overline{\langle(\boldsymbol{\beta}\widehat{\Pi})\rangle_t}$. Again all results coincide with the classical theory

$$\overline{\langle(\boldsymbol{\beta}\widehat{\Pi})\rangle_t} = \frac{\zeta_1\beta}{\gamma\sqrt{1-\beta^2\cos^2\alpha}\sqrt{1+\gamma^2\beta^2\sin^2\alpha}}\,(\cos^2\alpha + \gamma^2\sin^2\alpha\cos\omega t).$$

Similarly, one can consider a precession of the spin vectors $\overline{\langle\widehat{\pi}^\alpha\rangle_t}$ and $\overline{\langle\widehat{\boldsymbol{\zeta}}\rangle_t}$ too. In all cases at the identical initial conditions, the results coinciding with the solutions of the corresponding classical BMT-equations are obtained[136].

Thus, all obtained results are in complete agreement with the classical theory of spin precession.

What do the results obtained here have to do with the spin-flip radiation theory? It is known (see Chapter 4) that in the classical theory spin or, more precisely, the intrinsic magnetic moment, orientated towards or opposite the field direction does not radiate during uniform and rectilinear particle motion. This radiation is accounted for by the spin precession in the plane orthogonal to the magnetic field and all radiation characteristics completely coincide with the quantum theory of spin-flip radiation. It follows from the above that, in fact, there is no difference between classical and quantum interpretation of the radiation in the nonstationary representation of spin-flip transitions.

2.8. Spin Integrals of Motion and Invariant Description of Spin States in the External Electromagnetic Field

For practical purposes in research of the exact solutions of spin equations it is very important to know the relativistic spin invariants which would be motion integrals. Invariant integrals of motion are very convenient as they retain the proper values upon a transition from one inertial system to another. The usual spin operators $\widehat{\Pi}^{\mu\nu}$ and $\widehat{\pi}^\alpha$ do not possess this property. Moreover, in the presence of the external electromagnetic field they, generally speaking, stop being integrals of motion. It is easy to show this if the equation of motion of operators in terms of the proper time representation is used (see Eq. (2.97)). For $\widehat{Q} = \widehat{\Pi}^{\mu\nu}$ we obtain

$$\frac{d\widehat{\Pi}^{\mu\nu}}{d\tau} = \frac{e}{m_0c}\,H^{[\mu\alpha}\sigma_\alpha{}^{\nu]},$$

hence it follows that

$$\begin{aligned} \frac{d\boldsymbol{\Pi}}{dt} &= \frac{e}{m_0c}\left(\rho_3[\boldsymbol{\sigma H}] + \rho_2[\boldsymbol{\sigma E}]\right), \\ \frac{d\boldsymbol{\Phi}}{dt} &= \frac{e}{m_0c}\left(\rho_2[\boldsymbol{\sigma H}] - \rho_3[\boldsymbol{\sigma E}]\right). \end{aligned} \tag{2.152}$$

Here we have taken into account that[l] $d/dt \div \rho_3 d/d\tau$ and, besides that,

$$H^{[\mu\alpha}\sigma_{\alpha}{}^{\nu]} = (-i[\boldsymbol{\alpha}H] - [\boldsymbol{\sigma}E], [\boldsymbol{\sigma}H] - i[\boldsymbol{\alpha}E]).$$

Hence, in the same case of the existence of a magnetic field ($\boldsymbol{E} = 0$) alone with a fixed direction, the projection of the spin onto the direction of the magnetic field is an integral of motion. On the other hand, in the same case of a constant electric field ($\boldsymbol{H} = 0$), the projection of the spin onto the direction of the electric field is, correspondingly, an integral of motion.

By analogy for $\widehat{Q} = \widehat{\pi}^{\alpha}$ we obtain

$$\frac{d\widehat{\pi}^{\alpha}}{dt} = \frac{e}{m_0c} H^{\alpha\beta}\gamma_{\beta}\gamma^5$$

and

$$\frac{d\widehat{\boldsymbol{\pi}}}{dt} = \frac{e}{m_0c}\left(\rho_1\boldsymbol{E} + [\boldsymbol{\sigma H}]\right), \quad \frac{d\widehat{\pi}^0}{dt} = \frac{e}{m_0c}(\boldsymbol{\sigma E}).$$

Hence, in the same case when there is no electric field ($\boldsymbol{E} = 0$) and for a magnetic field with a fixed direction, the projections of the space component $\widehat{\boldsymbol{\pi}}$ onto the direction of the external magnetic field as well as the component $\widehat{\pi}^0$ are integrals of motion. If there is no magnetic field ($\boldsymbol{H} = 0$), the projection of the component $\widehat{\boldsymbol{\pi}}$ onto the direction along which the electric field vanishes is an integral of motion. However, the component $\widehat{\pi}^0$ stops being an integral of motion.

Taking account of the anomalous magnetic moment of an electron, the class of spin integrals of motion is some more reduced. In the uniform magnetic field $\boldsymbol{H} = (0, 0, H)$ only the projection $\widehat{\Pi}_z$ is an integral of motion (see also Sec. 2.7).

---

[l] This property follows from the usual rule of operators differentiation

$$\frac{d\widehat{Q}}{dt} = \frac{\partial\widehat{Q}}{\partial t} + \frac{1}{i\hbar}(\widehat{Q}\widehat{\mathcal{H}} - \widehat{\mathcal{H}}\widehat{Q}), \tag{2.153}$$

where for the Dirac particle Hamiltonian $\mathcal{H}$ has the form

$$\widehat{\mathcal{H}} = c(\widehat{\boldsymbol{\alpha}}\widehat{\boldsymbol{P}}) + e\varphi + \rho_3 m_0c^2 = i\hbar\frac{\partial}{\partial t} + c\rho_3\widehat{H} \div i\hbar\frac{\partial}{\partial t}. \tag{2.154}$$

Substituting this expression into Eq. (2.153) we obtain

$$\frac{d\widehat{Q}}{dt} \div \frac{ic}{\hbar}\rho_3[\widehat{H}, \widehat{Q}] = \rho_3\frac{d\widehat{Q}}{d\tau}.$$

Of course, using Eq. (2.153) instead of Eq. (2.97) we obtain the same result [see Eq. (2.152)].

Of course, the known spin operators and their components allow constructing the new operators which will be integrals of motion in the presence of the external field of a more complete configuration.

One of the first successful results of such attempts is the Dirac spin operator[137]

$$\widehat{J}_{\rm D} = \rho_3[(\boldsymbol{\sigma}\widehat{\boldsymbol{l}}) + \hbar],$$

where $\widehat{\boldsymbol{l}} = [\boldsymbol{r}\widehat{\boldsymbol{P}}]$ is the operator of the orbital angular momentum.

It is found that the Dirac operator $\widehat{J}_{\rm D}$ has a strict Poincaré–invariant origin. To show this, let us construct, first, a tensor dual to the tensor of the orbital angular momentum $L^{\mu\nu}$ given by the classical Eq. (2.1). According to Eq. (2.6), we have

$$K^{\alpha\beta} = \frac{1}{2}\,\varepsilon^{\alpha\beta\rho\sigma}L_{\rho\sigma} = \frac{1}{m_0c}\,\varepsilon^{\alpha\beta\rho\sigma}R_\rho P_\sigma,$$

where R^ρ are the coordinates of the mass center, $L^{\alpha\beta} = (\boldsymbol{K}, \boldsymbol{L})$ is the tensor of the orbital angular momentum of the mass center. Tensor $K^{\alpha\beta} = (-\boldsymbol{L}, \boldsymbol{K})$ is a spacelike one:

$$K^{\alpha\beta}v_\beta = 0.$$

Spin-orbital connection is ensured by the zero component of the four-dimensional vector

$$J^\mu = K^{\mu\nu}\pi_\nu = ((\boldsymbol{L}\boldsymbol{\pi}),\ \boldsymbol{L}\pi^0 - [\boldsymbol{K}\boldsymbol{\pi}]).$$

According to the definitions of $\widehat{L}^{\mu\nu}$ and $\widehat{R}^\mu$ [see Eqs (2.31) and (2.32)], we have in the operational form

$$\widehat{\boldsymbol{L}} = \widehat{\boldsymbol{l}} - \frac{\hbar}{2m_0c}\,\rho_2(\boldsymbol{\sigma}\widehat{\boldsymbol{P}}).$$

Taking into account the expression for $\widehat{\boldsymbol{\pi}}$ given by Eq. (2.42) [see also Eq. (2.29)] we obtain

$$\widehat{J}^0 = (\widehat{\boldsymbol{L}}\widehat{\boldsymbol{\pi}}) = \rho_3(\widehat{\boldsymbol{\sigma}}\widehat{\boldsymbol{l}}) - \frac{\hbar}{m_0c}\,(\widehat{\boldsymbol{\alpha}}\widehat{\boldsymbol{P}}) \div \widehat{J}_{\rm D} - \frac{\hbar}{m_0c}\,(\widehat{\mathcal{H}} - e\varphi)$$

and

$$\widehat{J}_{\rm D} \div J^0 + \frac{\hbar}{m_0c}\left(\widehat{\mathcal{H}} - e\varphi\right).$$

The Dirac operator $\widehat{J}_{\rm D}$ in the central force field commutates with Hamiltonian $\widehat{\mathcal{H}}$ of Eq. (2.154) and is used for quantization of the total angular momentum.

Another example of construction of the spin integrals of motion for mutually orthogonal fields of the plane electromagnetic wave type and Redmond's configuration (the plane wave and the uniform magnetic field are directed along the velocity of wave propagation) is the operator[138]

$$\widehat{\pi}_R = (\boldsymbol{n}\widehat{\boldsymbol{\pi}}) + \delta\widehat{\pi}^0,$$

where $\boldsymbol{n}$ is the constant unit vector, δ is any constant. It is possible to show that

$$\frac{d\widehat{\pi}_R}{dt} = \frac{e}{m_0 c}\,\left[\rho_1(\boldsymbol{nE}) - (\boldsymbol{\sigma}[\boldsymbol{nH}]) + \delta(\boldsymbol{\sigma E})\right]. \tag{2.155}$$

Hence it follows that in the fields of the type

$$\boldsymbol{H} = [\boldsymbol{nE}] + \boldsymbol{n}(\boldsymbol{nH}), \quad \boldsymbol{E} = -[\boldsymbol{nH}] + \boldsymbol{n}(\boldsymbol{nE}),$$

Eq. (2.155) goes to zero at $\delta = -1$ and $(\boldsymbol{nE}) = 0$.

An exhaustive analysis of the linear in momentum force integrals of motion has been made in Ref.[139]

Let us turn to consideration of the spin invariants which are integrals of motion. The possibility of building such integrals of motion was pointed out in Ref.[138] where the operator

$$\widehat{I} = \frac{1}{2}\,H_{\mu\nu}\widehat{\Pi}^{\mu\nu}$$

was constructed.

Later on the operator $\widehat{I}$ was used to calculate specific physical effects[140,141]. In Ref.[140] it also was shown that the operator $\widehat{I}$ has an equivalent representation in the form

$$\widehat{I'} = E^{\mu\nu}\widehat{\pi}_\mu\widehat{P}_\nu/m_0 c = \widehat{I},$$

where

$$E^{\mu\nu} = \frac{1}{2}\,\varepsilon^{\mu\nu\rho\sigma}H_{\rho\sigma}.$$

We shall consider here the properties of the spin invariants $\widehat{I}$ and $\widehat{I'}$ within the framework of approximation which is valid for the BMT-equations. This question has an important significance in connection with the problem of adequateness of relativistically invariant description of the spin states in the classical and quantum theories[142].

First, let us show that the spin invariants I and I' in the classical theory, as well as in the quantum theory, are equivalent. This is easy to do by means of correlation relations $\widehat{\Pi}^{\mu\nu}$ and $\widehat{\pi}^\alpha$ (see Eqs (2.8) and (2.9)), according to which

$$\pi^{[\mu}v^{\nu]} = -c\Phi^{\mu\nu},$$

where $\Phi^{\mu\nu}$ is the tensor dual to $\Pi^{\mu\nu}$. Taking into account this relation we obtain

$$I = \frac{1}{2}\,H_{\mu\nu}\Pi^{\mu\nu} = -\frac{1}{2}\,E_{\mu\nu}\Phi^{\mu\nu} = \frac{1}{c}\,E_{\mu\nu}\pi^\mu v^\nu = I'.$$

Now let us consider how the spin invariants I and I' vary in the course of time. If electromagnetic fields are constant in time, it is possible to determine that $I = I' = const$ applying the spin equations (2.61) and (2.62) as well as the equation of charge motion (2.60). In the proof the condition should be taken into account that $\Pi^{\mu\nu}$ and

π^α are the space-like tensor and vector, respectively, (see Eqs (2.4) and (2.10)) and the identities

$$H_{\mu\alpha}\Pi^{\alpha\beta}H_\beta{}^\mu = v_\mu H^{\mu\alpha}\Pi_{\alpha\beta}H^{\beta\nu}v_\nu = 0,$$

$$\pi_\alpha H^{\alpha\beta}E_{\beta\gamma}v^\gamma = v_\alpha H^{\alpha\beta}E_{\beta\gamma}\pi^\gamma = 0.$$

In the operational form the classical relations given by Eqs (2.8) and (2.9) take the form

$$\widehat{\Pi}^{\mu\nu} = \frac{1}{m_0 c}\,\varepsilon^{\mu\nu\alpha\beta}\widehat{\pi}_\alpha\widehat{P}_\beta + \frac{e\hbar}{m_0^2c^3}\,\gamma^5 E^{\mu\nu},$$

$$\widehat{\pi}^\alpha = \frac{1}{m_0 c}\,\varepsilon^{\alpha\beta\mu\nu}\widehat{\Pi}_{\mu\nu}\widehat{P}_\beta + i\,\frac{e\hbar}{m_0^2c^3}\,E^{\alpha\beta}\gamma_\beta.$$

However, in the BMT approximation the terms proportional to $\hbar$ can be omitted and then the classical Eqs (2.8) and (2.9) remain valid.[m]

The obvious expression for the spin invariant I in the quantum theory has the form

$$\widehat{I} = (\boldsymbol{H}\widehat{\boldsymbol{\Pi}}) + (\boldsymbol{E}\widehat{\boldsymbol{\Phi}}) = (\boldsymbol{\sigma}H) + \frac{1}{m_0 c}\left[\rho_2(\boldsymbol{H}[\boldsymbol{\sigma}\widehat{\boldsymbol{P}}]) - \rho_3(\boldsymbol{E}[\boldsymbol{\sigma}\widehat{\boldsymbol{P}}])\right] \div \widehat{I}'.$$

Thus, both the invariants describe a spin projection in the rest system onto the direction of the magnetic field.

Let us turn to consideration of spin invariants evolution in time. In terms of the quantum theory with definite parity of operators we have[86]

$$\frac{d\widehat{I}}{d\tau} = \frac{i}{\hbar}\,[\widehat{G}, I] = 0 \tag{2.156}$$

where $\widehat{G}$ is the squared Hamiltonian of the proper time Dirac equation (2.102). In the proof of Eq. (2.156) the commutators (to an accuracy necessary for the BMT-approximation) were used

$$[\widehat{P}_\alpha\widehat{P}^\alpha, P^\mu] = -2i\hbar\frac{e}{c}H^{\mu\nu}\widehat{P}_\nu,$$

$$[\sigma_{\alpha\beta}H^{\alpha\beta}, \gamma^\mu\gamma^5] = 4iH^{\alpha\beta}\gamma_\beta\gamma^5, \quad [H_{\alpha\beta}\gamma^\alpha\widehat{P}^\beta, \gamma^\mu\gamma^5] = 2H^{\alpha\beta}P_\beta\gamma^5,$$

---

[m] We note that in the symmetrized form (see footnote k)

$$\widehat{\Pi}^{\mu\nu} = \frac{1}{m_0 c}\varepsilon^{\mu\nu\alpha\beta}(\widehat{S}_\alpha, \widehat{P}_\beta), \quad \widehat{\pi}^\alpha = \frac{1}{2m_0 c}\varepsilon^{\alpha\beta\mu\nu}(\widehat{\Pi}_{\mu\nu}, \widehat{P}_\beta)$$

the classical Eqs (2.8) and (2.9) and also the Frenkel and Tamm's condition

$$(\widehat{\Pi}_{\mu\nu}, \widehat{P}^\nu) = (\widehat{\pi}_\alpha, \widehat{P}^\alpha) = 0$$

are valid.

and also the relations (to the same accuracy)

$$E^{\mu\alpha} P_\alpha H_{\mu\beta} \gamma^\beta = \frac{1}{4} E^{\alpha\beta} H_{\alpha\beta} \gamma_\mu \widehat{P}^\mu \div i m_0 c(\boldsymbol{EH}),$$

$$E^{\mu\alpha} P_\alpha H_{\mu\beta} P^\beta = \frac{1}{4} E^{\alpha\beta} H_{\alpha\beta} \widehat{P}_\mu \widehat{P}^\mu \div m_0^2 c^2 (\boldsymbol{EH}).$$

Thus, it is clear that in the BMT-approximation both the spin invariants $I = I'$ are integrals of motion in classical as well as in the quantum theory.

Bibliography

1. E. Wigner, *Ann. Math.* **50** (1939) 149.
2. V. Bargmann, *Ann. Math.* **48** (1947) 568.
3. V. Bargmann, E.P. Wigner, *Proc. NAS. USA* **34** (1948) 211.
4. Yu.M. Schirokov, *Zh. Exp. Teor. Fiz.* **21** (1951) 748; **33** (1957) 861, 1196, 1208 [in Russian].
5. E.M. Corson, *Introduction to Tensors, Spinors and Relativistic Wave-equations* (Blackie and Son, London, Glasgov, 1954).
6. H.J. Bhabha, *Rev. Mod. Phys.* **21** (1949) 451.
7. A.S. Wightman, in: *Dispersion Relations and Elementary Particles* (Wiley, New York, 1960), p.161-226.
8. S.S. Schweber, *An Introduction to Relativistic Quantum Field Theory* (Row, Peterson, Evanston, Ill.-Elmsford, N.Y., 1961).
9. G.E. Uhlenbeck, S. Goudsmit, *Naturwissenschaften* **13** (1925) 953; *Nature* 117 (1926) 264.
10. L.H. Thomas, *Phil. Mag.* **117** (1926) 514; (1927) 1.
11. J. Frenkel, *Zs. Phys.* **37** (1926) 243.
12. I. Tamm, *Zs. Phys.* **55** (1929) 199.
13. J.K. Lubański, *Physica* **9** (1942) 310.
14. H.A. Bethe, *Intermediate quantum mechanics* (Benjamin, N.Y.-Amsterdam, 1964).
15. F. Halbwachs, *Progr. Theor. Phys.* **24** (1960) 291.
16. J.A. deVos, *Ned. T. Natuurk* **27** (1961) 287.
17. H.C. Corben, *Proc. NAS USA* **48** (1962) 387.
18. Yu.M. Schirokov, *Dokl. Akad. Nauk SSSR* **49** (1954) 737 [in Russian].
19. L. Michel, S. Wightman, *Phys. Rev.* **98** (1955) 1190.
20. C. Bouchiat, L. Michel, *Nucl. Phys.* **5** (1958) 416.
21. J. Hilgewoord, S.A. Wouthuysen, *Nucl. Phys.* **40** (1963) 1.
22. F. Calogero, *Nuovo Cim.* **20** (1961) 280.
23. D.M. Fradkin, R.H. Good, Jr., *Rev. Mod. Phys.* **33** (1961) 343.
24. M. Kolsrud, *Phys. Norv.* **2** (1967) 141.
25. V. Fock, *Zs. Phys.* **55** (1929) 127; **68** (1931) 522.
26. M.H.L. Pryce, *Proc. Roy. Soc.* **195** (1948) 62.

27. A.A. Sokolov, *J. Phys. USSR* **9** (1945) 363.
28. J.M. Jauch and F. Rohrlich, *The Theory of Photons and Electrons* (Cambridge, 1959).
29. A.A. Sokolov, Yu.M. Loskutov, I.M. Ternov, *Quantum Mechanics* (Holt, Reinchart, and Winston, New York, 1966) [English translation].
30. W. Pauli, in: *Le magnétisme* (Bruxelles, 1930) p.175.
31. N.F. Mott, H.S. Massey, *The theory of atomic collisions* (Oxford, Clarendon Press, 1965).
32. R.P. Feynman, *Quantum electrodynamics* (Benjamin, N.Y., 1961).
33. M. Bunge, *Nuovo Cim.* **1** (1955) 577.
34. H.C. Corben, *Phys. Rev.* **121** (1961) 1833.
35. V.A. Bordovitsyn, I.M. Ternov, *Theor. Math. Phys.* **51** (1982) 327.
36. L.L. Foldy, S.A. Wouthuysen, *Phys. Rev.* **78** (1950) 29.
37. B. Stech, *Zs. Phys.* **144** (1956) 214.
38. V.N. Baier, V.M. Katkov, *Yadern. Fiz.* **3** 1 (1966) 81 [in Russian].
39. M. Rosen, *Nuovo Cim.* **23** (1964) 1664.
40. I.M. Ternov, V.G. Bagrov, V.A. Bordovitsyn, O.F. Dorofeev, *Zh. Eksp. Teor. Fiz.* **55** (1968) 2273 [*Sov. Phys.- JETP* **28** (1969) 1206]; *Dokl. Akad. Nauk SSSR* **183** (1968) 810 [in Russian].
41. J. Schwinger, *Phys. Rev.* **73** (1948) 416.
42. L.D. Landau and E.M. Lifshitz, *Classical Theory of Fields* (Addison-Wesly, Reading, Mass., 1971).
43. V. Bargmann, L. Michel, and V.L. Telegdi, *Phys. Rev. Lett.* **2** (1959) 435.
44. H. Bacry, *Theses Dr. Sc. Phys.* (Masson, Paris, 1963).
45. M. Kolsrud, *Phys. Norv.* **2** (1966) 51.
46. V.A. Bordovitsyn, S.V. Sorokin, *Izv. Vuz. Fiz.* **26** 8 (1983) 125 [in Russian].
47. H.C. Corben, *Classical and Quantum Theories of Spinning Particles* (Holden-Day, San-Francisko, 1968).
48. R.D. Sard, *Relativistic Mechanics* (Beniamin, N.Y., 1970).
49. R.A. Mann, *The Classical Dynamics of Particles Galilean and Lorentz Relativity* (Acad. Press, N.Y. e.a., 1974).
50. J.D. Jackson, *Classical Electrodynamics* (Wiley, N.Y., 1975).
51. H. Zatzkis, *J. Frankl. Inst.* **269** (1960) 268.
52. H. Bacry, *Nuovo Cim.* **26** (1962) 1664; *C.r. Acad. Sci.* **267** (1968) B 89.
53. A. Chakrabarti, *J. Math. Phys.* **5** (1964) 1747; *Nuovo Cim.* **A43** (1966) 576.
54. L.S. Schulman, *Nucl. Phys.* **B18** (1970) 595.
55. G.P. Fisher, *Am. J. Phys.* **40** (1972) 1772.
56. S.R. Barone, *Phys. Rev. D.* **8** (1973) 3492.
57. J. Schwinger, *Am. J. Phys.* **42** (1974) 507, 510.
58. P.F. Browne, *J. Phys.* **A10** (1977) 727.
59. G. Ziino, *Lett. Nuovo Cim.* 18 (1981) 629.
60. J. Bolog, P. Hrasko, *Found. Phys.* **11** (1981) 873.
61. H.C. Corben, *Am. J. Phys.* **61** (1993) 551.

62. W. Pauli, *Helv. Phys. Acta* **5** III (1932) 179.
63. R. Schiller, *Phys. Rev.* **128** (1962) 1402.
64. V.I. Maslov, *Usp. Matem. Nauk* **18** (1963) 220 [in Russian].
65. S.I. Rubinov, J.B. Keller, *Phys. Rev.* **131** (1963) 2789.
66. K. Rafanelli, R. Schiller, *Phys. Rev.* **B135** (1964) 279.
67. M. Kolsrud, *Nuovo Cim.* **39** (1965) 504.
68. H. Yamasaki, *Progr. Theor. Phys.* **36** (1966) 72.
69. J. Sniatycki, *Zesz. nauk. Univ. Jagell 176* (1967) 361.
70. M. Pardy, *Acta. Phys. Slov.* **23** (1973) 5.
71. V.Ch. Zhukovskii, N.S. Nikitina, *Vestn. Mosk. Univ., III* **14** 2 (1974) 189 [in Russian].
72. J. Stachel, J. Plebański, *J. Math. Phys.* **18** (1977) 2368.
73. V.D. Mur, V.S. Popov, *Yadern. Fiz.* **28** (1978) 837 [in Russian].
74. D. Bazeia, *Lett. Nuovo Cim.* **29** (1980) 228.
75. I.M. Ternov, V.A. Bordovitsyn, *Vestn. Mosk. Univ., III* **23** 6 (1982) 72 [in Russian].
76. V. Fock, *Phys. Zs. Sow.* **12** (1937) 404.
77. Y. Nambu, *Progr. Theor. Phys.* **5** (1950) 82.
78. R.P. Feynman, *Phys. Rev.* **84** (1951) 108.
79. H.E. Moses, *Ann. Phys.* **52** (1969) 444.
80. I. Saavedra, *Nucl. Phys.* **77** (1966) 673.
81. J.E. Johnson, K.K. Chang, *Phys. Rev.* **D10** (1974) 2421.
82. R.E.. Collins, J.R. Fanchi, *Nuovo Cim.* **A48** (1978) 314.
83. J.D. Bjorken, S.D. Drell, *Relativistic Quantum Mechanics* (Mc Graw-Hill, N.Y., 1964).
84. M. Alonzo and H. Valk, *Quantum Mechanics* (Addison-Wesley, Reading, MA, 1973), p.268.
85. Mc Graw-Hill, *Encyclopedia of Physics* (Mc Graw-Hill, N.Y., 1983), p.293.
86. V.A. Bordovitsyn, I.M. Ternov, *Teor. Matem. Fiz.* **54** (1983) 338 [in Russian].
87. R. Feynman, M. Gell-Mann, *Phys. Rev.* **109** (1958) 193.
88. A.S. Davydov, *Quantum Mechanics, 2nd ed.* (Pergamon Press, N.Y., 1976).
89. V.A. Bordovitsyn, in: *Quantum Systems: New Trends and Methods* eds. A.O. Barut et al., p.125 (World Scientific, Singapore, 1995).
90. V.A. Bordovitsyn, *Izv. Vuz. Fiz.* **38** 4 (1995) 72 [*Sov. Phys. J.* **38** (1995) 390].
91. V.A. Bordovitsyn, I.M. Ternov, *Izv. Vuz. Fiz.* **34** 2 (1991) 5 [*Sov. Phys. J.* **34** (1991) 91].
92. V.A. Bordovitsyn, N.N. Bysov, *Izv. Vuz. Fiz.* **20** 10 (1977) 36 [in Russian].
93. I.M. Ternov, V.A. Bordovitsyn, *Usp. Fiz. Nauk* **132** (1980) 345 [in Russian]; *Ann der Phys.* **39** (1982) 295.
94. W. Pauli, *Zs. Phys.* **43** (1927) 601.
95. P.A.M. Dirac, *Proc. Roy. Soc.* **A117** (1928) 610.
96. W. Pauli, *Rev. Mod. Phys.* **13** (1941) 203.
97. A.A. Schupp, R.W. Pidd, H.R. Crane, *Phys. Rev.* **121** (1961) 1.

98. G. Charpak, F.J.M. Farley, R.L. Garvin et al. *Phys. Rev. Lett.* **6** (1961) 128.
99. H.J. Meister, *Zs. Phys.* **166** (1962) 468.
100. V.N. Baier, *Usp. Fiz. Nauk* **105** (1971) 441 [in Russian].
101. Ya.S. Derbenev, A.M. Kondratenko, *Zh. Exp. Teor. Fiz.* **64** (1973) 1918 [in Russian].
102. B.W. Montague, *Physics Reports* **113** (1984) 1.
103. J.D. Jackson, *Rev. Mod. Phys.* **48** (1976) 417.
104. J.H. Field, E. Picasso, F. Combley, *Tests of Fundamental Physical Theories from Measurements of Free Charged Leptons* (CERN, Geneva, 28 February, 1978).
105. W.W. Mackay, J.F. Hassard, R.T. Giles et al. *Phys. Rev.* **D29** (1984) 2483.
106. A.N. Skrinsky, Yu.M. Shatunov, *Usp. Fiz. Nauk* **158** (1989) 315 [in Russian].
107. H.C. Corben, *Intern. J. Teor. Phys.* **34** (1995) 19.
108. R.H. Good, Jr. *Phys. Rev.* **125** (1962) 2112.
109. A.I. Solomon, *Nuovo Cim.* **26** (1962) 1320.
110. P. Nyborg, *Nuovo Cim.* **31** (1964) 1290; **32** (1964) 1131.
111. J. Weyssenhoff, A. Raabe, *Acta. Phys. Polon.* **9** (1947) 7.
112. K. Nagy, *Bull. Polsk. Akad. Nauk, III* **4** (1956) 333 [in Russian].
113. F. Halbwachs. *C. r. Acad. Sci.* **243** (1956) 1022, 1098; **245** (1957) 1298; *Acta. Phys. Polon.* **19** (1960) 93; *Nuovo Cim.* **36** (1965) 382.
114. D. Bohm, J.-P. Vigier, *Phys. Rev.* **109** (1958) 1822.
115. B. Linet, *Lett. Nuovo Cim.* **8** (1973) 63.
116. P. Grassberger, *J. Phys.* **A11** (1978) 1221.
117. W.G. Dixon, *Nuovo Cim.* **38** (1965) 1616.
118. E. Plathe, *Nuovo Cim. Suppl.* **4** (1966) 246, 291; **5** (1967) 944.
119. B.S. Scagerstam, A. Stern., *Phys. Scr.* **24** (1981) 493.
120. P.L. Nash, *J. Math. Phys.* **25** (1984) 2104.
121. A.O. Barut, *Electrodynamics and Classiacal Theory of Fields and Particles* (Mac Millan, N.Y., 1964).
122. S.R. de Groot, and L.G. Suttorp, *Foundations of Electrodynamics* (North Holland, Amsterdam, 1972).
123. V.A. Bordovitsyn, *Izv. Vuz. Fiz.* **25** 3 (1982) 109 [in Russian].
124. I.M. Ternov, *Fiz. Elementarnykh Chastits: Atomn. Yadra (EChAYa)* **17** (1986) 884 [in Russian].
125. J. Kessler, *Polarized Electrons* (Springer, Berlin-Heidelberg, 1985).
126. V.A. Bordovitsyn, *Izv. Vuz. Fiz.* **36** 11 (1993) 39 [*Sov. Phys. J.* **36** (1993) 1046].
127. I.M. Ternov, V.A. Bordovitsyn, G.K. Razina, *Izv. Vuz. Fiz.* **24** 1 (1981) 44 [*Sov. Phys. J.* **24** (1981) 42].
128. I.M. Ternov, V.A. Bordovitsyn, *Izv. Vuz. Fiz.* **22** 4 (1987) 120 [*Sov. Phys. J.* **22** (1987) 350].
129. L. Allen, J.H. Eberly, *Optical Resonance and Two-Lewel Atoms* (John Wiley and Sons, N.Y., 1975).
130. W.H. Louisell, *Radiation and Noise in Quantum Electronics* (Mc Graw-Hill, N.Y., 1964).

131. V.Ya. Epp, G.K. Razina, V.A. Bordovitsyn, *Izv. Vuz. Fiz.* **22** 3 (1979) 120 [*Sov. Phys. J.* **22** (1979) 329].
132. A.A. Sokolov, A.V. Borisov et al. *Izv. Vuz. Fiz.* **17** 12 (1974) 5 [in Russian].
133. I.M. Ternov, V.A. Bordovitsyn, S.V. Sorokin, *Izv. Vuz. Fiz.* **26** 12 (1983) 112 [in Russian].
134. I.M. Ternov, V.R. Khalilov, O.S. Pavlova, *Izv. Vuz. Fiz.* **21** 12 (1978) 89; **22** 2 (1979) 39 [in Russian].
135. I.M. Ternov, V.G. Bagrov, A.M. Khapaev, *Zh. Eksp. Teor. Fiz.* **48** (1965) 321 [in Russian].
136. V.A. Bordovitsyn, V.S. Gushchina, A.N. Myagkii, *Nucl. Instr. Meth. Phys. Res.* **A 405** (1998) 256.
137. P.A.M. Dirac, *The Principles of Quantum Mechanics* (Oxford, Clarendon, 1976).
138. I.M. Ternov, V.G. Bagrov, V.A. Bordovitsyn, *Izv. Vuz. Fiz.* **10** 4 (1967) 41 [in Russian].
139. V.G. Bagrov, D.M. Gitman, *Exact Solution of Relativistic Wave Equations* (Kluver, London, 1990).
140. V.Ch. Zhukovskii, *Zh. Eksp. Teor. Fiz.* (1973) 1169 [in Russian].
141. Yu.V. Klimenko, *Yadern. Fiz.* **27** (1978) 1677 [in Russian].
142. V.A. Bordovitsyn, I.M. Ternov, *Teor. Matem. Fiz.* **59** (1984) 220 [in Russian].

Chapter 3

Quantum Theory of Synchrotron Radiation

V.G. Bagrov
Tomsk State University 634050 Tomsk, Russia[a]

The concept of the trajectory is alien to the quantum mechanics. Therefore the classical model for a relativistic electron moving along a circumference is unfit for the construction of the quantum theory of synchrotron radiation. Some rather complex external electromagnetic fields are used in practice to provide the stability of electron orbits in accelerators and storage rings. However, the problem of the stability of electron motion concerns only slightly the synchrotron radiation. Abstracting, we have the uniform stationary magnetic field perpendicular to the orbital plane as a simplest one which ensures from the classical point of view the motion of the electron along the circular trajectory. Just this model is choosed in most cases to describe the sinchrotron radiation. We shall present below fundamental methods and physical results of the theoretical quantum-mechanical research of the radiation by electrons moving in a uniform stationary magnetic field.

3.1. Scalar Particles in a Magnetic Field. Klein-Gordon Equation

It is convenient to begin with spinless particles in a magnetic field so as to compare later a fermionic radiation with a bosonic one and thus, to draw some conclusions concerning the role of the spin in synchrotron radiation.

The behavior of scalar particles in an external field is described by the Klein-

[a]E-mail: bagrov@phys.tsu.tomsk.su

Gordon equation

$$(i\hbar\frac{\partial}{\partial t} - e\varphi)^2\Psi = \{c^2\hat{\mathcal{P}}^2 + (mc^2)^2\}\Psi, \tag{3.1}$$

where

$$\hat{\boldsymbol{\mathcal{P}}} = \hat{\boldsymbol{p}} - \frac{e}{c}\boldsymbol{A}, \quad \hat{\boldsymbol{p}} = -i\hbar\nabla, \tag{3.2}$$

$\boldsymbol{A}$ and φ are the external field potentials, e, m, $\hbar$, c are the charge of the particle, its mass, the Planck constant and the light speed, respectively.

We can choose the potentials of an uniform stationary magnetic field $\boldsymbol{H}$ in the following way

$$\boldsymbol{A} = -\frac{1}{2}\,[\boldsymbol{r}, \boldsymbol{H}], \quad \varphi = 0.$$

Pointing z-axes along the field we get

$$A_x = -\frac{y}{2}H, \quad A_y = \frac{x}{2}H, \quad A_z = 0, \quad \varphi = 0. \tag{3.3}$$

In terms of cylindrical coordinates r, φ, z, which are the most suitable ones in this case, Eq. (3.1) gets the form

$$\begin{aligned}&\left\{\frac{1}{r}\frac{\partial}{\partial r}(r\frac{\partial}{\partial r}) + \frac{1}{r^2}\frac{\partial^2}{\partial\varphi^2} + \frac{\partial^2}{\partial z^2} - \frac{1}{c^2}\frac{\partial^2}{\partial t^2} + \right.\\ &\left. + i\frac{eH}{\hbar c}\frac{\partial}{\partial\varphi} - \frac{1}{\hbar^2}m^2c^2 - \frac{e^2H^2}{4c^2\hbar^2}r^2\right\}\Psi = 0.\end{aligned} \tag{3.4}$$

We seek now a stationary solution

$$\Psi = e^{-i\frac{E}{\hbar}t}\psi(\boldsymbol{r}),$$

for a particle with the negative charge $e = -|e|$. Throughout what follows we shall consider particles with negative charges, the symbol e being assigned further to the absolute value of the charge. Assume the wave function to be the eigenfunction for the operators of z-components of the linear and angular momentum vectors

$$\hat{\mathcal{P}}_z\Psi = \hbar k_3\Psi,$$

$$\hat{L}_z\Psi = -i\hbar\frac{\partial}{\partial\varphi}\Psi = \hbar l\Psi.$$

Under such requirements we have the solutions of Eq. (3.4) in the form (see Refs[1,2])

$$\psi(\boldsymbol{r}) = \frac{e^{ik_3z}}{\sqrt{L}}e^{il\varphi}\sqrt{\frac{\gamma}{\pi}}I_{n,s}(\gamma r^2), \tag{3.5}$$

the energy being given by the expression

$$E_n = chK_n, \quad K_n = \sqrt{k_0^2 + k_3^2 + 4\gamma(n+1/2)}\,,$$
$$E_n = mc^2\sqrt{1 + (\mathcal{P}_z/mc)^2 + \frac{H}{H_*}(2n+1)}, \tag{3.6}$$

where

$$k_0 = \frac{mc}{\hbar}, \quad \gamma = \frac{eH}{2c\hbar}, \quad H^* = \frac{m^2c^3}{e\hbar} \tag{3.7}$$
$$I_{n,s}(x) = \sqrt{\frac{s!}{n!}} e^{-\frac{x}{2}} x^{\frac{n-s}{2}} L_s^{n-s}(x)\,, \quad n = l + s = 0, 1, 2, \ldots\,, \tag{3.8}$$

L_s^l are the Laguerre polynomials

$$L_s^l(x) = \frac{1}{s!} e^x x^{-l} \frac{d^s}{dx^s}(e^{-x} x^{s+l}). \tag{3.9}$$

The functions $I_{n,s}$ will be further called Laguerre functions. Later we shall consider their simplest properties. The quantum numbers n, l, s are called principal, azimuthal and radial ones respectively, and have the ranges of values

$$\begin{cases} 0 \le n < \infty \\ -\infty < l \le n \\ 0 \le s < \infty \end{cases}.$$

It is easy to prove that the wave functions of Eq. (3.5) are normalized to unity

$$\int |\Psi(\boldsymbol{r})|^2\, d^3x = 1$$

The quantum numbers n, l, s have a simple physical interpretation.

According to classical electrodynamics, the z-projection of the angular momentum of a negative charge is[3]

$$L_z^{\text{cl}} = \left[\boldsymbol{r}, \left(\frac{m\boldsymbol{v}}{\sqrt{1-\frac{v^2}{c^2}}} - \frac{e}{c}\boldsymbol{A}\right)\right]_z.$$

For a charge moving in a circle averaging over all directions results in

$$\langle \hat{L}_z^{\text{cl}} \rangle = \hbar\gamma(R^2 - a^2), \tag{3.10}$$

where R is the trajectory radius and $\boldsymbol{a}$ is the radius vector of the center of rotation. Here we have taken into account that the circle radius is connected with the magnetic field by the following relation[3]

$$R = \frac{E}{eH}\sqrt{1 - \left(\frac{mc^2}{E}\right)^2}. \tag{3.11}$$

For the quantum motion we have

$$\langle \hat{L}_z \rangle = \hbar(n - s).$$

Comparing Eq. (3.10) with Eq. (3.11) we conclude

$$R \approx \sqrt{\frac{n}{\gamma}}, \quad a \approx \sqrt{\frac{s}{\gamma}}, \tag{3.12}$$

that means the numbers n and s characterize the orbital radius and the deviation of the orbital center from the origin of the coordinates, respectively.

Let us find the quantum values $\langle r^2 \rangle$ and $\langle \hat{\mathcal{P}}_\varphi^2 \rangle = \langle \hat{\mathcal{P}}_x^2 \rangle + \langle \hat{\mathcal{P}}_y^2 \rangle$. Simple calculation results in

$$\langle r^2 \rangle = \frac{1}{\gamma}(n + s + 1) > 0 \tag{3.13}$$

$$\langle \hat{\mathcal{P}}_\varphi^2 \rangle = 4\gamma h^2(n + 1/2). \tag{3.14}$$

Eq. (3.13) confirms the relations of Eqs (3.12) and (3.14) and leads to the classical formula

$$E = c\sqrt{(mc)^2 + \langle \hat{\mathcal{P}}_3^2 \rangle + \langle \hat{\mathcal{P}}_\varphi^2 \rangle} = c\sqrt{(mc)^2 + \langle \hat{\mathcal{P}}^2 \rangle}. \tag{3.15}$$

Hence the number n characterizes the rotational energy of the particle. Note that the energy cannot vanish

$$E \geq \hbar c\sqrt{2\gamma}. \tag{3.16}$$

It means there are "zero oscillations" of the particle similar to those of a harmonic oscillator.

3.2. Electron in a Magnetic Field. Dirac Equation.

The behavior of an electron in an external field should be described by the Dirac equation

$$(i\hbar\frac{\partial}{\partial t} - \hat{H})\Psi = 0, \quad \hat{H} = c(\boldsymbol{\alpha}\hat{\boldsymbol{\mathcal{P}}}) + p_3 mc^2 + e\varphi, \tag{3.17}$$

where $\boldsymbol{\alpha}$, p_3 are the Dirac matrices,[4] $\hat{\boldsymbol{\mathcal{P}}}$ has been defined by Eq. (3.2), Ψ is a four-component spinor

$$\Psi = \begin{pmatrix} \Psi_1 \\ \Psi_2 \\ \Psi_3 \\ \Psi_4 \end{pmatrix}. \tag{3.18}$$

If H does not depend explicitly on time we can seek a stationary solution of Eq. (3.17) in the form

$$\Psi = e^{-i\frac{E}{\hbar}t}\psi(\boldsymbol{r}) \tag{3.19}$$

and then represent Eq. (3.17) as a system of four equations for four functions

$$\begin{aligned}
(E - mc^2)\psi_1 - c(\hat{\mathcal{P}}_x - i\hat{\mathcal{P}}_y)\psi_4 - c\hat{\mathcal{P}}_z\psi_3 &= 0\,,\\
(E - mc^2)\psi_2 - c(\hat{\mathcal{P}}_x + i\hat{\mathcal{P}}_y)\psi_3 + c\hat{\mathcal{P}}_z\psi_4 &= 0\,,\\
(E + mc^2)\psi_3 - c(\hat{\mathcal{P}}_x - i\hat{\mathcal{P}}_y)\psi_2 - c\hat{\mathcal{P}}_z\psi_1 &= 0\,,\\
(E + mc^2)\psi_4 - c(\hat{\mathcal{P}}_x + i\hat{\mathcal{P}}_y)\psi_1 + c\hat{\mathcal{P}}_z\psi_2 &= 0
\end{aligned} \tag{3.20}$$

Many authors[5-13] worked at the solution of Eqs (3.20) in the case of the uniform stationary magnetic field.

To solve Eq. (3.20) we need to take into account that in the case of a uniform stationary magnetic field, besides H there are two constants of motion $\hat{M}_z$ and $\hat{\mathcal{P}}_z$, since the corresponding operators $\hat{M}_z = -i\hbar\partial/\partial\varphi + \hbar\sigma_3/2$ and $\hat{\mathcal{P}}_z = i\hbar\partial/\partial z$ do not depend on the time explicitly and commute with H. Then the wave function is to be the eigenfunction for these operators.

$$\hat{M}_z\Psi = \hbar(l - 1/2)\Psi,$$

$$\hat{\mathcal{P}}_z\Psi = \hbar k_3\Psi. \tag{3.21}$$

On such conditions one has to look for a solution in cylindrical coordinates in the form

$$\begin{aligned}
\psi_{1,3}(\boldsymbol{r}) &= \frac{e^{ik_3z}}{\sqrt{L}}\frac{e^{i(l-1)\varphi}}{\sqrt{2\pi}}f_{1,3}(\boldsymbol{r})\,,\\
\psi_{2,4}(\boldsymbol{r}) &= \frac{e^{ik_3z}}{\sqrt{L}}\frac{e^{il\varphi}}{\sqrt{2\pi}}f_{2,4}(\boldsymbol{r}),
\end{aligned} \tag{3.22}$$

where

$$f(\boldsymbol{r}) = \sqrt{2\gamma} = \begin{pmatrix} c_1I_{n-1,s}(\gamma r^2)\\ ic_2I_{n,s}(\gamma r^2)\\ c_3I_{n-1,s}(\gamma r^2)\\ ic_4I_{n,s}(\gamma r^2)\end{pmatrix}$$

The coefficients c_i are to be found from the following real system of linear equations

$$\begin{aligned}
(K - k_0)c_1 - \sqrt{4\gamma n}c_4 - k_3c_3 &= 0\,,\\
(K - k_0)c_2 - \sqrt{4\gamma n}c_3 + k_3c_4 &= 0\,,\\
(K + k_0)c_3 - \sqrt{4\gamma n}c_2 - k_3c_1 &= 0\,,\\
(K + k_0)c_4 - \sqrt{4\gamma n}c_1 + k_3c_2 &= 0.
\end{aligned} \tag{3.23}$$

(Notation is clear from Eqs (3.6)–(3.9)). So that Eq. (3.23) has a solution we must equate the determinant to zero

$$\Delta = (K^2 - k_0^2 - k_3^2 - 4\gamma n)^2. \tag{3.24}$$

It follows from above that the energy spectrum is discrete, the energy levels being determined by the principal quantum number n

$$K = k_0\sqrt{1 + (\mathcal{P}_z/mc)^2 + 2\frac{H}{H_*}n} = \sqrt{k_0^2 + k_3^2 + 4\gamma n}. \tag{3.25}$$

If the functions of Eq. (3.22) are to be normalized to unity we need

$$\int \psi^+(\boldsymbol{r})\psi(\boldsymbol{r})d^3x = \sum_{i=1}^{4} c_i^+ c_i = 1 \tag{3.26}$$

If the determinant of Eq. (3.24) vanishes, then all determinants of the third rank turn out to be equal to zero. Therefore, even the requirement of Eq. (3.26) is not sufficient to provide the unique solution of Eq. (3.23). This is caused by the possibility of different orientations of the electron spin.

The solution of Eq. (3.23) together with the normalization condition of Eq. (3.26) results in the following coefficients[13]

$$c_1 = A\sqrt{\frac{1}{2}(1 + \frac{k_0}{K})},$$

$$c_2 = B\sqrt{\frac{1}{2}(1 + \frac{k_0}{K})},$$

$$c_3 = \frac{\sqrt{4\gamma n}B + k_3 A}{\sqrt{2K(K + k_0)}},$$

$$c_4 = \frac{\sqrt{4\gamma n}A - k_3 B}{\sqrt{2K(K + k_0)}},$$

$$AA^+ + BB^+ = 1 \tag{3.27}$$

Analogous wave functions have been obtained in terms of rectangular coordinates in Ref.[14]

In order to clarify the physical meaning of the quantities A and B, let us average σ_3 over the functions with these coefficients

$$\langle\sigma_3\rangle = [1 - \frac{4\gamma n}{K(K + k_0)}](AA^+ - BB^+) + \frac{k_3\sqrt{4\gamma n}}{K(K + k_0)}(AB^+ + A^+B). \tag{3.28}$$

In the nonrelativistic case we get

$$\langle\sigma_3\rangle = AA^+ - BB^+.$$

Thus, in the nonrelativistic case, the functions with the coefficients of Eq. (3.27) describe an electron with the spin pointing in the direction of the field when $A = 1$ and $B = 0$, and directed against the field when $A = 0$ and $B = 1$, whereas arbitrary A and B correspond to the state of partial polarization of the electron. An unpolarized electron is described by wave functions with

$$AA^+ = BB^+ = \frac{1}{2}, \quad AB^+ + A^+B = 0$$

(see Eq. (3.28). Note that $\langle\sigma_1\rangle = \langle\sigma_2\rangle = 0$ for any A and B, owing to the symmetry of the problem.

Note, finally, that the physical interpretation of quantum numbers n, l, s remains the same as was clarified in Sec. 3.1. Calculations lead to the expressions

$$\langle\hat{\mathcal{P}}_\varphi^2\rangle = 4\gamma\hbar^2(n - \frac{1}{2}\langle\sigma_3\rangle), \tag{3.29}$$

$$\langle r^2\rangle = \frac{1}{\gamma}(n + s + \frac{1}{2} - \frac{1}{2}\langle\sigma_3\rangle)$$

(compare with Eqs (3.13) and (3.14).

For an unpolarized electron we have

$$\langle\hat{\mathcal{P}}_\varphi^2\rangle = 4\gamma n\hbar^2, \quad \langle r^2\rangle = \frac{1}{\gamma}(n + s + \frac{1}{2})$$

returning again to the classical formula of Eq. (3.15). Taking into account Eq. (3.29) we can conclude that in the ground state $n = 0$ and $\sigma_3 = -1$, i.e. $A = 0$, $B = 1$, because of $\langle\hat{\mathcal{P}}_\varphi^2\rangle \geq 0$. Besides, the same result follow from Eqs (3.22) and (3.27). Taking into account $I_{-1,s}(x) = 0$ together with the requirement of Eq. (3.26), we have to set $A = 0$, $B = 1$. Therefore, in the ground state the electron is fully polarized, its wave function is the eigenfunction of the operator σ_3, and its spin points against the field.

$$\sigma_3\Psi = \lambda\Psi, \quad \lambda = -1.$$

The wave function has the form (compare with Eq. (3.22)

$$f(r) = \sqrt{\gamma} I_{0,s}(\gamma r^2) \begin{pmatrix} 0 \\ \sqrt{1 + \frac{k_0}{K}} \\ 0 \\ -\operatorname{sign} k_3 \sqrt{1 - \frac{k_0}{K}} \end{pmatrix},$$

where

$$\operatorname{sign} k_3 = \begin{cases} 1 & k_3 > 0 \\ 0 & k_3 = 0 \\ -1 & k_3 < 0 \end{cases}$$

Note, finally, that in the absence of motion along the field ($k_3 = 0$) the wave function does not vanish.

$$f(r) = \sqrt{2\gamma} I_{0,s}(\gamma r^2)$$

Although the total energy is equal to the rest energy, there are some "zero spin oscillations" of the electron with the energy

$$E = \sqrt{c^2 \langle \hat{\mathcal{P}}_\varphi^2 \rangle} = \hbar c \sqrt{2\gamma}$$

at the mean distance from the origin of the coordinates

$$\langle r^2 \rangle = \frac{s+1}{\gamma} > 0$$

(see Eqs (3.13) and (3.16)).

As in the case of the boson, we notice again the resemblance with the behavior of harmonic oscillator. However, as distinct from an electron, the boson energy cannot be equal to the electron energy: $K = \sqrt{k_0^2 + 2\gamma} > k_0$.

3.3. Solution of the Dirac Equation for an Electron with a Definite State of Spin Polarization.

Let us consider now the Dirac equation Eq. (3.20) together with the equation

$$\hat{\Lambda}\Psi = \lambda\Psi \ ,$$

where $\hat{\Lambda}$ is one of spin operators.

3.3.1. Longitudinally Polarised Electron

Let the polarization operator be the operator

$$\hat{\pi}^0 = \frac{(\boldsymbol{\sigma}\hat{\boldsymbol{\mathcal{P}}})}{mc}$$

(compare with Eq. (2.28)). This operator, the most often used before, was introduced first by A.A.Sokolov.[15] It is easy to check that for magnetic field with a fixed direction, operator $\hat{\pi}^0$ is an integral of motion. The equation

$$\hat{\pi}^0 \Psi = \zeta \frac{\lambda_1}{k^0} \Psi, \quad \zeta^2 = 1 \tag{3.30}$$

can be represented explicitly with the help of expressions of Eqs (3.21) and (3.22) for the functions Ψ. Now, besides Eqs (3.23) we have also the following additional relations between the coefficients c_i

$$\begin{aligned} \zeta\lambda_1 c_1 - (K + k_0) c_3 &= 0 \ , \\ \zeta\lambda_1 c_2 - (K + k_0) c_4 &= 0 \ , \\ (K - k_0) c_1 - \zeta\lambda_1 c_3 &= 0 \ , \\ (K - k_0) c_2 - \zeta\lambda_1 c_4 &= 0. \end{aligned} \tag{3.31}$$

Putting the determinant of the system of Eq. (3.31) equal to zero we obtain

$$\lambda_1 = \sqrt{K^2 - k_0^2} \tag{3.32}$$

with $\zeta = \pm 1$ describing the spin projection onto the direction of motion and opposite to it, respectively. On the condition of Eq. (3.26) we solve Eq. (3.31) together with Eq. (3.23) and find

$$\begin{aligned}
c_1 &= \frac{\zeta}{2}\sqrt{(1+\frac{k_0}{K})(1+\zeta\frac{k_3}{\lambda_1})}\,, \\
c_2 &= \frac{1}{2}\sqrt{(1+\frac{k_0}{K})(1-\zeta\frac{k_3}{\lambda_1})}\,, \\
c_3 &= \frac{1}{2}\sqrt{(1-\frac{k_0}{K})(1+\zeta\frac{k_3}{\lambda_1})}\,, \\
c_4 &= \frac{\zeta}{2}\sqrt{(1-\frac{k_0}{K})(1-\zeta\frac{k_3}{\lambda_1})}.
\end{aligned} \tag{3.33}$$

3.3.2. Vector Operator of Transverse Polarization

$$\hat{\pi}_z\Psi = \zeta\frac{\lambda_2}{k_0}\Psi,$$

where $\hat{\boldsymbol{\pi}}$ is defined by Eq. (2.28). It is possible to show (see Chapter 2, Sec.16) that the projection of $\hat{\boldsymbol{\pi}}$ onto the direction of magnetic field is an integral of motion. In the explicit form

$$\begin{aligned}
(k_0 - \zeta\lambda_2)c_1 + k_3c_3 &= 0\,, \\
(k_0 + \zeta\lambda_2)c_2 - k_3c_4 &= 0\,, \\
(k_0 + \zeta\lambda_2)c_3 - k_3c_1 &= 0\,, \\
(k_0 - \zeta\lambda_2)c_4 + k_3c_2 &= 0.
\end{aligned} \tag{3.34}$$

Putting the determinant of this system equal to zero we find

$$\lambda_2 = \sqrt{k_0^2 + k_3^2}.$$

The value $\zeta = \pm 1$ describes two possible directions of the spin relative to z-axis, i.e. to the field direction. Solving Eq. (3.34) together with Eq. (3.23) and taking into account the normalization condition we find

$$c_1 = \frac{\zeta}{2}\sqrt{\frac{1}{2}(1+\zeta\frac{\lambda_2}{K})}\left\{\sqrt{1+\frac{k_3}{\lambda_2}} + \zeta\sqrt{1-\frac{k_3}{\lambda_2}}\right\},$$

$$
\begin{aligned}
c_2 &= \frac{1}{2}\sqrt{\frac{1}{2}(1-\zeta\frac{\lambda_2}{K})}\left\{\sqrt{1+\frac{k_3}{\lambda_2}}-\zeta\sqrt{1-\frac{k_3}{\lambda_2}}\right\}, \\
c_3 &= \frac{1}{2}\sqrt{\frac{1}{2}(1+\zeta\frac{\lambda_2}{K})}\left\{\sqrt{1+\frac{k_3}{\lambda_2}}-\zeta\sqrt{1-\frac{k_3}{\lambda_2}}\right\}, \\
c_4 &= \frac{\zeta}{2}\sqrt{\frac{1}{2}(1-\zeta\frac{\lambda_2}{K})}\left\{\sqrt{1+\frac{k_3}{\lambda_2}}+\zeta\sqrt{1-\frac{k_3}{\lambda_2}}\right\}.
\end{aligned}
$$

3.3.3. Tensor Operator of Transverse Polarization

Let us consider a tensor operator of polarization (see also Eq. (2.28))

$$
\hat{\Pi}^{\mu\nu} = (\hat{\boldsymbol{\Phi}}, \hat{\boldsymbol{\Pi}}), \quad \hat{\boldsymbol{\Pi}} = \boldsymbol{\sigma} + \frac{\rho_2(\boldsymbol{\sigma}\hat{\boldsymbol{\mathcal{P}}})}{mc}, \quad \hat{\boldsymbol{\Phi}} = \frac{\rho_2(\boldsymbol{\sigma}\hat{\boldsymbol{\mathcal{P}}})}{mc}.
$$

The projection Π_z

$$
\hat{\Pi}_z = (\boldsymbol{\Pi}\boldsymbol{H})/H
$$

onto the direction of magnetic field is an integral of motion, and

$$
\hat{\Pi}_z\Psi = \zeta\frac{\lambda_3}{k_0}\Psi.
$$

We rewrite in the explicit form

$$
\begin{aligned}
(K-\zeta\lambda_3)c_1 - k_3c_3 &= 0\,, \\
k_3c_1 - (K+\zeta\lambda_3)c_3 &= 0\,, \\
(K+\zeta\lambda_3)c_2 + k_3c_4 &= 0\,, \\
k_3c_2 + (K-\zeta\lambda_3)c_4 &= 0.
\end{aligned} \tag{3.35}
$$

Equating the determinant to zero we get

$$
\lambda_3 = \sqrt{K^2 - k_3^2}
$$

and from Eqs (3.35), (3.23) and (3.26) we find

$$
\begin{aligned}
c_1 &= \frac{1}{2}\sqrt{\frac{1}{2}(1+\zeta\frac{k_0}{\lambda_3})}\left\{\sqrt{1+\frac{k_3}{K}}+\zeta\sqrt{1-\frac{k_3}{K}}\right\}, \\
c_2 &= -\frac{\zeta}{2}\sqrt{\frac{1}{2}(1-\zeta\frac{k_0}{\lambda_3})}\left\{\sqrt{1+\frac{k_3}{K}}-\zeta\sqrt{1-\frac{k_3}{K}}\right\}, \\
c_3 &= \frac{1}{2}\sqrt{\frac{1}{2}(1+\zeta\frac{k_0}{\lambda_3})}\left\{\sqrt{1+\frac{k_3}{K}}-\zeta\sqrt{1-\frac{k_3}{K}}\right\}, \\
c_4 &= \frac{\zeta}{2}\sqrt{\frac{1}{2}(1-\zeta\frac{k_0}{\lambda_3})}\left\{\sqrt{1+\frac{k_3}{K}}+\zeta\sqrt{1-\frac{k_3}{K}}\right\},
\end{aligned}
$$

$\zeta = \pm 1$ corresponds to two possible spin orientations relative to the field $\boldsymbol{H}$.

3.3.4. Radial Projection of the Spin

Finally, we consider the operator of the projection onto the radius of electron orbit which is also an integral of motion

$$\begin{aligned}
\hat{\Phi}_R &= -(\hat{\boldsymbol{\Phi}}\boldsymbol{H})/H = -\rho_3(\boldsymbol{\sigma}[\hat{\boldsymbol{\mathcal{P}}}\boldsymbol{H}])/H, \\
\hat{\Phi}_R\Psi &= \zeta\frac{\lambda_4}{k_0}\Psi.
\end{aligned}$$

We have the following system of equations for coefficients c_i

$$\begin{aligned}
(k_3 - i\zeta\lambda_4)c_1 - (K + k_0)c_3 &= 0\,, \\
(k_3 - i\zeta\lambda_4)c_2 + (K + k_0)c_4 &= 0\,, \\
(K - k_0)c_1 - (k_3 + i\zeta\lambda_4)c_3 &= 0\,, \\
(K + k_0)c_2 + (k_3 + i\zeta\lambda_4)c_4 &= 0.
\end{aligned} \tag{3.36}$$

Putting the determinant equal to zero and solving Eqs (3.36), (3.23) and (3.26) we find

$$\lambda_4 = \sqrt{4\gamma n},$$

$$\begin{aligned}
c_1 &= \frac{i}{2}\sqrt{1 + \frac{k_0}{K}}e^{i\zeta\frac{\theta}{2}}, \\
c_2 &= \frac{\zeta}{2}\sqrt{1 + \frac{k_0}{K}}e^{i\zeta\frac{\theta}{2}}, \\
c_3 &= \frac{i}{2}\sqrt{1 - \frac{k_0}{K}}e^{-i\zeta\frac{\theta}{2}}, \\
c_4 &= -\frac{\zeta}{2}\sqrt{1 - \frac{k_0}{K}}e^{-i\zeta\frac{\theta}{2}},
\end{aligned}$$

$$\cot\theta = \frac{\beta_3}{\sqrt{\beta^2 - \beta_3^2}}, \quad \beta^2 = 1 - \frac{k_0^2}{K^2}, \quad \beta_3 = \frac{k_3}{K}.$$

This is a unique case when the coefficients c_i are complex.

So we have found the wave functions for some possible operators of the electron spin polarization.

For the given energy and z-projection of the total angular and linear momentum vectors we have two possible spin orientations and, correspondingly, two linearly-independent wave functions which form a complete system of functions. Obviously, an eigenfunction of any polarization operator can be decomposed into two eigenfunctions of any other polarization operator. The corresponding coefficients can be easily calculated, but they are too bulky and we shall not write them out here. The possibility of such a decomposition shows the fact that any state of the electron spin polarization can be represented as a sum of states with partial polarization relative to another direction.

3.4. Matrix Elements of the Dirac Matrices. Some Properties of Laguerre Functions

As is known from quantum electrodynamics (e.g., see, Refs[4,14]) the power W of radiation during a transition from state a to the state b with account of radiation polarization in the first order of perturbation theory with respect to the fine structure constant is descried by the following expression

$$W_i = \frac{ce^2}{2\pi}\int d^3\kappa\,\delta(\kappa_{ab}-\kappa)S_i \quad \kappa_{ab} = K_a - K_b. \tag{3.37}$$

Here $\boldsymbol{\kappa} = \kappa\boldsymbol{\kappa}_0$ is the wave vector of the emitted photon, $c\kappa = \omega$ is the radiation frequency, and the unit vector

$$\boldsymbol{\kappa}_0 = (\sin\theta\cos\varphi', \sin\theta\sin\varphi', \cos\theta)$$

points to the directions of the photon escape and θ, φ' are the spherical angles. Obviously,

$$d^3\kappa = \kappa^2 d\kappa d\Omega, \quad d\Omega = \sin\theta d\theta d\varphi',$$
$$0 < \kappa < \infty, \quad 0 \le \theta < \pi, \quad 0 \le \varphi' < 2\pi.$$

For the electron quantities S_i are determined as follows

$$S_i = |((\boldsymbol{\alpha})\boldsymbol{\beta}_i)|^2 \tag{3.38}$$

Matrix elements $\langle\boldsymbol{\alpha}\rangle$ in this formula have the form

$$\langle\boldsymbol{\alpha}\rangle = \int \psi_b^+ exp[-i(\boldsymbol{\kappa z})]\boldsymbol{\alpha}\psi_a d^3x \tag{3.39}$$

and unit vectors $\boldsymbol{\beta}_i$ $(i = -1, 1, 2, 3)$ define the radiation polarization (see Chapter 1) in the following way. Choosing

$$\boldsymbol{\beta}_2 = \frac{[\boldsymbol{\kappa}_0\boldsymbol{j}]}{\sqrt{1-(\boldsymbol{\kappa}_0\boldsymbol{j})^2}},$$
$$\boldsymbol{\beta}_3 = \frac{\boldsymbol{j}\boldsymbol{\beta}_2][\boldsymbol{\kappa}_0(\boldsymbol{\kappa}_0\boldsymbol{j}) - \boldsymbol{j}}{\sqrt{1-(\boldsymbol{\kappa}_0\boldsymbol{j})^2}},$$

where $\boldsymbol{j}$ is an arbitrary constant polarization unit vector, we obtain two linear components of polarization called σ- and π-component, respectively. Choosing

$$\boldsymbol{\beta}_l = \frac{1}{\sqrt{2}}(\boldsymbol{\beta}_2 + il\boldsymbol{\beta}_3)\,, \quad l = \pm 1,$$

we have the radiation with the left-circular ($l = -1$) and right-circular ($l = +1$) polarization. It is convenient to rewrite the expression of Eq. (3.38) as follows

$$S_i = S = |l_2 S_2 + i l_3 S_3|^2. \tag{3.40}$$

If we put $l_2 = 1$, $l_3 = 0$ in this formula we have the σ-component of the linear polarization. If $l_2 = 0$, $l_3 = 1$ we have the π-component. If $l_2 = l l_3 = 1/\sqrt{2}$ we have the left-circular ($l = -1$) and right-circular polarizations. If $l_2^2 = l_3^2 = 1$, $l_2 \cdot l_3 = 0$ we get the total intensity of the unpolarized radiation. The polarization vector $\boldsymbol{j}$ should be choosen in some physically distinguished direction. In our problem such a direction is determined by an external magnetic field, hence $\boldsymbol{j} = (0, 0, 1)$.

In the case of a spinless particle, all the above formulas above hold with the substitution the matrix elements of $\langle \boldsymbol{\alpha} \rangle$ by the matrix elements of $\langle \boldsymbol{\mathcal{P}} \rangle$.

$$\langle \boldsymbol{\mathcal{P}} \rangle = \frac{1}{\hbar\sqrt{K_b K_a}} \int \psi_b^+ \exp[-i(\boldsymbol{\kappa z})] \hat{\boldsymbol{\mathcal{P}}} \psi \, d^3x.$$

Finally, if we wish to obtain the total power of radiation we must summarize over all final states in the formula of Eq. (3.37).

If we are interested in the probability (rathe than in power) of radiation we have to use the expression

$$w = \frac{e^2}{2\pi\hbar} \int \frac{d^3\kappa}{\kappa} \delta(\kappa_{ab} - \kappa) S. \tag{3.41}$$

As it follows from the above we need to calculate the matrix elements of matrices $\langle \boldsymbol{\alpha} \rangle$ or of operators $\langle \boldsymbol{\mathcal{P}} \rangle$ which are expressed in terms of the Laguerre functions given by the formulas of Eqs (3.8) and (3.9). The properties of Laguerre polynomials are well known (see Ref.[16], p. 1051). First of all, using these properties we lay down the following recurrence formulas connecting the functions, in which the indices differ in unity

$$\begin{aligned}
I_{n+1,n'}(x) &= \sqrt{\frac{x}{n+1}} \left[\frac{x+n-n'}{2x} I_{n,n'}(x) - I'_{n,n'}(x) \right], \\
I_{n,n'+1}(x) &= \sqrt{\frac{x}{n'+1}} \left[\frac{n-n'-x}{2x} I_{n,n'}(x) + I'_{n,n'}(x) \right], \\
I_{n-1,n'}(x) &= \sqrt{\frac{x}{n}} \left[\frac{x+n-n'}{2x} I_{n,n'}(x) + I'_{n,n'}(x) \right], \\
I_{n,n'-1}(x) &= \sqrt{\frac{x}{n'}} \left[\frac{n-n'-x}{2x} I_{n,n'}(x) - I'_{n,n'}(x) \right], \\
I_{n-1,n'-1}(x) &= \sqrt{\frac{x^2}{nn'}} \left[\frac{n+n'-x}{2x} I_{n,n'}(x) - I'_{n,n'}(x) \right], \\
I_{n+1,n'+1}(x) &= \frac{x}{\sqrt{(n+1)(n'+1)}} \left[\frac{n+n'+2-x}{2x} I_{n,n'}(x) + I'_{n,n'}(x) \right].
\end{aligned} \tag{3.42}$$

Since the Laguerre polynomials have the following property (see Ref.[4], p.152)

$$(-1)^l (s+l)! x^{-l} L_{s+l}^{-l}(x) = s! L_s^l(x),$$

the Laguerre functions satisfy the relation

$$I_{n.n'}(x) = (-1)^{n-n'} I_{n',n}(x).$$

The Laguerre polynomials may be expressed in terms of generating functions (see Ref.[16], p.1052)

$$n! L_n^{m-n}(x) = \frac{d^n}{dz^n} e^{-xz}(1+z)^m \Big|_{z=0}, \tag{3.43}$$

with the help of which it is easy to calculate, for example, a sum

$$\begin{aligned} J &= \sum_{n'=0}^{\infty} a^{n'} I_{n,n'}(x) I_{p,n'}(x) = \frac{(-1)^{n+p}}{\sqrt{n!p!}} e^{-x} x^{-\frac{n+p}{2}} \times \\ &\times \frac{d^n}{dt^n}\frac{d^p}{d\tau^p} \left(e^{-x(t+\tau)} \sum_{n'=0}^{\infty} \frac{[ax(1+t)(1+\tau)]^{n'}}{n'!} \right) \Big|_{t=\tau=0}. \end{aligned}$$

Assuming that $|t| < 1$ and $|\tau| < 1$ we ensure the equiconvergence of the series and thereby may interchange the order of summation and differentiation. Summarizing we find

$$J = \frac{(-1)^{n+p}}{\sqrt{n!p!}} e^{-x} x^{-\frac{n+p}{2}} \frac{d^n}{dt^n}\frac{d^p}{d\tau^p} e^{-x[t+\tau-a(1+t)(1+\tau)]} \Big|_{t=\tau=0},$$

and using again the formula of Eq. (3.43) we eventually obtain

$$J = i^{p-n} e^{\frac{a^2-1}{2a}x} a^{\frac{p+n}{2}} I_{p,n}(-\frac{(a-1)^2}{a} x).$$

Putting $a = 1$ we have (see also Ref.[4])

$$\sum_{n'=0}^{\infty} I_{p,n'}(x) = \delta_{p,n}.$$

Using the formulas of Eq. (3.42) we can easily derive the following relations for Laguerre functions

$$\begin{aligned} (\hat{\mathcal{P}}_x - i\hat{\mathcal{P}}_y) \frac{e^{il\varphi}}{\sqrt{2\pi}} I_{n,s}(\gamma r^2) &= -ih\sqrt{4\gamma n} \frac{e^{i(l-1)\varphi}}{\sqrt{2\pi}} I_{n-1,s}(\gamma r^2) \\ (\hat{\mathcal{P}}_x + i\hat{\mathcal{P}}_y) \frac{e^{i(l-1)\varphi}}{\sqrt{2\pi}} I_{n-1,s}(\gamma r^2) &= ih\sqrt{4\gamma n} \frac{e^{il\varphi}}{\sqrt{2\pi}} I_{n,s}(\gamma r^2), \end{aligned} \tag{3.44}$$

where $\hat{\mathcal{P}}$ has been determined by the formulas of Eqs (3.2) and (3.3).

In order to calculate the matrix elements of the Dirac matrices $\langle \boldsymbol{\alpha} \rangle$, we use the integral representation of the Bessel function $J_s(x)$ (see Refs,[16] p.966)

$$\frac{1}{2\pi}\int_0^{2\pi} d\varphi\, e^{-ix\sin\varphi+is\varphi} = J_s(x),$$

and also the integral

$$\int_0^{\infty} I_{n',s'}(x)I_{n,s}(x)J_{l-l'}(\sqrt{ax})dx = I_{n,n'}\left(\frac{a}{4}\right)I_{s,s'}\left(\frac{a}{4}\right), \quad n-s-l=n'-l'-s'.$$

Taking into account the relation of Eq. (3.44) we can calculate the matrix elements of $\langle \hat{\mathcal{P}} \rangle$. When we calculate the matrix elements in this way and substitute them in the formulas of Eqs (3.38) and (3.39), the values S_2, S_3, S_l do not depend on the angle φ', according to the axial symmetry of the problem. So we may put $\varphi' = \pi/2$ and get the result

$$S_2 = \langle \alpha_1 \rangle \langle \alpha_1 \rangle^+, \tag{3.45}$$

$$S_3 = |\langle \alpha_2 \rangle \cos\theta - \langle \alpha_3 \rangle \sin\theta|^2, \tag{3.46}$$

$$S^{(l)} = (\langle \alpha_1 \rangle^+ \langle \alpha_2 \rangle - \langle \alpha_1 \rangle \langle \alpha_2 \rangle^+)\cos\theta - (\langle \alpha_1 \rangle^+ \langle \alpha_3 \rangle - \langle \alpha_3 \rangle^+ \langle \alpha_1 \rangle)\sin\theta. \tag{3.47}$$

For a boson $\langle \boldsymbol{\alpha} \rangle$ is to be replaced by $\langle \hat{\mathcal{P}} \rangle$. The matrix elements of the Dirac matrices and those of the kinetic momentum operator have the form[17,18]

$$\begin{aligned}
\langle \alpha_1 \rangle &= iI_{s,s'}(x)\{(c_1^{+'}c_4 + c_3^{+'}c_2)I_{n,n'-1}(x) - (c_1c_4^{+'} + c_3c_2^{+'})I_{n-1,n'}(x)\}f\,, \\
\langle \alpha_2 \rangle &= I_{s,s'}(x)\{(c_1^{+'}c_4 + c_3^{+'}c_2)I_{n,n'-1}(x) + (c_1c_4^{+'} + c_3c_2^{+'})I_{n-1,n'}(x)\}f\,, \\
\langle \alpha_3 \rangle &= I_{s,s'}(x)\{(c_1^{+'}c_3 + c_1c_3^{+'})I_{n-1,n'-1}(x) - (c_2c_4^{+'} + c_2^{+'}c_4)I_{n,n'}(x)\}f,
\end{aligned} \tag{3.48}$$

$$\begin{aligned}
\langle \hat{\mathcal{P}} \rangle_1 &= i\sqrt{\frac{\gamma}{KK'}}I_{s,s'}(x)\{\sqrt{n+1}I_{n+1,n'}(x) - \sqrt{n}I_{n-1,n'}(x)\}f\,, \\
\langle \hat{\mathcal{P}} \rangle_2 &= \sqrt{\frac{\gamma}{KK'}}I_{s,s'}(x)\{\sqrt{n+1}I_{n+1,n'}(x) + \sqrt{n}I_{n-1,n'}(x)\}f\,, \\
\langle \hat{\mathcal{P}} \rangle_3 &= \frac{k_3}{\sqrt{KK'}}I_{s,s'}(x)I_{n,n'}(x)\}f,
\end{aligned} \tag{3.49}$$

where

$$x = \frac{k^2\sin^2\theta}{4\gamma} \qquad f = \delta_{k_3',k_3-\kappa\cos\theta}. \tag{3.50}$$

The quantities corresponding to a final state are denoted by primes. Now taking into account the energy conservation law

$$\kappa = K - K' \tag{3.51}$$

and the above obtained conservation condition for the momentum long the field

$$k_3' = k_3 - \kappa \cos\theta, \tag{3.52}$$

and solving together these equations for κ, we arrive at the following result

$$\kappa = \frac{K}{\sin^2\theta}\left[1 - \beta_3 \cos\theta - \sqrt{(1-\beta_3\cos\theta)^2 - (\beta^2 - \beta_3^2)\frac{\nu}{n}\sin^2\theta}\right],$$

where we denoted

$$1 - \beta^2 = \varepsilon_0 = (\frac{mc^2}{E})^2, \quad \beta_3 = \frac{k_3}{K} = \frac{v_z}{c}, \quad \nu = n - n'.$$

In particular, if $k_3 = 0$ we have a simpler result

$$\kappa = \frac{K}{\sin^2\theta}\left(1 - \sqrt{1 - \beta^2\frac{\nu}{n}\sin^2\theta}\right). \tag{3.53}$$

Finally, keeping in mind the integration rules for the expressions with δ-function and integrating over κ and φ' we get from Eqs (3.37) and (3.41)

$$w = \frac{e^2}{\hbar}\int \frac{\kappa \sin\theta d\theta}{1 + \frac{\partial K'}{\partial \kappa}} S, \quad \frac{\partial K'}{\partial \kappa} = -\frac{k_3 - \kappa\cos\theta)\cos\theta}{K - \kappa}, \tag{3.54}$$

$$W = ce^2 \int \frac{\kappa^2 \sin\theta d\theta}{1 + \frac{\partial K'}{\partial \kappa}} S, \tag{3.55}$$

where S was determined by Eq. (3.40). The summation over all final states n', s', k_3', ζ' in Eqs (3.54) and (3.55) would give us the total transition probability and radiation power.

3.5. Qualitative Estimations for Manifestation of Quantum Effects in SR

It is clear that characteristics of the synchrotron radiation calculated by quantum methods always differ from those yielded in the framework of the classical theory. From the physical point of view, there are three most significal factors which cause the difference. Firstly, it is the quantization of the particle energy in a magnetic field (see Eqs (3.6) and (3.25)). Secondly, the recoil effect due to the photon emission, which is implicated, in particular, in conservation laws of Eqs (3.51) and (3.52). Thirdly, spin properties of a particle causing, in particular, the difference between the radiation characteristics of scalar and spinor particles. All these factors are absent in the classical theory.

Now we wish to estimate qualitatively the role of the factors mentioned above, assuming there is no motion of the particle along the z-axis in the initial state.[b] For the estimation we will essentially use the constant field H^* defined by Eq. (3.7) and call it further the Schwinger field. Admitting that the rest mass of the particle under consideration coincides with one of the electrons, we have the numerical value

$$H^* \approx 4,41 \times 10^{13} \text{ Oe}.$$

From Eqs (3.6) and (3.25) it follows that both the discreteness of the energy spectrum of the particle and the photon recoil effect become very important when the strength of the external field H becomes comparable with or higher than H^*. In such cases, we will talk about a region of superstrong magnetic fields. The quantum effects play a leading part in this region. We will discuss the physical results of this region in another chapter.

Under conditions of a physical laboratory, magnetic fields of accelerators and storages, which are synchrotron radiation sources, are much weaker than the Schwinger field, usually $H \sim 10^3$ to 10^5 Oe. Then the discreteness of the energy levels may be important only for small values of the principal quantum number n. But in this case, provided

$$2nH \ll H^*$$

from Eqs (3.6) and (3.13) follows that

$$\beta^2 = 1 - \left(\frac{mc^2}{E}\right)^2 \ll 1,$$

which means we have a nonrelativistic particle and may use in calculations the expansion with respect to the small parameter β^2. Though the nonrelativistic approximation is for sure of theoretical interests, in practically important cases the energy of the particle is large when compared with the rest energy $mc^2/E \ll 1$. In the relativistic case with $H \ll H^*$ the initial number n is large and the energy difference between neighboring levels is small compared with the energy itself

$$\begin{aligned} \Delta E &= E_{n+1} - E_n = E_n \left(\sqrt{1 + 2\frac{H}{H^*}\left(\frac{mc^2}{E_n}\right)^2} - 1 \right), \\ \Delta E/E &\approx \frac{H}{H^*}\left(\frac{mc^2}{E}\right)^2 \ll 1, \end{aligned}$$

[b]We can take into account the motion of a particle along the z-axis with the aid of Lorentz transformation in final physical results.

and that means the energy spectrum becomes quasicontinuous, and in practice there is no need to take into consideration its discreteness. If in this case the energy of the emitted photon is small compared with that of the particle, the recoil factor is not essential and the greatest coincidence of the classical theoretical results with the quantum ones is to be expected. Otherwise, if the photon energy is comparable with the electron one the recoil factor will not be negligible and the quantum effects will manifest again.

A characteristic parameter which evaluates the possibility of manifestation of quantum effects may be found with the help of the following estimation. The requirement that the energy of the emitted photon E_{ph} is to be small relative to the energy of the particle

$$E_{ph} \ll E$$

in account of Eq. (3.53) takes the form

$$\frac{1 - \sqrt{1 - (\beta^2 \nu \sin^2\theta)/n}}{\sin^2\theta},$$

that is valid only if ($\beta^2 \sim 1$)

$$\nu \ll n.$$

If this condition holds we find from Eq. (3.53) for the energy of the emitted photon

$$E_{ph} = c\hbar k = c\hbar K \beta^2 \frac{\nu}{2n} = E\beta^2 \frac{\nu}{2n} = \hbar\omega\nu, \quad \omega = ceH/E,$$

where ω is the classical frequency of rotation of a particle along a circular orbit in the field H with the energy E, ν is the number of emitted harmonic. According to the classical theory, the main contribution to the radiation is made by harmonics with $\nu \sim \nu_{eff}$,

$$\nu_{eff} = \frac{3}{2}\left(\frac{E}{mc^2}\right)^3. \tag{3.56}$$

Assuming that this conclusion remains to be valid in our case too, we get

$$E_{ph} \approx \hbar\omega\nu_{\text{eff}} = E\xi, \quad \xi = \frac{3}{2}\frac{H}{H^*}\frac{E}{mc^2}.$$

The validity of the condition $E_{ph} \ll E$ is ensured by the condition $\xi \ll 1$. Otherwise, if $\xi \gg 1$ the quantum effects in the synchrotron radiation of a relativistic particle in the field $H \ll H^*$ become essential. it is in this case that the dimensionless parameter ξ becomes characteristic.

3.6. Quantum Effects in Radiation of a Nonrelativistic Electron

From a practical point of view the radiation of nonrelativistic particles is of rather a limited interest. However, the discovery of distinctions between the classical theory and the quantum one would be very instructive. Expanding the matrix elements in powers of small a parameter β^2 we get the following result for the total radiation intensity

$$W = RW^{\rm cl}, \quad W^{\rm cl} = \frac{2e^4H^2\beta^2}{3m^2c^3(1-\beta^2)},$$

where $W^{\rm cl}$ is the corresponding classical value and the dimensionless constant R depends on a number of energy level of the particle in the initial state. For the electron

$$R = R^l = \frac{2n-1}{2n}[1 - \frac{11n-8}{2n(2n-1)}\beta^2 + \cdots]$$

and for a spinless particle we have another expression

$$R = R^b = \frac{2n}{2n+1}[1 - \frac{11}{2(2n+1)}\beta^2].$$

The following inequalities are valid for any finite n

$$1 > R^b > R^l.$$

The quantities R^b and R^l monotonously increase with n, tending to unity. So the quantum effects decrease the emitted intensity when compared with the classical value, the decrease being more significant for an electron than for a spinless particle[c] The greatest difference of R^b from R^l (and also the greatest difference from the classical theory) take place when $n = 1$. In this case

$$R^b = \frac{2}{3} - \frac{11}{9}\beta^2, \quad R^l = \frac{1}{2} - \frac{3}{4}\beta^2.$$

The total probability radiation in the first nonzero approximation in β^2 has the form

$$w^l = \frac{2n-1}{T^{\rm cl}}, \quad w^b = \frac{2n}{T^{\rm cl}}, \quad T^{\rm cl} = \frac{3}{2}\frac{m^3c^5}{e^4H^2}.$$

It is not difficult now to calculate the total time of radiation of a particle. In the nonrelativistic approximation the transitions have a dipole character ($\nu = 1$), and the particle emitting photons successively proceeds from a state n to a state $n-1$ until it finds itself in a state of minimally possible energy, i.e. it will be "illuminated out". The total time of transition is $T(n)$

$$T(n) = \sum_{k=0}^{n}[w(k)]^{-1} = T^{\rm cl}\delta.$$

[c]We fix here the energy level (of number n) and the energy itself (more exactly, the quantity β).

For an electron, e.g., δ is equal to

$$\delta = \sum_{k=1}^{n} \frac{1}{2k-1}.$$

For large n we get

$$\delta \approx c/2 + \ln 2\sqrt{n} + \frac{1}{48n^2} + \cdots,$$

where C is the Euler constant. Since the quantity δ increases slowly with n (according to the logarithmic law), and remains practically always of the order of unity, the total "depletion" occurs during the time $T \sim T^{\rm cl}$ which can be obtained by means of integration of the classical expression

$$\begin{aligned} W^{\rm cl} = -\frac{dE}{dt} &= \frac{2}{3}\frac{e^4 H^2}{m^4 c^7}(E^2 - m^2 c^4), \\ \frac{E}{mc^2} &= \coth\frac{t + t_0}{T^{\rm cl}}. \end{aligned}$$

So the time $T^{\rm cl}$ plays a part of the life time of the exited state, depending only on the value of the external field. In the case of $H \sim 10^4$ gauss we have $T^{\rm cl} \sim$ sec.

Let us consider in more detail the role of the spin properties of the particles in the radiation. Considering, for example, the case of the transversely oriented electron spin, we find that the probability of transition with the spin-flip is by factor of β^2 less than that of transition without a flip. Namely, for the probability w_1 of transitions with the flip of transverse spin we have

$$w_1 = \frac{2}{5}\frac{1}{T_0}\left(\frac{H}{H^*}\right)^3 \left(\frac{1+\xi}{2} + \frac{1-\xi}{2}\beta^4 a\right), \quad T_0 = \frac{\hbar^2}{mce^2},$$

where the numerical coefficient a is of the order of unity. We see that in the case $\zeta = 1$ (the initial field is parallel to the field), the probability of the spin-flip is much more than the corresponding probability in the case of $\zeta = -1$ (the initial spin against the field). The radiation process occurring on the whole without a spin-flip, leads to self-polarization of the spin, namely, to the predominant orientation of the spin against the field. Later we shall consider this effect in more detail noting only the following fact. If we have a system of nonrelativistic noninteracting electrons in a magnetic field and the electrons are unpolarized on the average, after a time $T \sim T^{\rm cl}$ all electrons with spin orientation against the field will be in the ground state $n = 0$ due to the radiation during transitions without a spin-flip. After the same time all electrons with a spin oriented along the field will occur in the first exited state $n = 1$, and it would be difficult for them to achieve the ground state because of the low probability of transitions with a spin-flip. So a two-phase system will arise containing electrons in the ground state with a spin against the field and those in the first exited state with a spin along the field.

3.7. Mathematical Features of the Problem of Radiation of Ultrarelativistic Particles

From the theoretical and practical points of view, it is of great interest and importance to analyse the problem of SR by ultrarelativistic particles in weak fields

$$\varepsilon_0 = 1 - \beta^2 \ll 1, \quad H \ll H^*.$$

In this case, the maximum radiation power is emitted in high harmonics as it folows from Eq. (3.56, and the energy spectrum of the particle may be regarded as quasicontinuous (as well as that of the emitted photons). Upto now it seemed impossible to obtain in this case correct analytical results concerning the total power of radiation and its other characteristics, using the exact expressions for the matrix elements of Eqs (3.48) and (3.49). However, it is possible to get asymptotic expansions with respect to the small parameter ε_0.

First of all, just as in the classical theory we regard the numbers of harmonics ν to be quasicontinuous and replace summation by integration introducing a new variable of integration in the following way

$$\kappa = K\frac{\xi y}{1+\xi y}, \quad 0 < y < \infty. \tag{3.57}$$

After the change of the variables we get

$$\frac{ce^2\kappa^2 d\nu}{1+\frac{\partial K'}{\partial \kappa}} = \frac{81}{16} W_0\, \xi^2 \varepsilon_0^{-5/2} \frac{y^2 dy}{(1+\xi y)^5}, \tag{3.58}$$

$$K' = \frac{K}{1+\xi y}, \quad \frac{k_3'}{K} = -\frac{\xi y}{1+\xi y}\cos\theta, \tag{3.59}$$

$$W_0 = \frac{8}{27}\frac{e^2 m^2 c^3}{\hbar^2}. \tag{3.60}$$

Further, assuming the radiation to be localized mainly in the orbital plane ($\theta = \pi/2$) with a small opening of an angle $\Delta\theta \sim \sqrt{\varepsilon}$ (as it takes place in the classical theory) we find

$$\varepsilon = 1 - \beta^2 \sin^2\theta = \varepsilon_0 + \beta^2\cos^2\theta \approx \varepsilon_0 + \beta^2(\Delta\theta)^2 \approx 2\varepsilon_0 \ll 1.$$

So ε and ε_0 are the quantities of the same order. Finally, we take the known asymptotic formulas for the Laguerre functions

$$\begin{aligned} I_{n,n'}(x) &= \frac{(1+\xi y)^{1/2}}{\pi\sqrt{3}}\,\varepsilon^{1/2} K_{1/3}(z_0), \\ I'_{n,n'}(x) &= \frac{(1+\xi y)^{3/2}}{\pi\sqrt{3}\xi y}\,\varepsilon K_{2/3}(z_0), \\ z_0 &= \frac{4}{2}\left(\frac{\varepsilon}{\varepsilon_0}\right)^{3/2}, \end{aligned} \tag{3.61}$$

where x has been determined by Eq. (3.50). From Eq. (3.42) and in view of Eqs (3.50), (3.57), (3.58), and (3.59) we obtain the following relations for the Laguerre functions with shifted indices

$$\begin{aligned} I_{n-1,n'}(x) &= I_{n,n'}(x) + \frac{\xi y}{1+\xi y} I'_{n,n'}(x), \\ I_{n'-1,n}(x) &= I_{n,n'}(x) - \xi y I'_{n,n'}(x), \\ I_{n-1,n'-1}(x) &= I_{n,n'}(x) - \frac{\xi^2 y^2}{1+\xi y} I'_{n,n'}(x), \\ I_{n+1,n'}(x) &= I_{n,n'}(x) - \frac{\xi y}{1+\xi y} I'_{n,n'}(x) \end{aligned} \qquad (3.62)$$

Now we are to use the formulas of Eq. (3.61).

3.8. Calculation of Angle Integrals

The integration over angles in the expressions for the radiation characteristics is now reduced to the calculation of integrals containing the products of the Macdonald functions and the powers of ε.

A method of calculating the integrals of this kind was proposed by N.P. Klepikov.[19] Such integrals are calculated in the same way, so we can illustrate it with only one example. For instance, let us take

$$J = \int_0^{\pi} d\theta \sin\theta \varepsilon^2 K^2_{2/3}\left[\frac{y}{2}\left(\frac{\varepsilon}{\varepsilon_0}\right)^{3/2}\right]$$

Using a change of variables

$$\cos\theta = t\sqrt{\frac{\varepsilon_0}{1-\varepsilon_0}}$$

and admitting the condition of $\varepsilon_0 \ll 1$ we reduce the integral to the form

$$J \approx 2\varepsilon_0^{5/2}\int_0^{\infty}(1+t^2)^2 K^2_{2/3}\left[\frac{y}{2}(1+t^2)^{3/2}\right] dt$$

Further we use the known relations for Γ- and Macdonald functions[20,16]

$$K_\mu(x)K_\nu(x) = 2\int_0^{\infty} K_{\mu-\nu}(2xcht)ch(\mu+\nu)tdt \ ,$$

$$K_\nu(x) = \frac{1}{8\pi i}\int_{K-i\infty}^{K+i\infty} dS\left(\frac{x}{2}\right)^{-S}\Gamma\left(\frac{S}{2}-\frac{\nu}{2}\right)\Gamma\left(\frac{S}{2}+\frac{\nu}{2}\right) ,$$

$$\int_{2y}^{\infty} \frac{K_\nu(x)}{x} dx = \frac{1}{8\pi i} \int_{K-i\infty}^{K+i\infty} dS \, \frac{y^{-S}}{S} \Gamma\left(\frac{S}{2} - \frac{\nu}{2}\right) \Gamma\left(\frac{S}{2} + \frac{\nu}{2}\right) ,$$

$$\int_{x}^{\infty} K_\nu(y) dy = 2K_{\nu-1}(x) - \int_{x}^{\infty} K_{\nu-2}(y) dy ,$$

$$\int_{0}^{\infty} \frac{ch 2px}{ch^{2q} x} dx = 2^{2(q-1)} \, \frac{\Gamma(q+p)\Gamma(q-p)}{\Gamma(2q)} ,$$

$$\int_{0}^{\infty} \frac{dt}{(1+t^2)^S} = \frac{\sqrt{\pi}}{2} \, \frac{\Gamma(S - \frac{1}{2})}{\Gamma(S)} ,$$

$$\Gamma(2x) = \frac{2^{2x-1/2}}{\sqrt{2\pi}} \, \Gamma(x) \Gamma\left(x + \frac{1}{2}\right), \quad \Gamma(S+1) = S\Gamma(S) ,$$

$$\Gamma(3x) = \frac{3^{3x-1/2}}{2\pi} \, \Gamma(x) \Gamma\left(x + \frac{1}{3}\right) \Gamma\left(x + \frac{2}{3}\right) .$$

Using these formulas we can easily represent the integral in the form

$$J = \frac{4\varepsilon_0^{5/2}}{8\pi i} \int_0^\infty dt \int_0^\infty dx \int_{k-i\infty}^{k+i\infty} ds \left(\frac{y}{2}\right)^{-s} \frac{\Gamma^2(\frac{s}{2})}{(1+t^2)^{3s/2-2}} \, \frac{ch \frac{4x}{3}}{ch^s x}$$

and after integration over x and t we arrive at

$$J = \frac{\varepsilon_0^{5/2}}{8\pi i} \, \frac{\sqrt{\pi}}{2} \int_{k-i\infty}^{k+i\infty} ds \left(\frac{y}{2}\right)^{-s} \frac{\Gamma^{2s}(\frac{s}{2})\Gamma(\frac{3s}{2} - \frac{5}{2})\Gamma(\frac{s}{2} + \frac{2}{3})\Gamma(\frac{s}{2} - \frac{2}{3})}{\Gamma(\frac{3s}{2} - 2)\Gamma(s)}$$

Now, using the properties of Γ-functions we get

$$J = \frac{\varepsilon_0^{5/2}}{4i\sqrt{3}} \int_{k-i\infty}^{k+i\infty} ds \left(\frac{y}{2}\right)^{-s} \frac{3s-2}{6} \, \frac{\Gamma(\frac{s}{2} - \frac{5}{6})\Gamma(\frac{s}{2} - \frac{1}{6})}{s-1}$$

and, shifting $s \to s+1$, we find

$$\begin{aligned} J &= \frac{\varepsilon_0^{5/2}}{4i\sqrt{3}y} \int_{k-i\infty}^{k+i\infty} ds \left(\frac{y}{2}\right)^{-s} \frac{\frac{s}{2} + \frac{1}{6}}{s} \Gamma\left(\frac{s}{2} + \frac{1}{3}\right) \Gamma\left(\frac{s}{2} - \frac{1}{3}\right) = \\ &= \frac{2\pi\varepsilon_0^{5/2}}{\sqrt{3}y} \left(K_{2/3}(y) + \frac{1}{3} \int_y^\infty \frac{K_{2/3}(x)}{x} dx \right) = \\ &= \frac{\pi\varepsilon_0^{5/2}}{\sqrt{3}y} \left(3K_{2/3}(y) - \int_y^\infty K_{1/3}(x) dx \right) . \end{aligned}$$

Using the method shown above we can calculate all the integrals we need

$$\int_0^\pi \varepsilon^2 K_{1/3}^2(z_0)\sin\theta d\theta = \frac{\pi\varepsilon_0^{5/2}}{\sqrt{3}y}(K_{2/3}(y) + \int_y^\infty K_{1/3}(x)dx),$$

$$\int_0^\pi \varepsilon^2 K_{2/3}^2(z_0)\sin\theta d\theta = \frac{\pi\varepsilon_0^{5/2}}{\sqrt{3}y}(3K_{2/3}(y) - \int_y^\infty K_{1/3}(x)dx),$$

$$\int_0^\pi \varepsilon^{3/2} K_{1/3}(z_0)K_{1/3}(z_0)\sin\theta d\theta = \frac{2\pi\varepsilon_0^2}{\sqrt{3}y}(K_{1/3}(y),$$

$$\int_0^\pi \varepsilon K_{1/3}^2(z_0)\sin\theta d\theta = \frac{2\pi\varepsilon_0^{3/2}}{\sqrt{3}y}\int_y^\infty K_{1/3}(x)dx, \tag{3.63}$$

$$\int_0^\pi \varepsilon\cos^2\theta K_{1/3}^2(z_0)\sin\theta d\theta = \frac{\pi\varepsilon_0^{5/2}}{\sqrt{3}y}(K_{2/3}(y) - \int_y^\infty K_{1/3}(x)dx),$$

$$\int_0^\pi \varepsilon^{3/2}\cos^2\theta K_{1/3}(z_0)K_{1/3}(z_0)\sin\theta d\theta = \frac{2\pi\varepsilon_0^3}{3\sqrt{3}y^2}\int_y^\infty K_{1/3}(x)dx,$$

$$\int_0^\pi \varepsilon^{5/2} K_{1/3}(z_0)K_{2/3}(z_0)\sin\theta d\theta = \frac{2\pi\varepsilon_0^3}{3\sqrt{3}y^2}(3yK_{1/3}(y) + \int_y^\infty K_{1/3}(x)dx.$$

With the help of above expressions we can obtain the spectral distribution of the radiation intensity in the explicit form.

3.9. Calculation of Integrals over Spectrum. Functions $\mathcal{L}_\mu$ and $\mathcal{D}_\mu$

The correct calculation of integrals over spectrum is one of the difficult problems of the radiation theory. The integrals have been calculated in Ref.[19], but the authors succeeded only in getting expressions of the integrals in terms of products of generalized hypergeometrical functions. These expressions are too cumbersome and cannot be used in the practical analysis. It is not clear, in particular, how one could derive from them the asymptotic formulas for $\xi \ll 1$ in a simple way[21]. Moreover, there are no tables of generalized hypergeometrical functions, which makes it especially difficult to use these functions in numerical estimations of results. Here we propose a new simple method of calculating integrals over a spectrum introducing for this purpose new special functions $\mathcal{L}_\mu$ and $\mathcal{D}_\mu$ which are related to the Bessel functions. We will study their properties and show how the integrals over a spectrum are expressed in terms of these functions.

It is easy to see that after the integration over angles the radiation intensity is expressed in the form of a linear combination of integrals of the kind

$$\int_0^\infty \frac{y^m}{(1+\xi y)^n} K_s(y)dy,$$

$$\int_0^\infty \frac{y^m dy}{(1+\xi y)^n} \int_y^\infty K_s(x)dx = \int_0^\infty dy K_s(y) \int_0^y \frac{x^m dx}{(1+\xi x)^n}.$$

Expanding the algebraic fractions under integration into simple ones,we can represent the radiation intensity as a linear combination of the following integrals

$$\int_0^\infty \ln(1+\xi y) K_s(y)dy,$$

$$\int_0^\infty \frac{K_s(y)dy}{(1+\xi y)^n} \qquad (n = 0, 1, 2, 3, \ldots) \tag{3.64}$$

Let us choose a new variable

$$x = \frac{1}{\xi} \tag{3.65}$$

Since below the symbol x will denominate only this variable, no ambiguity will arise. Now we introduce a function[18]

$$\mathcal{L}_\mu(x) = \int_0^\infty \frac{K_\mu(y)dy}{x+y} = \int_0^\infty \frac{\xi K_\mu(y)dy}{1+\xi y}. \tag{3.66}$$

It is easy to see that due to Eqs (3.64) and (3.65) we have

$$\begin{aligned} \mathcal{D}_\mu(x) &= \int_0^\infty \ln\left(1+\frac{y}{x}\right) K_\mu(y)dy = \int_0^\infty \ln(1+\xi y)K_\mu(y)dy = \\ &= \int_0^x \mathcal{L}_\mu(t)dt - \frac{\pi}{2\cos\frac{\mu\pi}{2}} \ln x - c_\mu, \end{aligned} \tag{3.67}$$

where

$$c_\mu = -\int_0^\infty \ln y K_\mu(y)dy.$$

As is known (see Ref.[16], p.691), the function $\mathcal{L}_\mu(x)$ can be expressed in terms of the Bessel functions of an imaginary argument $I_\mu(x)$ and Anger functions $J_\mu(x)$. Namely, we have

$$\mathcal{L}_\mu = \frac{\pi^2}{2\sin^2\mu\pi} \left\{ I_\mu(x) + I_\mu(x) - e^{-i\frac{\mu\pi}{2}} y_\mu(ix) - e^{-i\frac{\mu\pi}{2}} y_{-\mu}(ix) \right\}.$$

Using the known expansions of $I_\mu(x)$ and $J_\mu(x)$ in series we get

$$\begin{aligned}\mathcal{L}_\mu &= \frac{\pi^2}{2\sin^2\mu\pi}\left\{\sum_{n=0}^{\infty}\frac{1}{n!\Gamma(n+1+\mu)}\left(\frac{x}{2}\right)^{\mu+2n}+\right.\\ &+\left.\sum_{n=0}^{\infty}\frac{1}{n!\Gamma(n+1-\mu)}\left(\frac{x}{2}\right)^{-\mu+2n}\right\}.\end{aligned}$$

So we see that $\mathcal{L}_\mu(x)$, and hence $\mathcal{D}_\mu(x)$ too, are represented everywhere by convergent power series. Using the known integral representations[16]

$$I_\mu(x)+I_{-\mu}(x)=\frac{2}{\pi}\left\{\int_0^\pi e^{x\cos\theta}\cos\mu\theta d\theta+\sin\mu\pi\int_0^\infty e^{-xcht}sh\mu t\,dt\right\},$$

$$J_\mu(x)=\frac{1}{\pi}\int_0^\pi\cos(\mu\theta-x\sin\theta)d\theta$$

we get the following significant integral representations

$$\begin{aligned}\mathcal{L}_\mu(x) &= \frac{\pi}{\sin\mu\pi}\left\{\int_0^{\frac{\pi}{2}}e^{-x\cos\theta}\sin\mu\theta d\theta+\int_0^\infty e^{-xcht}8h\mu tdt\right\},\\ \mathcal{D}_\mu(x) &= \frac{\pi}{\sin\mu\pi}\left\{\int_0^{\frac{\pi}{2}}\frac{1-e^{-x\cos\theta}}{\cos\theta}\sin\mu\theta d\theta+\right.\\ &+\left.\int_0^\infty\frac{1-e^{-xcht}}{cht}8h\mu tdt\right\}-\frac{\pi}{2\cos\frac{\mu\pi}{2}}\ln x-c_\mu.\end{aligned}$$

We imply here that $|\mu|\neq 0,1,2,\ldots$ Some important properties follow from the integral representations. First of all, it is obvious that

$$\mathcal{L}_\mu(x)=\mathcal{L}_{-\mu}(x). \tag{3.68}$$

Differentiating Eq. (3.68) we get

$$2\mathcal{L}'_\mu(x)=\mathcal{L}_{1-\mu}(x)+\mathcal{L}_{1+\mu}(x) \tag{3.69}$$

and in addition

$$\mathcal{L}_{1-\mu}(x)-\mathcal{L}_{1+\mu}(x)=\frac{2\mu}{x}\mathcal{L}_\mu(x)+\frac{\pi}{x\sin\frac{\mu\pi}{2}}. \tag{3.70}$$

Combining Eqs (3.68), (3.69) and (3.70) we get the following significant relations

$$\begin{aligned}\mathcal{L}'_\mu(x) &= \mathcal{L}_{1-\mu}(x)-\frac{\mu}{x}\mathcal{L}_\mu(x)-\frac{\pi}{2\sin\frac{\mu\pi}{2}}\frac{1}{x},\\ \mathcal{L}'_{1-\mu}(x) &= \mathcal{L}_\mu(x)-\frac{1-\mu}{x}\mathcal{L}_{1-\mu}(x)-\frac{\pi}{2\cos\frac{\mu\pi}{2}}\frac{1}{x}\end{aligned} \tag{3.71}$$

and, finally, the following differential equation for $\mathcal{L}_\mu(x)$

$$\mathcal{L}''_\mu(x) + \frac{1}{x}\mathcal{L}'_\mu(x) - \left(1 + \frac{\mu^2}{x^2}\right)\mathcal{L}_\mu(x) = \frac{\mu\pi}{2\sin\frac{\mu\pi}{2}}\frac{1}{x^2} - \frac{1}{2\cos\frac{\mu\pi}{2}}\frac{1}{x}$$

which can be regarded as a definition of the function $\mathcal{L}_\mu(x)$.

Using Eqs (3.71), (3.66) and (3.67) it is easy to obtain the fundamental relations[18]

$$\begin{aligned}
&\int_0^\infty \ln(1+\xi y)K_\mu(y)dy = \mathcal{D}_\mu(x),\\
&\int_0^\infty \frac{K_\mu(y)dy}{1+\xi y} = x\mathcal{L}_\mu(x),\\
&\int_0^\infty \frac{K_\mu(y)dy}{(1+\xi y)^2} = -x^2\mathcal{L}'_\mu(x) = \mu x\mathcal{L}_\mu(x) - x^2\mathcal{L}_{1-\mu}(x) + \frac{\pi}{2\sin\frac{\mu\pi}{2}}x,\\
&\int_0^\infty \frac{K_\mu(y)dy}{(1+\xi y)^3} = \frac{x^3}{2}\mathcal{L}''_\mu(x) = \frac{1}{2}[x^3 + \mu(1+\mu)]\mathcal{L}_\mu(x) -\\
&\qquad - \frac{x^2}{2}\mathcal{L}_{1-\mu}(x) + \frac{\pi(1+\mu)}{4\sin\frac{\mu\pi}{2}}x - \frac{\pi}{4\cos\frac{\mu\pi}{2}}x^2,\\
&\int_0^\infty \frac{K_\mu(y)dy}{(1+\xi y)^4} = -\frac{x^4}{6}\mathcal{L}'''_\mu(x) = \frac{1}{6}[(1+\mu)x^3 + \mu(1+\mu)(2+\mu)]\mathcal{L}_\mu(x) -\\
&\qquad - \frac{1}{6}[x^4 + (2+\mu^2)x^2]\mathcal{L}_{1-\mu}(x) + \frac{(1+\mu)(2+\mu)\pi}{12\sin\frac{\mu\pi}{2}}x -\\
&\qquad - \frac{\pi x^2}{6\cos\frac{\mu\pi}{2}} + \frac{\pi x^3}{12\sin\frac{\mu\pi}{2}}.
\end{aligned} \tag{3.72}$$

With the help of these relations it is possible to represent the radiation intensity as a linear combination of the functions $\mathcal{L}_{1/3}$, $\mathcal{L}_{2/3}$, $\mathcal{D}_{1/3}$, $\mathcal{D}_{2/3}$. However, the function $\mathcal{D}_{2/3}$ has never been found in calculations, and $\mathcal{D}_{1/3}$ occurs seldom.

Now we seek asymptotic representations of $\mathcal{L}_\mu(x)$ and $D_\mu(x)$ for large x (i.e. for small ξ). For this purpose, let us consider the integrals

$$I = \int_0^{\pi/2} e^{-x\cos\theta}\sin\mu\theta d\theta \quad J = \int_0^\infty e^{-x\cosh t}\sinh\mu t dt$$

which after a change of the variables

$$\cos\theta = t\ , \quad \cosh t - 1 = \tau^2$$

have the form

$$I = \int\limits_0^1 e^{-xt} \frac{\sin \mu \arccos t}{\sqrt{1-t^2}}\,, \quad J = \sqrt{2}e^{-x} \int\limits_0^\infty e^{-x\tau^2} \frac{\sinh 2\mu \operatorname{arcsinh} \frac{\tau}{\sqrt{2}}}{1+\frac{\tau^2}{2}}.$$

For the functions

$$f_1 = \cos \mu \arccos t\,, \quad f_2 = \cosh 2\mu \operatorname{arcsinh} y$$

we have

$$f_1' = \frac{\mu \sin \mu \arccos t}{\sqrt{1-t^2}}, \quad f_2' = \frac{2\mu \sinh 2\mu \operatorname{arcsinh} y}{\sqrt{1+y^2}}. \tag{3.73}$$

One can make sure by direct verification that the functions f_1 and f_2 obey the equations

$$\begin{aligned} (1-t^2)f_1'' - tf_1' = \mu^2 f_1 = 0, \\ (1+y^2)f_2'' + yf_2' - 4\mu^2 f_2 = 0. \end{aligned}$$

Looking for solutions of these equations in the form of series and taking into account Eq. (3.73) we find

$$\frac{\sin \mu \arccos t}{\sqrt{1-t^2}} = \frac{\sin \mu}{\pi} \sum_{n=0}^\infty \frac{2^{n-1}\Gamma(\frac{n+1+\mu}{2})\Gamma(\frac{n+1-\mu}{2})}{\Gamma(n+1)}(-t)^n, \quad |t| < 1,$$

$$\frac{\sqrt{2}\sinh 2\mu \operatorname{arcsinh}\left(\frac{\tau}{\sqrt{2}}\right)}{\sqrt{1+\frac{\tau^2}{2}}} = \frac{2\sin \mu\pi}{\pi}\tau \sum_{n=0}^\infty (-1)^n \frac{2^n\Gamma(n+1-\mu)\Gamma(n+1+\mu)}{\Gamma(2n+2)}\tau^{2n},$$

$$|\tau| < 1.$$

Now it is not difficult, using the saddle-point method, to find the asymptotics of $\mathcal{L}_\mu(x)$ for the case of large x (small ξ) (see Ref.[22], Lemma on p.447)

$$\begin{aligned} \mathcal{L}_\mu(x) &= \xi \sum_{n=0}^\infty \left\{ 2^{n-1}\Gamma\left(\frac{n+1+\mu}{2}\right)\Gamma\left(\frac{n+1-\mu}{2}\right) + \right. \\ &+ \left. e^{-\frac{1}{\xi}} \frac{\Gamma(n+1+\mu)\Gamma(n+1-\mu)}{(2n+1)!!} \right\} (-\xi)^n. \end{aligned}$$

Taking into account the relation

$$\frac{d}{d\xi}\,\mathcal{D}_\mu(x) = \frac{\pi}{2\cos\frac{\mu\pi}{2}}\,\frac{1}{\xi} - \frac{1}{\xi^2}\,\mathcal{L}_\mu(x)$$

it is easy to find the asymptotics

$$\mathcal{D}_\mu(x) = \xi \sum_{n=0}^{\infty} \left\{ \frac{2^n}{n+1} \Gamma\left(\frac{n+2+\mu}{2}\right) \Gamma\left(\frac{n+2-\mu}{2}\right) - e^{-\frac{1}{\xi}} d_n^\mu \right\} (-\xi)^n,$$

where d_n^μ are defined by the recurrence formulas

$$\begin{aligned} d_0^\mu &= \frac{\mu\pi}{\sin \mu\pi}, \\ d_n^\mu &= n d_{n-1}^\mu + \frac{\Gamma(n+1+\mu)\Gamma(n+1-\mu)}{(2n+1)!!}. \end{aligned}$$

Introducing the quantities

$$\begin{aligned} a_n &= 2^n \Gamma\left(\frac{3n+4}{6}\right) \Gamma\left(\frac{3n+8}{6}\right), \quad a_{n+1} = \frac{(3n+8)(3n+4)}{9} a_{n-1}, \\ b_n &= \frac{\Gamma(\frac{3n+8}{3})\Gamma(\frac{3n+4}{3})}{(2n+3)!!}, \quad b_{n+1} = \frac{(3n+8)(3n+4)}{9(2n+5)} b_n, \end{aligned}$$

and the quantities $d_n^{2/3}$ and $d_n^{1/3}$, defined by the recurrence formulas

$$\begin{aligned} d_0^{2/3} &= \frac{4\sqrt{3}\pi}{9}, \quad d_0^{1/3} = \frac{2\sqrt{3}\pi}{9}, \\ d_n^{2/3} &= n d_{n-1}^{2/3} + b_{n-1}, \quad d_n^{1/3} = n d_{n-1}^{1/3} + \frac{3n+1}{3n+2} b_{n-1}, \end{aligned}$$

we can write asymptotical formulas in the form

$$\begin{aligned} \mathcal{L}_{2/3}(x) &= \xi \sum_{n=0}^{\infty} \left(a_{n-1} + e^{-\frac{1}{\xi}} b_{n-1}\right) (-\xi)^n, \\ \mathcal{L}_{1/3}(x) &= \xi \sum_{n=0}^{\infty} \left(\frac{3n-2}{3} a_{n-2} + e^{-\frac{1}{\xi}} \frac{3n+1}{3n+2} b_{n-1}\right) (-\xi)^n, \\ \mathcal{D}_{1/3}(x) &= \xi \sum_{n=0}^{\infty} \left(\frac{3n+1}{3n+2} a_{n-1} - e^{-\frac{1}{\xi}} d_n^{1/3}\right) (-\xi)^n, \\ \mathcal{D}_{2/3}(x) &= \xi \sum_{n=0}^{\infty} \left(\frac{a_n}{n+1} - e^{-\frac{1}{\xi}} d_n^{2/3}\right) (-\xi)^n. \end{aligned} \tag{3.74}$$

Here we write some first coefficients for the reference use

$$\left\{ \begin{array}{l} a_{-2} = -\frac{\sqrt{3}\pi}{2} \\ a_{-1} = \pi \\ a_0 = \frac{2\sqrt{3}\pi}{9} \\ a_1 = \frac{5\pi}{9} \\ a_2 = \frac{64\sqrt{3}\pi}{81} \end{array} \right., \quad \left\{ \begin{array}{l} b_{-1} = \frac{4\sqrt{3}\pi}{9} \\ b_0 = \frac{20\sqrt{3}\pi}{243} \\ b_1 = \frac{\sqrt{3}\pi}{2} \\ b_2 = \frac{1408\sqrt{3}\pi}{19683} \end{array} \right., \quad \left\{ \begin{array}{l} d_0^{1/3} = \frac{2\sqrt{3}\pi}{9} \\ d_1^{1/3} = \frac{178\sqrt{3}\pi}{243} \\ d_2^{1/3} = \frac{3316\sqrt{3}\pi}{2187} \\ d_3^{1/3} = \frac{90812\sqrt{3}\pi}{19683} \end{array} \right., \quad \left\{ \begin{array}{l} d_0^{2/3} = \frac{4\sqrt{3}\pi}{9} \\ d_1^{2/3} = \frac{56\sqrt{3}\pi}{243} \\ d_2^{2/3} = \frac{1136\sqrt{3}\pi}{2187} \\ d_3^{2/3} = \frac{32080\sqrt{3}\pi}{19683} \end{array} \right. .$$

We note that in the case of small x (large ξ), the functions $\mathcal{L}_\mu(x)$ are approximated as follows

$$\mathcal{L}_\mu(x) \approx \frac{2^\mu \pi \Gamma(\mu)}{2\sin\mu\pi}\, x^{-\mu}. \tag{3.75}$$

Thus, we have studied the functions $\mathcal{L}_\mu(x)$ and $D_\mu(x)$ to an extent sufficient for our purposes, and from now on one can use them for qualitative investigation and also for quantitative estimations of radiation intensity.

3.10. Radiation of the Electron with the Longitudinally Polarized Spin

In the case of an electron with a longitudinally polarized spin, there are two possible directions of the spin, namely, along the motion and against it. The wave functions of such an electron are defined by the formulas of Eqs (3.22) and (3.33). Here we will carry out a rather detailed calculation of radiation intensity and transition probabilities, so as to omit further the insignificant calculating details.

Let us consider a σ-component of linear polarization which was determined by formulas of Eqs (3.45) and (3.48). It is easy to find (the factor $I_{ss'}x$ is omitted here)

$$\begin{aligned}\alpha_1 &= \frac{i}{4}\left\{\sqrt{\left(1-\frac{k_0}{K'}\right)(1+\sqrt{\varepsilon_0})} + \zeta\zeta'\sqrt{\left(1+\frac{k_0}{K'}\right)(1+\sqrt{1-\sqrt{\varepsilon_0}})}\right\}\\ &\times \left\{\sqrt{1+\zeta'\frac{k_3'}{\lambda_1'}}\, I_{n,n'-1} - \zeta\zeta'\sqrt{1-\zeta'\frac{k_3'}{\lambda_1'}}\, I_{n-1,n'}\right\},\end{aligned}$$

where λ is defined by Eq. (3.32). Squaring we get

$$\begin{aligned}S_2 &= \frac{1}{8}\left\{1-\frac{k_0}{K'}\sqrt{\varepsilon_0} + \zeta\zeta'\sqrt{(1-\varepsilon_0)\left(1-\frac{k_0^2}{K'^2}\right)}\right\}\left\{(I_{n,n'-1} - \zeta\zeta' I_{n-1,n'})^2\right.\\ &+ \left.\zeta'\frac{k_3'}{\lambda_1'}(I^2_{n,n'-1} - I^2_{n-1,n'}) - 2\zeta\zeta'\sqrt{1-\frac{k_3'}{l_1'^2}}\, I_{n-1,n'}I_{n,n'-1}\right\}.\end{aligned}$$

Changing the variable (see Eqs (3.57)—(3.61)) and expanding in series with respect to ε_0 and $\cos^2\theta$ (in fact, to ε) we have

$$\begin{aligned}S_2 &= \frac{1}{4}\frac{1+\zeta\zeta'}{2}\{(I_{n,n'-1} - I_{n-1,n'})^2 + \xi^2 y^2\cos^2\theta I^2_{n,n'} - \zeta\xi y\cos\theta(I^2_{n,n'-1} - I^2_{n-1,n'})\}\\ &+ \frac{1}{4}\frac{1-\zeta\zeta'}{2}\varepsilon_0\xi y I^2_{n,n'}.\end{aligned}$$

It remains now to use the formulas of Eq. (3.62).

In a quite analogous manner calculations are carried out for the other polarization components. As a result, we have expressions characterizing the spectral-angular distribution of the radiation intensity

$$W_i = \frac{27}{64\pi^2} W_0 \xi^2 \varepsilon_0^{-5/2} \int_0^\infty \int_0^\pi \frac{y^2 \sin\theta d\theta dy}{(1+\xi y)^4} \Phi_i(\xi, y, \theta),$$

$$\Phi_i(\xi, y, \theta) = \frac{1+\zeta\zeta'}{2} \Phi_i^{(1)}(\xi, y, \theta) + \frac{1-\zeta\zeta'}{2} \Phi_i^{(2)}(\xi, y, \theta),$$

where the functions $\Phi_i^{(j)}(\xi, y, \theta)$ have the following form

$$\begin{aligned}
\Phi_2^{(1)} &= \{(2+\xi y)\varepsilon K_{2/3}(z_0) + \zeta\sqrt{\varepsilon}\xi y \cos\theta K_{1/3}(z_0)\}^2, \\
\Phi_3^{(1)} &= \{(2+\xi y)\sqrt{\varepsilon}\cos\theta K_{1/3}(z_0) + \zeta\xi y \varepsilon K_{2/3}(z_0)\}^2, \\
\Phi_l^{(1)} &= \frac{1}{2}(2+\xi y + l\zeta\xi y)^2(\varepsilon\cos^2\theta K_{1/3}^2(z_0) + \varepsilon^2 K_{2/3}^2(z_0)) + \\
&+ 2\zeta[\xi y(2+\xi y) + \zeta l(2+2\xi y+\xi^2 y^2)]\varepsilon^{3/2}\cos\theta K_{1/3}(z_0)K_{2/3}(z_0),
\end{aligned}$$

$$\Phi_l^{(2)} = (1+l\zeta)\Phi^{(1)}, \quad \Phi_2^{(2)} = \Phi_3^{(2)} = \Phi' = \xi^2 y^2 \varepsilon_0 \varepsilon K_{1/3}^2(z_0).$$

Integrating over angles with the help of Eq. (3.63) we get the spectral distribution of the intensity

$$W_i = \frac{9\sqrt{3}}{64\pi} W_0 \xi^2 \int_0^\infty \frac{y dy}{(1+\xi y)^4} \left\{ \frac{1+\zeta\zeta'}{2} F_i^{(1)}(\xi, y) + \frac{1-\zeta\zeta'}{2} F_i^{(2)}(\xi, y) \right\},$$

where the functions $F_i^{(j)}(\xi, y)$ have the form

$$\begin{aligned}
F_2^{(1)} &= [3(2+\xi y)^2 + \xi^2 y^2]K_{2/3}(y) - [(2+\xi y)^2 + \xi^2 y^2]\int_y^\infty K_{1/3}(x)dx, \\
F_3^{(1)} &= [(2+\xi y)^2 + 3\xi^2 y^2]K_{2/3}(y) - [(2+\xi y)^2 + \xi^2 y^2]\int_y^\infty K_{1/3}(x)dx, \\
F_l^{(1)} &= (2+\xi y + l\zeta\xi y)^2(2K_{2/3}(y) - \int_y^\infty K_{1/3}(x)dx),
\end{aligned}$$

$$F_l^{(2)} = (1+l\zeta)F', \quad F_2^{(2)} = F_3^{(2)} = F' = 2\xi^2 y^2 \int_y^\infty K_{1/3}(x)dx.$$

The probability of transitions is obtained from the radiation intensity if we divide the spectral-angular density by $\hbar c\kappa$. The angular distribution of the probability of transition is not of great interest, so we immediately write the spectral distribution formula

$$w = \frac{18}{T_1}\int\limits_0^\infty \frac{dy}{(1+\xi y)^3}\left\{\frac{1+\zeta\zeta'}{2}\psi_1(\xi,y) + \frac{1-\zeta\zeta'}{2}\psi_2(\xi,y)\right\},$$

where

$$\psi_1 = (2+2\xi y+\xi^2y^2)(2K_{2/3}(y) - \int\limits_y^\infty K_{1/3}(x)dx),$$

$$\psi_2 = \xi^2y^2\int\limits_y^\infty K_{1/3}(x)dx,$$

and we have denoted

$$T_1 = T_0\,\frac{H^*}{H},\quad T_0 = 36\sqrt{3}\pi\,\frac{hc}{l^2}\,\frac{h}{mc^2} = 3.457\dot{1}0^{-17}\text{sec.}$$

Now if we average now over the initial spins and sum over the final ones, we get the radiation intensity and the probability of transition for the unpolarized electron

$$W_2 = \frac{9\sqrt{3}}{16\pi}W_0\xi^2\int\limits_0^\infty \frac{ydy}{(1+\xi y)^4}\Big\{(3+3\xi y+\xi^2y^2)K_{2/3}(y) - - (1+\xi y)\int\limits_y^\infty K_{1/3}(x)dx\Big\}, \tag{3.76}$$

$$W_3 = \frac{9\sqrt{3}}{16\pi}W_0\xi^2\int\limits_0^\infty \frac{ydy}{(1+\xi y)^4}\Big\{(1+\xi y+\xi^2y^2)K_{2/3}(y) - - (1+\xi y)\int\limits_y^\infty K_{1/3}(x)dx\Big\}, \tag{3.77}$$

$$W = \frac{9\sqrt{3}}{18\pi}W_0\xi^2\int\limits_0^\infty \frac{ydy}{(1+\xi y)^4}\Big\{(2+2\xi y+\xi^2y^2)K_{2/3}(y) - - (1+\xi y)\int\limits_y^\infty K_{1/3}(x)dx\Big\}, \tag{3.78}$$

$$w = \frac{36}{T_1}\int\limits_0^\infty \frac{dy}{(1+\xi y)^3}\left\{(2+2\xi y+\xi^2y^2)K_{2/3}(y) - (1+\xi y)\int\limits_y^\infty K_{1/3}(x)dx\right\} \tag{3.79}$$

Finally, the integration over a spectrum with the help of Eq. (3.72) results in

$$W_2 = q_1W_0\Big\{(36x^2+82)\mathcal{L}_{2/3}(x) + (27x^3+201x)\mathcal{L}_{1/3}(x) - - 9\pi\sqrt{3}x^2 - 27\pi x - 67\pi\sqrt{3}\Big\}, \tag{3.80}$$

$$W_3 = q_1 W_0 \left\{(198x^2+46)\mathcal{L}_{2/3}(x) + (27x^3+363x)\mathcal{L}_{1/3}(x) - \right.$$
$$\left. - 9\pi\sqrt{3}x^2 - 189\pi x - 85\pi\sqrt{3}\right\}, \qquad (3.81)$$

$$w = \frac{x}{T_1}\left\{\frac{1+\zeta\zeta'}{2}\left[(9x^2+56)\mathcal{L}_{2/3}(x) - 15x\mathcal{L}_{1/3}(x) - 9\pi x + 7\pi\sqrt{3} - 18\mathcal{D}_{1/3}(x)\right] + \right.$$
$$\left. + \frac{1-\zeta\zeta'}{2}\left[9x^2\mathcal{L}_{2/3}(x) + 33x\mathcal{L}_{1/3}(x) - 9\pi x - 9\pi\sqrt{3} + 18\mathcal{D}_{1/3}(x)\right]\right\}. \qquad (3.82)$$

If we sum the radiation intensity over the final spin states but do not average over the initial ones, we have

$$\begin{aligned} W_l &= q_1 W_0 \left\{[(117-99l\zeta)x^2 + (64-44l\zeta)]\mathcal{L}_{2/3}(x) + \right. \\ &+ [27x^3 + (282-297l\zeta)x]\mathcal{L}_{1/3}(x) - 9\pi\sqrt{3}x^2 - \\ &- \left.(108-99l\zeta)\pi x - (76-77l\zeta)\pi\sqrt{3}\right\}. \end{aligned} \qquad (3.83)$$

The transitions occurring due to the spin-flip give the following contribution to the radiation

$$\begin{aligned} W^{(2)} &= 3q_1 W_0 \left\{72x^2\mathcal{L}_{2/3}(x) + (9x^3+139x)\mathcal{L}_{1/3}(x) - \right. \\ &- \left. 3\pi\sqrt{3}x^2 - 69\pi x - 33\pi\sqrt{3} + 54\mathcal{D}_{1/3}(x)\right\} \end{aligned} \qquad (3.84)$$

and the total probability of a transition for the unpolarized electron is equal to

$$w = \frac{2x}{T_1}\left\{(9x^2+28)\mathcal{L}_{2/3}(x) + 9x\mathcal{L}_{1/3}(x) - 9\pi x - \pi\sqrt{3}\right\}, \qquad (3.85)$$

where

$$q_1 = \frac{\sqrt{3}}{288\pi}.$$

Finally, we can find the asymptotics of the expressions for the case of small ξ. Using Eq. (3.74) and rejecting the terms containing $e^{-1/\xi}$ being small in the case of $\xi \ll 1$, we get

$$W_2 = 27q_1 W^{\text{cl}} \sum_{n=0}(n+1)(n^2+5n+14)a_n(-\xi)^n, \qquad (3.86)$$

$$W_3 = 27q_1 W^{\text{cl}} \sum_{n=0}(n+1)(n^2-n+2)a_n(-\xi)^n, \qquad (3.87)$$

$$W = 54q_1 W^{\text{cl}} \sum_{n=0}(n+1)(n^2+2n+8)a_n(-\xi)^n, \qquad (3.88)$$

$$w = \frac{3}{T_1}\sum_{n=0}\left\{\frac{1+\zeta\zeta'}{2}\left(3n^2+11n+22-\frac{2(3n+1)}{n+1}\right)+\right.$$
$$\left.+\ \frac{1-\zeta\zeta'}{2}\frac{n(n-1)(3n+1)}{n+1}\right\}a_n(-\xi)^n.$$

The transition probability averaged over spin states is equal to

$$w = \frac{6}{T_1}\sum_{n=0}(3n^2+3n+10)a_{n-1}(-\xi)^n.$$

The transitions accompanied by a spin-flip are described by the asymptotic expansion of the form

$$W^{(2)} = 27q_1W^{\rm cl}\sum_{n=0}\frac{(3n+4)(n+1)(n-1)n}{n+2}a_n(-\xi)^n,$$

where we designate

$$W^{\rm cl} = W_0\xi^2 = \frac{2}{3}\frac{ce^2}{R^2}\left(\frac{E}{mc^2}\right)^4.$$

In particular, the first three terms give

$$W_2 = W^{\rm cl}\left\{\frac{1+\zeta\zeta'}{2}(\frac{7}{8}-\frac{25\sqrt{3}}{12}\xi+\frac{163}{9}\xi^2)+\frac{1-\zeta\zeta'}{2}\frac{5}{9}\xi^2\right\},$$

$$W_3 = W^{\rm cl}\left\{\frac{1+\zeta\zeta'}{2}(\frac{1}{8}-\frac{5\sqrt{3}}{24}\xi+\frac{19}{9}\xi^2)+\frac{1-\zeta\zeta'}{2}\frac{5}{9}\xi^2\right\}, \tag{3.89}$$

$$W = W^{\rm cl}\left(1-\frac{55\sqrt{3}}{24}\xi+\frac{64}{3}\xi^2\right). \tag{3.90}$$

The result of Eq. (3.90), has first been obtained in Ref.[23] (see also Refs[4,17,24,25]). In contrast to the results obtained in Refs[24,25,23] we have also terms containing $\exp(-1/\xi)$ which we rejected as small. In particular, the contribution of the first one to W is

$$W^{\rm cl}\frac{e^{-1/\xi}}{8\xi^4}$$

It has a maximum at $\xi = 1/4$, reaching the value of

$$W^{\rm cl}\frac{1}{8}\left(\frac{4}{e}\right)^4 \approx 0.59W^{\rm cl}$$

The asymptotic formulas, however, are reliable only in the region of small ξ (they give an error in the third sign even when $\xi = 0.1$), and in this region the above term is really small. For example, it equals $\approx W^{\rm cl}\cdot 10^{-35}$ at $\xi = 0.01$.

For W_l the asymptotics has the form

$$W_l = 27q_1W^{\rm cl}\sum_{n=0}(n+1)(n^2+2n+8+\frac{11}{3}nl\zeta)a_n(-\xi)^n,$$

$$W_l = W^{\rm cl}\left[\frac{1}{2}-\frac{55\sqrt{3}}{144}(3+l\zeta)\xi+\frac{4}{9}(24+11l\zeta)\xi^2\right].$$

In the opposite limit of $\xi \gg 1$ due to Eq. (3.75) we have[17,26]

$$\begin{aligned} W_2 &= \left(\frac{1+\zeta\zeta'}{2}\frac{41}{64}+\frac{1-\zeta\zeta'}{2}\frac{9}{32q_2}\frac{ln\xi}{\xi^{2/3}}\right)W^{\text{total}}, \\ W_3 &= \left(\frac{1+\zeta\zeta'}{2}\frac{23}{64}+\frac{1-\zeta\zeta'}{2}\frac{9}{32q_2}\frac{ln\xi}{\xi^{2/3}}\right)W^{\text{total}}, \end{aligned}$$

where we denote

$$\begin{aligned} W^{\text{total}} &= q_2 W_0 \xi^{2/3}, \\ q_2 &= 2^{2/3}\Gamma(\frac{2}{3})\frac{4}{9}. \end{aligned}$$

The circular polarization has the form

$$W_l = \frac{W^{\text{total}}}{2}\left[\frac{1+\zeta\zeta'}{2}\left(1+\frac{11}{16}l\zeta\right)+\frac{1-\zeta\zeta'}{2}\left(1+l\zeta\right)\frac{9}{32q_2}\frac{ln\xi}{\xi^{2/3}}\right]. \tag{3.91}$$

The results of the numerical calculations for intermediate values are represented in Fig. 3.1–3.9. Fig. 3.2, 3.4, 3.7 give the functions

$$\frac{W}{W^{\text{total}}}, \quad \frac{W_2}{W^{\text{cl}}}, \quad \frac{W_3}{W^{\text{cl}}}$$

for ξ between 0 and 1 and also

$$\frac{W}{W^{\text{total}}}, \quad \frac{W_2}{W^{\text{total}}}, \quad \frac{W_3}{W^{\text{total}}}$$

for $0.01 \le \xi \le 1$, and Fig. 3.3, 3.6 and 3.8 give

$$\frac{W}{W^{\text{total}}}, \quad \frac{W_2}{W^{\text{total}}}, \quad \frac{W_3}{W^{\text{total}}}$$

for $0 \le \xi \le 0.01$. In the region of $\xi < 1$ we notice that all quantities are less than their limiting classical values and decrease rather rapidly with ξ. In the region of $\xi > 1$ (i.e. $x < 1$) the corresponding quantities tend from below to their limits when $\xi \to \infty$. In Fig. 3.9 shown are the circular polarization components as functions of ξ, $l\xi = 1$ corresponding to W_1 and $l\xi = -1$ to W_{-1}. It is worth noting that in the case $\xi \gg 1$, the value W_{-1} tends to its limit $(5/32)W^{\text{total}}$ from above, as it can be seen from Eq. (3.83). However, the maximum of this function falls on the region of $\xi > 1000$ (i.e. $x < 0.001$) and we have not succeeded in calculating it numerically. In connection with the circular polarization we mark the following two significant facts. Firstly, there is a correlation between the spin of the emitted photon and that of the electron in the region of high electron energies (see Eq. (3.91), where the correlation term is $(11/16)l\xi$). Secondly, the difference between the components of circular polarization increases with electron energy. Also, it is significant that averaging over the electron

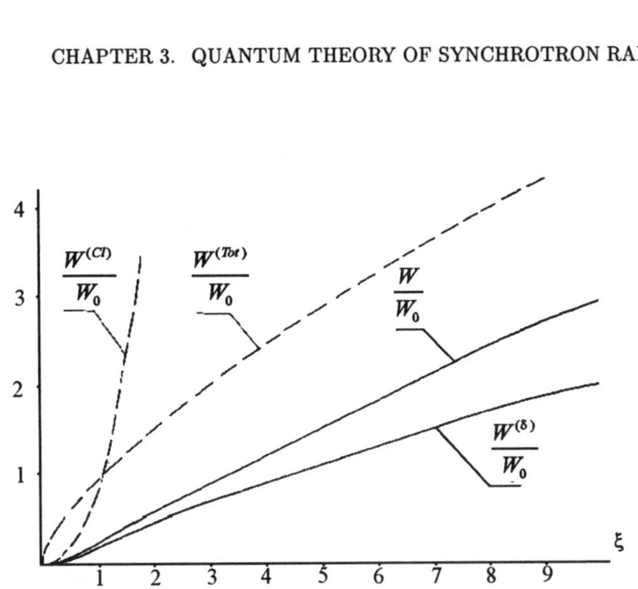

Fig. 3.1. Total radiation intensity of the unpolarized electron and the boson as a function of ξ, and also the functions $W^{\rm cl}(\xi)$ and $W^{\rm total}(\xi)$ for comparison.

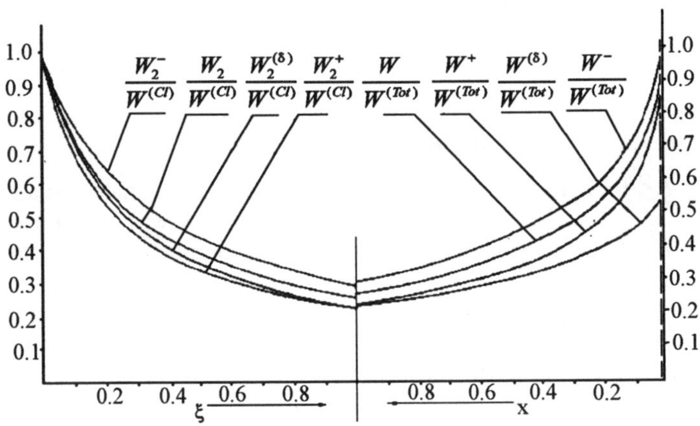

Fig. 3.2. Total radiation of the particles depending on $\xi = 1/x$.

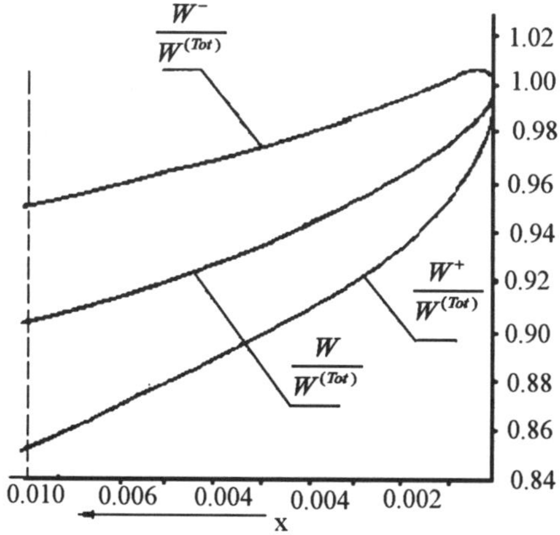

Fig. 3.3. Total radiation of the particles depending on $\xi = 1/x$.

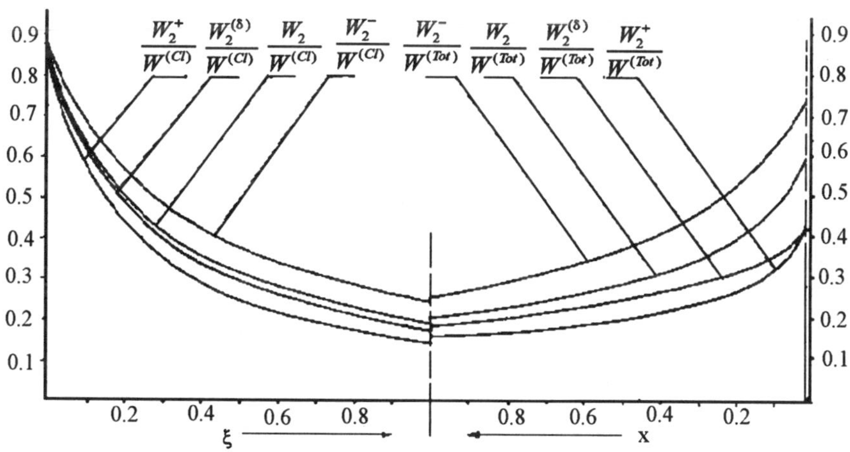

Fig. 3.4. W_2–component of linear polarization versus ξ or x

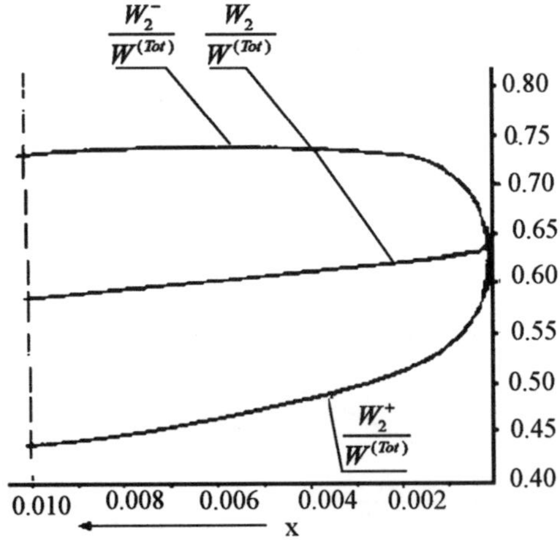

Fig. 3.5. W_2–component of linear polarization at small x

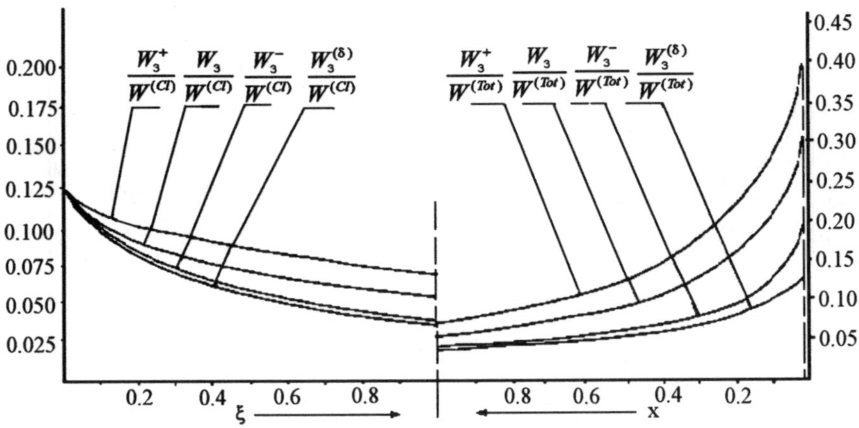

Fig. 3.6. W_3–component of linear polarization versus ξ or x

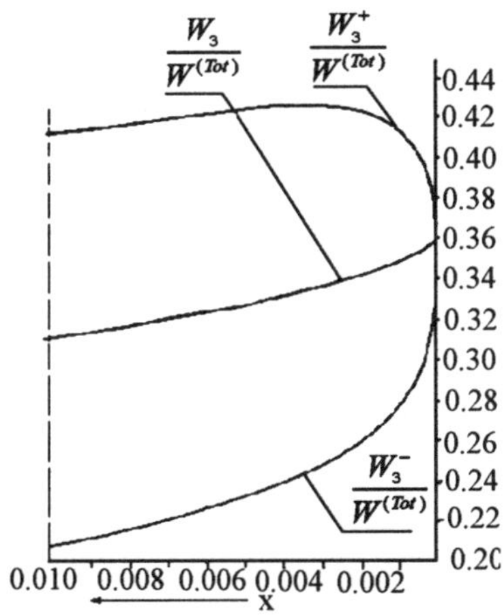

Fig. 3.7. W_3–component of linear polarization at small x

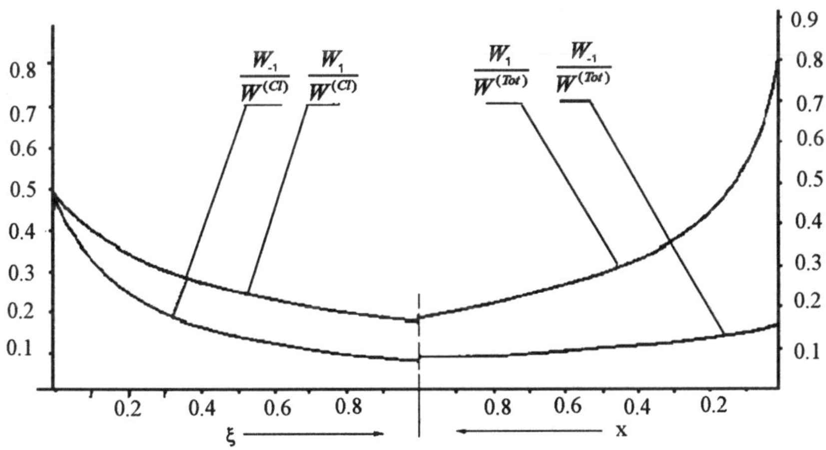

Fig. 3.8. Circular polarization components with a longitudinally oriented spin ($W_1 : \ l\zeta = 1$, $W_{-1} : \ l\zeta = -1$).

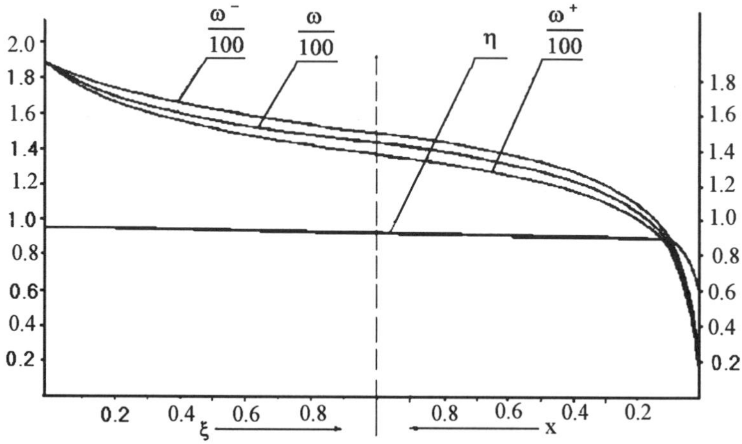

Fig. 3.9. Total probabilities of transitions with the radiation of the unpolarized and transversely polarized electron ($\omega = wT_1$), of the electron polarized along the field ($\omega^+ = w^+T_1$, $\zeta = \pm 1$) and against the field ($\omega^- = w^-T_1$, $\zeta = -1$).

spin results in the disappearance of the circular polarization of radiation in accordance with the general rule, namely, the radiation with the circular polarization can be emitted by particles with an oriented spin. In conclusion, we note that the initial orientation of spin against or along the field does not affect on the total radiation and the linear components of polarization.

We will undertake a more detailed analysis of the radiation properties after consideration of the other electron spin orientations and of the boson radiation, so as to have the possibility of comparing different cases and distinguishing general rules and peculiar features in each case.

3.11. Radiation of the Electron with a Spin Oriented to the Magnetic Field

The electron transverse polarization may be described either by the polarization vector operator or by the polarization tensor operator. However, the vector operator is inconvenient in definite calculations, since in the relativistic case, it changes into an operator $\widehat{T} \approx \widehat{P}_3$ containing no spin matrix. Meanwhile when we have to deal with the polarization tensor there arises no disadvantage of the kind, hence, this operator should be used to describe transverse polarization.

By the same method we obtain spectral-angular distribution in the following form

$$\begin{aligned}
\Phi_2^{(1)} &= \{(2+\xi y)\varepsilon K_{2/3}(z_0) - \zeta\sqrt{\varepsilon_0\varepsilon}\,\xi y K_{1/3}(z_0)\}^2, \\
\Phi_3^{(1)} &= (2+\xi y)^2 \varepsilon \cos^2\theta K_{1/3}^2(z_0),
\end{aligned}$$

$$\Phi_2^{(2)} = \xi^2 y^2 \varepsilon \cos^2\theta K_{1/3}^2(z_0),$$
$$\Phi_3^{(2)} = \xi^2 y^2 (\varepsilon K_{2/3}(z_0) + \zeta\sqrt{\varepsilon_0\varepsilon}\, K_{1/3}(z_0))^2.$$

The circular polarization is absent. Integrating over angles we have[17]

$$\begin{aligned}
F_2^{(1)} &= 3(2+\xi y)^2 K_{2/3}(y) - 4\zeta\xi y(2+\xi y)^2 K_{1/3}(y) + \\
&+ (\xi^2 y^2 - 4\xi y - 4)\int_y^\infty K_{1/3}(x)dx, \\
F_3^{(1)} &= (2+\xi y)^2 (K_{2/3}(y) - \int_y^\infty K_{1/3}(x)dx), \\
F_2^{(2)} &= \xi^2 y^2 (K_{2/3}(y) - \int_y^\infty K_{1/3}(x)dx), \\
F_3^{(2)} &= \xi^2 y^2 (K_{2/3}(y) + 4\zeta K_{1/3}(y) + \int_y^\infty K_{1/3}(x)dx).
\end{aligned}$$

Integrating over spectrum and presenting the results in the form

$$W_i = \frac{1+\zeta\zeta'}{2}\, W_i^{(1)} + \frac{1-\zeta\zeta'}{2}\, W_i^{(2)}$$

we obtain for our case

$$\begin{aligned}
W_2^{(1)} &= \frac{q_1}{2} W_0 \Big\{(54\zeta)x^2 + 81x^2 - 372\zeta x + 150)\mathcal{L}_{2/3}(x) + \\
&+ (54x^3 + 90\zeta x + 456x - 200\zeta)\mathcal{L}_{1/3}(x) - \\
&- 18\pi(\sqrt{3} + 3\zeta)x^2 - (7 + 2\sqrt{3}\zeta)9\pi x - \\
&- 6\pi(25\sqrt{3} - 62\zeta) + 81\mathcal{D}_{1/3}(x)\Big\}, \\
W_3^{(1)} &= \frac{q_1}{2} W_0 \Big\{(-9x^2 + 50)\mathcal{L}_{2/3}(x) + \\
&+ 54x\mathcal{L}_{1/3}(x) + 9\pi x - 20\sqrt{3}\pi - 81\mathcal{D}_{1/3}(x)\Big\}, \\
W_2^{(2)} &= \frac{q_1}{2} W_0 \Big\{(-9x^2 + 14)\mathcal{L}_{2/3}(x) - 54x\mathcal{L}_{1/3}(x) + \\
&+ 16\sqrt{3}\pi - 81\mathcal{D}_{1/3}(x)\Big\}, \\
W_3^{(2)} &= \frac{q_1}{2} W_0 \Big\{(54\zeta)x^3 + 405x^2 + 600\zeta x + 42)\mathcal{L}_{2/3}(x) + \\
&+ (54x^3 + 414\zeta x + 672x + 160\zeta)\mathcal{L}_{1/3}(x) - \\
&- 18\pi(\sqrt{3} + 3\zeta)x^2 - 9\pi(43 + 14\sqrt{3}\zeta)x - \\
&- 6\pi(25\sqrt{3} + 82\zeta) + 81\mathcal{D}_{1/3}(x)\Big\}.
\end{aligned}$$

For transition probabilities we get the following spectral distribution

$$\begin{aligned}\psi_1 &= (2+\xi y)^2 K_{2/3}(y) - 2(1+\zeta\xi y)\int_y^\infty K_{1/3}(x)dx - \zeta\xi y(2+\xi y)K_{1/3}(y),\\ \psi_2 &= \xi^2 y^2 (K_{2/3}(y) - \zeta K_{1/3}(y)).\end{aligned}$$

The integration over spectrum gives

$$\begin{aligned}w &= \frac{x}{T_1}\frac{1+\zeta\zeta'}{2}\Big[(9x^2-9\zeta x+52)\mathcal{L}_{2/3}(x) + (9\zeta x^2 - 9x - 14\zeta)\mathcal{L}_{1/3}(x) -\\ &- 3\pi(3+\zeta\sqrt{3})x + \pi(5\sqrt{3}+12\zeta)\Big] + \frac{x}{T_1}\frac{1-\zeta\zeta'}{2}\Big[(9x^2+27\zeta x+4)\mathcal{L}_{2/3}(x)\\ &+ (9\zeta x^2+27x+10\zeta)\mathcal{L}_{1/3}(x) - 3\pi(3+\zeta\sqrt{3})x - \pi(7\sqrt{3}+24\zeta)\Big]. \qquad (3.92)\end{aligned}$$

The asymptotics for small ξ has the form

$$\begin{aligned}W_2^{(1)} &= \frac{9q_1}{2}W^{\mathrm{cl}}\sum_{n=0}\Big\{\frac{n+1}{n-1}\left[(n+3)(n+8)(3n+7)+n(n-1)(3n+4)\right]a_n +\\ &+ \zeta 2n(n+1)(n+5)(3n+1)a_{n-1}\Big\}(-\xi)^n,\\ W_3^{(1)} &= \frac{9q_1}{2}W^{\mathrm{cl}}\sum_{n=0}\frac{(n+1)(n+3)(n+8)}{(n+2)}a_n(-\xi)^n,\\ W_2^{(2)} &= \frac{9q_1}{2}W^{\mathrm{cl}}\sum_{n=0}\frac{n(n-1)(n+1)}{(n+2)}a_n(-\xi)^n,\\ W_3^{(2)} &= \frac{9q_1}{2}W^{\mathrm{cl}}\sum_{n=0}\frac{n(n-1)(n+1)}{(n+2)}\left[(6n+11)a_n + 2\zeta(n+2)(3n+1)a_{n-1}\right](-\xi)^n,\end{aligned}$$

$$\begin{aligned}w &= \frac{3}{T_1}\sum_{n=0}\Big\{\frac{1+\zeta\zeta'}{2}\left[(3(n+1)(n+2)+14)a_n + \zeta(n+3)(3n-2)na_{n-2}\right] +\\ &+ \frac{1-\zeta\zeta'}{2}n(n-1)\left[3a_{n-1}+\zeta(3n-2)a_{n-2}\right]\Big\}(-\xi)^n. \qquad (3.93)\end{aligned}$$

The first terms of expansion give

$$\begin{aligned}W_2 &= W^{\mathrm{cl}}\Big\{\frac{1+\zeta\zeta'}{2}\left[\frac{7}{8} - \left(\frac{25\sqrt{3}}{12}+\zeta\right)\xi + \left(\frac{335}{18}+\frac{245\sqrt{3}}{48}\zeta\right)\xi^2\right] +\\ &+ \frac{1-\zeta\zeta'}{2}\frac{1}{18}\xi^2\Big\},\\ W_3 &= W^{\mathrm{cl}}\Big\{\frac{1+\zeta\zeta'}{2}\left[\frac{1}{8} - \frac{5\sqrt{3}}{24}\xi + \frac{25}{18}\xi^2\right] + \frac{1-\zeta\zeta'}{2}\left(\frac{23}{18}+\frac{35\sqrt{3}}{48}\zeta\right)\xi^2\Big\} \qquad (3.94)\end{aligned}$$

In the opposite case of $\xi \gg 1$ we have limiting expansions

$$\begin{aligned} W_2^{(1)} &= \frac{75}{128} W^{\rm total}, \quad W_3^{(1)} = \frac{25}{128} W^{\rm total}, \\ W_2^{(2)} &= \frac{7}{128} W^{\rm total}, \quad W_3^{(2)} = \frac{21}{128} W^{\rm total}. \end{aligned}$$

In Figs 3.1–3.9 the results of numerical calculations for intermediate values are represented, with W_i^+ corresponding to $\xi = +1$ and W_i^- to $\xi = -1$. Note that W^- (the case of $\xi = -1$) in Fig. 3.2, 3.3 is the most intensive in comparison with all other functions, which means that the radiation by the electron with a spin oriented against the field is the most intensive. As it follows from Figs 3.3, 3.6 and 3.8, the functions W^-, W_2^-, W_3^- in the region of $\xi \gg 1$ approach their limiting values from the above.

The absence of the circular polarization of radiation is of significance. It can be caused by the fact that the spin of the emitted photon is directed nearly along an instantaneous velocity of the electron, whereas the electron spin is orthogonal to it , thereby the correlation is absent.

The peculiarity of this electron polarization is that both the radiation and the probability of transitions depend explicitly on the spin, this dependence appearing in the case of $\xi \ll 1$ owing to the term proportional to $\hbar$ (see Eq. (3.94). In this connection, the transition with a spin-flip can have the directional character already established in Refs[27,28,29] It is interesting to note that if we introduce noncovariant transverse polarization of the electron spin, for example, putting $A = \frac{1+\zeta}{2}$ and $B = \frac{1-\zeta}{2}$ in Eq. (3.27), we arrive, nevertheless at the same result. In particular, in Ref.[27] the noncovariantly defined spin was used and only in further works[28,29] it was shown that the choice of covariant electron transverse polarization or noncovariant one does not matter. The dependence of the probability of the spin flip (see Eq. (3.92)) on the initial spin orientation causes the electron spin to be oriented preferably against the field due to the radiation. Thus, for a sufficiently long time the majority of electrons of an electron beam rotating in a magnetic field acquires spins oriented against the field. The function $\eta(\xi)$ represented in Fig. 3.10, shows the part of electrons with a spin directed against the field. As we see, the percentage of such electrons is very high, slowly changing with ξ.

Only in the case of very large ξ, part of the electrons with $\zeta = 1$ becomes nearly equal to the part with $\zeta = -1$, i.e. $\eta \to \frac{1}{2}$ with $\xi \to \infty$.

Finally note that in this case, averaging over initial spin orientations leads to the same results as in the case of the longitudinal polarization, namely, the averaged radiation is described by Eqs (3.76)–(3.78), (3.80), (3.81) and the probability by Eqs (3.79), (3.82).

3.12. Radiation of a Spinless Particle in a Magnetic Field

The radiation of a spinless particle (boson) can be studied in the same manner as that of an electron. One should replace $\langle \boldsymbol{\alpha} \rangle$ by $\langle \boldsymbol{P} \rangle$ in the expressions of Eqs (3.45)–

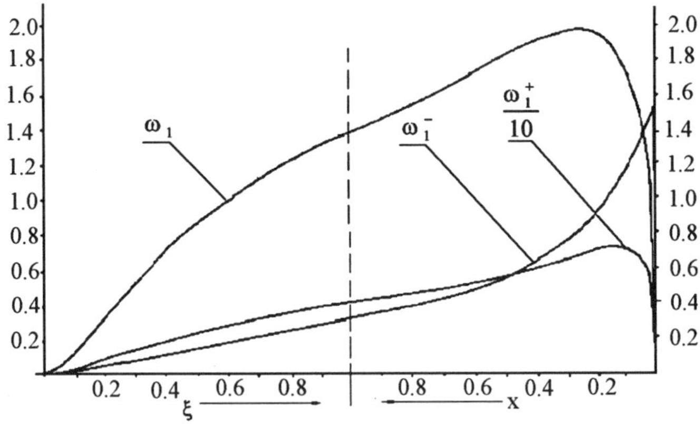

Fig. 3.10. Analogous probabilities as in Fig. 3.9 but only for the transition accompanied by the change of spin orientation, ω_1 corresponding to the longitudinally polarized electron and $\omega_1^{\pm}$ to two transverse polarizations. The function $\eta(\xi)$ shows the part of electrons with a spin directed against the field.

(3.47) and use the resulting formulas together with the formulas of Eq. (3.49) for the matrix elements. Then we can obtain spectral-angular distribution in the form[18]

$$W_i^{\delta} = \frac{27}{16\pi^2} W_0 \xi^2 \varepsilon_0^{-5/2} \int_0^{\pi} \int_0^{\infty} \frac{y^2 \sin\theta dy d\theta}{(1+\xi y)^3} \Phi_i^{(\delta)}(\xi, y, \theta),$$

$$\Phi_2^{(\delta)} = \varepsilon^2 K_{2/3}^2(z_0), \quad \Phi_3^{(\delta)} = \varepsilon \cos^2\theta K_{1/3}^2(z_0).$$

The integration over angles results in the expressions

$$W_i^{\delta} = \frac{9\sqrt{3}}{16\pi} W_0 \xi^2 \int_0^{\infty} \frac{y dy}{(1+\xi y)^3} F_i^{(\delta)}(\xi, y),$$

$$\begin{aligned} F_2^{(\delta)} &= 3K_{2/3}(y) - \int_y^{\infty} K_{1/3}(x) dx, \\ F_3^{(\delta)} &= K_{2/3}(y) - \int_y^{\infty} K_{1/3}(x) dx. \end{aligned}$$

There is no circular polarization. The integrals over spectrum lead to the relations

$$\begin{aligned} W_2^{(\delta)} &= 54 q_1 W_0 \left\{ (1-3x^2)\mathcal{L}_{2/3}(x) - 2x\mathcal{L}_{1/3}(x) + 3\pi x \right\}, \\ W_3^{(\delta)} &= 18 q_1 W_0 \left\{ \mathcal{L}_{2/3}(x) + 3x\mathcal{L}_{1/3}(x) - \pi\sqrt{3} \right\}. \end{aligned}$$

In the case of small ξ, we have asymptotics (compare with Eqs ((3.86) and (3.87))

$$\begin{aligned} W_2^{(\delta)} &= 54q_1 W^{\text{cl}} \sum_{n=0} (n+1)(3n+7)a_n(-\xi)^n, \\ W_3^{(\delta)} &= 54q_1 W^{\text{cl}} \sum_{n=0} (n+1)a_n(-\xi)^n. \end{aligned}$$

The first terms of expansion give

$$\begin{aligned} W_2 &= W^{\text{cl}} \left\{ \frac{7}{8} - \frac{25\sqrt{3}}{12}\xi + \frac{162 \pm 6}{9}\xi^2 \right\}, \\ W_3 &= W^{\text{cl}} \left\{ \frac{1}{8} - \frac{5\sqrt{3}}{24}\xi + \frac{6 \pm 2}{3}\xi^2 \right\}, \\ W &= W^{\text{cl}} \left\{ 1 - \frac{55\sqrt{3}}{24}\xi + \frac{60 \pm 4}{3}\xi^2 \right\}. \end{aligned}$$

Here the sign "+" is related to electrons and the sign "-" to bosons. So we see that in the case of $\xi \ll 1$, a boson radiates somewhat weaker than an electron and this difference appears only in terms of the order of $\hbar^2$, both polarization components decreasing by the same quantity. Moreover, it is easy to see that the relation

$$W_2 - W_2^{(\delta)} = W_3 - W_3^{(\delta)}$$

holds, that means both the boson polarization components equally differ from the electron ones.

In the opposite limiting case of $\xi \gg 1$ we have

$$W_2^{(\delta)} = \frac{27}{64} W^{\text{total}}, \quad W_3^{(\delta)} = \frac{9}{64} W^{\text{total}}, \quad W^{(\delta)} = \frac{9}{16} W^{\text{total}}.$$

We conclude that a boson always emits somewhat less than an electron, this difference being essential in the case of $\xi \gg 1$ (see also Refs[21,24]) The results of numerical calculations for intermediate values are represented in Figs 3.1–3.9).

Note here one significant fact, namely, the difference between the radiation of an unpolarized electron and a boson occurs in the case of $\xi \ll 1$ in terms of the order of $\hbar^2$, which has been earlier marked by Matveyev (see Refs[21,28]). If the electron is polarized this difference appears in terms of the order of $\hbar$ (see Eq. (3.94).

3.13. Characteristic Features of Electron Radiation in a Magnetic Field

Here we analyze represented results of numerical calculations.

Let us begin with the analysis of the characteristic features of electron and boson radiation (see Figs 3.1 – 3.9). The components of the radiation are designated by W_i for an unpolarized electron and for an electron with a longitudinally oriented spin. W_i^+ corresponds to a transversally polarized electron with $\zeta = +1$, W_i^- to the same electron with $\zeta = -1$ and $W_i^{(b)}$ to a boson. First of all, note one significant common feature of all graphs, namely, in the region of $\xi < 1$ both the radiation of an electron with an arbitrary spin orientation and that of a boson are less than their limiting classical values. In Fig. 3.1 the total radiation intensity of the unpolarized electron and the boson is represented as a function of ξ, and also the functions $W^{\rm cl}(\xi)$ and $W^{\rm total}(\xi)$ for comparison. Both in the case of $\xi < 1$ and of $\xi > 1$ the intermediate values differ essentially from the limiting values $W^{\rm cl}$ ($\xi \to 0$) and $W^{\rm total}$ ($\xi \to \infty$) (for an electron). It is significant that the electron radiation corresponds to the classical model to a greater extent than the boson one.

Now look at Figs 3.2 and 3.3 representing the total radiation of the particles depending on $\xi = 1/x$. We see that the electron with a spin opposite to the field emits most intensively, whereas the least radiation intensity corresponds to the electron with a spin oriented along the field when $\xi < 1$ and to boson when $\xi > 1$. Note again that W^- in Fig. 3.3 approaches its limiting value from the above when $\xi \to \infty$, i.e. the radiation is stronger than $W^{\rm total}$ beginning approximately with $\xi \geq 700$.

Figs 3.3 – 3.8 show graphical representation of the polarization components. It is interesting to note that the component W_2 is the least for the electron in the case of the spin orientation along the field and the greatest in the case of the spin orientation against the field. On the contrary, the component W_3 is the least for the electron with a spin against the field and the greatest for the electron with a spin along the field. Their behavior in the region of small x (i.e. $\xi \to \infty$) is analogous to that of the total intensity (see Figs 3.6 and 3.8).

In Fig. 3.8 we have circular polarization components with a longitudinally oriented spin ($W_1 : l\zeta = 1$, $W_{-1} : l\zeta = -1$). We see that the maximal difference between the components appears in the region of large ξ (of small x), W_{-1} reaching its limiting value in the region of small x from the above. This fact can be easily found with the help of Eq. (3.83).

In Fig. 3.9 we present the total probabilities of transitions with the radiation of the unpolarized and transversely polarized electron ($\omega = wT_1$, Eq. (3.14), of the electron polarized along the field ($\omega^+ = w^+T_1$, $\zeta = \pm 1$, Eq. (3.92)) and against the field ($\omega^- = w^-T_1$, $\zeta = -1$, Eq. (3.92)). As we see, the electron with a spin against the field emits more photons per unit time than the unpolarized electron or the electron with a spin along the field.

Fig. 3.10 shows the analogous probabilities but only for the transition accompanied by the change of spin orientation, ω_1 corresponding to the longitudinally polarized electron and $\omega_1^\pm$ to two transverse polarizations. As we see from these graphs, the spin oriented along the field is the most unstable one and, on the contrary, the orientation opposite to the field is the most stable (when $\xi < 10$). So, as it has been noted above, spontaneous polarization of an electron bunch rotating in the magnetic field for a sufficiently long time is possible, the percentage of the electrons with a spin against the field being remarkably high[29] and weakly changing with ξ (see the function η in Fig. 3.9).

Note one interesting peculiarity, namely, each type of the electron spin polarization has the point of maximal instability (see Fig. 3.10), although all these points are in the region of large ξ ($\xi > 4$). It should also be noted that the total probability of radiation proves to be the largest for an electron with a spin oriented against the field. However, spin-flip transitions for such an electron are the least probable compared with the other spin orientations.

Fig. 3.11, which is of great interest, demonstrates the functions characterizing the relative contribution to the radiation intensity by transitions with a spin flip.

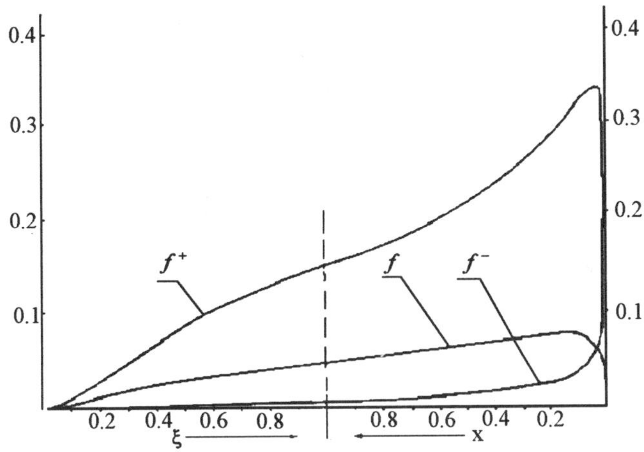

Fig. 3.11. Functions characterizing the relative contribution to the radiation intensity by transitions with a spin flip.

For example, the function f is $W^{(2)}/W$, where $W^{(2)}$ (see Eq. (3.84) is a part of the radiation of longitudinally polarized electron related with the transitions accompanied by a spin flip. The functions $f^{\pm}$ have the same meaning for a transversally polarized electron.

We see that in the case of longitudinal polarization, the contributions of transitions with a spin flip are maximum, even though the maximal contribution does not reach 8 p.c., whereas in the case of the spin oriented along the field ($\zeta = +1$), the transitions with a spin flip give the maximal contribution of nearly 34 p.c. of radiation intensity. In the case of a spin oriented opposite to the field, the maximal value of 21.9 p.c. is achieved only in the limit of $\xi \to \infty$.

Now let us consider a degree of radiation polarization as a function of the parameter ξ, the polarization degree being determined by the formula

$$P = (W_\alpha - W_\beta)(W_\alpha + W_\beta)^{-1}. \tag{3.95}$$

The dependence of radiation polarization on ξ is shown in Fig. 3.13. Here P is the degree of the linear polarization of radiation of the unpolarized electron and of the electron with a longitudinally oriented spin, $P^{(b)}$ is the degree of the linear polarization of boson radiation, P_c is the degree of the circular polarization of radiation of the longitudinally polarized electron, $P^{\pm}$ is the polarization of radiation of the transversally polarized electron ($\zeta = \pm 1$, respectively). We see that the degree of the linear polarization of radiation of arbitrary polarized electron uniformly decreases with ξ until $\xi = 33$. In the case of an electron with a spin along the field there appears an essential peculiarity, i.e. the minimum of the linear polarization of radiation. The degree of the circular polarization of radiation P_c of the longitudinally polarized electron uniformly increases with ξ. The strongest polarization occurs in the case of the electron with a spin directed against the field ($\zeta = -1$).

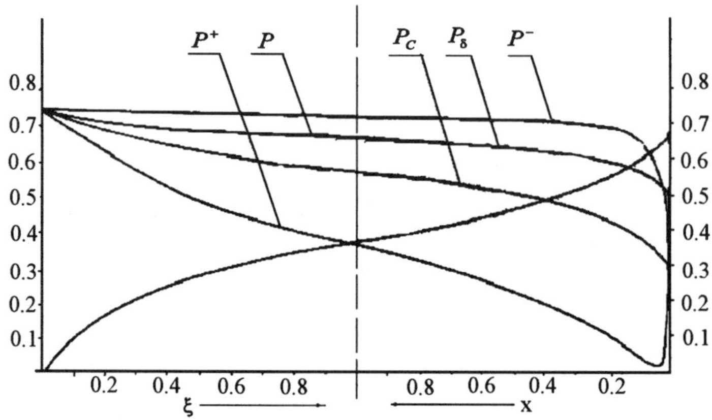

Fig. 3.12.

Boson radiation is considerably polarized. It is interesting to note that $P^{(b)} = \frac{1}{2}$ in the limit of $\xi \to \infty$, which coincides with the polarization degree of radiation of a nonrelativistic particle.

We remind that in the region of $\xi \ll 1$ the difference between the radiations of the unpolarized electron and the boson manifests itself only in terms of the order of $\hbar^2$, whereas the difference between boson radiation and radiation of an electron with an oriented spin may appear in terms of the order of $\hbar$.

We see that the study of the polarization properties of the radiation of an electron with an oriented spin is, undoubtedly, of interest due to many fine peculiarities which can be clearly detected in exact numerical calculations.

Now let us determine the wavelength on which the maximum of the electrons radiation falls.

Using the relations of Eqs (3.76) and (3.78) for an unpolarized electron we can

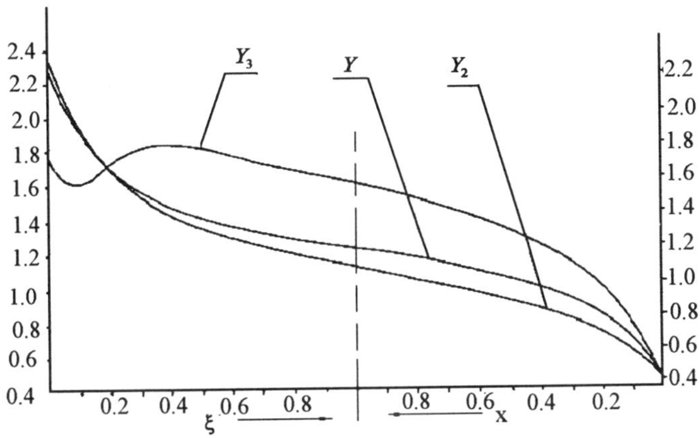

Fig. 3.13. Dependence of radiation polarization on ξ.

find

$$W_i = \frac{9\sqrt{3}}{16\pi} W_0 \xi^4 \int_{\lambda m}^{\infty} d\,\frac{\lambda}{\lambda_0}\, f_i(\xi, y),$$

where we have to put down

$$\begin{aligned} f_2 &= \frac{y^3}{(1+\xi y)^4}\left\{(3+3\xi y+\xi^2y^2)K_{2/3}(y) - (1+\xi y)\int_y^{\infty} K_{1/3}(x)dx\right\}, \\ f_3 &= \frac{y^3}{(1+\xi y)^4}\left\{(1+\xi y+\xi^2y^2)K_{2/3}(y) - (1+\xi y)\int_y^{\infty} K_{1/3}(x)dx\right\}. \end{aligned}$$

for the total intensity $f = f_2 + f_3$ on the condition that

$$y = \frac{\lambda_c}{\lambda - \lambda_m}, \quad \lambda = \frac{2\pi}{\kappa}, \quad \lambda_c = \lambda_0 \xi^{-2}, \quad \lambda_m = \lambda_0 \xi, \quad \lambda_0 = \frac{2\pi\hbar}{mc}\frac{H}{H^*}.$$

The maximum of the intensity in the component f_i is determined from the condition

$$\frac{df_i}{d\lambda} = -\frac{\xi^2 y^2}{\lambda_0}\frac{df_i}{dy} = 0$$

and hence, from the condition

$$\frac{df_i}{dy} = 0.$$

After differentiation we arrive at the following relations

$$\begin{aligned} \psi_2(\xi, y_2) &= \psi_3(\xi, y_3) = \psi(\xi, y) = \\ &= \psi_2(\xi, y) + \psi_3(\xi, y) = 0 \end{aligned}$$

for W_2, W_3 and W, respectively, where we denoted;

$$\begin{aligned} \psi_2(\xi, y) &= \Big[(21+15\xi y+7\xi^2y^2+\xi^3y^3)K_{2/3}(y) - (6+12\xi y+9\xi^2y^2+ \\ &+ 3\xi^3y^3)yK_{1/3}(y) - 9(1+\xi y)\int_y^{\infty} K_{1/3}(x)dx\Big], \\ \psi_3(\xi, y) &= (7+5\xi y+11\xi^2y^2+\xi^3y^3)K_{2/3}(y) - 9(1+\xi y)\int_y^{\infty} K_{1/3}(x)dx - \\ &- 3\xi^2y^2(1+\xi y)yK_{1/3}(y). \end{aligned}$$

These equations have been solved numerically yielding the functions $y_i = y_i(\xi)$. The results are illustrated in Fig. 3.12. The radiational maximum falls either on $\lambda \sim \lambda_c$ when $\xi < 1$ or on $\lambda \sim \lambda_m$ when $\xi > 1$. In both cases it falls on small wavelengths (high frequencies) in accordance with the assumptions made above. The spectral properties of the radiation have been investigated experimentally[30] for the case of $\xi \ll 1$ displaying good agreement between theory and experiment.

Extensive information on the results of numerical calculations of the characteristics of synchrotron radiation can be found by V.G. Bagrov, O.F. Dorofeev, Yu.N. Kargin, G.F. Kopytov, G.K. Razina[31] ["The numerical analysis of spectral distribution of the synchrotron radiation. (Quantum theory)." Deposited in VINITI 12.11.87, N8002-V87, 98 p.].

See also the list of references.

3.14. Radiative Self-Polarization of an Electron Spin

Since the probability of transitions with a spin flip for transversally polarized electrons depends on the initial spin orientation, the stability of spin states turns out to be different, and the predominant (i.e. the most stable) orientation of an electron spin will arise in an initially unpolarized electron beam due to radiation.

The possibility of self-polarization of the electron beam has been first pointed out in I.M.Ternov's dissertation,[32] the first exact quantitative predictions were made in Refs[28,29] and at present we have extensive literature on this subject (see Ref.[33] and the references cited there).

The dynamics of radiation self-polarization can be investigated in the following way. Acts of radiation accompanied by a spin flip occur rather seldom, so they can be considered independent. Let n_1 be a number of electrons with a given spin orientation (say along the field) and n_2 a number of electrons with the opposite spin orientation. Let w_{12} be the probability of transition from state 1 to state 2 per unit time and w_{21} the probability of the reverse event. The dynamics (i.e. the dependence of n_1 and n_2 on time) of statistically independent transitions is determined by the following system of equations

$$\dot{n}_1 = -w_{12}n_1 + w_{21}n_2, \quad \dot{n}_2 = -w_{21}n_2 + w_{12}n_1$$

having the first obvious integral of motion

$$\begin{aligned} n_1(t) + n_2(t) &= n = n_1^0 + n_2^0 = const, \\ n_1^0 &= n_1(0), \quad n_2^0 = n_2(0) \end{aligned}$$

and the following general solution

$$\begin{aligned} n_1(t) &= \tau[nw_{21} + (n_1^0 w_{12} - n_2^0 w_{21}) \exp(-t/\tau)], \\ n_2(t) &= \tau[nw_{21} + (n_2^0 w_{21} - n_1^0 w_{12}) \exp(-t/\tau)], \\ \tau &= (w_{12} + w_{21})^{-1}. \end{aligned} \tag{3.96}$$

As it follows from Eq. (3.96), the equilibrium distribution of electrons in the beam

$$\begin{aligned} \frac{n_1}{n} &= 1 - \eta = \frac{w_{21}}{w_{21} + w - 12}, \\ \frac{n_2}{n} &= \eta = \frac{w_{12}}{w_{21} + w - 12} \end{aligned}$$

is established when $\tau \gg t$, τ being the relaxation time. Assuming that state 1 corresponds to the spin orientation along the field and state 2 to the orientation against the field, we find with the help of Eq. (3.93) for the case of $\xi \ll 1$

$$\tau = \frac{8}{5\sqrt{3}} \frac{\hbar^2}{mce^2} \left(\frac{H^*}{H}\right)^3 \left(\frac{mc^2}{E}\right)^2, \quad \eta = \frac{15 + 8\sqrt{3}}{30} \approx 0,962.$$

When $E = 1$ Gev and $H \approx 10^4$ gauss, the relaxation time $\tau \sim 1$ hour, so we can observe the radiative self-polarization effect in modern electron storage rings. This effect has been experimentally discovered first in 1968 in France (Orsay) and later in many accelerator centers. The experimental peculiarities of the observation were discussed, e.g., in Ref.[33] (see also literature cited there).

The polarization degree η of the electron beam is very high (more than 96 p.c. in the case of $\xi \ll 1$), weakly depending on ξ, as it can be seen in Fig. 3.9.

Finally, one could raise the following question. The description of the transverse spin orientation by means of the tensor spin operator is not the only possible way. Moreover, we can take a linear combination of spin operators which are the constants of motion and try to enlarge the polarization degree η or to diminish the relaxation time τ by a proper selection of the coefficients of such a combination. Such an attempt was made in Ref.[34] It was shown that the polarization degree can be actually enlarged up to 99 p.c. and the relaxation time be diminished nearly by 20 p.c.

In the world outlook aspect, the phenomenon of radiative self-polarization of the electron spin is of considerable interest as an explicit manifestation of the quantum-law-governed effects at the macroscopic level. This phenomenon is not only of theoretical interest but also of great practical importance remaining nowadays an only experimentally elaborated method of obtaining beams of relativistic particles with the oriented spin.

3.15. Radiation of Uncharged Particles with Magnetic and Electric Moments Moving in Uniform Stationary Electromagnetic Fields

Existence of magnetic moments and maybe electric moments of elementary particles (even uncharged ones, e.g. neutrons) has to cause interaction between these particles and an external electromagnetic field and, as a result, lead to a spontaneous electromagnetic radiation. The possibility of such radiation was shown first in Ref.[35] on the basis of the Dirac-Pauli equation and, namely, the radiation of a particle with an anomalous magnetic moment moving in a constant magnetic field was considered. Later the results were generalized for the case of an electric moment of a particle and constant electromagnetic fields (see Refs[36–38]).

We shall consider here the radiation of au neutral particle having the anomalous magnetic moment $\mu_1 = \mu \cos\alpha$ and the electric one $\mu_2 = \mu \sin\alpha$ and moving in stationary uniform electromagnetic fields $\boldsymbol{E}$ and $\boldsymbol{H}$. As is well known, there is a Lorentz system of reference in which the fields are parallel provided that both the invariants of electromagnetic field

$$I_1 = \boldsymbol{E}^2 - \boldsymbol{H}^2 \,, \quad I_2 = (\boldsymbol{E}\boldsymbol{H})$$

are not simultaneously equal to zero, which is assumed be valid. Let us choose a system of coordinates and direct the z-axis along the direction of fields

$$\boldsymbol{H} = (0, 0, Q\cos\eta), \quad \boldsymbol{E} = (0, 0, Q\sin\eta), \quad Q = \sqrt{E^2 + H^2}. \tag{3.97}$$

Let us assume that the wave functions of a particle obey the Dirac-Pauli equations

$$\begin{aligned} i\hbar\partial_t\Psi &= \mathcal{H}\Psi, \\ \mathcal{H} &= c(\boldsymbol{\alpha}\boldsymbol{p}) + \rho_3 mc^2 + c\hbar\mu_1[\rho_3(\boldsymbol{\sigma}\boldsymbol{H}) + \rho_2(\boldsymbol{\sigma}\boldsymbol{E})] + \\ &+ c\hbar\mu_2[\rho_3(\boldsymbol{\sigma}\boldsymbol{E}) - \rho_2(\boldsymbol{\sigma}\boldsymbol{H})]. \end{aligned} \tag{3.98}$$

In the case of the fields of Eq. (3.97), stationary solutions of Eq. (3.98), corresponding to the values of the energy $c\hbar K$, the momentum $\hbar\boldsymbol{k}$ and a spin $\zeta = \pm 1$, have the form

$$\psi = Nb\exp[-icKt + -(\boldsymbol{k}\boldsymbol{r})], \tag{3.99}$$

where the constant spinor b may be written in the form

$$b = \begin{pmatrix} c_1 \\ c_2 e^{i\varphi} \\ c_3 \\ c_4 e^{i\varphi} \end{pmatrix}. \tag{3.100}$$

The coefficients c_k are to be found from the solution of Eq. (3.98)

$$\begin{aligned} c_1 &= \frac{1}{2}\sqrt{\frac{1}{2}(1 + \zeta\frac{k_0}{d}\cos\Phi_0)}\left(\sqrt{1+\beta\cos\theta}e^{i\gamma} + \zeta\sqrt{1-\beta\cos\theta}e^{-i\gamma}\right), \\ c_2 &= -\frac{\zeta}{2}\sqrt{\frac{1}{2}(1 - \zeta\frac{k_0}{d}\cos\Phi_0)}\left(\sqrt{1+\beta\cos\theta}e^{i\delta} - \zeta\sqrt{1-\beta\cos\theta}e^{-i\delta}\right), \\ c_3 &= \frac{1}{2}\sqrt{\frac{1}{2}(1 + \zeta\frac{k_0}{d}\cos\Phi_0)}\left(\sqrt{1+\beta\cos\theta}e^{i\gamma} - \zeta\sqrt{1-\beta\cos\theta}e^{-i\gamma}\right), \\ c_4 &= \frac{\zeta}{2}\sqrt{\frac{1}{2}(1 - \zeta\frac{k_0}{d}\cos\Phi_0)}\left(\sqrt{1+\beta\cos\theta}e^{i\delta} + \zeta\sqrt{1-\beta\cos\theta}e^{-i\delta}\right). \end{aligned} \tag{3.101}$$

Here for the momentum vector we assume

$$\boldsymbol{k} = k(\sin\theta\cos\varphi, \sin\theta\sin\varphi, \cos\theta).$$

Let us introduce the following notation

$$d = \sqrt{k_0^2\cos^2\Phi_0 + k^2\sin^2\theta}\,, \quad 2\gamma = \Phi_0 - \zeta\Phi\,, \quad 2\delta = -\Phi_0 - \zeta\Phi\,,$$

$$\Phi_0 = \eta - \alpha, \quad \sin\Phi = (K^2 - k^2\cos^2\theta)^{-1/2} k_0 \sin\Phi_0\,, \quad \beta = k/K.$$

The energy K is defined by the following expression

$$K = [k_0^2 \sin^2\Phi_0 + k^2\cos^2\theta + (d + \zeta\mu Q)^2]^{1/2}, \quad k_0 = mc/\hbar.$$

It follows from Eqs (3.99) and (3.101) that

$$\Psi^+\Psi = |N|^2.$$

Since

$$\langle\sigma_3\rangle = \Psi^+\sigma_3\Psi = \zeta|N|^2(k_0/d)\cos\Phi_0, \quad |\Phi_0| \le \frac{\pi}{2}$$

the spin is oriented along the z-axis when $\zeta = 1$ and against it when $\zeta = -1$.

The radiation power W and the probability w are expressed as follows

$$\begin{aligned} W &= \frac{c^3\hbar^2\mu^2}{2\pi}\int \frac{\kappa^4 \sin\theta' d\theta' d\varphi'}{1 + \partial K'/\partial\kappa} S_i, \\ w &= \frac{c^2\hbar\mu^2}{2\pi}\int \frac{\kappa^3 \sin\theta' d\theta' d\varphi'}{1 + \partial K'/\partial\kappa} S, \\ S &= S_2 + S_3, \quad \kappa = K - K'. \end{aligned}$$

We choose a system of spherical coordinates for the photon momentum vector $\boldsymbol{\kappa}$

$$\boldsymbol{\kappa} = \kappa \cdot \boldsymbol{\kappa}_0 \quad \boldsymbol{\kappa}_0 = (\sin\theta'\cos\varphi', \sin\theta'\sin\varphi', \cos\theta'),$$

S_i are determined by expressions of Eqs (3.38)–(3.40) in which the matrices $\boldsymbol{\alpha}$ are to be replaced by the matrices $\boldsymbol{B}$

$$\boldsymbol{B} = \{\rho_3[\boldsymbol{\sigma\kappa}_0] + \rho_2\boldsymbol{\sigma}\}\cos\alpha + \{\rho_3\boldsymbol{\sigma} - \rho_2[\boldsymbol{\sigma\kappa}_0]\}\sin\alpha.$$

The calculation of the matrix elements of Eq. (3.39), which is not so much difficult as cumbersome, leads to the following results. The radiation is possible only due to the transition from the state $\zeta = 1$ to the state $\zeta' = -1$, all other transitions being strongly forbidden. Thereby, as in the case of the electron, the radiation results in radiative self-polarization of the beam. We should carry out numerical calculations using expansion with respect to the parameter μ, since the Hamiltonian as defined by Eq. (3.98) is valid only in this approximation. For the radiation frequency κ we obtain the following expression

$$\begin{aligned} \kappa &= 2\mu Q(1 - \beta\cos\Omega)^{-1}[(1 - \beta^2)\cos^2\Phi_0 + \beta^2\sin^2\theta]^{1/2}, \\ \cos\Omega &= \sin\theta\sin\theta'\cos\varphi_0 + \cos\theta\cos\theta', \quad \varphi_0 = \varphi - \varphi', \end{aligned}$$

where Ω is the angle between the particle velocity and the photon escape direction.

The angular distribution of radiation can be represented as follows

$$\begin{aligned}
\frac{dW}{d\Omega'} &= \frac{8c^2\hbar^2\mu^6 Q^4(1-\beta^2\cos^2\theta)^2(1-\sin^2\Phi_0)|l_2^0F_2+il_3^0F_3|^2}{\pi(1-\beta\cos\Omega)^5},\\
F_2 &= i\beta\sin\theta\sin\theta'\sin\Phi_0 - i(1-S)(1-\beta\cos\theta\cos\theta')\sin\Phi_0\cos\varphi_0 -\\
&- (\beta\cos\theta-\cos\theta')[\sqrt{S(1-S\sin^2\Phi_0)}\cos\varphi_0 + i\cos\Phi_0\sin\varphi_0],\\
F_3 &= i\beta\sin\theta\sin\theta'\cos\Phi_0 + i(1-S)(\beta\cos\theta-\cos\theta')\sin\Phi_0\cos\varphi_0 +\\
&+ (1-\beta\cos\theta\cos\theta')[\sqrt{S(1-S\sin^2\Phi_0)}\sin\varphi_0 - i\cos\Phi_0\cos\varphi_0],\\
l_2^0 &= l_2\cos\alpha - il_3\sin\alpha, \quad l_3^0 = l_3\cos\alpha - il_2\sin\alpha,\\
S &= (1-\beta^2)/(1-\beta^2\cos^2\theta), \quad d\Omega' = \sin\theta' d\theta' d\varphi'. \qquad (3.102)
\end{aligned}$$

Integrating over the angles in Eq. (3.102) we find the following expressions or the total emitted power and radiation probability

$$W = W_0\left(\frac{Q}{Q_0}\right)^3\left(\frac{\cos^2\Phi_0+\gamma^2\sin^2\Phi_0}{1-\gamma^2}\right)^2, \qquad (3.103)$$

$$w = \frac{1}{T_0}\left(\frac{Q}{Q_0}\right)^4\frac{(1-\beta^2\cos^2\theta)^{3/2}}{1-\beta^2}(\cos^2\Phi_0+\gamma^2\sin^2\Phi_0)^{1/2}. \qquad (3.104)$$

Here we denoted

$$W_0 = \frac{128}{3}c^3\hbar^2\mu^2k_0^4\,, \quad T_0^{-1} = \frac{64}{3}c^2\mu^2k_0^3\hbar\,, \quad Q_0 = \frac{k_0}{\mu}, \quad \gamma^2 = \frac{\beta^2\sin^2\theta}{1-\beta^2\cos^2\theta}.$$

From Eq. (3.102) we find for the components of linear and circular polarizations $(l=\pm1)$

$$W_2 = \frac{1-p}{2}W, \quad W_3 = \frac{1+p}{2}, \quad p = \frac{W_3-W_2}{W},$$

$$p = \frac{[2S+(3-2s+S^2)\cos2\Phi_0]\cos2\alpha + (1-S)(3+S)\sin2\Phi_0\sin2\alpha}{4(2-S+S\cos2\Phi_0)},$$

$$W_l = \frac{W}{2}\{1+\frac{1}{2}l\beta\cos\theta\cos\Phi_0[(1-\gamma^2)/(\cos^2\Phi_0+\gamma^2\sin^2\Phi_0)]^{1/2}\}. \qquad (3.105)$$

Now let us briefly analyze the results. In the case of the relativistic particle $(\beta\sim1)$, the radiation is localized in the narrow cone with the angle opening of the order of $\sqrt{1-\beta^2}$, the axis of the cone coinciding with the direction of the velocity vector. The above feature is caused by the factor of $(1-\beta\cos\Omega)^s$ in the denominator of the formula of Eq. (3.102). In the nonrelativistic approximation $(\beta\ll1)$ the radiation is uniformly distributed over the space

$$\pi dW = 8c^2\hbar^2\mu^6(Q\cos\Phi_0)^4|l_2^0e^{i\varphi_0}\cos\theta' + il_3^0e6-i\varphi_0|^2d\Omega'.$$

In this approximation the total emitted power coincides with the power of a particle moving along the z-axis and is equal to

$$W = W_0 \left(\frac{Q \cos \Phi_0}{Q_0} \right)^4 .$$

In the ultrarelativistic approximation under the condition of $\theta \neq 0$ (more exactly, $\sin^2 \theta > 1 - \beta^2$) from Eqs (3.103), (3.104) it follows

$$W = W_0 \left(\frac{Q}{Q_0} \right)^4 \left(\frac{\sin^2 \theta}{1 - \beta^2} \right)^2 , \quad w = \frac{1}{T_0} \left(\frac{Q}{Q_0} \right)^3 \frac{\sin^3 \theta}{1 - \beta^2} .$$

There is no radiation of a particle being at rest or moving along the z-axis, if

$$\cos \Phi_0 = 0 \rightarrow \quad \eta = \alpha \pm \pi/2,$$

i.e. there is a relation between the moments μ_1, μ_2 and the fields E, H

$$\mu_1 H = \mu_2 E.$$

A particle moving at a nonzero angle to the z-axis always radiates. As it follows from Eqs (3.102) and (3.105) the existence of the electric moment affects essentially the radiation polarization, namely, the orientation of axes of polarization ellipse depends on α (see Eq. (3.102)) and the linear polarization degree essentially changes with α.

The possibility of neutral particles generating electromagnetic radiation is of great interest. For instance, one could suppose that the radio-frequency radiation of the Galaxy is caused if only partly by the mechanism considered above. The identification of this part of the cosmic radio-frequency radiation could give us some information on the existence of neutral particles in the space. At present we have no information of this kind.

Bibliography

1. L. Page, *Phys. Rev.* **36** (1930) 444.
2. M.S. Pleesset, *Phys.Rev.* **36** (1930) 1728.
3. D.D. Ivanenko and A.A. Sokolov, *The Classical Theory of Fields* (Gostekhizdat, Moskwa-Leningrad, 1949) [in Russian].
4. A.A. Sokolov, I.M. Ternov, *Relativistic Electron* (Moscow, Nauka, 1983).
5. I.I. Rabi, *Zs. f. Phys.* **49** (1928) 507.
6. L.D. Huff, *Phys. Rev.* **38** (1931) 501.
7. V.A. Fock, *Beginning of Quantum Mechanics* (Leningrad,1932) [in Russian].
8. V.S. Fursov, *Zh. Exp. Teor. Fiz.* **4** (1934) 767 [Sov. Phys.– JETP].
9. I.M. Luttinger, *Phys. Rev.* **74** (1948) 893.
10. M.N. Johnson and B.A. Lippman, *Phys. Rev.* **76** (1949) 828.

11. M.N. Johnson and B.A. Lippman, *Phys. Rev.* **77** (1950) 702.
12. I.M. Ternov and V.S. Tumanov, *Dokl. Akad. Nauk. SSSR* **124** (1959) 1038 [Sov. Phys. Dokl.].
13. A.A. Sokolov and I.M. Ternov, *Dokl. Akad. Nauk. SSSR* **92** (1953) 537 [Sov. Phys. Dokl.].
14. G.I. Ahiezer and V.B. Berestetskiy, *Quantum Electrodynamics* (Fismatgiz, 1953) [in Russian].
15. A.A. Sokolov, *Journ. of Phys. USSR* **9** (1945) 303.
16. I.S. Gradsteyn and I.M. Ryzhik, *Tables of Integrals, Series, and Products* (Academic Press, New York, 1965).
17. I.M. Ternov, V.G. Bagrov and R.A. Rzaev, *Zh. Exp. Teor. Fiz.* **46** (1964) 374 [Sov. Phys.- JETP].
18. V.G. Bagrov, *Izv. Vuz. Fiz.* N 5 (1965) 121 [Sov. Phys. J.].
19. N.P. Klepikov, Doctor Thesis, Moscow State University 1952.
20. G.N. Watson, *A Treatise on the Theory of Bessel Functions* (Cambridge University Press, London, 1958).
21. A.N. Matveev, Doctor Thesis, Moscow State University 1958.
22. M.A. Lavrent'ev and B.V. Shabat, *Methods of Function Theory of Complex Variaty* (Fizmatgiz, 1958).
23. A.A. Sokolov, I.P. Klepikov and I.M. Ternov, *Zh. Exp. Teor. Fiz.* **24** (1953) 249 [Sov. Phys.- JETP].
24. A.A. Sokolov, A.I. Matveev and I.M. Ternov, *Dokl. Akad. Nauk. SSSR* **102** (1955) 65 [Sov. Phys. Dokl.].
25. A.I. Matveev, *Zh. Exp. Teor. Fiz.* **31** (1956) 479 [Sov. Phys.- JETP].
26. I.M. Ternov, V.G. Bagrov and R.A. Rzaev, *Izv. Vuz. Fiz.* N 5 (1963) 127 [Sov. Phys. J.].
27. I.M. Ternov, Yu.M. Loskutov and L.I. Korovina, *Zh. Exp. Teor. Fiz.* **41** (1961) 1924 [Sov. Phys.- JETP].
28. A.A. Sokolov and I.M. Ternov, *Dokl. Akad. Nauk. SSSR* **153** (1963) 1052 [Sov. Phys. Dokl.].
29. I.M. Ternov, V.G. Bagrov and R.A. Rzaev, *Vestn. Moskov. Univ.* N 4 (1964) 62 [in Rassian].
30. D. Tomboulian and P. Hartman, *Phys. Rev.* **102** (1956) 1423.
31. V.G. Bagrov, O.F. Dorofeev, Yu.N. Kargin, G.F. Kopytov, G.K. Razina, Deposited in VINITI 12.11.87, N8002-V87, 98 p. [in Rassian].
32. I.M. Ternov, Doctor Thesis, Moscow State University 1961.
33. I.M. Ternov, V.V. Mikhailin and V.R. Khalilov, *Synchrotron Radiation and its Application* (Harwood Acad. Publ., Chur, Switzerland, 1986).
34. V.G. Bagrov, O.F. Dorofeev, A.A. Sokolov, I.M. Ternov, V.P. Khalilov, *Dokl. Akad. Nauk. SSSR* **221** (1975) 312 [Sov. Phys. Dokl.].
35. I.M. Ternov, V.G. Bagrov, A.M. Khapaev, *Zh. Exp. Teor. Fiz.* **48** (1965) 921 [Sov. Phys.- JETP].
36. I.M. Ternov, V.G. Bagrov, G.M. Kruzhkov, A.M. Khapaev, *Izv. Vuz. Fiz.* N4

(1967) 10 [Sov. J. Phys.].
37. V.L. Luboshits. Nuclear Physics, **4** (1966) 269 [In Russian].
38. V.G. Bagrov, P.V. Bozrikov, D.M. Gitman, Yu.I. Klimenko, A.I. Khudomyasov. *Izv. Vuz. Fiz.* N6 (1974) 17 [Sov.J.Phys.].

Chapter 4

Spin Light

V.A. Bordovitsyn, V.S. Gushchina
Physics Faculty, Tomsk State University
634050 Tomsk, GSP-14, Russia [a]

4.1. Introduction

The spin magnetic moment of an electron moving in a magnetic field is a source of electromagnetic radiation. This radiation (spin light, or μ-radiation) shows up at high energies and it is available for measurements in contemporary electron accelerators. Here the spin light identification problem is considered when the spin radiation proceeds against the background of powerful synchrotron radiation (SR), recoil effects, and other relativistic phenomena. A relativistic neutron is considered to be a source of pure spin radiation.

For a long time the μ-radiation problem of elementary particles (electron, neutron) was little studied not only in quantum but also in classical electrodynamics. Although there were some papers in this area, but no systematic and comprehensive investigation using the correspondence principle was carried out. Perhaps, this situation was due to the fact that the spin magnetic moment as a physical value possesses more complicated tensor properties as compared with the charge. As a result, all the calculations are considerably complicated.

The relativistic classical theory of μ-radiation has been worked out for years. At first, the potentials of radiation fields were obtained by J. Frenkel[1]. Then the fields themselves and the integral responses of radiation were found by H.J. Bhabha and H.C. Corben[2]. Later these results were repeated to certain extents by other authors (A. Bialas[3], M. Kolsrud and E. Leer[4], J. Cohn and H. Wiebe[5], and others[6–8]), but definite applications of theory were not considered.

The point is that the magnetic dipole μ-radiation which accompanies spin transitions is negligible in the background of electron-charge SR. In its pure form, the spin

[a]E-mail: bord@urania.tomsk.su

magnetic moment radiation (spin light) contributes to SR power as a small correction proportional to $\hbar$ (see Sec. 4.3.3)

$$W^{\mu}/W_{\rm SR} \approx \xi^2,$$

where ξ is the well-known parameter in the relativistic quantum theory of radiation [b]

$$\xi = \frac{3}{2}\,\frac{\hbar\gamma^2}{m_0 c\rho} = 3\,\frac{\mu_0 H}{m_0 c^2}\,\gamma = \frac{3}{2}\,\frac{H}{H^*}\,\gamma = \frac{3}{2}\,\frac{\hbar}{m_0 c^3}\sqrt{w_\mu w^\mu},$$

$\gamma = 1/\sqrt{1-\beta^2}$ is the relativistic Lorentz factor, $\mu_0 = e_0\hbar/(2m_0 c)$ is the Bohr magneton, $H^* = m_0^2 c^3/(e\hbar) \approx 4.4\cdot 10^{13}$ Oe is Schwinger's magnetic field, w^μ is the four-dimensional acceleration of an electron. In the nonrelativistic case of $\gamma \approx 1$, this value is negligible ($\xi \approx 3.4\cdot 10^{-7}$) even for superstrong magnetic fields ($H \approx 10^7$ Oe).

At the present time the problem of radiation of the relativistic magnetic moment is particularly urgent in connection with the construction of ultrahigh energy accelerators. Therefore, the power of the magnetic moment radiation of a relativistic electron becomes an experimentally attainable value. For the above example, one can take on $\gamma \approx 10^5$, and even if $H \approx 10^5$ Oe then $\xi \approx 10^{-4}$. Now it is possible to carry out more detailed investigation of SR which, as it was found, has a complex structure. Side by side with recoil effects, it is specified by a number of fundamental phenomena including the radiation of the electron spin magnetic moment.

Although the quantum corrections to the SR power have already been known for a long time[9,10,11],

$$W = W_{\rm SR}\left[1 - \zeta\xi - \frac{55\sqrt{3}}{24}\,\xi + \frac{64}{3}\,\xi^2 + \zeta\,\frac{35\sqrt{3}}{6}\,\xi^2\right],$$

their physical interpretation remained obscure.

In 1983 the first quantum spin-orientation-dependent correction to SR power $-\zeta\xi$ was experimentally identified at the Institute of Nuclear Physics, Siberian Branch of the Academy of Sciences (Novosibirsk)[19].[c] In this experiment, synchrotron radiation was for the first time observed to be dependent on the spin orientation of a free electron moving in a macroscopic magnetic field.

It will be shown below that this radiation is due to the interference of charge radiation and the spin magnetic moment radiation of the electron (mixed $e\mu$-radiation), and this gives rise to the complex radiation proportional to $\hbar$.

Note that the problem of the electron spin observability has a long and intricate history. As Bohr showed, a direct measurement of free electron spin is impossible in

---

[b] See also footnote j in Chapter 1.

[c] The procedure for experimental observation of spin dependence of SR power was proposed in 1977[20], and the experiment itself was described in detail in Ref.[21] The authors are very grateful to A.E. Bondar and V.N. Zhilich, scientists from the Novosibirsk Institute of Nuclear Physics, for the exhaustive information on the experiment.

experiments like those made by Stern and Gerlach, as the spin observability condition comes into conflict with the uncertainty relation and the wave nature of an electron when the Lorentz force comes into play. A comprehensive analysis of this problem can be found in the paper written by Pauli in 1930[23].

Since the direct method of measuring the spin (by Stern and Gerlach) proved to be inconsistent, other methods were proposed to produce and observe free polarized electrons. Some of them are successfully realized[24,25]. However, it was not clear yet how the spin could be observed in a macroscopic magnetic field[26]. It is possible that the Bohr-Pauli 'taboo' syndrome is responsible for the fact that the cited experiment and its fundamental significance were hardly noticed.

The interpretation of spin corrections to SR power is not simple since spin corrections have an equal competitor, namely, the recoil effects accompanying the radiation. They also make contribution to the SR power, proportional to $\hbar$ as well as $\hbar^2$, and, moreover, the recoil effects are essentially different for an electron and a charged spinless particle (boson) in the second order in $\hbar$.

Considered here is the contribution of various fundamental factors (charge radiation, recoil effects, radiation of the magnetic moment and spin effects, radiation connected with Larmor and Thomas precession of the spin, mixed $e\mu$-radiation, and radiation from the anomalous part of magnetic moment of an electron) into the spectral-angular distribution and the total power of SR.

4.2. Classical Theory of μ-Radiation

In this section we shall present the fields of a point-like magnetic moment (magneton). The covariant retarded solution of the wave equation for the Hertzian tensor-potential is obtained by complete analogy with the solution for the Liénard-Wiechert potentials for a point charge. The electric dipole case would then be described by dual formalism. The formulas obtained were applied for computing the power of radiation in uniform motion of magneton. More complete information can be found in our works[60] (see also Refs[12,13,14]).

4.2.1. Equations of an Electromagnetic Field

Let us introduce the Hertzian tensor-potential of a relativistic magneton $Q^{\mu\nu} = (\boldsymbol{Z}, \boldsymbol{Q})$, linked with the usual vector potential $A^\mu = (\varphi, \boldsymbol{A})$ by the relation[12]

$$A^\mu = \partial_\nu Q^{\mu\nu}, \tag{4.1}$$

that is

$$A^\mu = \left(-div\boldsymbol{Q},\ rot\boldsymbol{Z} + \frac{1}{c}\frac{\partial \boldsymbol{Q}}{\partial t}\right).$$

The tensors of the magnetic polarization density $M^{\mu\nu} = (\boldsymbol{N}, \boldsymbol{M})$ and the density of the "magnetic current" j^μ_{mag} satisfy the relation (see also Ref.[6])

$$j^\mu_{\text{mag}} = c\partial_\nu M^{\mu\nu} = \left(div\boldsymbol{N}, c\,rot\boldsymbol{M} + \frac{\partial \boldsymbol{N}}{\partial t} \right). \tag{4.2}$$

It is easy to see that the Lorentz condition and the condition of continuity of the current density are satisfied automatically

$$\partial_\nu A^\nu = 0, \quad \partial_\nu j^\nu_{\text{mag}} = 0.$$

The vector A^μ satisfies the wave equations

$$\Box A^\mu = -\frac{4\pi}{c}(j^\mu + c\partial_\rho M^{\mu\rho}),$$

and the tensor-potential $Q^{\mu\nu}$ is governed by the tensor equation

$$\Box Q^{\mu\nu} = -4\pi M^{\mu\nu}, \tag{4.3}$$

which can be obtained from the variational principle of $\delta S = 0$ for an action function

$$S = \int \mathcal{L}(Q^{\mu\nu}, \partial_\rho Q^{\mu\rho}; r^\lambda) d\Upsilon,$$

if

$$\mathcal{L} = M_{\mu\nu}Q^{\mu\nu} - \frac{1}{8\pi}\partial_\rho Q^{\mu\nu}\partial^\rho Q^{\mu\nu} \tag{4.4}$$

is taken as the density of the Lagrange function.

Further, acting in the usual manner we shall obtain the Eiler-Lagrange equations

$$\partial_\rho \frac{\partial \mathcal{L}}{\partial_\rho Q^{\mu\nu}} - \frac{\partial \mathcal{L}}{\partial Q^{\mu\nu}} = 0,$$

which after substitution of Eq. (4.4) give Eq. (4.3).

Differentiating Eq. (4.3) with respect to the coordinates, one can obtain the usual wave equation for a vector potential with the "current" of j^μ_{mag}.

Let us represent the tensor-potential $Q^{\mu\nu}$ in a form analogous with the four-dimensional Liénard-Wiechert potential

$$Q^{\rho\nu} = -\frac{\mu c}{\tilde{r}_\sigma v^\sigma}\,\Pi^{\rho\nu}, \tag{4.5}$$

where μ is the magnetic moment, $v^\rho = dr^\rho/d\tau$ is the four-dimensional velocity of a magneton, $\tilde{r}^\rho = R^\rho - r^\rho$ is the four-dimensional vector drawn from the world point of the magneton to the world point of the observer. The mutual orientation of the vectors $\tilde{\boldsymbol{r}}$, $\boldsymbol{R}$, $\boldsymbol{r}$ in three-dimensional space is shown in Fig. 1.1.

The dimensionless spacelike tensor $\Pi^{\mu\nu} = (\boldsymbol{\Phi}, \boldsymbol{\Pi})$ (see Chapter 2) is connected with the magnetic polarization tensor by the relation

$$M^{\rho\nu}(t) = \mu\pi^{\rho\nu}\delta\left[\boldsymbol{\xi} - \boldsymbol{r}(t)\right],$$

where $\pi^{\rho\nu} = \Pi^{\rho\nu}/\gamma$, $\gamma = 1/\sqrt{1-\beta^2}$ is the Lorentz-factor, $\boldsymbol{\xi}$ is the vector of the "smeared" magneton at the time t connected with the moment of observation of the radiation $\tilde{t}$ by the relation $\tilde{t} = t - \tilde{r}/c$.

We shall exactly describe the electric dipole moment by means of the polarization tensor potential $Z^{\mu\nu} = (-\boldsymbol{Q}, \boldsymbol{Z})$ which is dual to the tensor potential of magnetic moment

$$Z^{\mu\nu} = \frac{1}{2}\,\varepsilon^{\mu\nu\alpha\beta} Q_{\alpha\beta}.$$

Tensors of the electric and magnetic polarization densities are related in the corresponding way

$$N^{\mu\nu} = \frac{1}{2}\,\varepsilon^{\mu\nu\alpha\beta} M_{\alpha\beta} = (-\boldsymbol{M}, \boldsymbol{N}).$$

The electromagnetic fields of the electric dipole moment are described by

$$\Box Z^{\mu\nu} = -4\pi N^{\mu\nu}.$$

The solution of this equation for a point electric dipole moment can be written as (cf. Eq. (4.5))

$$Z^{\alpha\beta} = -\frac{\nu c}{\tilde{r}_\rho v^\rho}\,\Phi^{\alpha\beta},$$

where ν is the electric dipole moment.

The dimensionless space-like tensor $\Phi^{\mu\nu} = (-\boldsymbol{\Pi}, \boldsymbol{\Phi})$ satisfies the condition

$$v_\mu \Phi^{\mu\nu} = 0,$$

from which it follows that

$$\boldsymbol{\Pi} = -[\boldsymbol{\beta}\boldsymbol{\Phi}].$$

These relations are formally similar to the Frenkel condition of Eq. (2.4) for relativistic magneton. However, in connection with that the vector potential A^μ is again given by Eq. (4.1), the electromagnetic field tensor of the electric dipole moment is more complicated. The reason is that the Frenkel condition of $v_\mu \Pi^{\mu\nu} = 0$ does not hold in the present case. As a result, additional terms appear in all structural elements of the electromagnetic field tensor $H^{\mu\nu} = (-\boldsymbol{E}, \boldsymbol{H})$.

However, through a duality transformation of the electromagnetic field tensor

$$E^{\mu\nu} = \frac{1}{2}\,\varepsilon^{\mu\nu\alpha\beta} H_{\alpha\beta} = (-\boldsymbol{H}, -\boldsymbol{E})$$

in the course of which we also use a similar transformation for a spin tensor on the right side

$$\Pi^{\mu\nu} = -\frac{1}{2}\,\varepsilon^{\mu\nu\alpha\beta}\Phi_{\alpha\beta}$$

we can find the tensor $E^{\mu\nu}$ which is formally identical to $H^{\mu\nu}$ for a relativistic magneton, except that we have

$$H^{\alpha\beta} \to E^{\alpha\beta}, \quad \mu\Pi^{\alpha\beta} \to \nu\Phi^{\alpha\beta}.$$

Noting that we have

$$\Pi^{\alpha\beta} = ([\boldsymbol{\beta\Pi}],\ \boldsymbol{\Pi}) \to \Phi^{\alpha\beta} = ([\boldsymbol{\beta\Phi}],\ \boldsymbol{\Phi}),$$

we can make the transformation of the fields in the switch from the magnetic to the electric dipole moment by means of replacements

$$\mu \to \nu, \quad -\boldsymbol{E} \to \boldsymbol{H}, \quad \boldsymbol{H} \to \boldsymbol{E}.$$

4.2.2. Potentials and Fields of a Relativistic Magneton

According to Eqs (4.1) and (4.5), the vector potential of magneton A^σ at the observation point has the form

$$A^\sigma = -\mu c \widetilde{D}_\nu \frac{\Pi^{\sigma\nu}}{\tilde{r}_\rho v^\rho}, \tag{4.6}$$

where the derivative $\widetilde{D}_\nu$ is determined by the rules of differentiation of functions with a retarded argument (see Sec. 1.2)

$$\widetilde{D}_\nu = \tilde{\partial}_\nu + \frac{\tilde{r}_\nu}{\tilde{r}_\rho v^\rho} \frac{d}{d\tau}, \tag{4.7}$$

where $\tilde{\partial}_\nu = \partial/\partial\tilde{r}^\nu$.

Differentiating in Eq. (4.6) according to the rules given by Eq. (4.7), we obtain

$$A^\sigma = -\mu c \left[\frac{\overset{\circ}{\Pi}{}^{\sigma\nu}\tilde{r}_\nu}{(\tilde{r}_\rho v^\rho)^2} - \frac{\Pi^{\sigma\nu}\tilde{r}_\nu}{(\tilde{r}_\rho v^\rho)^3}\left(c^2 + \tilde{r}_\rho w^\rho\right) \right], \tag{4.8}$$

where $w^\rho = dv^\rho/d\tau$ is the four-dimensional acceleration of magneton, the derivative with respect to the proper time is written with a circle. The components $A^\sigma = (\varphi, \boldsymbol{A})$ are represented in the form

$$\begin{aligned} \boldsymbol{A} &= -\frac{\mu}{\gamma^2[1-(\boldsymbol{n\beta})]^2} \left\{ \frac{1}{\tilde{r}c}\left[(\boldsymbol{n}-\boldsymbol{\beta})\,\overset{\circ}{\boldsymbol{\Pi}}\right] + \frac{1}{\tilde{r}^2\gamma}\left[(\boldsymbol{n}-\boldsymbol{\beta})\boldsymbol{\Pi}\right] J \right\}, \\ \varphi &= \frac{\mu}{\gamma^2[1-(\boldsymbol{n\beta})]^2} \left\{ \frac{1}{\tilde{r}c}\left(\boldsymbol{n}\,\overset{\circ}{\boldsymbol{\Phi}}\right) + \frac{1}{\tilde{r}^2\gamma}(\boldsymbol{n\Phi})\, J \right\}, \end{aligned} \tag{4.9}$$

where $\boldsymbol{n} = \tilde{\boldsymbol{r}}/\tilde{r}^0 = \tilde{\boldsymbol{r}}/c(\tilde{t}-t)$ is the unit vector directed to the observation point,

$$J = 1 + \frac{1}{c^2}\tilde{r}_\rho w^\rho = 1 + \frac{1}{c^2}\tilde{r}\gamma^2\left\{(\boldsymbol{na}) - \gamma^2(\boldsymbol{\beta a})[1-(\boldsymbol{n\beta})]\right\}.$$

The potentials of Eq. (4.9) can be also obtained by means of the Lorentz transformations. Actually, in the system of rest ($\beta = 0$) the potentials of Eq. (4.8) take on the form

$$\boldsymbol{A}' = \frac{\mu}{\tilde{r}'c^2}\left\{[\boldsymbol{a}'\boldsymbol{\zeta}] - c[\boldsymbol{n}'\dot{\boldsymbol{\zeta}}] - (\boldsymbol{n}'\boldsymbol{a}')[\boldsymbol{n}'\boldsymbol{\zeta}]\right\} - \frac{\mu}{\tilde{r}'^2}[\boldsymbol{n}'\boldsymbol{\zeta}],$$

$$\varphi' = \frac{\mu}{\tilde{r}'c}\,(\boldsymbol{n}'[\boldsymbol{a}'\boldsymbol{\zeta}]).$$

Considering that the stroked and unstroked variables are connected by the relations

$$\begin{aligned}\boldsymbol{\zeta} &= \frac{\boldsymbol{\Pi}}{\gamma} + \frac{\gamma}{\gamma+1}\,\boldsymbol{\beta}(\boldsymbol{\beta}\boldsymbol{\Pi}),\\ \dot{\boldsymbol{\zeta}} &= \frac{\overset{\circ}{\boldsymbol{\Pi}}}{\gamma} + \frac{\gamma}{\gamma+1}\,\boldsymbol{\beta}(\boldsymbol{\beta}\,\overset{\circ}{\boldsymbol{\Pi}}) + \gamma[\boldsymbol{\beta}[\overset{\circ}{\boldsymbol{\beta}}\,\boldsymbol{\Pi}]],\end{aligned} \tag{4.10}$$

(see also Eqs (1.3) and (1.6)) and taking into account the Lorentz transformations

$$\boldsymbol{A} = \gamma(\boldsymbol{A}' + \boldsymbol{\beta}\varphi') + \frac{\gamma^2}{\gamma+1}[\boldsymbol{\beta}[\boldsymbol{\beta}\boldsymbol{A}']],$$

$$\varphi = \gamma[\varphi' + (\boldsymbol{\beta}\boldsymbol{A}')],$$

after simple but quite laborious calculations one can obtain Eq. (4.9).

From the physical point of view, this result is not unusual; however, the Lorentz transformations allow us to more deeply understand the character of the behavior and the transformed properties of different physical variables.

Using the standard definition of the electromagnetic field tensor, on the basis of the potentials of Eq. (4.8) and differentiation rules of Eq. (4.7) we obtain

$$H^{\alpha\beta} = H_1^{\alpha\beta} + H_2^{\alpha\beta} + H_3^{\alpha\beta} \tag{4.11}$$

$$\begin{aligned}
H_1^{\alpha\beta} &= \frac{\mu c}{\tilde{r}_\rho v^\rho}\left\{\frac{\overset{\circ\circ}{\Pi}{}^{[\alpha\sigma}\tilde{r}_\sigma\tilde{r}^{\beta]}}{(\tilde{r}_\rho v^\rho)^2} - \frac{3\tilde{r}_\rho w^\rho\,\overset{\circ}{\Pi}{}^{[\alpha\sigma}\tilde{r}_\sigma\tilde{r}^{\beta]} + \tilde{r}_\rho\,\overset{\circ}{w}{}^\rho\Pi^{[\alpha\sigma}\tilde{r}_\sigma\tilde{r}^{\beta]}}{(\tilde{r}_\rho v^\rho)^3} + \right.\\
&+ \left.\frac{3(\tilde{r}_\rho w^\rho)^2\Pi^{[\alpha\sigma}\tilde{r}_\sigma\tilde{r}^{\beta]}}{(\tilde{r}_\rho v^\rho)^4}\right\},\\
H_2^{\alpha\beta} &= \frac{\mu c}{(\tilde{r}_\rho v^\rho)^2}\left\{2\,\overset{\circ}{\Pi}{}^{\alpha\beta} - \frac{\tilde{r}_\rho w^\rho\Pi^{\alpha\beta}}{\tilde{r}_\rho v^\rho} - \right.\\
&- \frac{\overset{\circ}{\Pi}{}^{[\alpha\sigma}v_\sigma\tilde{r}^{\beta]} + 2\,\overset{\circ}{\Pi}{}^{[\alpha\sigma}\tilde{r}_\sigma v^{\beta]} + \Pi^{[\alpha\sigma}\tilde{r}_\sigma w^{\beta]}}{\tilde{r}_\rho v^\rho} - \\
&- \left. 3\,\frac{c^2\,\overset{\circ}{\Pi}{}^{[\alpha\sigma}\tilde{r}_\sigma\tilde{r}^{\beta]} - \tilde{r}_\rho w^\rho\Pi^{[\alpha\sigma}\tilde{r}_\sigma v^{\beta]}}{(\tilde{r}_\rho v^\rho)^2} + 6\frac{c^2\tilde{r}_\rho w^\rho\Pi^{[\alpha\sigma}\tilde{r}_\sigma\tilde{r}^{\beta]}}{(\tilde{r}_\rho v^\rho)^3}\right\},\\
H_3^{\alpha\beta} &= -\frac{\mu c}{(\tilde{r}_\rho v^\rho)^3}\left\{2\Pi^{\alpha\beta} - 3\frac{\Pi^{[\alpha\sigma}\tilde{r}_\sigma v^{\beta]}}{\tilde{r}_\rho v^\rho} - 3c^2\frac{\Pi^{[\alpha\sigma}\tilde{r}_\sigma\tilde{r}^{\beta]}}{(\tilde{r}_\rho v^\rho)^2}\right\}.
\end{aligned}$$

One can show by direct calculations taking into account Eq. (4.7) that the tensor $H^{\alpha\beta}$ satisfies Maxwell's equation $\partial_\beta H^{\alpha\beta} = 0$.

In a general case, the magnetic and the electric field intensities are very complicated expressions. However, in the absolute rest system ($\beta = \dot{\beta} = \ddot{\beta} = 0$) they are considerably simplified, and from Eq. (4.11) the expressions known in electrodynamics for the electric and magnetic fields of an absolutely precessing magnetic moment follow[13]

$$\boldsymbol{E}' = \frac{[\boldsymbol{n}'\dot{\boldsymbol{\zeta}}]}{\tilde{r}'^2 c} + \frac{[\boldsymbol{n}'\ddot{\boldsymbol{\zeta}}]}{\tilde{r}' c^2},$$

$$\boldsymbol{H}' = \frac{[\boldsymbol{n}'[\boldsymbol{n}'\ddot{\boldsymbol{\zeta}}]]}{\tilde{r}' c^2} + \frac{3\boldsymbol{n}'(\boldsymbol{n}'\dot{\boldsymbol{\zeta}}) - \dot{\boldsymbol{\zeta}}}{\tilde{r}'^2 c} + \frac{3\boldsymbol{n}'(\boldsymbol{n}'\boldsymbol{\zeta}) - \boldsymbol{\zeta}}{\tilde{r}'^3}.$$

Setting in Eq. (4.9) $\dot{\beta} = 0$, we find the potentials of the fields emitted by a magneton in a uniform rectilinear motion to be

$$\boldsymbol{A} = -\frac{\mu}{\gamma^2[1-(\boldsymbol{n}\boldsymbol{\beta})]^2}\left\{\frac{1}{\tilde{r}c}\,[(\boldsymbol{n}-\boldsymbol{\beta})\,\overset{\circ}{\boldsymbol{\Pi}}] + \frac{1}{\tilde{r}^2}\,\frac{[(\boldsymbol{n}-\boldsymbol{\beta})\boldsymbol{\Pi}]}{\gamma[1-(\boldsymbol{n}\boldsymbol{\beta})]}\right\},$$

$$\varphi = (\boldsymbol{\beta}\boldsymbol{A}).$$

Hence, it follows that in this case, the vector potential A^μ is a space-like vector and satisfies the condition

$$v_\mu A^\mu = 0.$$

In precisely the same way taking into account that

$$\begin{aligned}
\Pi^{[\alpha\sigma}\tilde{r}_\sigma\tilde{r}^{\beta]} &= -\tilde{r}^2\,\left(-[\boldsymbol{n}[\boldsymbol{n}[(\boldsymbol{n}-\boldsymbol{\beta})\boldsymbol{\Pi}]]],\ [\boldsymbol{n}[(\boldsymbol{n}-\boldsymbol{\beta})\boldsymbol{\Pi}]]\right),\\
\Pi^{[\alpha\sigma}\tilde{r}_\sigma v^{\beta]} &= -\gamma c\tilde{r}\,\left(\boldsymbol{\beta}(\boldsymbol{n}[\boldsymbol{\beta}\boldsymbol{\Pi}]) + [(\boldsymbol{n}-\boldsymbol{\beta})\boldsymbol{\Pi}],\ [\boldsymbol{\beta}[(\boldsymbol{n}-\boldsymbol{\beta})\boldsymbol{\Pi}]]\right),\\
\overset{\circ}{\Pi}{}^{[\alpha\sigma}v_\sigma\tilde{r}^{\beta]} &= 0,
\end{aligned}$$

we find the fields from Eq. (4.11) ($\dot{\beta} = \ddot{\beta} = 0$)

$$\begin{aligned}
\boldsymbol{H}_1 &= \frac{\mu}{\tilde{r}c^2\gamma^3[1-(\boldsymbol{n}\boldsymbol{\beta})]^3}\,[\boldsymbol{n}[(\boldsymbol{n}-\boldsymbol{\beta})\,\overset{\circ\circ}{\boldsymbol{\Pi}}]], \qquad (4.12)\\
\boldsymbol{H}_2 &= -\frac{\mu}{\tilde{r}^2 c\gamma^2[1-(\boldsymbol{n}\boldsymbol{\beta})]^2}\left\{2\,\overset{\circ}{\boldsymbol{\Pi}} - 2\frac{[\boldsymbol{\beta}[(\boldsymbol{n}-\boldsymbol{\beta})\,\overset{\circ}{\boldsymbol{\Pi}}]]}{1-(\boldsymbol{n}\boldsymbol{\beta})} + 3\frac{[\boldsymbol{n}[(\boldsymbol{n}-\boldsymbol{\beta})\,\overset{\circ}{\boldsymbol{\Pi}}]]}{\gamma^2[1-(\boldsymbol{n}\boldsymbol{\beta})]^2}\right\},\\
\boldsymbol{H}_3 &= \frac{\mu}{\tilde{r}^3\gamma^3[1-(\boldsymbol{n}\boldsymbol{\beta})]^3}\left\{2\boldsymbol{\Pi} - 3\frac{[\boldsymbol{\beta}[(\boldsymbol{n}-\boldsymbol{\beta})\boldsymbol{\Pi}]]}{1-(\boldsymbol{n}\boldsymbol{\beta})} + 3\frac{[\boldsymbol{n}[(\boldsymbol{n}-\boldsymbol{\beta})\boldsymbol{\Pi}]]}{\gamma^2[1-(\boldsymbol{n}\boldsymbol{\beta})]^2}\right\},
\end{aligned}$$

$$\begin{aligned}
\boldsymbol{E}_1 &= -\frac{\mu}{\tilde{r}c^2\gamma^3[1-(\boldsymbol{n}\boldsymbol{\beta})]^3}\,[\boldsymbol{n}[\boldsymbol{n}[(\boldsymbol{n}-\boldsymbol{\beta})\,\overset{\circ\circ}{\boldsymbol{\Pi}}]]], \qquad (4.13)\\
\boldsymbol{E}_2 &= -\frac{\mu}{\tilde{r}^2 c\gamma^2[1-(\boldsymbol{n}\boldsymbol{\beta})]^2}\left\{2[\boldsymbol{\beta}\,\overset{\circ}{\boldsymbol{\Pi}}] + \right.\\
&+\left. 2\frac{[(\boldsymbol{n}-\boldsymbol{\beta})\,\overset{\circ}{\boldsymbol{\Pi}}] + \boldsymbol{\beta}(\boldsymbol{n}[\boldsymbol{\beta}\,\overset{\circ}{\boldsymbol{\Pi}}])}{1-(\boldsymbol{n}\boldsymbol{\beta})} + 3\frac{[\boldsymbol{n}[\boldsymbol{n}[(\boldsymbol{n}-\boldsymbol{\beta})\,\overset{\circ}{\boldsymbol{\Pi}}]]]}{\gamma^2[1-(\boldsymbol{n}\boldsymbol{\beta})]}\right\},
\end{aligned}$$

$$\boldsymbol{E}_3 = -\frac{\mu}{\tilde{r}^3\gamma^3[1-(\boldsymbol{n\beta})]^3}\{2[\boldsymbol{\beta\Pi}] + + 3\frac{[(\boldsymbol{n}-\boldsymbol{\beta})\boldsymbol{\Pi}]+\boldsymbol{\beta}(\boldsymbol{n}[\boldsymbol{\beta\Pi}])}{1-(\boldsymbol{n\beta})} + 3\frac{[\boldsymbol{n}[\boldsymbol{n}[(\boldsymbol{n}-\boldsymbol{\beta})\boldsymbol{\Pi}]]]}{\gamma^2[1-(\boldsymbol{n\beta})]}\right\}.$$

The fields of the electric dipole moment can be obtained on the basis of dual formalism[14].

4.2.3. Wave Zone and Radiation Field

The wave zone is dominated by the field which falls off as $1/\tilde{r}$, where $\tilde{r}$ is the distance from the radiating particle (the charge or magneton) to the observation point.

Differentiation of A^μ with respect to the proper time in the wave zone (see Chapter 1) yields

$$\overset{\circ}{\tilde{A}}{}^{\mu} = -\frac{\mu c}{(\tilde{r}_\rho v^\rho)^2}\left\{\overset{\circ\circ}{\Pi}{}^{\mu\lambda}\tilde{r}_\lambda - \frac{3\,\overset{\circ}{\Pi}{}^{\mu\lambda}\tilde{r}_\lambda\tilde{r}_\rho w^\rho + \Pi^{\mu\lambda}\tilde{r}_\lambda\tilde{r}_\rho\,\overset{\circ}{w}{}^\rho}{\tilde{r}_\rho v^\rho} + 3\frac{\Pi^{\mu\lambda}\tilde{r}_\lambda(\tilde{r}_\rho w^\rho)^2}{(\tilde{r}_\rho v^\rho)^2}\right\}. \quad (4.14)$$

Substituting this expression into Eq. (1.20), we find $\widetilde{H}^{\mu\nu} = H_1^{\mu\nu}$, where $H_1^{\mu\nu}$ is given in Eq. (4.11) and corresponds to the field which falls off as $1/\tilde{r}$.

It is convenient to write the tensor $\widetilde{H}^{\mu\nu}$ in a more compact form

$$\widetilde{H}^{\mu\nu} = \frac{\mu c}{(\tilde{r}_\rho v^\rho)^3}\,Q^{[\mu}\tilde{r}^{\nu]},$$

where, according to Eq. (1.20), Q^μ is the expression in the curly brackets in Eq. (4.14).

The antisymmetric combination which appear in the tensor $\widetilde{H}^{\mu\nu}$ can be written

$$\begin{aligned}
\Pi^{[\mu\sigma}\tilde{r}_\sigma\tilde{r}^{\nu]} &= \tilde{r}^2(\boldsymbol{K},[\boldsymbol{Kn}]),\\
\overset{\circ}{\Pi}{}^{[\mu\sigma}\tilde{r}_\sigma\tilde{r}^{\nu]} &= \tilde{r}^2(\boldsymbol{K}_1,[\boldsymbol{K}_1\boldsymbol{n}]),\\
\overset{\circ\circ}{\Pi}{}^{[\mu\sigma}\tilde{r}_\sigma\tilde{r}^{\nu]} &= \tilde{r}^2(\boldsymbol{K}_2,[\boldsymbol{K}_2\boldsymbol{n}]),
\end{aligned}$$

where

$$\begin{aligned}
\boldsymbol{K} &= -[\boldsymbol{n}[\boldsymbol{n}[(\boldsymbol{n}-\boldsymbol{\beta})\boldsymbol{\Pi}]]],\\
\boldsymbol{K}_1 &= -[\boldsymbol{n}[\boldsymbol{n}[(\boldsymbol{n}-\boldsymbol{\beta})\,\overset{\circ}{\boldsymbol{\Pi}}]]] + [\boldsymbol{n}[\boldsymbol{n}[\overset{\circ}{\boldsymbol{\beta}}\,\boldsymbol{\Pi}]]],\\
\boldsymbol{K}_2 &= -[\boldsymbol{n}[\boldsymbol{n}[(\boldsymbol{n}-\boldsymbol{\beta})\,\overset{\circ\circ}{\boldsymbol{\Pi}}]]] + 2[\boldsymbol{n}[\boldsymbol{n}[\overset{\circ}{\boldsymbol{\beta}}\overset{\circ}{\boldsymbol{\Pi}}]]] + [\boldsymbol{n}[\boldsymbol{n}[\overset{\circ\circ}{\boldsymbol{\beta}}\,\boldsymbol{\Pi}]]].
\end{aligned}$$

Introducing the representation

$$\Pi^{\mu\nu} = \gamma\pi^{\mu\nu} = \gamma(\boldsymbol{\varphi},\boldsymbol{\pi}), \quad \boldsymbol{\varphi} = [\boldsymbol{\beta\pi}],$$

we find[12]

$$\widetilde{\boldsymbol{H}} = \frac{\mu}{\tilde{r}c^2[1-(\boldsymbol{n\beta})]^3}\,[\boldsymbol{n}[(\boldsymbol{n}-\boldsymbol{\beta})\boldsymbol{h}]], \quad \widetilde{\boldsymbol{E}} = -[\boldsymbol{n}\widetilde{\boldsymbol{H}}], \tag{4.15}$$

where

$$\boldsymbol{h} = \ddot{\boldsymbol{\pi}} + \frac{(\boldsymbol{n}\dot{\boldsymbol{\beta}})\dot{\boldsymbol{\pi}} + 2(\boldsymbol{n}\dot{\boldsymbol{\pi}})\dot{\boldsymbol{\beta}} + (\boldsymbol{n\pi})\ddot{\boldsymbol{\beta}}}{1-(\boldsymbol{n\beta})} + 3\,\frac{(\boldsymbol{n\pi})(\boldsymbol{n}\dot{\boldsymbol{\beta}})\dot{\boldsymbol{\beta}}}{[1-(\boldsymbol{n\beta})]^2}.$$

It is easy to see that the radiation electromagnetic fields automatically satisfy the conditions of orthogonality and transversality

$$(\boldsymbol{nH}) = (\boldsymbol{nE}) = (\boldsymbol{EH}) = 0.$$

In the rest frame ($\beta = 0$) we have

$$\widetilde{\boldsymbol{H}}' = \frac{\mu}{\tilde{r}'c^2}\,[\boldsymbol{n}'[\boldsymbol{n}'\boldsymbol{h}']] = [\boldsymbol{n}'\widetilde{\boldsymbol{E}}'], \tag{4.16}$$

$$\boldsymbol{h}' = \ddot{\boldsymbol{\zeta}} + (\boldsymbol{n}'\dot{\boldsymbol{\beta}}')\dot{\boldsymbol{\zeta}} + (\boldsymbol{n}'\boldsymbol{\zeta})\ddot{\boldsymbol{\beta}}' + [3(\boldsymbol{n}'\boldsymbol{\zeta})(\boldsymbol{n}'\dot{\boldsymbol{\beta}}') + 2(\boldsymbol{n}'\dot{\boldsymbol{\zeta}}) - (\dot{\boldsymbol{\beta}}'\boldsymbol{\zeta})]\dot{\boldsymbol{\beta}}'.$$

It was considered here that according to Eq. (4.10)

$$\boldsymbol{\pi} = \boldsymbol{\zeta} - \frac{\gamma}{\gamma+1}\,\boldsymbol{\beta}(\boldsymbol{\beta\zeta}).$$

Proceeding from some conclusions drawn previously, we can also write the expressions for the fields radiated by a relativistic electric dipole moment

$$\widetilde{\boldsymbol{E}}' = \frac{\nu}{\tilde{r}'c^2}\,[\boldsymbol{n}'[\boldsymbol{n}'\boldsymbol{e}']], \quad \widetilde{\boldsymbol{H}}' = [\boldsymbol{n}'\widetilde{\boldsymbol{E}}'],$$

where $\boldsymbol{e}'$ is formally the same as $\boldsymbol{h}'$ if we make the substitution $\boldsymbol{\zeta} \to \boldsymbol{\delta}$ in the latter. The unit vector along the electric dipole moment $\boldsymbol{\delta}$ is related to the spatial component $\boldsymbol{\varphi}$ in the tensor $\pi^{\mu\nu}$ by the Lorentz transformation

$$\boldsymbol{\delta} = \boldsymbol{\varphi} + \frac{\gamma^2}{\gamma+1}\,\boldsymbol{\beta}(\boldsymbol{\beta\varphi}).$$

4.2.4. Total Radiation Power in a Uniform Motion of a Magneton

In a general case, the total power of instantaneous radiation of a magneton is a highly difficult expression (see also Ref.[60]). However, for the uniform motion of magneton in constant and homogeneous fields this expression is considerably simplified

$$W^{\mu} = \frac{1}{3}\,\frac{\mu^2}{c^3}\,\overset{\circ\circ}{\Pi}{}^{\mu\nu}\,\overset{\circ\circ}{\Pi}_{\mu\nu}\,. \tag{4.17}$$

This result can be found in Ref.[66]

Eq. (4.17) can be also obtained by means of the Lorentz transformation for radiation power (see Chapter 1).

In the present case, $\dot{\boldsymbol{\beta}}' = \ddot{\boldsymbol{\beta}}' = 0$ and according to Eq. (4.16), the fields in the system of rest have the form

$$\widetilde{\boldsymbol{H}}' = \frac{\mu}{\widetilde{r}' c^2} [\boldsymbol{n}'[\boldsymbol{n}'\ddot{\boldsymbol{\zeta}}']] = [\boldsymbol{n}'\widetilde{\boldsymbol{E}}'].$$

These expressions differ from the respective terms in Eq. (1.5) for charge in the replacement

$$e\boldsymbol{a}' \to \mu\ddot{\boldsymbol{\zeta}}, \quad \widetilde{\boldsymbol{E}}' \to \widetilde{\boldsymbol{H}}', \quad \widetilde{\boldsymbol{H}}' \to -\widetilde{\boldsymbol{E}}'.$$

Therefore, the invariant functional I_a in Eq. (1.37) takes on the form

$$I_a = \overset{\circ\circ}{\pi}_\alpha \overset{\circ\circ}{\pi}{}^\alpha - (e_\alpha \overset{\circ\circ}{\pi}{}^\alpha)^2 = -e_\alpha \overset{\circ\circ}{\Pi}{}^{\alpha\beta} \overset{\circ\circ}{\Pi}_{\beta\gamma} e^\gamma,$$

where e_α is the unit spacelike four-vector (see Chapter 1), $\overset{\circ\circ}{\pi}{}^\alpha$ and $\overset{\circ\circ}{\Pi}{}^{\alpha\beta}$ are the second derivatives of the covariant vector and the tensor of spin with respect to the proper time.

Now it is easy to find the angular distribution of radiation

$$\frac{dW^\mu}{d\Omega} = -\frac{\mu^2}{4\pi\gamma} \frac{\overset{\circ\circ}{\pi}_\alpha \overset{\circ\circ}{\pi}{}^\alpha - (e_\alpha \overset{\circ\circ}{\pi}{}^\alpha)^2}{(n_\rho v^\rho)^3} = \frac{\mu^2}{4\pi\gamma} \frac{e_\alpha \overset{\circ\circ}{\Pi}{}^{\alpha\beta} \overset{\circ\circ}{\Pi}_{\beta\gamma} e^\gamma}{(n_\rho v^\rho)^3}.$$

To carry out the covariant integration with respect to the angles, we rewrite these expressions in the form (cf. Eq. (1.28))

$$\frac{d\mathcal{P}^\rho}{d\tau} = \frac{\mu^2}{4\pi c^3} \oint v^\rho \frac{\overset{\circ\circ}{\pi}_\alpha \overset{\circ\circ}{\pi}{}^\alpha - (e_\alpha \overset{\circ\circ}{\pi}{}^\alpha)^2}{(n_\rho v^\rho)^2} d\Omega = -\frac{\mu^2}{4\pi c^3} \oint v^\rho \frac{e_\alpha \overset{\circ\circ}{\Pi}{}^{\alpha\beta} \overset{\circ\circ}{\Pi}_{\beta\gamma} e^\gamma}{(n_\rho v^\rho)^2} d\Omega.$$

After integration, according to Eq. (1.30) we obtain

$$\frac{d\mathcal{P}^\rho}{d\tau} = \frac{2}{3} \frac{\mu^2}{c^5} v^\rho \overset{\circ\circ}{\pi}_\alpha \overset{\circ\circ}{\pi}{}^\alpha = \frac{1}{3} \frac{\mu^2}{c^5} v^\rho \overset{\circ\circ}{\Pi}{}^{\alpha\beta} \overset{\circ\circ}{\Pi}_{\alpha\beta}.$$

It is easy to obtain Eq. (4.17) from this relation.

By the use of the tensor BMT-equation for the uniform motion of a magneton (see Chapter 2), after averaging over the period of spin precession it is possible to find that

$$\overline{\overset{\circ\circ}{\Pi}_{\mu\nu} \overset{\circ\circ}{\Pi}{}^{\mu\nu}} = \lambda^4 \Pi_{\mu\nu}\Pi^{\mu\nu} - \frac{1}{2}\lambda^2 (h_{\mu\nu}\Pi^{\mu\nu})^2, \tag{4.18}$$

where

$$\lambda = \gamma\sqrt{(\boldsymbol{h} - [\boldsymbol{\beta}\boldsymbol{e}])^2 - (\boldsymbol{\beta}\boldsymbol{h})^2} = \gamma\omega,$$

ω is the spin precession frequency,

$$h^{\mu\nu} = \frac{ge}{2m_0 c} H^{\mu\nu}.$$

Taking into account that

$$\Pi_{\mu\nu}\Pi^{\mu\nu} = 2\zeta^2, \quad h_{\mu\nu}\Pi^{\mu\nu} = h_{\mu\nu}\Pi_1^{\mu\nu},$$

where $\Pi_1^{\mu\nu}$ are the initial components of the spin tensor, Eq. (4.17) can be represented in the form

$$\overline{W}^\mu = \frac{2}{3}\frac{\mu^2}{c^3}\,\lambda^2\left[\lambda^2\zeta^2 - \frac{1}{4}\,(h_{\mu\nu}\Pi_1^{\mu\nu})^2\right]. \tag{4.19}$$

This formula is good for the description of the radiation of a magneton in a uniform motion in constant and homogeneous electromagnetic fields of various configurations. Let us consider some of them.

1. Uniform magnetic field: $\boldsymbol{H} = (0,0,H), \quad \boldsymbol{E} = 0.$

Let the magneton velocity be given by the components $\boldsymbol{\beta} = \beta(\sin\alpha, 0, \cos\alpha)$, which corresponds to the motion in the XZ plane. Therefore,

$$\lambda = \gamma h\sqrt{1-\beta^2\cos^2\alpha}.$$

The invariant initial condition according to Eq. (2.131) has the form

$$\boldsymbol{\Pi}_1 = \zeta\left(-\frac{\sin\varphi_1}{\sqrt{1-\beta^2\cos^2\alpha}}, \gamma\cos\varphi_1, 0\right), \quad h_{\mu\nu}\Pi_1^{\mu\nu} = 0.$$

The radiation power averaged over the precession period is represented in the form

$$\overline{W}^\mu = \frac{2}{3}\frac{\mu^2}{c^3}\,\omega^4\gamma^4\zeta^2 \tag{4.20}$$

with the spin precession frequency

$$\omega = h\sqrt{1-\beta^2\cos^2\alpha} = \frac{2|\mu|H}{\hbar}\sqrt{1-\beta^2\cos^2\alpha}. \tag{4.21}$$

It is interesting that the same result has been obtained for neutron radiation in the quantum theory on the basis of the Dirac-Pauli equation[59], with the only difference which can be eliminated if one takes on $\zeta = 2$ [d]

At the initial orientation of $\boldsymbol{\zeta}$ along the Z axis according to Eqs (4.10) and (2.129) the spin precession is absent and

$$\boldsymbol{\Pi} = \zeta_{\parallel}\left(-\frac{\gamma\beta^2\sin\alpha\cos\alpha}{\sqrt{1-\beta^2\cos^2\alpha}},\ 0,\ \gamma\sqrt{1-\beta^2\cos^2\alpha}\right) = \boldsymbol{\Pi}_1,$$

where $\zeta_{\parallel} = \pm 1$. Since

$$h_{\mu\nu}\Pi_1^{\mu\nu} = 2h\Pi_{1z} = 2\zeta_{\parallel}h\gamma\sqrt{1-\beta^2\cos^2\alpha},$$

[d] This basic difference between the classical and quantum radiation theories will be discussed in detail further.

then for the given λ and $\zeta^2 = \zeta_{\|}^2 = 1$ we obtain $\overline{W}^\mu = 0$. As it should be expected, the magneton in a uniform motion with a constant spin orientation does not emit in the classical theory.

2. Mutually orthogonal fields of different magnitudes.

First of all, let us consider the translational motion of a magneton in the plane of the orthogonal fields

$$\boldsymbol{H} = (0, 0, Q\cos\eta), \quad \boldsymbol{E} = (0, Q\sin\eta, 0), \quad \boldsymbol{\beta} = \beta(0, \sin\alpha, \cos\alpha),$$

where $Q = \sqrt{E^2 + H^2}$, $\tan\eta = E/H$.

According to Eq. (2.134) for spin tensor, the condition $h_{\mu\nu}\Pi_1^{\mu\nu} = 0$ is carried out again. Therefore, the formula of radiation power has the same form as in Eq. (4.20). Only the spin precession frequency $\omega = \lambda/\gamma$ is varied

$$\omega = q\sqrt{\cos^2\eta(1 - \beta^2\cos^2\alpha) + \beta^2\sin^2\eta\cos^2\alpha}, \tag{4.22}$$

where $q = (2\mu Q)/\hbar$.

In the particular case at $\eta = 0$ from Eq. (4.22) a result follows for homogeneous magnetic field, and at $\eta = \pi/4$ it is possible to obtain the radiation power for the orthogonal fields of equal magnitude

$$\boldsymbol{H} = (0, 0, H), \quad \boldsymbol{E} = (0, E, 0).$$

In this case, the formula for the radiation power has the same form as Eq. (4.20) but the spin precession frequency is $\omega = h$.

One should also consider the motion of a magneton along the direction orthogonal to the plane of the fields

$$\boldsymbol{H} = (0, 0, Q\cos\eta), \quad \boldsymbol{E} = (0, Q\sin\eta, 0), \quad \boldsymbol{\beta} = (\beta, 0, 0).$$

The corresponding solution of the spin precession equation is given by Eq. (2.135). The initial conditions are specified by the expression

$$\boldsymbol{\Pi}_1 = \zeta(\cos\varphi_1, \gamma\sin\varphi_1, 0), \quad h_{\mu\nu}\Pi_1^{\mu\nu} = 0.$$

The radiation power is also determined by Eq. (4.20), but the spin precession frequency has another form

$$\omega = \lambda/\gamma = q(\cos\eta - \beta\sin\eta). \tag{4.23}$$

In the real form the corresponding expression for radiation power is as follows

$$\overline{W}^\mu = \frac{128}{3}\frac{\mu^6 Q^4}{c^3\hbar^4}\left(\frac{(\cos\eta - \beta\sin\eta)^2}{1 - \beta^2}\right)^2. \tag{4.24}$$

Eq. (4.24) was also obtained in the quantum theory of neutron radiation[64] (at $\zeta = 2$).

If we put $\beta = \tan\eta = \sin\chi$ in the above-mentioned expressions for this case, then we obtain the results for the motion of a magneton in the Wien-filter-type fields. The motion in these fields is of interest because the charged particle (electron) moves uniformly and rectilinearly when its velocity is equal to $\beta = \sin\chi$. Earlier, it has been shown (see Chapter 2) that the spin precession frequency rapidly decreases with increasing particle velocity: $\omega = h/\gamma^2$.

3. Parallel fields of different magnitudes: $\boldsymbol{H} = (0, 0, Q\cos\eta)$, $\boldsymbol{E} = (0, 0, Q\sin\eta)$. We shall consider that the magneton moves in the XZ plane at the angle α with respect to the field directions

$$\boldsymbol{\beta} = \beta(\sin\alpha, 0, \cos\alpha).$$

The spin precession is described by the expressions given by Eq. (2.136). One can show that the condition $h_{\mu\nu}\Pi_1^{\mu\nu} = 0$ is valid again. Therefore, the radiation power is described by Eq. (4.20) as before. The only difference is that the spin precession frequency is specified by the expression

$$\omega = q\sqrt{\cos^2\eta(1 - \beta^2\cos^2\alpha) - \beta^2\sin^2\eta\sin^2\alpha}. \tag{4.25}$$

In the real form we obtain the radiation power as follows

$$\overline{W^\mu} = \frac{128}{3}\frac{\mu^6 Q^4}{\hbar^4 c^3}\left(\frac{\cos^2\eta + \delta^2\sin^2\eta}{1 - \delta^2}\right)^2, \tag{4.26}$$

where

$$\delta^2 = \frac{\beta^2\sin^2\alpha}{1 - \beta^2\cos^2\alpha}, \quad 1 - \delta^2 = \frac{1 - \beta^2}{1 - \beta^2\cos^2\alpha}.$$

For $\zeta = 2$, this formula is in complete agreement with that for the radiation power from a neutron in parallel fields obtained by I.M. Ternov, V.G. Bagrov, and others[63] using the quantum theory based on the Dirac-Pauli equation (see Chapter 3).

At $\eta = 0$ we obtain the radiation power in a uniform magnetic field. Setting $\eta = \pi/2$, one can obtain the radiation power in the case of magneton motion in a uniform electric field with the precessional frequency $\omega = q\beta\cos\alpha$.

Thus, the radiation of a magneton (spin light) has a great number of diversities even in the simplest case of uniform motion. Further, we shall see that in the presence of a charge the problem becomes complicated due to interference of charge radiation and the spin magnetic moment radiation.

4.3. Radiation of the Relativistic Magnetic Moment in Uniform Fields

Let us consider the radiation of a magneton moving uniformly and rectilinearly in constant and homogeneous electromagnetic fields of various configurations ($\boldsymbol{E} \perp \boldsymbol{H}$, $\boldsymbol{E} \parallel \boldsymbol{H}$ where $E \neq H$) as a supplement to the classical theory of μ-radiation from a magneton developed by the authors.

4.3.1. Angular Distribution and Frequency of Radiation

The fields of a magneton in a uniform rectilinear motion in the wave zone are equal, according to Eq. (4.15) ($\mu < 0$)

$$\widetilde{\boldsymbol{E}} = \frac{|\mu|}{\tilde{r}c^2\gamma^3}\,\frac{[\boldsymbol{n}[\boldsymbol{n}[(\boldsymbol{n}-\boldsymbol{\beta})\,\overset{\circ\circ}{\boldsymbol{\Pi}}]]]}{[1-(\boldsymbol{n}\boldsymbol{\beta})]^3}, \quad \widetilde{\boldsymbol{H}} = [\boldsymbol{n}\widetilde{\boldsymbol{E}}]. \tag{4.27}$$

The time-average radiation power is given by ($\boldsymbol{\Pi} = \gamma\boldsymbol{\pi}$)

$$\overline{W^\mu} = \frac{\mu^2}{4\pi c^3}\oint\left\{\frac{\overline{\ddot{\boldsymbol{\pi}}^2}}{[1-(\boldsymbol{n}\boldsymbol{\beta})]^3} + 2\frac{\overline{(\boldsymbol{n}\ddot{\boldsymbol{\pi}})(\boldsymbol{\beta}\ddot{\boldsymbol{\pi}})}}{[1-(\boldsymbol{n}\boldsymbol{\beta})]^4} - \frac{(1-\beta^2)\overline{(\boldsymbol{n}\ddot{\boldsymbol{\pi}})^2}}{[1-(\boldsymbol{n}\boldsymbol{\beta})]^5}\right\}d\Omega, \tag{4.28}$$

where $d\Omega$ is the solid angle, and the superior bar means an average over the precession period of the magneton.

It is interesting that the substitution $|\mu|\ddot{\boldsymbol{\pi}} \to e\boldsymbol{a}$ formally converts this expression into the known expression for the power radiated by charge. Let us consider the angular distribution and frequency of the radiation in the above particular cases.

1. Uniform magnetic field.

As before, we assume that the point magneton is moving in the XZ-plane at the angle α with respect to the magnetic field $\boldsymbol{H} = (0, 0, H)$, i.e.

$$\boldsymbol{\beta} = \beta(\sin\alpha, 0, \cos\alpha).$$

The frequency of the radiation of the magneton as determined through the Lorentz transformation is

$$\widetilde{\omega} = \frac{\omega}{1-\beta\cos\theta} = \frac{\omega_0}{\gamma(1-\beta\cos\theta)}, \quad \omega_0 = \frac{2|\mu|H_0}{\hbar},$$

where $\omega_0 = \gamma\omega$ and $H_0 = \gamma H\sqrt{1-\beta^2\cos^2\theta}$ are the frequency and intensity of the magnetic field in the proper frame of the magneton.

This expression is exactly the same as that for the frequency of the radiation of a neutron in the relativistic quantum theory[59].

After evaluating the expectated values and substituting them into Eq. (4.28), we find

$$\begin{aligned}\overline{W^\mu} &= \frac{\mu^2\zeta^2}{8\pi c^3}\,\omega^4\oint\left\{\frac{[1-\beta(\sin\alpha\sin\theta\cos\varphi+\cos\alpha\cos\theta)]^2}{[1-\beta(\sin\alpha\sin\theta\cos\varphi+\cos\alpha\cos\theta)]^5} + \right.\\ &\left. + \frac{h^2}{\lambda^2}\frac{(\beta\cos\alpha-\cos\theta)^2}{[1-\beta(\sin\alpha\sin\theta\cos\varphi+\cos\alpha\cos\theta)]^5}\right\}d\Omega,\end{aligned} \tag{4.29}$$

where $\lambda = \gamma\omega$, ω is the spin precession frequency specified by Eq. (4.21), and the spherical angles correspond to the following definition of the unit vector $\boldsymbol{n}$

$$\boldsymbol{n} = (\sin\theta\cos\varphi, \sin\theta\sin\varphi, \cos\theta).$$

The integration over the angles can be carried out most simply by rotating the X- and Z-axes through an angle α around the Y-axis

$$\begin{aligned} \boldsymbol{n} &= (\cos\alpha\sin\theta'\cos\varphi' + \sin\alpha\cos\theta', \sin\theta'\sin\varphi' \,, \\ & \quad -\sin\alpha\sin\theta'\cos\varphi' + \cos\alpha\cos\theta') \,. \end{aligned}$$

Then Eq.(4.29) is simplified becoming

$$\overline{W^{\mu}} = \frac{\mu^2\zeta^2}{8\pi c^3}\,\omega^4 \oint \left\{ \frac{1}{(1-\beta\cos\theta')^3} + \frac{h^2}{\lambda^2}\, \frac{[\cos\alpha(\beta-\cos\theta') + \sin\alpha\sin\theta'\cos\varphi']^2}{(1-\beta\cos\theta')^5} \right\} d\Omega',$$

where $d\Omega' = \sin\theta' d\theta' d\varphi'$. After integration over the angles, we obtain Eq. (4.20).

We can see in Fig. 4.1 the indicatrix of radiation and its XZ-plane section in this case, when $\alpha = 0$.

It is interesting that the power of radiation when the magneton moves in a direction transverse with respect to the magnetic field is greater by a factor of γ^4 than that in the case of motion in the longitudinal direction

$$\overline{W^{\mu}_{\perp}} = \gamma^4 \overline{W^{\mu}_{\parallel}}.$$

For the motion along the field, the power of radiation is completely independent of the velocity

$$\overline{W^{\mu}_{\parallel}} = \frac{2^5}{3}\,\frac{\mu^6\zeta^2 H^4}{c^3\hbar^4}.$$

In the nonrelativistic appoximation we have the following familiar result from classical[16] ($\zeta = 1$) and quantum[61] ($\zeta = 2$) mechanics (see also Sec. 4.8)

$$\overline{W^{\mu}_{0}} = \frac{2\mu^2\omega_0^4}{3c^3}\zeta^2, \quad \omega = \frac{2|\mu|H_0}{\hbar}.$$

Thus, the adequate classical model of radiation with spin-flip is found.

2. Mutually orthogonal fields of different magnitudes: $\boldsymbol{H} = (0, 0, Q\cos\eta)$, $\boldsymbol{E} = (0, Q\sin\eta, 0)$.

First, let us consider the linear motion of a magneton in the plane of the fields

$$\boldsymbol{\beta} = \beta(0, \sin\alpha, \cos\alpha).$$

The angular distribution of the radiation, averaged over a spin precession period, is determined by Eq. (4.28). It is worth simplifying the calculations by rotating the coordinate system about the X axis through an angle α. Then

$$\begin{aligned} \boldsymbol{n} &= (\sin\theta'\cos\varphi', \cos\alpha\sin\theta'\sin\varphi' + \sin\alpha\cos\theta', \\ & \quad -\sin\alpha\sin\theta'\sin\varphi' + \cos\alpha\cos\theta'), \end{aligned}$$

$$1 - (\boldsymbol{n\beta}) = 1 - \beta\cos\theta'.$$

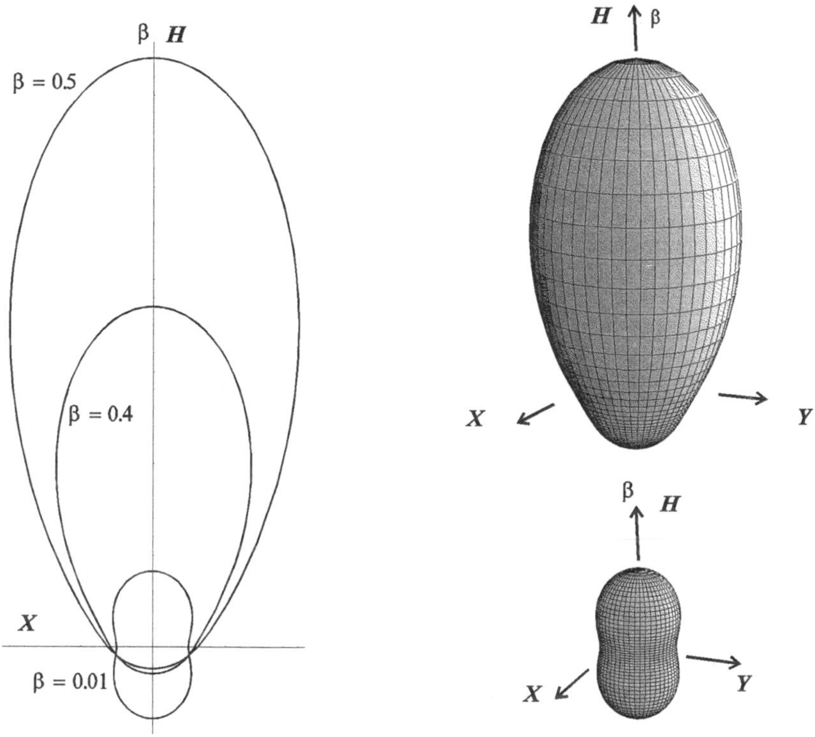

Fig. 4.1. Space indicatrix of radiation and its XZ plane section when $\boldsymbol{\beta} \parallel \boldsymbol{H}$.

The angular distribution of the total radiation power is rather difficult in this case. But it can be considerably simplified by integration over the polar angle φ'. Then we obtain

$$\overline{W}^{\mu} = \frac{2}{3}\frac{\mu^2\zeta^2\lambda^4}{c^3}\int\limits_0^{\pi} F(\theta')\sin\theta' d\theta', \tag{4.30}$$

where

$$\int\limits_0^{\pi} F(\theta')\sin\theta' d\theta' = 1, \quad F(\theta') = \kappa f(\theta'), \quad \kappa = \frac{3}{16\gamma^4},$$

$$f(\theta') = 2\frac{1+(q/\lambda)^2\cos^2\eta\cos^2\alpha}{(1-\beta\cos\theta')^3} + \frac{1}{\gamma^2}\sin^2\theta'\,\frac{1-3(q/\lambda)^2\cos^2\eta\cos^2\alpha}{(1-\beta\cos\theta')^5}. \tag{4.31}$$

The radiation frequency is

$$\widetilde{\omega} = \omega/(1-\beta\cos\theta'),$$

where the precessional frequency of the magneton is determined by Eq. (4.22).

Now let us consider the radiation emitted by a magneton moving along the X-axis perpendicular to the plane of the fields

$$\boldsymbol{H} = (0,0,Q\cos\eta), \quad \boldsymbol{E} = (0,Q\sin\eta,0), \quad \boldsymbol{\beta} = (\beta,0,0).$$

The parameter η affects the radiation power through the precessional frequency only (see Eq. (4.23)).

The angular distribution of the radiation determined from Eq. (4.28) is of the form

$$\overline{W}^{\mu} = \frac{\mu^2\zeta^2\omega^4}{8\pi c^3}\oint\left\{\frac{1}{(1-\beta\sin\theta\cos\varphi)^3} + \frac{(1-\beta^2)\cos^2\theta}{(1-\beta\sin\theta\cos\varphi)^5}\right\} d\Omega. \tag{4.32}$$

Formally, we have the same expression in a purely magnetic field when a magneton moves perpendicular to the magnetic field ($\alpha = \pi/2$) (see Eq. (4.29)), although here the magnitude of $\lambda = \omega/\gamma$ is different.

The indicatrix of radiation and its XZ plane section are shown in Fig. 4.2.

3. Parallel fields of different magnitudes: $\boldsymbol{H} = (0,0,Q\cos\eta)$, $\boldsymbol{E} = (0,0,Q\sin\eta)$.

We consider the motion of a magneton in the XZ-plane at the angle α with respect to the field directions

$$\boldsymbol{\beta} = \beta(\sin\alpha, 0, \cos\alpha).$$

To simplify further integration, we rotate the coordinate system through an angle α about the Y axis

$$\begin{aligned}\boldsymbol{n} \;=\;& (\cos\alpha\sin\theta'\cos\varphi' + \sin\alpha\cos\theta', \sin\theta'\sin\varphi',\\ & -\sin\alpha\sin\theta'\cos\varphi' + \cos\alpha\cos\theta').\end{aligned}$$

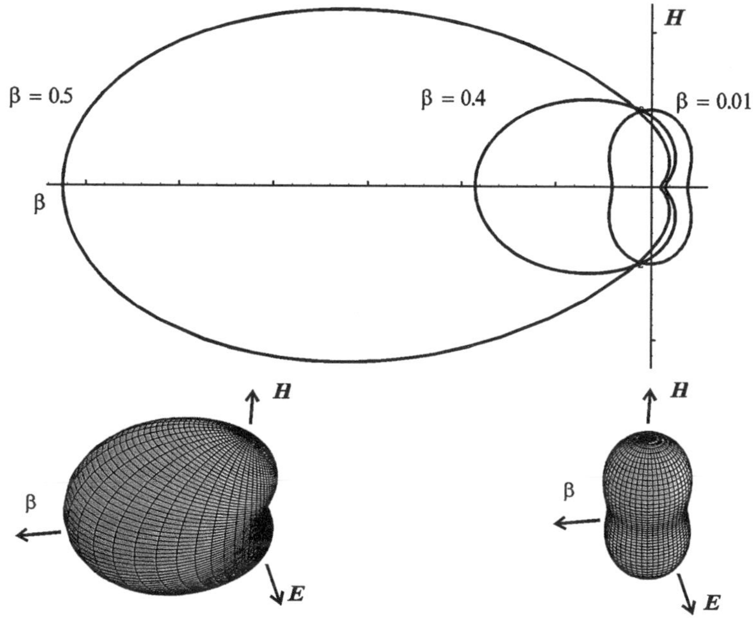

Fig. 4.2. Space indicatrix of radiation and its XZ plane section in the case of magneton motion perpendicular to the plane of the mutually orthogonal fields.

Then using Eq. (4.28) after averaging over the period of spin precession and integration over the angle φ, we obtain the angular distribution of the radiation power determined by Eqs (4.30) and (4.31). The difference is associated with the precession frequency (see Eq. (4.25)).

4.3.2. Linear and Circular Polarization of the Radiation

We define radiation polarization by means of the unit vectors (see Fig. 1.4a)

$$\boldsymbol{e}_\pi = (\cos\theta\cos\varphi, \cos\theta\sin\varphi, -\sin\theta), \quad \boldsymbol{e}_\sigma = (-\sin\varphi, \cos\varphi, 0),$$

which form with the unit vector $\boldsymbol{n}$ along the direction to the observer a triad of orthogonal vectors

$$(\boldsymbol{n}\boldsymbol{e}_\pi) = (\boldsymbol{n}\boldsymbol{e}_\sigma) = (\boldsymbol{e}_\sigma\boldsymbol{e}_\pi) = 0, \quad \boldsymbol{e}_\pi = [\boldsymbol{e}_\sigma\boldsymbol{n}].$$

According to Eq. (4.27) we have ($s = \pi, \sigma$)

$$\widetilde{E}_s = (\boldsymbol{e}_s\widetilde{\boldsymbol{E}}) = \frac{|\mu|}{\tilde{r}c^2\gamma^3}\,\frac{([(\boldsymbol{n}-\boldsymbol{\beta})\boldsymbol{e}_s]\,\overset{\circ\circ}{\boldsymbol{\Pi}})}{[1-(\boldsymbol{n}\boldsymbol{\beta})]^3},$$

where

$$([(\boldsymbol{n}-\boldsymbol{\beta})\boldsymbol{e}_s]\,\overset{\circ\circ}{\boldsymbol{\Pi}}) = \left\{\begin{array}{l} \left(\{\boldsymbol{e}_\sigma - [\boldsymbol{\beta}\boldsymbol{e}_\pi]\}\,\overset{\circ\circ}{\boldsymbol{\Pi}}\right) \\ -\left(\{\boldsymbol{e}_\pi + [\boldsymbol{\beta}\boldsymbol{e}_\sigma]\}\,\overset{\circ\circ}{\boldsymbol{\Pi}}\right) \end{array}\right. ,$$

and also

$$\begin{aligned} \{\boldsymbol{e}_\sigma - [\boldsymbol{\beta}\boldsymbol{e}_\pi]\} &= (-\sin\varphi(1-\beta\cos\alpha\cos\theta), \beta\sin\alpha\sin\theta + \cos\varphi(1-\beta\cos\alpha\cos\theta), \\ &\quad -\beta\sin\alpha\cos\theta\sin\varphi), \\ -\{\boldsymbol{e}_\pi + [\boldsymbol{\beta}\boldsymbol{e}_\sigma]\} &= (\cos\varphi(\beta\cos\alpha - \cos\theta), \sin\varphi(\beta\cos\alpha - \cos\theta), \sin\theta - \beta\sin\alpha\cos\varphi). \end{aligned}$$

1. Uniform magnetic field.

For the general case of an arbitrary direction of the magneton with respect to the external magnetic field, the projections $\widetilde{E}_\pi$ and $\widetilde{E}_\sigma$ are given by some rather complex expressions

$$\begin{aligned} \widetilde{E}_\pi &= -\frac{|\mu|\zeta h\lambda}{\tilde{r}c^2\gamma^2}\left\{\frac{(1-\beta\cos\alpha\cos\theta)\sin\varphi\sin(\omega t+\varphi_1)}{[1-\beta(\sin\alpha\sin\theta\cos\varphi+\cos\alpha\cos\theta)]^3} + \right. \\ &+ \left.\frac{\lambda}{h}\frac{[\cos\varphi - \beta(\cos\alpha\cos\theta\cos\varphi+\sin\alpha\sin\theta)]\cos(\omega t+\varphi_1)}{[1-\beta(\sin\alpha\sin\theta\cos\varphi+\cos\alpha\cos\theta)]^3}\right\}, \\ \widetilde{E}_\sigma &= -\frac{|\mu|\zeta h\lambda}{\tilde{r}c^2\gamma^2}(\cos\theta-\beta\cos\alpha)\left\{\frac{\cos\varphi\sin(\omega t+\varphi_1)}{[1-\beta(\sin\alpha\sin\theta\cos\varphi+\cos\alpha\cos\theta)]^3} + \right. \\ &+ \left.\frac{\lambda}{h}\frac{\sin\varphi\cos(\omega t+\varphi_1)}{[1-\beta(\sin\alpha\sin\theta\cos\varphi+\cos\alpha\cos\theta)]^3}\right\}. \end{aligned}$$

However, in the case of primary interest, $\alpha = 0$ (motion along the magnetic field) and $\alpha = \pi/2$ (motion across this field), these expressions can be simplified and written in the form

$$\widetilde{E}_\pi = -b_\pi \cos(\omega t + \varphi_1 - \varphi), \quad \widetilde{E}_\sigma = -b_\sigma \sin(\omega t + \varphi_1 - \varphi),$$

where

$$b_\pi = \frac{|\mu|\zeta h^2}{\tilde{r}c^2\gamma^2} \frac{1}{(1-\beta\cos\theta)^2}, \quad b_\sigma = \frac{|\mu|\zeta h^2}{\tilde{r}c^2\gamma^2} \frac{\cos\theta - \beta}{(1-\beta\cos\theta)^3} \tag{4.33}$$

in the case of $\alpha = 0$ and

$$\begin{aligned} b_\pi &= \frac{|\mu|\zeta h^2}{\tilde{r}c^2} \frac{\sqrt{(1-\beta\sin\theta\cos\varphi)^2 - \beta^2\cos^2\theta\sin^2\varphi}}{(1-\beta\sin\theta\cos\varphi)^3}, \\ b_\sigma &= \frac{|\mu|\zeta h^2}{\tilde{r}c^2} \frac{\cos\theta\sqrt{1-\beta^2\cos^2\varphi}}{(1-\beta\sin\theta\cos\varphi)^3} \end{aligned}$$

in the case of $\alpha = \pi/2$.

We write the polarization components of the radiation power in the form

$$\overline{W}_s^\mu = \frac{2}{3}\frac{\mu^2\zeta^2}{c^3}\lambda^4 \oint F_s(\theta,\varphi)d\Omega.$$

The total angular distribution of the radiation, $F = F_\pi + F_\sigma$, is normalized (to unity)

$$\oint F(\theta,\varphi)d\Omega = 1.$$

For further discussion it is convenient to single out the normalization factor κ from F

$$\oint F(\theta,\varphi)d\Omega = \kappa \oint f(\theta,\varphi)d\Omega = 1, \quad \kappa = \frac{3}{16\pi\gamma^4}.$$

The functions $F_s = \kappa f_s$ have the same factor.

Here are expressions for the functions f_s in the cases of interest

$$f_\pi = \frac{1}{(1-\beta\cos\theta)^3}, \quad f_\sigma = \frac{(\cos\theta-\beta)^2}{(1-\beta\cos\theta)^5}$$

for $\alpha = 0$ and

$$\begin{aligned} f_\pi &= \frac{(1-\beta\sin\theta\cos\varphi)^2 - \beta^2\cos^2\theta\sin^2\varphi}{(1-\beta\sin\theta\cos\varphi)^5}, \\ f_\sigma &= \frac{\cos^2\theta(1-\beta^2\cos^2\varphi)}{(1-\beta\sin\theta\cos\varphi)^5} \end{aligned}$$

for $\alpha = \pi/2$.

We see that in the case of motion along the magnetic field the polarization components, like the total angular distribution of the radiation, are independent on φ and

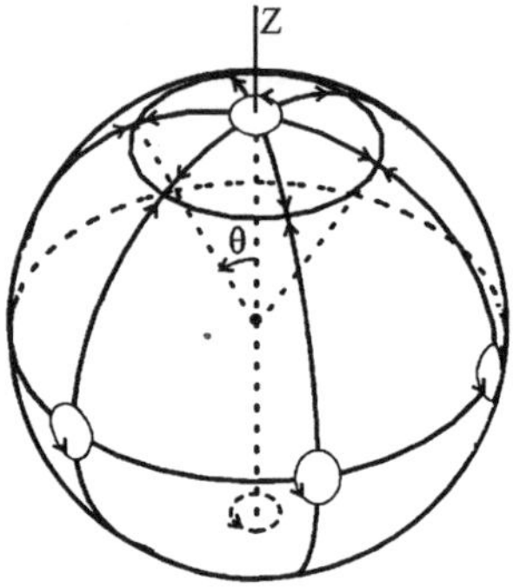

Fig. 4.3. Magneton radiation polarization in the motion along the magnetic field direction.

have the azimuthal symmetry. Completely polarized radiation (the π- component) is observed at the angle $\theta = \arccos\beta$ (see Fig. 4.3). After the integration over θ we finally obtain

$$\overline{W}^{\mu}_{\pi} = \frac{3}{4}\overline{W}^{\mu}, \quad \overline{W}^{\mu}_{\sigma} = \frac{1}{4}\overline{W}^{\mu}, \quad \overline{W}^{\mu} = \overline{W}^{\mu}_{\pi} + \overline{W}^{\mu}_{\sigma} = \overline{W}^{\mu}_{\|}. \tag{4.34}$$

For the case of transverse motion we find

$$\overline{W}^{\mu}_{\pi} = \frac{3}{4}\left(1 - \frac{1}{6}\beta^2 + \frac{1}{12}\beta^4\right)\gamma^4\overline{W}^{\mu}_{\|}, \quad \overline{W}^{\mu}_{\sigma} = \frac{1}{4}\left(1 + \frac{1}{2}\beta^2 - \frac{1}{4}\beta^4\right)\gamma^4\overline{W}^{\mu}_{\|}, \tag{4.35}$$

$$\overline{W}^{\mu} = \overline{W}^{\mu}_{\pi} + \overline{W}^{\mu}_{\sigma} = \gamma^4\overline{W}^{\mu}_{\|} = \overline{W}^{\mu}_{\perp}.$$

It is interesting that in the first case, when $\alpha = 0$, neither the degree of linear polarization nor the total radiation power depends on the velocity of the magneton.

Circular polarization is of greatest interest in the case of $\alpha = 0$, when the radiation has azimuthal symmetry. The circular-polarization amplitudes $a_{\pm 1}$ are related to the linear polarization amplitudes b_π and b_σ by the formulas

$$a_{+1} = \frac{b_\pi + b_\sigma}{\sqrt{2}}, \quad a_{-1} = \frac{b_\pi - b_\sigma}{\sqrt{2}}.$$

Using Eq. (4.33) we find the following expressions for these amplitudes

$$a_{\pm 1} = \frac{21}{\sqrt{2}}\frac{|\mu|h^2\zeta}{\tilde{r}c^2\gamma^2}\frac{(1 \pm \beta)(1 \mp \cos\theta)}{(1 - \beta\cos\theta)^3}.$$

The radiation power averaged over the spin precession period is

$$\overline{W}^{\mu}_{\pm 1} = \frac{\mu^2\omega^4\zeta^2}{8\pi c^3}\oint F_{\pm 1}d\Omega,$$

where

$$F_{\pm 1} = \kappa f_{\pm 1}, \quad f_{\pm 1} = \frac{1}{2}(1 \pm \beta)^2 \frac{(1 \mp \cos\theta)^2}{(1-\beta\cos\theta)^5}. \tag{4.36}$$

It follows that only the right-hand-polarized radiation will be observed in the direction of $\theta = 0$, and only the left-hand-polarized radiation in the opposite direction of $\theta = \pi$. At $\cos\theta = \beta$, the difference between the circular polarization of the radiation disappears, and, as we have already mentioned, the radiation is completely linearly polarized in this case.

Note that the function $F = F_{+1} + F_{-1}$ is normalized to unity, as in the linear-polarized case.

After an integration over the angles we have

$$\overline{W}^{\mu}_{\pm 1} = \frac{1}{2}\left(1 \pm \frac{1}{2}\beta\right)\overline{W}^{\mu}, \quad \overline{W}^{\mu} = \overline{W}^{\mu}_{+1} + \overline{W}^{\mu}_{-1} = \overline{W}^{\mu}_{\parallel}. \tag{4.37}$$

All the results found agree completely with the quantum theory[59,63].

2. Mutually orthogonal fields of different magnitudes.

We shall define the linearly polarized radiation relative to the vector $\boldsymbol{\beta}' = (0,0,\beta)$. The corresponding unit vectors of polarization are then of the form

$$\boldsymbol{e}_{\pi} = (\cos\theta'\cos\varphi', \cos\theta'\sin\varphi', -\sin\theta'), \quad \boldsymbol{e}_{\sigma} = (-\sin\varphi', \cos\varphi', 0).$$

In this case we have

$$\begin{aligned} \{\boldsymbol{e}_{\sigma} - [\boldsymbol{\beta}\boldsymbol{e}_{\pi}]\} &= (-\sin\varphi'(1-\beta\cos\theta'), \cos\varphi'(1-\beta\cos\theta'), 0), \\ -\{\boldsymbol{e}_{\pi} + [\boldsymbol{\beta}\boldsymbol{e}_{\sigma}]\} &= (\cos\varphi'(\beta-\cos\theta'), \sin\varphi'(\beta-\cos\theta'), \sin\theta'). \end{aligned}$$

We first consider the motion of a magneton in the plane of fields. For further calculations we have to transform the components of the spin precession vector (see Sec. 2.7) to a new coordinate system. Acting in the standard way after integration with respect to φ' we obtain ($s = \pi, \sigma$)

$$\overline{W}^{\mu}_{s} = \overline{W}^{\mu} \int F_s(\theta')\sin\theta' d\theta', \quad F_s(\theta') = \kappa f_s(\theta'), \tag{4.38}$$

where

$$\begin{aligned} f_{\pi}(\theta') &= \frac{1 + (q/\lambda)^2\cos^2\eta\cos^2\alpha}{(1-\beta\cos\theta')^3}, \\ f_{\sigma}(\theta') &= f_{\pi}(\theta') + \frac{1}{\gamma^2}\sin^2\theta' \frac{1 - 3(q/\lambda)^2\cos^2\eta\cos^2\alpha}{(1-\beta\cos\theta')^5}, \end{aligned} \tag{4.39}$$

$$f_{\pi}(\theta') + f_{\sigma}(\theta') = f(\theta').$$

Subsequent integration with respect to θ' gives

$$\overline{W}^{\mu}_{\pi} = \frac{3}{8}\left(1 + \frac{q^2}{\lambda^2}\cos^2\eta\cos^2\alpha\right)\overline{W}^{\mu}, \quad \overline{W}^{\mu}_{\sigma} = \frac{5}{8}\left(1 - \frac{3}{5}\frac{q^2}{\lambda^2}\cos^2\eta\cos^2\alpha\right)\overline{W}^{\mu},$$

$$\overline{W}^\mu_\pi + \overline{W}^\mu_\sigma = \overline{W}^\mu.$$

In the ultrarelativistic case when $\gamma \gg 1$, the polarization components are independent of the angle α.

By analogy we consider the linear polarization of radiation emitted by a magneton moving perpendicularly to the plane of fields. As a result, we obtain Eq. (4.38). But the functions $f_s(\theta')$ differ from Eq. (4.39) in this case

$$f_\pi(\theta') = \frac{1}{(1-\beta\cos\theta')^3}, \quad f_\sigma(\theta') = f_\pi(\theta') + \frac{1}{\gamma^2}\frac{\sin^2\theta'}{(1-\beta\cos\theta')^5}.$$

It should be noted that along the direction of motion of the magneton ($\theta' = 0$) the functions f'_π and f'_σ are the same and, consequently, the radiation is purely circularly polarized. The same situation also occurred in the previous case [see Eq. (4.39)]. This phenomenon can evidently be used to determine the direction of motion of the magneton.

After integration with respect to θ' we obtain

$$\overline{W}^\mu_\pi = \frac{3}{8}\,\overline{W}^\mu, \quad \overline{W}^\mu_\sigma = \frac{5}{8}\,\overline{W}^\mu.$$

3. Mutually parallel fields of different magnitudes.

Linear polarization can be calculated in a way similar to that in the previous case. We obtain the same expressions as Eqs (4.38) and (4.39). Only the precession frequencies are different.

4.3.3. Ultrarelativistic Case

First we consider the motion along the magnetic field ($\alpha = 0$). In this case, with $\beta \to 1$ it is convenient to introduce the variable $z = \gamma\sin\theta \approx \gamma\theta$. Since the radiation is azimuthally symmetric, we have $d\Omega = 2\pi\sin\theta d\theta \approx 2\pi\gamma^{-2}zdz$. Thus, we have

$$\oint(\ldots)d\Omega \to \frac{2\pi}{\gamma^2}\int\limits_0^\infty(\ldots)zdz. \tag{4.40}$$

For the functions f_s we have

$$f_\pi = \frac{8\gamma^6}{(1+z^2)^3}, \quad f_\sigma = \frac{8\gamma^6(1-z^2)^2}{(1+z^2)^5}, \quad f = f_\pi + f_\sigma = \frac{16\gamma^6(1+z^4)}{(1+z^2)^5}.$$

An integration over the angles carried out in accordance with Eq. (4.40) gives us polarization in Eq. (4.34) again. As we have already mentioned, the polarization is independent of the velocity.

In the other extreme case of motion across the magnetic field ($\alpha = \pi/2$), the most convenient variable is $x = \gamma\cos\theta \approx \gamma\psi$, where $\psi = \pi/2 - \theta$. Now we have

$d\Omega = \sin\theta d\theta d\varphi \approx -\gamma^{-2} d\varphi dx$ and

$$\oint (\ldots) d\Omega \to \frac{2}{\gamma} \int_0^\infty (\ldots) d\varphi dx.$$

After the integration over φ we find

$$\int_0^{2\pi} f_\pi d\varphi = \frac{5\pi}{2}\,\gamma^4\,\frac{(8+x^2)x^2}{(1+x^2)^{9/2}}, \quad \int_0^{2\pi} f_\sigma d\varphi = \frac{\pi}{2}\,\gamma^4\,\frac{12+7x^2}{(1+x^2)^{7/2}},$$

$$\int_0^{2\pi} f d\varphi = \frac{5\pi}{2}\,\gamma^4\,\frac{12+59x^2+12x^4}{(1+x^2)^{9/2}}.$$

Now integrating over x we obtain

$$\overline{W}^\mu_\pi = \frac{11}{16}\,\gamma^4 \overline{W}^\mu_\parallel = \frac{11}{16}\,\overline{W}^\mu_\perp, \quad \overline{W}^\mu_\sigma = \frac{5}{16}\,\gamma^4 \overline{W}^\mu_\parallel = \frac{5}{16}\,\overline{W}^\mu_\perp.$$

This result also follows from Eq. (4.35) if we put $\beta \to 1$.

The circular polarization ($\alpha = 0$) is given by the functions

$$f_{+1} = \frac{16\gamma^6}{(1+z^2)^5}, \quad f_{-1} = \frac{16\gamma^6 z^4}{(1+z^2)^5},$$

$$f_{+1} + f_{-1} = f,$$

where $f_{\pm 1}$ were determined in Eq. (4.36). After the integration over z we find

$$\overline{W}^\mu_{+1} = \frac{3}{4}\,\overline{W}^\mu_\parallel, \quad \overline{W}^\mu_{-1} = \frac{1}{4}\,\overline{W}^\mu_\parallel,$$

which could also be derived from Eq. (4.37) in the ultrarelativistic limit.

We note in conclusion that the power of the radiation from a relativistic magneton which we have been discussing here is much lower than the power radiated by a relativistic electron. Nevertheless, in the case of transverse motion of the Bohr magneton we obtain the following ratio with the synchrotron radiation of an electron ($\zeta = 2$)

$$\frac{\overline{W}^\mu_\perp}{W_{\rm SR}} = \left(\frac{\mu}{\mu_0}\right)^6 \left(\frac{\gamma}{\beta}\,\frac{H}{H^*}\right)^2 = \left(\gamma\,\frac{H}{H^*}\right)^2 = \frac{4}{9}\,\xi^2,$$

where ξ is the invariant relativistic parameter introduced above. We find also that $\overline{W}^\mu_\perp / W_{\rm SR}$ is of the second order of the quantum corrections to the power of SR (see Secs 4.7, 4.8).

4.3.4. Spectral Composition of the Radiation

Let us now consider the spectral properties of the radiation emitted by a relativistic magneton moving uniformly in a straight line in constant and homogeneous electromagnetic fields of various configurations. For this purpose we shall first write down the spectral-angular disrtibution of the radiation in the magneton rest system, averaged over a spin precession period

$$\overline{d^2W_0^\mu} = J_0(\omega_0;\theta_0,\varphi_0)\delta(\omega_0-\widetilde{\omega}_0)d\Omega_0 d\widetilde{\omega}_0. \tag{4.41}$$

The spherical angles θ_0 and φ_0 characterize the direction of the radiation, $d\Omega_0 = \sin\theta_0 d\theta_0 d\varphi_0$ is the corresponding solid angle, δ-function distinguishes the fundamental frequency $\widetilde{\omega}_0$ of the radiation which coincides with the spin precession frequency ω_0 in the rest system, and the specific form of the function J_0 is determined by the choice of the coordinate system and the configuration of the external fields.

Then, using the known relationships (see Sec. 4.2) we perform the Lorentz transformations

$$\overline{d^2W_0^\mu} = \overline{d^2W^\mu}\gamma^2(1-\beta\cos\theta), \tag{4.42}$$

$$\omega_0 = \gamma\omega = \widetilde{\omega}_0, \quad \widetilde{\omega}_0 = \widetilde{\omega}\gamma(1-\beta\cos\theta), \quad d\Omega_0 = d\Omega/[\gamma(1-\beta\cos\theta)]^2.$$

In addition, if θ is the angle between the direction of the magneton velocity $\boldsymbol{\beta}$ and the direction of the radiation, then

$$\cos\theta_0 = \frac{\cos\theta-\beta}{1-\beta\cos\theta}, \quad \sin\theta_0 = \frac{1}{\gamma}\frac{\sin\theta}{1-\beta\cos\theta}. \tag{4.43}$$

Since the ultrarelativistic case is of the greatest interest, for $\beta\to 1$ we have

$$1-\beta\cos\theta = \frac{1+x^2}{2\gamma^2}, \quad \cos\theta-\beta = \frac{1-x^2}{2\gamma^2},, \quad x=\gamma\theta \tag{4.44}$$

The field intensities entering into ω_0 are also transformed

$$\begin{aligned} H_0 &= \gamma\sqrt{H^2-(\boldsymbol{\beta}\boldsymbol{H})^2+[\boldsymbol{\beta}\boldsymbol{E}]^2-2(\boldsymbol{H}[\boldsymbol{\beta}\boldsymbol{E}])}, \\ E_0 &= \gamma\sqrt{E^2-(\boldsymbol{\beta}\boldsymbol{E})^2+[\boldsymbol{\beta}\boldsymbol{H}]^2-2(\boldsymbol{E}[\boldsymbol{\beta}\boldsymbol{H}])}. \end{aligned} \tag{4.45}$$

According to Eqs (4.41)—(4.45), the spectral-angular distribution assumes the form

$$\overline{d^2W^\mu} = J(\omega;\theta,\varphi)\,\frac{\delta[\gamma\omega-\gamma\widetilde{\omega}(1-\beta\cos\theta)]}{\gamma^3(1-\beta\cos\theta)^2}\,d\Omega d\widetilde{\omega}. \tag{4.46}$$

Now let us consider specific cases.

1. Uniform magnetic field: $\boldsymbol{H}=(0,0,H)$, $\boldsymbol{\beta}=\beta(\sin\alpha,0,\cos\alpha)$.

The spectral-angular distribution of the magneton radiation in the rest system, averaged over a spin precession period, is determined by the expression (see Eq. (4.29))

$$\overline{d^2W_0^\mu} = \frac{2}{\pi}\frac{\mu^6\zeta^2H_0^4}{\hbar^4c^3}(1+\cos^2\theta_0)\delta(\omega_0-\widetilde{\omega}_0)d\Omega_0 d\widetilde{\omega}_0. \tag{4.47}$$

Hence, it follows that

$$J(\omega_0; \theta_0) = \frac{2}{\pi} \frac{\mu^6 \zeta^2 H_0^4}{\hbar^4 c^3} (1 + \cos^2 \theta_0).$$

Before applying the Lorentz transformations the coordinate system must be turned through an angle α about the Y-axis. Then

$$\cos \theta_0 = \sin \alpha \sin \theta_0' \sin \varphi_0' - \cos \alpha \cos \theta_0'.$$

One can now use Eq. (4.46) and the transforms of Eqs (4.41)–(4.45). In the ultrarelativistic case we obtain

$$\overline{d^2 W^\mu} = \frac{16}{\pi} \frac{\mu^6 \zeta^2 H^4 \gamma^6}{\hbar^4 c^3} (1 - \beta^2 \cos^2 \alpha)^2 \times \tag{4.48}$$

$$\times \frac{(1 + x^2)^2 + [2x \sin \alpha \sin \varphi' - \cos \alpha (1 - x^2)]^2}{(1 + x^2)^4} \frac{1}{\widetilde{\omega}} \delta \left(1 + x^2 - \frac{\omega_{\max}}{\widetilde{\omega}}\right) d\Omega d\widetilde{\omega},$$

where $d\Omega = x dx d\varphi' / \gamma^2$.

The maximum radiation frequency (see below) is $\omega_{\max} = 2\gamma^2 \omega$. Substituting here the spin precession frequency $\omega = h\sqrt{1 - \beta^2 \cos^2 \alpha}$, we obtain

$$\omega_{\max} = \frac{4|\mu| H}{\hbar} \gamma^2 \sqrt{1 - \beta^2 \cos^2 \alpha}.$$

It is interesting that in the ultrarelativistic case, for $\alpha = \pi/2$ (motion perpendicular to the magnetic field) the maximum frequency is shifted to shorter wavelengths.

Using the properties of δ-function, after integration over the angle φ' one can obtain the spectral-angular distribution of the radiation power in the form

$$\overline{d^2 W^\mu} = \overline{W}^\mu \int_0^1 d\xi \int_0^\infty f x, \xi x dx,$$

$$f(x, \xi) = 3 \frac{(1 + x^2)^2 + 2x^2 \sin^2 \alpha + \cos^2 \alpha (1 - x^2)^2}{(1 + x^2)^5} \delta \left(\xi - \frac{1}{1 + x^2}\right),$$

where $\xi = \widetilde{\omega} / \omega_{\max}$.

Then, integrating with respect to x $(0 \leq x < \infty)$ we obtain the spectral distribution of the radiation in the form

$$\overline{dW^\mu} = \overline{W}^\mu F(\xi) d\xi,$$

where $F(\xi)$ is the spectral distribution function normalized by unity (see Fig. 4.4).

$$F(\xi) = \frac{3}{2} [\xi(1 + \cos^2 \alpha) + 2(1 - 3\cos^2 \alpha)(\xi^2 - \xi^3)], \tag{4.49}$$

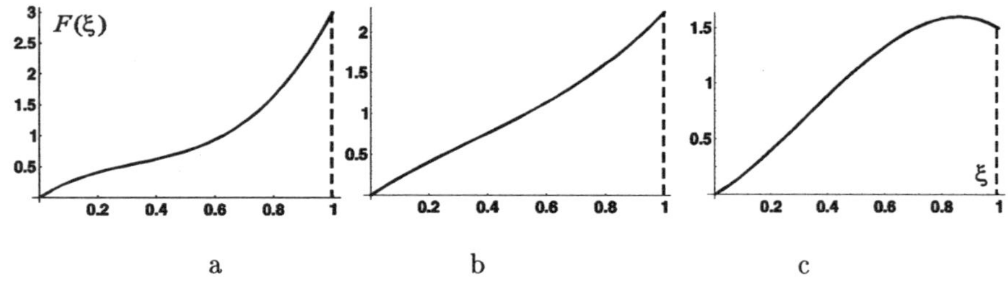

Fig. 4.4. The normalized spectral functions $F(\xi)$ for $\alpha = 0$ (a); $\alpha = \pi/4$ (b); $\alpha = \pi/2$ (c).

$$\int_0^1 F(\xi) d\xi = 1.$$

After integration with respect to ξ we obtain the total radiation power

$$\overline{W}^\mu = \frac{32}{3} \frac{\mu^6 \zeta^2 H^4}{\hbar^4 c^3} \left(\frac{1 - \beta^2 \cos^2 \alpha}{1 - \beta^2} \right)^2 . \tag{4.50}$$

This result is already well-known from the classical (see Eq. (4.19)) and quantum[59] theories of radiation.

The specral functions $F(\xi)$ for $\alpha = 0$ (motion along the field) and $\alpha = \pi/2$ (motion perpendicular to the field) differ considerably from each other. It should be borne in mind [see Eq. (4.50)] that in these extreme cases, the dimentional factors $\overline{W}^\mu$ in the spectral distribution differ by the factor of γ^4!

2. Mutually orthogonal fields

$$\boldsymbol{H} = (0, 0, Q \cos \eta), \quad \boldsymbol{E} = (0, Q \sin \eta, 0), \quad Q = \sqrt{E^2 + H^2}.$$

First let us consider the motion of the magnetic moment in the plane of the fields $\boldsymbol{\beta} = \beta(0, \sin \alpha, \cos \alpha)$.

According to Eq. (4.28), with account of the components of the spin precession vector (see Sec. 2.7) $\overline{d^2 W^\mu_0}$, in this case, is of exactly the same form as in Eq. (4.47) (where $H_0 \to Q_0 = \sqrt{E_0^2 + H_0^2}$), differing only in the frequency

$$\omega_{\max} = \frac{4\mu Q}{\hbar} \gamma^3 \sqrt{\cos^2 \eta (1 - \beta^2 \cos^2 \alpha) + \beta^2 \sin^2 \eta \cos^2 \alpha}.$$

Let us turn our attention to the fact that $\omega_{\max}$ is independent of the angle α in the case of the fields of an equal magnitude.

Further transformations and integration with respect to the angles will evidently give the same spectrum as in the preceding case (see Eq. (4.49)).

In the other extreme case, when the magnetic moment moves along the X axis perpendicularly to the plane of the fields, a similar situation occurs with

$$\omega_{\max} = \frac{4\mu Q}{\hbar} \frac{1}{1+\beta}$$

and the spectral distribution of Eq. (4.49), where one must take $\alpha = \pi/2$.

3. Parallel fields

$$\boldsymbol{H} = (0, 0, Q\cos\eta), \quad \boldsymbol{E} = (0, 0, Q\sin\eta), \quad \boldsymbol{\beta} = \beta(\sin\alpha, 0, \cos\alpha)$$

In the rest system the spectral-angular distribution of the radiation agrees formally with Eq. (4.47) where, instead of H_0, one must put $Q_0 = \sqrt{H_0^2 + E_0^2}$. Thus, that is all reduced to a difference in the frequency

$$\omega_{\max} = \frac{4\mu Q}{\hbar} \gamma^3 \sqrt{\cos^2\eta(1 - \beta^2\cos^2\alpha) + \beta^2 \sin^2\eta \sin^2\alpha}.$$

We have considered the simplest types of radiation of the point intrinsic magnetic moment (magneton). The interesting fact is that the results of the quantum and classical theories are similar in spite of the different approaches in description of μ-radiation. In the quantum theory the radiation of the magnetic moment (neutron) in a uniform motion is accounted for by spin-flip transitions, while in the classical theory by spin precession with respect to the magnetic field direction. The obtained results can be used in the analysis of the SR properties with regard for the intrinsic magnetic moment of an electron. Of course, the radiation features of the intrinsic magnetic moment of the electron differ from the above model of pure μ-radiation. But it is highly important to be able to separate the pure spin light from a great number of accessory phenomena among the quantum corrections proportional to $\hbar$. The quasiclassical (or semiclassical) approach is a reliable method for comparison of classical and quantum theories.

4.4. Relativistic Semiclassical Radiation Theory

The progress in the quantum theory of SR[9] based on the exact solutions of the relativistic Dirac equation and the rigorous methods of quantum electrodynamics made it possible to describe the spectral-angular distribution of SR power with consideration for all quantum features, including spin evolution[9,27,30]. However, the physical interpretation of the quantum corrections to the classical theory remained obscure since there was no simple way of transition to the classical theory. It was not until relatively recently that SR has been found to be a complex structure depending on a number of fundamental phenomena: the charge radiation itself (conventional SR); recoil effects accompanying the radiation; interference of fields of charge radiation and of the spin magnetic moment radiation (mixed $e\mu$-radiation); the spin magnetic moment radiation itself (spin light); and the radiation associated with the anomalous magnetic moment of the electron[15,18].

The interpretation of spin corrections to the SR power is not simple since spin corrections have an equal competitor, namely, the recoil effects accompanying the radiation. They also make contribution to the SR power, proportional to $\hbar$ as well as $\hbar^2$, and, moreover, the recoil effects are essentially different for an electron and a charged spinless particle (boson) in the second order in $\hbar$.

In this situation, the relativistic quasiclassical theory proposed by Schwinger[31] and developed by Baier[32-34] proved to be a handy instrument for research. The basic conclusions of the strict quantum theory[9,27] were replicated in the quasiclassical approach for high-energy particles following macroscopic paths. The full mathematical justification of the quasiclassical approach is given in Ref.[35] Here we use a simplified semiclassical version of this theory (Jackson's semiclassical method[36]) which we used earlier in Refs[10,37-39]. The following approximations are the basis of the semiclassical theory of SR.

1. The electron radiation is formed by the classical trajectory of motion. The quantum nature of motion (discreteness of energy states) is unimportant.

According to the formula of quantization of the electron energy in a uniform magnetic field (see Chapter 3), the number of energy state n is specified by the expression

$$n = \frac{m_0 c \rho \gamma}{2\hbar} = \frac{\rho P}{2\hbar},$$

where ρ is the orbital radius, P is the momentum of an electron. Hence, it follows that

$$\frac{\Delta n}{n} = \frac{\Delta P}{P} + \frac{\Delta \rho}{\rho}.$$

Since the maximum of radiation falls on the frequency

$$\widetilde{\omega} \approx \frac{1}{3}\,\widetilde{\omega}_{\max} = \frac{1}{2}\,\frac{c}{\rho}\,\gamma^3,$$

then

$$\Delta P = \frac{\hbar\widetilde{\omega}}{c} \approx \frac{\hbar\gamma^3}{2\rho}.$$

The quantity of $\Delta\rho$ can be found from the indeterminacy relation

$$\Delta\rho\Delta P \approx \hbar.$$

Finally, we obtain

$$\frac{\Delta n}{n} = \frac{1}{2}\,\gamma\,\frac{H}{H*} + \frac{2}{\gamma^3},$$

where $H^* = m_0^2 c^3 / e_0 \hbar = 4.41 \cdot 10^{13}\ Oe$ is Schwinger's magnetic field.

The condition of smallness of the first term

$$\frac{\Delta P}{P} = \frac{1}{2}\gamma\frac{H}{H^*} = \frac{1}{2}\frac{\hbar\gamma^2}{m_0 c\rho} = \frac{1}{3}\xi \ll 1, \tag{4.51}$$

coincides with the criterion of validity of the classical radiation theory of Eq. (1.95).

The second term monotonously tends to zero with increasing γ.

For typical parameters of synchrotron

$$E = m_0c^2 = 2.5\ GeV, \quad \gamma = 4.9\cdot 10^3, \quad H_\rho = 8.34\cdot 10^6\ Oe\cdot cm, \tag{4.52}$$

$$\xi = 1.07\cdot 10^{-6}$$

we have

$$n = 8.25\cdot 10^{16}, \quad \frac{\Delta n}{n} \approx \frac{\Delta P}{P} = 3.56\cdot 10^{-7}, \quad \frac{\Delta\rho}{\rho} = 1.68\cdot 10^{-11} \ll \frac{\Delta P}{P}. \tag{4.53}$$

Thus, one can assume with a great degree of accuracy that $\Delta n/n \ll 1$. This means that with increasing energy the trajectory of electron becomes a classical one.

2. The purely quantum process of emission itself must not influence the radiation formation.

For realization of this condition it is necessary that the time of emission $\Delta\tilde{t}$ is much less than the time of radiation formation Δt_{rad}, i.e.

$$\frac{\Delta\tilde{t}}{\Delta t_{rad}} = \frac{\Delta\tilde{l}}{\Delta l_{rad}} \ll 1,$$

where

$$\Delta\tilde{l} = c\Delta\tilde{t} = \frac{\hbar c}{\Delta E} = \frac{c}{\tilde{\omega}} \approx \frac{2c}{\omega\gamma^3} = \frac{2\rho}{\gamma^3},$$

$\Delta l_{rad} = \rho/\gamma$ is the length of radiation formation (see Sec. 1.7.4). For typical parameters of Eq. (4.52)

$$\frac{\Delta\tilde{l}}{\Delta l_{rad}} = \frac{2}{\gamma^2} = 0.84\cdot 10^{-7} \ll 1.$$

Thus, one can assume that the process of emission is almost inertialess.

3. The quantum effects connected with the response of electron-positron vacuum (Zitterbewegung, pair production) are negligible. This condition can be derived from the equation of motion of an electron with account of Zitterbewegung (see Chapter 2)

$$\frac{\hbar}{2m_0c^2}\sqrt{\left|\frac{\overset{\circ}{w}_\mu \overset{\circ}{w}{}^\mu}{w_\mu w^\mu}\right|} \ll 1.$$

Substituting there the four-dimensional acceleration and hyperacceleration of an electron in a uniform magnetic field ($e = -e_0 < 0$)

$$w^\mu = -c\gamma^2\omega(0, \cos\omega t, \sin\omega t, 0), \quad \overset{\circ}{w}{}^\mu = c\gamma^3\omega^2(0, \sin\omega t, -\cos\omega t, 0)$$

we obtain the condition of one-particle approximation in quantum electrodynamics

$$\frac{1}{2}\,\frac{H}{H^*} \ll 1.$$

It is clear that if the condition of Eq. (4.51) is realized, then this condition should be realized too.

4. The radiative damping must not influence the validity of the laws of classical electrodynamics. This problem has been discussed in Sec. 1.7 where it was shown that even in the case when the radiative damping force is greater than the Lorentz force, classical electrodynamics with account of radiative damping force is valid in some fields

$$\frac{1}{\alpha\gamma^2}\,H^* \ll H \ll \frac{1}{\gamma}\,H^*. \tag{4.54}$$

The radiation damping force can be neglected on the condition that

$$\alpha\gamma^2\,\frac{H}{H^*} \ll 1.$$

Taking into account for above numerical example the fine structure constant $\alpha = e^2/\hbar c \approx 1/137$ we have

$$\alpha\gamma^2\,\frac{H}{H^*} = 2.54 \cdot 10^{-5} \ll 1.$$

As compared with Eqs (4.51) and (4.53), this limitation is the most strict but at the same time easily realized.

We note that the condition of Eq. (4.51) also means that the electron remains a relativistic particle after the emission of a photon

$$\frac{\Delta P}{P} = \frac{\hbar\tilde{\omega}}{m_0 c p \gamma} \ll 1.$$

Further, we shall not take into account the radiative damping force although the semiclassical method considered here allows such generalization.

Thus, in the relativistic semiclassical radiation theory all features of motion of a relativistic electron are considered to be classical along the trajectory and the quantum processes associated with the radiation are accounted for by the replacement of the electron velocity with the transition matrix element

$$\boldsymbol{\beta} \to \langle f \mid \widehat{\boldsymbol{\alpha}} \mid i \rangle, \tag{4.55}$$

where $\boldsymbol{\beta} = \boldsymbol{u}/c$ is the dimensionless velocity of an electron, $\widehat{\boldsymbol{\alpha}} = \widehat{\boldsymbol{\alpha}}(t)$ are the well-known Dirac matrices in the Heisenberg representation, $\mid i\rangle$ is the ket-vector of the initial state, and $\langle f \mid$ is the bra-vector of the final state. Since the inertiality of an electron motion is not essential for quantum transitions in the range of temperatures typical of SR, the matrix element of Eq. (4.55) is calculated for the wave functions of a free electron (quantum kinematics) with the ket-vector

$$| \, i \rangle = \frac{1}{\sqrt{2}} \begin{pmatrix} \sqrt{\frac{\gamma+1}{2}} \\ \\ \sqrt{\frac{\gamma}{\gamma+1}} (\widehat{\boldsymbol{\sigma}} \boldsymbol{\beta}) \end{pmatrix} | \, \zeta \rangle, \tag{4.56}$$

where $\gamma = 1/\sqrt{1-\beta^2}$ is the relativistic Lorentz factor, $\widehat{\boldsymbol{\sigma}}$ is the Pauli matrix, and the spin states in projection onto the arbitrary direction $\boldsymbol{\zeta} = (\sin\nu\cos\lambda, \sin\nu\sin\lambda, \cos\nu)$ are given by the spin ket-vector

$$| \, \zeta \rangle = \frac{1}{\sqrt{2}} \begin{pmatrix} \zeta\sqrt{1+\zeta\cos\nu} \\ \\ \sqrt{1-\zeta\cos\nu} \quad e^{i\lambda} \end{pmatrix}, \qquad \zeta = \pm 1. \tag{4.57}$$

where λ and ν are the direction angles of the spin projection. The normalization is chosen so that

$$\langle i \mid \widehat{\boldsymbol{\alpha}} \mid i \rangle = \boldsymbol{\beta}, \quad \langle \zeta \mid \widehat{\boldsymbol{\sigma}} \mid \zeta \rangle = \boldsymbol{\zeta}. \tag{4.58}$$

Inhomogeneities in the leading magnetic field are assumed to have no effect on the electron polarization process[40,41].

The matrix element of Eq. (4.55) should be calculated with the laws of conservation of energy and momentum

$$m_0 c^2 \gamma = m_0 c^2 \gamma\,' + \hbar\widetilde{\omega}, \quad \gamma\boldsymbol{\beta} = \gamma\,'\boldsymbol{\beta}\,' + \hbar\widetilde{\omega}\boldsymbol{n}. \tag{4.59}$$

The Fourier transform of the radiation field is specified in the same way as in the classical electrodynamics (cf. Eq. (1.58)).

$$\boldsymbol{E}^s_{\widetilde{\omega}} = i\frac{e\widetilde{\omega}}{Rc} \int \left(\langle f \mid \widehat{\boldsymbol{\alpha}} \mid i \rangle \boldsymbol{n}^s \right) e^{i\widetilde{\omega}\left[t - \frac{1}{c}(\boldsymbol{n}\boldsymbol{r})\right]} dt, \tag{4.60}$$

where $\boldsymbol{n}^s$ are the basis vectors of the linear radiation polarization ($s = \sigma, \pi$), $\boldsymbol{n}$ is the unit vector from the charge towards the point of observation, $\widetilde{\omega}$ is the radiation frequency, and all the other symbols are a conventional notation.

In the ultrarelativistic approximation, all variables are expanded into series in terms of small parameters γ^{-1}, ct/ρ, and ψ. We have (see Sec. 1.7)

$$\begin{aligned} \boldsymbol{\beta} &= \left(1 - \frac{1}{2\gamma^2} - \frac{\psi^2}{2} - \frac{1}{2}\frac{c^2t^2}{\rho^2}\right)\boldsymbol{n} + \frac{ct}{\rho}\boldsymbol{n}_\sigma - \psi\boldsymbol{n}_\pi, \\ \boldsymbol{n} &= \left(1 - \frac{\psi^2}{2}, 0, \psi\right), \quad \boldsymbol{n}_\sigma = (0,1,0), \quad \boldsymbol{n}_\pi = \left(-\psi, 0, 1 - \frac{\psi^2}{2}\right). \end{aligned} \tag{4.61}$$

It is correct to the second order of smallness ($e = -e_0 < 0$, the magnetic field is parallel to the Z-axis).

The spectral-angular distribution of the radiation is specified by the usual expression[9]

$$\frac{d^2W^s}{d\tilde{\omega}d\Omega} = \frac{c^2R^2}{8\pi^3\rho} \mid \widetilde{\boldsymbol{E}}^s_{\tilde{\omega}} \mid^2. \tag{4.62}$$

4.5. Recoil Effects and Mixed $e\mu$-Radiation

We introduce the recoil parameter $\varepsilon = \hbar\tilde{\omega}/m_0c^2\gamma\,'$. Then, instead of Eq. (4.59) we have

$$\gamma\,' = \frac{\gamma}{1+\varepsilon}, \quad \boldsymbol{\beta}\,' = (1+\varepsilon)\boldsymbol{\beta} - \varepsilon\boldsymbol{n}. \tag{4.63}$$

In the linear approximation in $\hbar$, the matrix element of Eq. (4.55) is calculated to be

$$\langle f \mid \hat{\boldsymbol{\alpha}} \mid i\rangle = \left(1+\frac{\varepsilon}{2}\right)\boldsymbol{\beta}\cdot\langle\zeta' \mid \zeta\rangle - i\frac{\varepsilon}{2}\left[\langle\zeta' \mid \hat{\boldsymbol{\sigma}} \mid \zeta\rangle\cdot\left(\boldsymbol{n} - \frac{\gamma}{\gamma+1}\boldsymbol{\beta}\right)\right]. \tag{4.64}$$

In this formula, the first term with $\boldsymbol{\beta}$ describes the radiation of a charge and the recoil effects associated with ε, whereas the second term with the Pauli matrices is responsible for the spin magnetic moment radiation (μ-radiation).

By substituting the matrix element of Eq. (4.64) into Eq. (4.60) we find

$$\begin{aligned} \widetilde{\boldsymbol{E}}^s_{\tilde{\omega}} &= -\frac{ie_0\tilde{\omega}\rho}{Rc^2\gamma^2}\left\{\left(1+\frac{\varepsilon}{2}\right)\langle\zeta' \mid \zeta\rangle U^s_0 + \frac{\varepsilon}{2}\langle\zeta' \mid \hat{\sigma}_z \mid \zeta\rangle U^s_1 + \right. \\ &+ \left. \varepsilon\left[\langle\zeta' \mid \hat{\sigma}_+ \mid \zeta\rangle U^s_2 + \langle\zeta' \mid \hat{\sigma}_- \mid \zeta\rangle U^s_3\right]\right\}, \end{aligned} \tag{4.65}$$

where

$$\begin{aligned} U^\sigma_0 &= \frac{2i}{\sqrt{3}}(1+\chi^2)K_{2/3}(x), \quad U^\pi_0 = -\frac{2}{\sqrt{3}}\chi(1+\chi^2)^{1/2}K_{1/3}(x), \\ U^\sigma_1 &= -\frac{2i}{\sqrt{3}}(1+\chi^2)^{1/2}K_{1/3}(x), \quad U^\pi_1 = 0, \\ U^\sigma_{2,3} &= \frac{i}{\sqrt{3}}\chi(1+\chi^2)^{1/2}K_{1/3}(x), \\ U^\pi_{2,3} &= \pm\frac{1}{\sqrt{3}}(1+\chi^2)^{1/2}\left[K_{1/3}(x) \mp (1+\chi^2)^{1/2}K_{2/3}(x)\right], \end{aligned}$$

and, in addition,

$$x = \frac{1}{3}\frac{\tilde{\omega}}{\omega}\eta^3, \quad \eta = \frac{1}{\gamma}\,(1+\chi^2)^{1/2}, \quad \omega = \frac{e_0H}{m_0c\gamma}, \quad \chi = \gamma\psi.$$

The modified Bessel functions $K_{1/3}(x)$ and $K_{2/3}(x)$ appear as a result of expansion of the pre-exponential factor in Eq. (4.60) with retention of the linear terms in γ^{-1}, ct/ρ, and ψ. In general, the time dependences of the $\hat{\boldsymbol{\sigma}}(t)$ operators should be considered. These operators are solutions to the Heisenberg spin precession equation for a uniform magnetic field ($\mu < 0$)

$$\widehat{\sigma}_{\pm}(t) = \sigma_{\pm} e^{\pm i\omega t}, \quad \widehat{\sigma}_z(t) = \sigma_z, \tag{4.66}$$

where $\sigma_{\pm} = (\sigma_x \pm i\sigma_y)/2$. However, it suffices to consider the dominant term in the expansion $\exp(\pm i\omega t) \approx 1 \pm ict/\rho$ because all the other terms in $(\langle f \mid \boldsymbol{\beta} \mid i\rangle \boldsymbol{n}^s)$ are of the first order of smallness. This correction will be considered in more detail for the 'true' magnetic moment of an electron (see Sec. 4.7).

If we then use the standard procedure of classical electrodynamics one obtains, according to Eqs (4.62) and (4.65), the spectral-angular radiation distribution in the form

$$\begin{aligned} \frac{d^2W^s}{d\tilde{\omega}d\Omega} &= \frac{e^2\tilde{\omega}^2}{8\pi^3\gamma^4} \left\{ \left| \left(1 + \frac{\varepsilon}{2}\right) \langle\zeta' \mid \zeta\rangle U_0^s + \frac{\varepsilon}{2} \langle\zeta' \mid \sigma_z \mid \zeta\rangle U_1^s \right|^2 + \right. \\ &+ \left. \varepsilon^2 |\langle\zeta' \mid \sigma_+ \mid \zeta\rangle U_2^s + \langle\zeta' \mid \sigma_- \mid \zeta\rangle U_3^s|^2 \right\}. \end{aligned} \tag{4.67}$$

Here we account for the fact that the terms like the products $U_0^s U_{2,3}^{*s}$ and $U_1^s U_{2,3}^{*s}$ are the odd function of χ and vanish upon integration with respect to angles.

If we restrict ourselves to the linear approximation in ε, then only the charge radiation, spinless recoil effects, and mixed $e\mu$-radiation of the "charge+electron' spin magnetic moment" system remain in Eq. (4.67)

$$\frac{d^2W^s}{d\tilde{\omega}d\Omega} = \frac{e^2\tilde{\omega}^2}{8\pi^3\gamma^4} \frac{1+\zeta\zeta'}{2} \left[(1+\varepsilon) |\, U_0^s \,|^2 + \zeta \cos\nu \; \varepsilon U_0^s U_1^{*s} \right]. \tag{4.68}$$

To deduce the last formula we used the relationships

$$\begin{aligned} \langle\zeta' \mid \zeta\rangle &= \frac{1+\zeta\zeta'}{2} = \delta_{\zeta,\zeta'}, \\ \langle\zeta' \mid \sigma_z \mid \zeta\rangle &= \frac{1}{2}\left[(1+\zeta\zeta')\zeta\cos\nu - (1-\zeta\zeta')\sin\nu \right]. \end{aligned} \tag{4.69}$$

Here ν is the spin orientation angle given by Eq. (4.57).

If we then integrate Eq. (4.68) with respect to the spectrum and angles in the usual way, we obtain the overall polarized radiation power

$$W^s = W_{\rm SR} \frac{1+\zeta\zeta'}{2} f^s(\xi), \tag{4.70}$$

where

$$W_{\rm SR} = \frac{2}{3} \frac{e^2 c\gamma^4}{\rho^2}$$

is the synchrotron radiation power and the function $f^s(\xi)$ specifies the linear polarization of radiation

$$\begin{aligned} f^\sigma(\xi) &= \frac{7}{8} - \frac{25\sqrt{3}}{12}\xi - \xi\zeta\cos\nu, \\ f^\pi(\xi) &= \frac{1}{8} - \frac{5\sqrt{3}}{24}\xi, \end{aligned}$$

where $\xi = 3\hbar\gamma^2/(2m_0 c\rho)$ (sometimes another parameter, $\chi = 2\xi/3$, is used instead of ξ, see Ref.[31]). The presence of the coefficient $(1+\zeta\zeta')/2$ in Eqs (4.68) and others indicates that this radiation is not related to the spin-flip since Eq. (4.68) equals zero for $\zeta' = -\zeta$.

Summation over $s = \sigma, \pi$ yields

$$W = W_{\rm SR}\,\frac{1+\zeta\zeta'}{2}\,f(\xi), \tag{4.71}$$

where

$$f(\xi) = 1 + \frac{55\sqrt{3}}{24}\xi - \xi\zeta\cos\nu.$$

Eqs (4.70) and (4.71) are obtained on the assumption that $\xi \ll 1$. In a real situation, this condition is satisfied for contemporary accelerators. There is no principal difficulty in extending the result to $\xi \gg 1$ (see, for example, Ref.[29]).

The quantum corrections in Eq. (4.71) were calculated earlier by the methods adopted in quantum electrodynamics. The first quantum correction $-(55\sqrt{3}/24)\xi$ was determined in Ref.[42] (later this result was replicated in Ref.[31]). The other quantum correction $-\zeta\xi$, which depends on spin orientation, was obtained in Ref.[30] (see also Ref.[43]). However, the physical meaning of these corrections remained unexplained for a long time. Today we know that the first describes the recoil effects and the second is the mixed $e\mu$-radiation[10], the interpretation of which is not, however, so simple as it may seem at first.

Let us consider the features of mixed $e\mu$-radiation in more detail. It follows from the common formulas for the radiation power [Eqs (4.70) and (4.71)] that this radiation is due to the quantum transitions without a spin flip. Mixed $e\mu$-radiation is completely linearly polarized in the orbital plane (σ-component). Electrons polarized along the magnetic field yield the largest contribution to the radiation when $\nu = 0$. Mixed $e\mu$-radiation power increases smoothly with the degree of polarization of the electron beam[19]. Mixed $e\mu$-radiation disappears for nonpolarized electrons when $\nu = \pi/2$; it is also absent for electrons polarized along the velocity or the orbital radius[44].

The angular mixed $e\mu$-radiation distribution is, in contrast to SR, more pointed (see Fig. 4.5). The peak radiation frequency $\omega^{e\mu}_{\rm max}$ is clearly shifted towards high frequencies in comparison with $\omega^{e}_{\rm max}$ (see Fig. 4.6).

The recoil effects can be demonstrated clearly and convincingly by the example of the radiation of a spinless particle (boson). In this case, the radiation associated with the spin is absent and quantum corrections proportional to $\hbar$ and $\hbar^2$ are reduced to the recoil effects.

The matrix element of Eq. (4.55) is calculated for a boson in the same semiclassical method for spin functions with the ket-vector

$$\mid i\rangle = \frac{1}{\sqrt{\gamma}}.$$

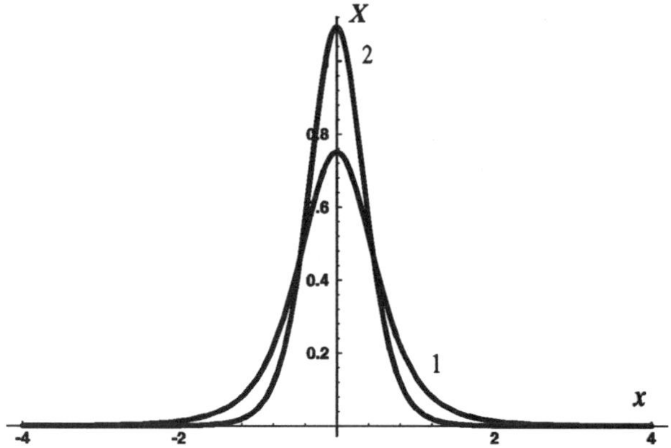

Fig. 4.5. Normalized angular distribution: synchrotron radiation $X_e^\sigma = 3/4(1+\chi^2)^{5/2}$ (1) and mixed $e\mu$-radiation $X_{e\mu}^\sigma = 35/32(1+\chi^2)^{9/2}$ (2).

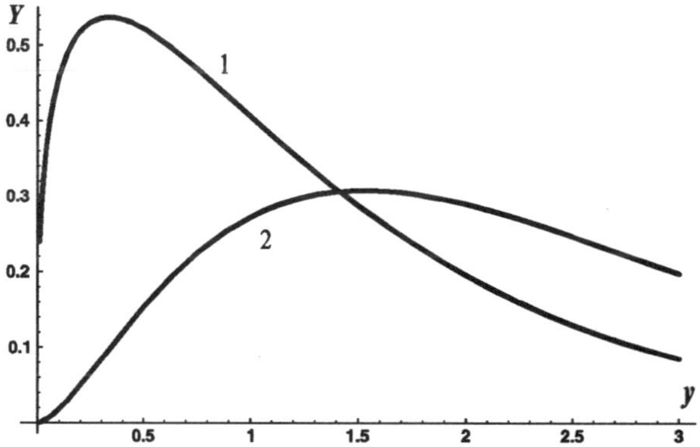

Fig. 4.6. Normalized spectral distribution: synchrotron radiation $Y_e^\sigma = (9\sqrt{3}/14\pi)\cdot y[\int_y^\infty K_{5/3}(x)dx + K_{2/3}(y)]$ (1) and mixed $e\mu$-radiation $Y_{e\mu}^\sigma = (9\sqrt{3}/8\pi)y^2K_{1/3}(y)$ (2).

The boson norming is chosen so that (see Ref.[9])

$$\langle i \mid \gamma\boldsymbol{\beta} \mid i\rangle = \boldsymbol{\beta}.$$

The formal difference from the first condition given by Eq. (4.58) is that $\widehat{\boldsymbol{\alpha}}$ is changed for $\gamma\boldsymbol{\beta}$. It is easy to see that

$$\langle f \mid \gamma\boldsymbol{\beta} \mid i\rangle = \sqrt{1+\varepsilon}\,\boldsymbol{\beta}. \tag{4.72}$$

In the linear approximation in ε or in $\hbar$, this expression coincides with the spinless term in Eq. (4.64), which is not zero for $\zeta = \zeta'$ according to Eq. (4.69). The recoil effects for an electron and a boson are different only for the terms proportional to $\hbar^2$. This fact is also known in the exact quantum theory[44] (see also Sec. 4.6).

The fact that the term with the Pauli matrices in Eq. (4.64) is related to the spin magnetic moment radiation may be shown by considering the features of the spin precession for a Dirac's particle in electromagnetic fields. We shall present the interaction between the electron spin magnetic moment and the field in the form

$$H_{\mu}^{int} = -\frac{\mu_0}{2\gamma}\,\widetilde{H}^{\text{eff}}_{\alpha\beta}\Pi^{\alpha\beta} = -\frac{\mu_0}{\gamma}\left(\boldsymbol{\zeta}\widetilde{\boldsymbol{H}}^{\text{eff}}_0\right), \tag{4.73}$$

where $\widetilde{H}^{\text{eff}}_{\alpha\beta}$ is some effective radiation field, $\Pi^{\alpha\beta}$ is the dimensionless spin tensor ($\mu_0\Pi^{\alpha\beta} = \mu^{\alpha\beta}$ is Frenkel's tensor[1]), $\boldsymbol{\zeta}$ and $\widetilde{\boldsymbol{H}}^{\text{eff}}_0$ are, respectively, the unit vector and the effective magnetic field in the system of an electron at rest. The field $\widetilde{\boldsymbol{H}}^{\text{eff}}_0$ can be determined from the BMT spin precession equation [45,46]. Here we rewrite this equation in the form (see Sec. 2.7)

$$\frac{d\boldsymbol{\zeta}}{dt} = [\boldsymbol{\Omega}\boldsymbol{\zeta}] = \frac{e}{m_0c\gamma}\left[\boldsymbol{\zeta}\boldsymbol{H}^{\text{eff}}_0\right], \tag{4.74}$$

where

$$\boldsymbol{\Omega} = -\frac{e}{m_0c\gamma}\,\boldsymbol{H}^{\text{eff}}_0,$$

$$\boldsymbol{H}^{\text{eff}}_0 = \gamma\left\{\left(a+\frac{1}{\gamma}\right)\boldsymbol{H} - \left(a+\frac{1}{\gamma+1}\right)[\boldsymbol{\beta}\boldsymbol{E}] - a\frac{\gamma}{\gamma+1}\boldsymbol{\beta}(\boldsymbol{\beta}\boldsymbol{H})\right\},$$

and the quantity $a = (g-2)/2$ characterizes the anomalous magnetic moment of the electron $\mu_a = \mu_0 a$. The presence of the Lorentz factor in the denominator of Eq. (4.74) may be explained by the way in which the variation of $\boldsymbol{\zeta}$ with time t is determined in the laboratory frame of reference.

If we neglect the anomalous magnetic moment of the electron in Eq. (4.74) and assume that $g = 2$, then

$$\boldsymbol{H}^{\text{eff}}_0 = \boldsymbol{H} - \frac{\gamma}{\gamma+1}\,[\boldsymbol{\beta}\boldsymbol{E}]. \tag{4.75}$$

The effective interaction between the spin magnetic moment of an electron and the radiation field is specified, according to Eqs (4.73) and (4.75), by the expression

$$H_\mu^{int} = -\frac{\mu_0}{\gamma}\left(\boldsymbol{\zeta}\left\{\widetilde{\boldsymbol{H}} - \frac{\gamma}{\gamma+1}[\boldsymbol{\beta}\widetilde{\boldsymbol{E}}]\right\}\right),$$

and the overall interaction of the charge and the magnetic moment of the electron with the radiation field has the form

$$H^{int} = H_e^{int} + H_\mu^{int} = -e(\boldsymbol{\beta}\widetilde{\boldsymbol{A}}) - \frac{\mu_0}{\gamma}\left(\boldsymbol{\zeta}\left\{\widetilde{\boldsymbol{H}} - \frac{\gamma}{\gamma+1}[\boldsymbol{\beta}\widetilde{\boldsymbol{E}}]\right\}\right). \tag{4.76}$$

With regard for

$$\widetilde{\boldsymbol{E}} = -i\kappa\widetilde{\boldsymbol{A}}, \quad \widetilde{\boldsymbol{H}} = [\boldsymbol{n}\widetilde{\boldsymbol{E}}],$$

where $\kappa = \widetilde{\omega}/c$ is the wave number, Eq. (4.76) can be rewritten in the form

$$H^{int} = -e\left(\boldsymbol{\beta}^{\text{eff}}\widetilde{\boldsymbol{A}}\right),$$

where

$$\boldsymbol{\beta}^{\text{eff}} = \boldsymbol{\beta} - i\frac{\hbar\kappa}{2m_0c\gamma}\left[\boldsymbol{\zeta}\left(\boldsymbol{n} - \frac{\gamma}{\gamma+1}\boldsymbol{\beta}\right)\right] \tag{4.77}$$

Clearly, Eq. (4.77) corresponds to the matrix element of Eq. (4.64) in the relativistic semiclassical theory of radiation but without recoil effects which have not been considered in the above purely classical analysis. It is remarkable that $\langle\zeta \mid \boldsymbol{\sigma} \mid \zeta\rangle = \boldsymbol{\zeta}$ for radiation without a spin-flip, and Eqs (4.77) and (4.64) coincide one-to-one when the recoil effects are not taken into account; thus the origin of mixed $e\mu$-radiation becomes clear.

4.6. Mixed $e\mu L$- and $e\mu Th$-Radiation of Relativistic Electron

Radiation of the intrinsic magnetic moment of an electron (μ-radiation) is decomposed into two basically different μL- and μTh- radiation types. μL-radiation is conditioned by the Larmor precession of the intrinsic magnetic moment of an electron (and its anomalous part), while μTh- radiation is related to the Thomas precession of the electron spin and has a purely kinematic origin. In the relativistic quantum theory of radiation based on the Dirac equation, μL- and μTh-types of radiation are not separated from each other and from the rest quantum effects of the electron radiation. A separate consideration is possible only within the framework of the relativistic semiclassical radiation theory. Of course, thereat all the integral responses of radiation coincide with the results of a more strict quantum theory.

The total Hamiltonian of interaction of an electron with the radiation field has the form

$$\widehat{\widehat{H}} = \widehat{\widehat{H}}_e + \widehat{\widehat{H}}_\mu^{\text{eff}} = \widehat{\widehat{H}}_e + \widehat{\widehat{H}}_\mu^L + \widehat{\widehat{H}}_\mu^{Th} = -e(\boldsymbol{\beta}^{\text{eff}}\widetilde{\boldsymbol{A}}),$$

where

$$\boldsymbol{\beta}^{\text{eff}} = \boldsymbol{\beta} + \boldsymbol{\beta}^{L} + \boldsymbol{\beta}^{Th}, \tag{4.78}$$

$$\begin{aligned}
\boldsymbol{\beta}^{L} &= -i\,\frac{\mu\widetilde{\omega}}{ec}\left\{[\boldsymbol{\sigma}(\boldsymbol{n}-\boldsymbol{\beta})] - \frac{\gamma}{\gamma+1}(\boldsymbol{\sigma\beta})[\boldsymbol{\beta n}]\right\}, \\
\boldsymbol{\beta}^{Th} &= -i\,\frac{\hbar\widetilde{\omega}}{2m_0c^2}\,\frac{\gamma}{\gamma+1}\,\{[\boldsymbol{\sigma\beta}](1-(\boldsymbol{n\beta})) + \boldsymbol{\beta}(\boldsymbol{\sigma}[\boldsymbol{\beta n}])\}\,.
\end{aligned}$$

Note that we do not take into account the recoil effects during radiation in Eq. (4.78), although it is not difficult to do this (as stated above).

The quantum nature of the spin behavior during radiation is taken into account by the matrix element $(\zeta, \zeta' = \pm 1)$

$$\langle \zeta' \mid \boldsymbol{\sigma} \mid \zeta \rangle = \frac{1-\zeta\zeta'}{2}\,(\boldsymbol{i} + i\zeta\boldsymbol{j}) + \frac{1+\zeta\zeta'}{2}\,\zeta\boldsymbol{k} = \overline{\boldsymbol{\zeta}}. \tag{4.79}$$

This formula can be interpreted as decomposition of the complex spin-vector $\overline{\boldsymbol{\zeta}}$ into parallel and orthogonal components

$$\overline{\boldsymbol{\zeta}} = \overline{\boldsymbol{\zeta}}_{\perp} + \overline{\boldsymbol{\zeta}}_{\|}, \quad \mid \overline{\boldsymbol{\zeta}} \mid = \frac{1+\zeta\zeta'}{2}\,\delta_{\zeta\zeta'} = \zeta_{\|}.$$

Further, we shall assume that the quantum transitions take place along the direction of the magnetic field, although a more general consideration of this question is possible. Moreover, we put $g = 2$ for simplification.

Earlier we have shown that the spectral-angular and polarization responses of radiation in the semiclassical theory can be obtained on the basis of the Fourier transform $(e = -e_0 < 0)$

$$\widetilde{E}_{s\widetilde{\omega}} = -i\,\frac{e_0\widetilde{\omega}}{\tilde{r}c}\int\limits_{-\infty}^{\infty}(\boldsymbol{\beta}^{\text{eff}}\boldsymbol{n}_s)e^{i\widetilde{\omega}\left[t-\frac{1}{c}(\boldsymbol{nr})\right]}dt, \tag{4.80}$$

where $\boldsymbol{n}_s$ are the polarization unit vectors $(s = \sigma, \pi)$, $\tilde{r}$ is the distance from the charge to the observation point.

We shall speak of the mixed radiation of the intrinsic magnetic moment of an electron conditioned by the Larmor or Thomas precession as $e\mu L$- and $e\mu Th$- radiation, respectively.

With regard for these remarks we obtain, according to Eqs (4.78) — (4.80), that

$$\widetilde{E}_{s\widetilde{\omega}} = -\frac{ie_0\widetilde{\omega}\rho}{\tilde{r}c^2\gamma^2}\left[\left(1+\frac{\varepsilon}{2}\right)\langle U_s^e\rangle + \frac{\varepsilon}{2}\left(\langle U_s^{\mu L}\rangle + \langle U_s^{\mu Th}\rangle\right)\right], \tag{4.81}$$

where

$$\langle U_s^e\rangle = \frac{1+\zeta\zeta'}{2}\,U_0^s,$$

$$\begin{aligned}
\langle U_\sigma^{\mu L}\rangle &= \frac{2i}{\sqrt{3}}\,\chi\,(1+\chi^2)^{1/2}\left\{\frac{1+\zeta\zeta'}{2}\,\zeta\chi K_{1/3}(x)+\right.\\
&\quad\left.+\,\frac{1-\zeta\zeta'}{2}\left[K_{1/3}(x)+\zeta(1+\chi^2)^{1/2}K_{2/3}(x)\right]\right\},\\
\langle U_\pi^{\mu L}\rangle &= -\frac{2}{\sqrt{3}}\,(1+\chi^2)\left\{\frac{1+\zeta\zeta'}{2}\,\zeta\chi K_{2/3}(x)+\right.\\
&\quad\left.+\,\frac{1-\zeta\zeta'}{2}\left[K_{2/3}(x)+\zeta(1+\chi^2)^{1/2}K_{1/3}(x)\right]\right\},\\
\langle U_\sigma^{\mu Th}\rangle &= -\frac{2i}{\sqrt{3}}\,\zeta\,(1+\chi^2)\left\{\frac{1+\zeta\zeta'}{2}\,(1+\chi^2)^{1/2}K_{1/3}(x)+\frac{1-\zeta\zeta'}{2}\,\chi K_{2/3}(x)\right\},\\
\langle U_\pi^{\mu Th}\rangle &= \frac{2}{\sqrt{3}}\,\zeta\chi\,(1+\chi^2)^{1/2}\left\{\frac{1+\zeta\zeta'}{2}\,(1+\chi^2)^{1/2}K_{2/3}(x)+\frac{1-\zeta\zeta'}{2}\,\chi K_{1/3}(x)\right\}.
\end{aligned}$$

We are not going to consider the SR of an electron in this section. Principal attention will be paid to the mixed $e\mu$-radiation only, which is associated with the interference terms

$$\widetilde{E}^{e}_{s\widetilde{\omega}}\widetilde{E}^{\mu L}_{s\widetilde{\omega}},\quad \widetilde{E}^{e}_{s\widetilde{\omega}}\widetilde{E}^{\mu Th}_{s\widetilde{\omega}}\quad \textit{and}\quad \textit{c.c.}$$

Integrating the spectral-angular distribution we take into account that

$$\frac{1}{T}\,\widetilde{\omega}^3 d\widetilde{\omega}=\frac{3^4c^5}{2\pi\rho^5\gamma^{12}}\,(1+\chi^2)^{-6}\xi^3 d\xi,\quad d\Omega=\frac{2\pi}{\gamma}\,d\chi.$$

As a result of the integration we obtain

$$W_\sigma^{e\mu L}=\frac{1+\zeta\zeta'}{2}\,W_{\rm SR}\,\frac{1}{6}\,\xi\zeta=W_\pi^{e\mu L},$$

$$W_\sigma^{e\mu Th}=-\frac{1+\zeta\zeta'}{2}\,W_{\rm SR}\,\frac{7}{6}\,\xi\zeta,\quad W_\pi^{e\mu Th}=-\frac{1+\zeta\zeta'}{2}\,W_{\rm SR}\,\frac{1}{6}\,\xi\zeta,$$

where, as usual, $W_{\rm SR}$ is the total power of SR of the electronic charge.

We draw attention to the fact that all the mixed $e\mu$-radiation is related to the transitions without a spin-flip.

Summing $e\mu L$- and $e\mu Th$-radiation we obtain the known result (see Eq. (4.70) without recoil effects and for $\nu=0$)

$$\begin{aligned}
W^{e\mu L} &= W_\sigma^{e\mu L}+W_\pi^{e\mu L}=\frac{1+\zeta\zeta'}{2}\,W_{\rm SR}\,\frac{1}{3}\,\xi\zeta,\\
W^{e\mu Th} &= W_\sigma^{e\mu Th}+W_\pi^{e\mu Th}=-\frac{1+\zeta\zeta'}{2}\,W_{\rm SR}\,\frac{4}{3}\,\xi\zeta,\\
W^{e\mu} &= W^{e\mu L}+W^{e\mu Th}=-\frac{1+\zeta\zeta'}{2}\,W_{\rm SR}\,\xi\zeta.
\end{aligned}\tag{4.82}$$

Moreover,

$$\begin{aligned} W_\sigma^{e\mu} &= W_\sigma^{e\mu L} + W_\sigma^{e\mu Th} = -\frac{1+\zeta\zeta'}{2} W_{\rm SR}\,\xi\zeta, \\ W_\pi^{e\mu} &= W_\pi^{e\mu L} + W_\pi^{e\mu Th} = 0. \end{aligned}$$

Thus, the π- components of $e\mu L$- and $e\mu Th$- radiation completely make up for each other at $g = 2$.

Now we show that the relativistic classical theory of $e\mu$-radiation is the theory of $e\mu L$-radiation.

Really, according to the general classical theory of μ-radiation[13] the radiation loss of energy for $e\mu$-radiation can be determined by the expression

$$\left(\frac{d\tilde{P}^\alpha}{d\tau}\right)^{e\mu} = \frac{2}{3}\frac{e\mu}{c^4}\left(\overset{\circ\circ}{\Pi}{}^{\alpha\beta} w_\beta - \frac{2}{c^2} v^\alpha\, \overset{\circ}{w}_\rho\, \Pi^{\rho\sigma} w_\sigma - \frac{1}{c^2}\Pi^{\alpha\beta} w_\beta w_\rho w^\rho\right).$$

Here P^α is the four-dimentional momentum of radiation, the circle means the derivative with respect to the proper time τ. The radiation power can be obtained from the zero component

$$W^{e\mu} = \frac{c}{\gamma}\left(\frac{d\tilde{P}^0}{d\tau}\right)^{e\mu}.$$

Substituting here the solutions of the Lorentz equations for a charge motion and the BMT-equation for spin (see Sec. 2.7) we obtain after averaging over the period of motion and spin precession in the case of $g = 2$ that ($\zeta_{\|} = \Pi_{1z}/\gamma = \pm 1$) (see Chapter 1)

$$\overline{W}^{e\mu} = W_{\rm SR}\left(1 + \frac{1}{3}\xi\zeta_{\|}\right) = W^{e\mu L},$$

where $W^{e\mu L}$ is the sum of σ- and π-components of $e\mu L$-radiation determined in Eq. (4.82) (multiplier $(1+\zeta\zeta')/2 = \delta_{\zeta\zeta'}$ is unimportant in this case). It is possible to show that the spectral-angular and polarization characteristics of $e\mu L$-radiation also coincide.

Of course, this coincidence is not casual. It is known that only in the ultimate nonrelativistic case when the Thomas precession can be neglected does Eq. (4.74) coincide with the equation for the magnetic moment precession in conventional classical electrodynamics ($g = 2$)

$$\frac{d\boldsymbol{\zeta}}{dt} = \frac{e}{m_0 c}\left[\boldsymbol{\zeta}\boldsymbol{H}_0\right].$$

It could be said that in this case, the spin magnetic moment of an electron exhibits the properties of a 'true' magnetic moment (see the papers by I.E. Tamm[47] and V.L. Ginzburg[48]).

The 'true' character of the magnetic moment of an electron manifests itself also in the hypothetical case of $g \gg 1$, when the entire magnetic moment is anomalous and Eq. (4.74) takes the form

$$\frac{d\boldsymbol{\zeta}}{dt} = \frac{ge}{2m_0c\gamma}\left[\boldsymbol{\zeta}\boldsymbol{H}_0\right]. \tag{4.83}$$

In fact Eq. (4.83) describes the spin precession for a neutron, the magnetic moment of which is entirely anomalous. This equation can also be used in the case of the uniform rectilinear motion of an electron in the Wien-filter-type fields when the Thomas precession disappears[49,50,51]. It is very interesting that the situation with the 'true' magnetic moment is the same in radiation theory (see Secs 4.7,4.8).

4.7. The Structure of Quantum Corrections to SR

Let us now study synchrotron radiation with regard for the quantum corrections of the second order in $\hbar$. The recoil effects and mixed $e\mu$-radiation are also present here. However, there is a new specific type of radiation due to the intrinsic magnetic moment of an electron (spin light). The $\hbar$-quadratic radiation is accompanied by transitions both with and without a spin-flip and it has a very complex nature. In this approximation the radiation from the anomalous magnetic moment of an electron becomes perceptible and it is, as we shall see, principally different from the spin magnetic moment radiation (the Bohr magneton).

In the quantum theory of radiation which is based on the Dirac-Pauli equation, the anomalous magnetic moment is taken into account through the following change[52]

$$\widehat{\boldsymbol{\alpha}} \to \widehat{\boldsymbol{\alpha}}^{\mathrm{eff}} = \widehat{\boldsymbol{\alpha}} - i\frac{\hbar\tilde{\omega}}{2mc^2}a\left(\rho_3[\widehat{\boldsymbol{\sigma}}\boldsymbol{n}] + \rho_2\widehat{\boldsymbol{\sigma}}\right).$$

Conversion to the semiclassical theory of radiation yields

$$\boldsymbol{\beta} \to \langle f \mid \widehat{\boldsymbol{\alpha}}^{\mathrm{eff}} \mid i\rangle. \tag{4.84}$$

instead of Eq. (4.55).

One must bear in mind that $(\hbar\tilde{\omega}/2mc^2)a = \varepsilon\gamma' a/2$ and $\gamma' a \approx \gamma a(1-\varepsilon)$ in the range of velocities typical for SR. Therefore, the matrix element can be calculated without taking into account the recoil effects ($\boldsymbol{\beta}\,' = \boldsymbol{\beta}$) in the linear approximation in ε. Otherwise we will follow the method described in Sec. 4.5 and obtain

$$\begin{aligned}\langle f \mid \widehat{\boldsymbol{\alpha}}^{\mathrm{eff}} \mid i\rangle &= \left(1+\frac{\varepsilon}{2}\right)\boldsymbol{\beta}\langle\zeta' \mid \zeta\rangle - i\frac{\varepsilon}{2}\left[\langle\zeta' \mid \widehat{\boldsymbol{\sigma}} \mid \zeta\rangle\cdot\left(\boldsymbol{n} - \frac{\gamma}{\gamma+1}\boldsymbol{\beta}\right)\right] - \\ &- i\frac{\varepsilon}{2}\gamma a\left\{[\langle\zeta' \mid \widehat{\boldsymbol{\sigma}} \mid \zeta\rangle\cdot(\boldsymbol{n}-\boldsymbol{\beta})] - \frac{\gamma}{\gamma+1}(\langle\zeta' \mid \widehat{\boldsymbol{\sigma}} \mid \zeta\rangle\boldsymbol{\beta})\cdot[\boldsymbol{\beta}\boldsymbol{n}]\right\}.\end{aligned}$$

In comparison with Eq. (4.64) there are additional terms with the anomalous part of the magnetic moment of the electron. In semiclassical interpretation, the energy of interaction of the magnetic moment and field is now presented in the form

$$\begin{aligned} H_\mu^{int} &= -\mu_0 \left\{ \left(\boldsymbol{\zeta} \left[\widetilde{\boldsymbol{H}} - \frac{\gamma}{\gamma+1} [\boldsymbol{\beta} \widetilde{\boldsymbol{E}}] \right] \right) - \right. \\ &\left. - \gamma a \left\{ \left(\boldsymbol{\zeta} \left[\widetilde{\boldsymbol{H}} - [\boldsymbol{\beta} \widetilde{\boldsymbol{E}}] \right] \right) - \frac{\gamma}{\gamma+1} (\boldsymbol{\zeta\beta})(\boldsymbol{\beta} \widetilde{\boldsymbol{H}}) \right\} \right\}. \end{aligned}$$

As in Sec. 4.5, it then follows that

$$\begin{aligned} \boldsymbol{\beta}^{\text{eff}} &= \left(1 + \frac{\varepsilon}{2}\right) \boldsymbol{\beta} - i \frac{\varepsilon}{2} \left\{ \left[\boldsymbol{\zeta} \left(\boldsymbol{n} - \frac{\gamma}{\gamma+1} \boldsymbol{\beta} \right) \right] + \right. \\ &+ \left. \gamma a \left\{ [\boldsymbol{\zeta} (\boldsymbol{n} - \boldsymbol{\beta})] - \frac{\gamma}{\gamma+1} (\boldsymbol{\zeta\beta}) [\boldsymbol{\beta n}] \right\} \right\}. \end{aligned}$$

This formula is an extension of Eq. (4.77) and an analogue of Eq. (4.64).

If the terms in the Fourier transform of the field are grouped by the same criterion as in Eq. (4.65), then the spectral-angular distribution of the radiation power does not formally differ from that given by Eq. (4.67). However, the functions U_i^s $(i = 1, 2, 3)$ have more complex forms

$$\begin{aligned} U_1^\sigma &= -\frac{2i}{\sqrt{3}} (1+\chi^2)^{1/2} (1 - a\chi^2) K_{1/3}(x), \\ U_1^\pi &= \frac{2}{\sqrt{3}} \chi (1+\chi^2) a K_{2/3}(x), \\ U_{2,3}^\sigma &= \frac{i}{\sqrt{3}} \chi (1+\chi^2)^{1/2} \left\{ (1+a) K_{1/3}(x) \mp 2a (1+\chi^2)^{1/2} K_{2/3}(x) \right\}, \\ U_{2,3}^\pi &= \pm \frac{1}{\sqrt{3}} \left\{ [1 + 2a(1+\chi^2)] K_{1/3}(x) \mp (1+2a)(1+\chi^2)^{1/2} K_{2/3}(x) \right\}. \end{aligned} \tag{4.85}$$

The functions U_0^s remain unchanged. In addition to Eq. (4.69), there is the relationship

$$\langle \zeta' \mid \sigma_\pm \mid \zeta \rangle = \frac{1}{2} \begin{pmatrix} \zeta' \\ \zeta \end{pmatrix} [1 \mp (\zeta - \zeta') \cos\nu - \zeta\zeta' \cos^2\nu]^{1/2} e^{\pm i\lambda}. \tag{4.86}$$

Further manipulations are performed in an ordinary fashion. Omitting insignificant details we present the final expressions for the linearly polarized radiation power in the linear approximation in a (in which the Dirac-Pauli equation is also true)

$$\begin{aligned} W^\sigma &= W_{\text{SR}} \left\{ \frac{1+\zeta\zeta'}{2} \left\langle \frac{7}{8} - \frac{25\sqrt{3}}{12} \xi + \frac{325}{18} \xi^2 - \right. \right. \\ &- \zeta \cos\nu \left[\xi - \frac{245\sqrt{3}}{48} \xi^2 - \frac{1}{6} a \left(\xi - \frac{245\sqrt{3}}{72} \xi^2 \right) \right] + \\ &+ \left. \cos^2\nu \frac{5-a}{9} \xi^2 - \sin^2\nu \cos^2\lambda \frac{1+2a}{18} \xi^2 \right\rangle + \end{aligned} \tag{4.87}$$

$$+ \quad \frac{1-\zeta\zeta'}{2}\left\langle\frac{5-a}{9}\sin^2\nu + (\cos^2\nu + \sin^2\nu\sin^2\lambda)\frac{1+2a}{18} + \right.$$
$$+ \quad \left.\zeta\cos\nu\frac{35\sqrt{3}}{216}a\right\rangle\xi^2\bigg\},$$

$$\begin{aligned} W^\pi &= W_{\rm SR}\left\{\frac{1+\zeta\zeta'}{2}\left\langle\frac{1}{8} - \frac{5\sqrt{3}}{24}\xi + \frac{25}{18}\xi^2 + a\zeta\cos\nu\left(\frac{1}{6}\xi - \frac{245\sqrt{3}}{432}\xi^2\right) + \right.\right. \\ &+ \left.\left[\frac{5+22a}{9}\sin^2\lambda + \frac{13}{18}(1+4a)\cos^2\lambda\right]\sin^2\nu\xi^2\right\rangle + \\ &+ \frac{1-\zeta\zeta'}{2}\left\langle\frac{5+22a}{9}(\cos^2\nu + \sin^2\nu\cos^2\lambda) + \right. \\ &+ \left.\left.\frac{13}{18}(1+4a)(\cos^2\nu + \sin^2\nu\sin^2\lambda) + \zeta\cos\nu\frac{35\sqrt{3}}{48}\left(1 + \frac{38}{9}a\right)\right\rangle\xi^2\right\}. \end{aligned} \tag{4.88}$$

The total radiation power is

$$\begin{aligned} W &= W_{\rm SR}\left\{\frac{1+\zeta\zeta'}{2}\left\langle 1 - \frac{55\sqrt{3}}{24}\xi + \frac{175}{9}\xi^2 - \zeta\cos\nu\left(\xi - \frac{245\sqrt{3}}{48}\xi^2\right) + \right.\right. \\ &+ \frac{1}{3}a\zeta\cos\nu\left(\xi - \frac{245\sqrt{3}}{72}\xi^2\right) + \cos^2\nu\frac{5-a}{9}\xi^2 + \\ &+ \left.\sin^2\nu\left(\frac{5+22a}{9}\sin^2\lambda + \frac{7+27a}{9}\cos^2\lambda\right)\xi^2\right\rangle + \\ &+ \frac{1-\zeta\zeta'}{2}\left\langle\frac{5-a}{9}\sin^2\nu + \frac{5+22a}{9}(\cos^2\nu + \sin^2\nu\cos^2\lambda) + \right. \\ &+ \left.\left.\frac{7+27a}{9}(\cos^2\nu + \sin^2\nu\sin^2\lambda) + \zeta\cos\nu\frac{35\sqrt{3}}{48}\left(1 + \frac{40}{9}a\right)\right\rangle\xi^2\right\}. \end{aligned} \tag{4.89}$$

In these equations the spin orientation can be averaged over the azimuthal angle λ

$$\begin{aligned} \overline{\sin^2\lambda} &= \overline{\cos^2\lambda} = \frac{1}{2}, \\ \cos^2\nu &+ \sin^2\nu\overline{\sin^2\lambda} = \cos^2\nu + \sin^2\nu\overline{\cos^2\lambda} = \frac{1}{2}(1+\cos^2\nu), \end{aligned} \tag{4.90}$$

and then they coincide with the results given in Ref.[38] (see also Ref.[53]). If instead of averaging over λ we consider the spin to be oriented along the velocity of the electron ($\nu = \pi/2, \lambda = 0$) or along the orbital radius (orthogonal to the magnetic field) ($\nu = \pi/2, \lambda = \pi/2$), then Eqs (4.87)—(4.89) go for $a = 0$ into the results true in quantum electrodynamics.

In the most interesting case of transversely polarized electrons when ($\nu = o$), Eqs (4.87)—(4.89) are simplified and upon averaging over λ take the form

$$W^s = W_{\rm SR}f(\xi), \tag{4.91}$$

$$
\begin{array}{l|c|c|}
f^\sigma(\xi) = \frac{1+\zeta\zeta'}{2}\left[\frac{7}{8}\right. & -\frac{25\sqrt{3}}{12}\xi + \frac{325}{18}\xi^2 & -\zeta\left(\xi - \frac{245\sqrt{3}}{48}\xi^2\right) \\
f^\pi(\xi) = \frac{1+\zeta\zeta'}{2}\left[\frac{1}{8}\right. & -\frac{5\sqrt{3}}{24}\xi + \frac{25}{18}\xi^2 & \\
\qquad (1) & (2) & (3)
\end{array}
$$

$$
\begin{array}{l|l|}
+\zeta\frac{a}{6}\left(\xi - \frac{245\sqrt{3}}{72}\xi^2\right) \; -\frac{a}{9}\xi^2 & \left. + \frac{5}{9}\xi^2\right] \; +\frac{1-\zeta\zeta'}{2}\left[\frac{1}{18} +\right. \\
\left. + \, \zeta\frac{a}{6}\left(\xi - \frac{245\sqrt{3}}{72}\xi^2\right)\right] & \qquad\qquad +\frac{1-\zeta\zeta'}{2}\left[\frac{23}{18} + \zeta\frac{35\sqrt{3}}{48}\right. \\
\qquad\qquad (5) & \qquad\qquad (4)
\end{array}
$$

$$
\begin{array}{|l}
\left. + \, \frac{a}{9}\left(1 + \zeta\frac{35\sqrt{3}}{24}\right)\right]\xi^2, \\
\left. + \, \frac{a}{9}\left(48 + 19\zeta\frac{35\sqrt{3}}{24}\right)\right]\xi^2. \\
\qquad\qquad (5)
\end{array}
$$

We pay our attention that the radiation related to the anomalous magnetic moment in the linear approximation in ξ coincides with classical $e\mu L$-radiation (see Sec. 4.6). Obviously, it is accounted for by the fact that the anomalous magnetic moment does not undergo the Thomas precession.

The total radiation power is specified by the expression

$$W = W^\sigma + W^\pi = W_{\mathrm{SR}} f(\xi), \tag{4.92}$$

$$
\begin{array}{l|c|c|}
f(\xi) = \frac{1+\zeta\zeta'}{2}\,[1 & -\frac{55\sqrt{3}}{24}\xi + \frac{175}{9}\xi^2 & -\zeta\left(\xi - \frac{245\sqrt{3}}{48}\xi^2\right) \\
\qquad (1) & (2) & (3)
\end{array}
$$

$$
\begin{array}{c|c|c|}
+\zeta\frac{a}{3}\left(\xi - \frac{245\sqrt{3}}{72}\xi^2\right) - \frac{a}{9}\xi^2 & \left. + \frac{5}{9}\xi^2\right] + \frac{1-\zeta\zeta'}{2}\left[\frac{4}{3} + \frac{35\sqrt{3}}{48}\right. & \left. + \, \frac{a}{9}\left(49 + \zeta\frac{175\sqrt{3}}{6}\right)\right]\xi^2. \\
(5) & (4) & (5)
\end{array}
$$

The vertical lines separate the charge radiation (1) from the recoil effects (2), mixed $e\mu$-radiation (3), spin magnetic moment radiation (μ-radiation) (4), and the radiation associated with the anomalous magnetic moment (5).

If we do not separate the structural elements of SR and neglect the anomalous magnetic moment of the electron, we arrive at the familiar expression for the radiation power [see, for example, Eq. (4.71)], where

$$f(\xi) = 1 - \zeta\xi - \frac{55\sqrt{3}}{24}\xi + \frac{64}{3}\xi^2 + \zeta\frac{35\sqrt{3}}{6}\xi^2.$$

A similar calculation for a spinless particle on the basis of the matrix element of Eq. (4.72) yields

$$f_0(\xi) = 1 - \frac{55\sqrt{3}}{24}\xi + \frac{56}{3}\xi^2.$$

instead of $f(\xi)$. This expression was first obtained in Ref.[55] (see also Refs[44,56,57]). Note that the 'role of spin' in SR was earlier limited to the quantum correction $(8/3)\xi^2$[258]. Here it is obtained as the averaged difference

$$\frac{1}{2}\sum_{\zeta=\pm1}[f(\xi) - f_0(\xi)] = \frac{8}{3}\xi^2.$$

However, the structural role and physical meaning of this and other quantum corrections have been revealed only because of Eq. (4.89).

It is of interest to write down the functions $f^s(\xi)$ for μ-radiation where μL- and μTh-components are separated. Neglecting the anomalous magnetic moment we obtain, according to Eq. (4.80) for $\nu = 0$,

$$\begin{array}{r|c|c|c|c|c}
f^\sigma_\mu = \left[\frac{1+\zeta\zeta'}{2}\left(\frac{1}{48}\right.\right. & +\frac{11}{16} & \left.-\frac{11}{72}\right) & +\frac{1-\zeta\zeta'}{2}\left(\frac{7}{48}+\frac{35\sqrt{3}}{432}\zeta\right. & +\frac{13}{144}- & \left.\left.-\frac{13}{72}-\frac{35\sqrt{3}}{432}\zeta\right)\right]\xi^2 \\
f^\pi_\mu = \left[\frac{1+\zeta\zeta'}{2}\left(\frac{13}{144}\right.\right. & +\frac{13}{144} & \left.-\frac{13}{72}\right) & +\frac{1-\zeta\zeta'}{2}\left(\frac{203}{144}+\frac{175\sqrt{3}}{216}\zeta\right. & +\frac{1}{48}- & \left.\left.-\frac{11}{72}-\frac{35\sqrt{3}}{432}\zeta\right)\right]\xi^2 \\
f_\mu = \left[\frac{1+\zeta\zeta'}{2}\left(\frac{1}{9}\right.\right. & +\frac{7}{9} & \left.-\frac{1}{3}\right) & +\frac{1-\zeta\zeta'}{2}\left(\frac{14}{9}+\frac{385\sqrt{3}}{432}\zeta\right. & +\frac{1}{9}- & \left.\left.-\frac{1}{3}-\frac{35\sqrt{3}}{216}\zeta\right)\right]\xi^2 \\
(L) & (Th) & (LTh) & (L) & (Th) & (LTh)
\end{array}$$

where the vertical lines separate the L-, Th-, and mixed LTh-radiation of the magnetic moment. Note that all properties of μL-radiation without a spin-flip were obtained earlier on the basis of the classical theory of μ-radiation[66].

The total power of μ-radiation is determined by the function

$$\begin{aligned} f^\mu(\xi) &= f^{\mu L}(\xi) + f^{\mu Th}(\xi) + f^{\mu LTh}(\xi) = \\ &= \left[\frac{1+\zeta\zeta'}{2}\frac{5}{9} + \frac{1-\zeta\zeta'}{2}\left(\frac{4}{3} + \frac{35\sqrt{3}}{48}\right)\right]\xi^2 \end{aligned}$$

and gets into the structure of Eq. (4.92).

Undoubtely, the above method enables the photon emission probability to be calculated as well. Although there is little sense in quoting these equations, we shall write down the expression for the probability of radiation with a spin-flip, averaged over the azimuthal angle λ and summed over linear polarizations of photons[38]

$$w = \frac{1}{2T_{\rm ST}}\left[1 + \frac{37}{9}a - \left(\frac{2}{9} - \frac{13}{9}a\right)\frac{\sin^2\nu}{2} + \zeta\frac{8}{5\sqrt{3}}\left(1 + \frac{14}{3}a\right)\cos\nu\right], \qquad (4.93)$$

where

$$T_{\rm ST} = \frac{8\sqrt{3}}{15}\frac{\hbar^2}{m_0 c e_0^2 \gamma^2}\left(\frac{H^*}{H}\right)^3$$

is the polarization time in the Sokolov-Ternov effect[9] (see also Chapter 3).

The following formula is more convenient for calculations

$$T_{\rm ST} = \frac{16}{15}\sqrt{3}\,\frac{(H^*/H)^3}{\alpha\gamma^2\omega_Z}, \tag{4.94}$$

where $\alpha = e_0^2/\hbar c \approx 1/137$ is the fine-structure constant, $\omega_Z = \hbar/2m_0c^2 \approx 1.6 \cdot 10^{21}\ sec^{-1}$ is the frequency of Zitterbewegung, $H^* = m_0^2c^3/e_0\hbar \approx 4.4 \cdot 10^{13}\ Oe$ is Schwinger's magnetic field.

Since the magnetic field is about $10^4\ Oe$ in a storage ring, the polarization time takes a value, which is accessible for observation of the polarization effect only in the high-energy range. In particular, this time is approximately equal to an hour for an electron with an energy about 1 GeV ($\gamma = 1.96 \cdot 10^3$).

According to Eq. (4.93), spin-flip transitions do not depend on the initial orientation of a spin for $\nu = \pi/2$ when the spin oriented in the orbital plane, whereas spin-flip transitions depend essentially on the orientation of the spin for $\nu = 0$ when the spin is oriented along the magnetic field and, thus, self-polarization of the electron beam occurs. The spin relaxation time is specified by the expression

$$[w(\zeta = 1) + w(\zeta = -1)]\, T = 1.$$

It follows that (see also Ref.[40])

$$T = T_{\rm ST}\left(1 + \frac{37}{9}a\right)^{-1} \approx T_{\rm ST}\left(1 - \frac{37}{9}a\right).$$

Thus the anomalous magnetic moment contributes only to ordering the spin orientation of the electron during radiation.

4.8. 'True' Magnetic Moment Radiation

We have seen (see Sec. 4.6) that spin exhibits the features of the 'true' magnetic moment only in the relativistic limit or during uniform rectilinear motion when the Thomas precession can be neglected. The case of uniform rectilinear motion when the electron charge radiation and mixed radiation are absent can be obtained from Eq. (4.85) in the limit of $\rho \to \infty$. Although the method of the relativistic semiclassical theory does not undergo any change in this case, the functions $U_{2,3}^s$ responsible for the 'true' magnetic moment radiation change considerably. In what follows we do not want to restrict ourselves by the assumption that the anomalous magnetic moment is small and proceed to the consideration of an arbitrary a. Eqs (4.85) are no longer applicable in this case. In a more general notation they are written as[37]

$$\begin{aligned} U^{\sigma}_{2,3} &= i\pi q\chi \left\{(1+a)Ai(z_{\pm}) \pm qaAi'(z_{\pm})\right\}, \\ U^{\pi}_{2,3} &= \pm\pi q \left\{[1+(1+\chi^2 \pm aq^3)a]Ai(z_{\pm}) \pm (1+a)qAi'(z_{\pm})\right\}, \end{aligned} \tag{4.95}$$

where

$$z_{\pm} = z_0 \pm aq, \quad z_0 = \frac{1+\chi^2}{q^2}, \quad q = \left(\frac{2c}{\widetilde{\omega}\rho}\right)^{1/3} \gamma.$$

The appearance of the Airy functions Ai and their derivatives are explained by the fact that the argument depends on the anomalous magnetic moment and it is no longer possible to employ the modified Bessel functions in $U^{s}_{2,3}$[36]. It may be shown that the previous equations for $U^{s}_{2,3}$ in Eq. (4.85) follow from Eq. (4.95) in the particular case of $a \ll 1$.

In the limit of $\rho \to \infty$ the functions $U^{s}_{2,3}$ are simplified

$$\begin{aligned} U^{\sigma}_{2,3} &= i\pi q\chi a Ai(z_{\pm}), \\ U^{\pi}_{2,3} &= \pm\pi q \left(1+\chi^2 \pm \frac{1}{2}aq^3\right) Ai(z_{\pm}). \end{aligned}$$

Moreover, as $\rho \to \infty$, with regard for Eq. (4.66),

$$Ai(z_{\pm}) \to \frac{1}{2\pi}\int\limits_{-\infty}^{\infty} dx\, e^{iz_{\pm}x} = \left(\frac{\widetilde{\omega}_{\max}}{a\widetilde{\omega}}\right)^{2/3} \begin{cases} 0 \\ \delta\left(1+\chi^2 - \frac{\widetilde{\omega}_{\max}}{\widetilde{\omega}}\right) \end{cases},$$

where $\widetilde{\omega}_{\max} = 2a\omega\gamma^3$, instead of Eq. (4.84) we have

$$\begin{aligned} U^{\sigma}_{2} &= 0, \\ U^{\sigma}_{3} &= i\pi\, \frac{\widetilde{\omega}_{\max}}{\widetilde{\omega}}\, \chi\, \delta\left(1+\chi^2 - \frac{\widetilde{\omega}_{\max}}{\widetilde{\omega}}\right), \\ U^{\pi}_{2,3} &= \pm\pi\, \frac{\widetilde{\omega}_{\max}}{\widetilde{\omega}}\, \frac{1}{2}\, (1+\chi^2)\delta\left(1+\chi^2 - \frac{\widetilde{\omega}_{\max}}{\widetilde{\omega}}\right). \end{aligned} \tag{4.96}$$

If the magnetic moment is anomalous (neutron), it can be assumed that $g = 2a$, and then

$$\widetilde{\omega}_{\max} = \frac{4 \mid \mu \mid H}{\hbar}\gamma^2, \quad \mid \mu \mid = \frac{g\mu_0}{2},$$

where μ_0 is the Bohr nuclear magneton. The frequency of spin precession mainly depends on the direction of the neutron motion in the magnetic field[59]

$$\omega_{\mu} = \frac{2 \mid \mu \mid H}{\hbar}[1 - \beta^2\cos^2\alpha]^{1/2},$$

where α is the angle between $\boldsymbol{\beta}$ and $\boldsymbol{H}$. In the case of interest, the neutron (as the electron) moves along the X, the magnetic field is parallel to the Z axis and, therefore, $\alpha = \pi/2$ and $\widetilde{\omega}_{\rm max} = 2\gamma^2/\omega_\mu$. According to Eq. (4.96) the frequency $\widetilde{\omega}_{\rm max}$ plays the part of the maximal radiation frequency.

The total spectral-angular distribution of the 'true' magnetic moment radiation (of a neutron, in our case) is calculated by means of Eqs (4.62), (4.67), and (4.96). As a result, we have

$$\frac{d^2W^\mu}{d\widetilde{\omega}d\Omega} = \frac{64\mu^6H^4}{\pi c^3\hbar^4} \mid \langle\zeta' \mid \sigma_- \mid \zeta\rangle \mid^2 \frac{4\chi^2 + (1+\chi^2)^2}{(1+\chi^2)^5}\,\delta\left(\widetilde{\omega} - \frac{\widetilde{\omega}_{\rm max}}{1+\chi^2}\right).$$

It then follows that the radiation proceeds at the frequency

$$\widetilde{\omega} = \frac{\widetilde{\omega}_{\rm max}}{1+\chi^2}.$$

Further integration with respect to the angles and frequency relative to Eq. (4.86) for $\langle\zeta' \mid \sigma_- \mid \zeta\rangle$, yields the total radiation power

$$W^\mu = \frac{128}{3}\,\frac{\mu^6H^4}{\hbar^4c^3}\,\gamma^4\,\frac{(1+\zeta\cos\nu)(1-\zeta'\cos\nu)}{4}. \tag{4.97}$$

In the quantum theory of radiation, spin transitions are considered only for time-invariant spin projections. In our case, adaption to such transitions corresponds to $\nu = 0$ when the spin of a neutron is oriented along the field. At $\nu = 0$ the spin transitions occur, according to Eq. (4.97), in only one direction — opposite to the magnetic field when $\zeta = 1$ and $\zeta' = -1$ ($\mu < 0$). The total power

$$W^\mu_{\uparrow\downarrow} = \frac{128}{3}\,\frac{\mu^6H^4}{\hbar^4c^3}\,\gamma^4,$$

coincides entirely with the spin-flip radiation of a neutron, calculated from the Dirac-Pauli equation[59]. At $\nu = \pi/2$ when the spin rotates in the plane orthogonal to the magnetic field vector, the total radiation power of the neutron decreases by the factor of 4 but all qualitative features of the radiation are the same. This case is adequate for the classical statement of the problem on the magnetic moment radiation and it is described fully within the scope of classical electrodynamics[58]. Note that the 'true' magnetic moment does not radiate at all in the classical theory at $\nu = 0$ when $\mu = const$. This radiation is associated with spin-flip transitions and is purely quantum in nature.

In a particular case, all the above results can be obtained by direct calculations or the Lorentz transformations as in Ref.[60], or by the quantum theory in Ref.[60] as well as by the classical theory in Ref.[61].

Eq. (4.97) can also be obtained by Jackson's method[36] in the limit of $a \to \infty$. For an arbitrary a the power of μ-radiation has the form

$$W^\mu = W_{\rm SR}\,\frac{4}{3}\,\xi^2\left\{\frac{1+\zeta\zeta'}{2}\left[\frac{\sin^2\nu}{2}\Phi_1(a) + \cos^2\nu\Phi_3(a)\right] + \right.$$

$$+\frac{1-\zeta\zeta'}{2}\left[\frac{1+\cos^2\nu}{2}\Phi_1(a)+\sin^2\nu\Phi_3(a)+\zeta\cos\nu\left(\frac{a}{|a|}\Phi_1(a)+e^{-\sqrt{12}|a|}\Phi_2(a)\right)\right]\right\},$$

where

$$\begin{aligned}
\Phi_1(a) &= 1+\frac{49}{12}a+\frac{89}{12}a^2+7a^3+\frac{49}{12}a^4+\frac{5}{4}a^5+\frac{1}{6}a^6,\\
\Phi_2(a) &= \frac{35\sqrt{3}}{64}\left(1+\frac{248}{315}a-\frac{19}{15}a^2-\frac{73}{315}a^3+\frac{76}{105}a^4-\frac{4}{15}a^5\right)-\\
&- \frac{a}{|a|}\left(1+\frac{77}{96}a-\frac{7}{6}a^2-\frac{7}{32}a^3+\frac{65}{96}a^4-\frac{1}{4}a^5\right),\\
\Phi_3(a) &= \frac{1}{12}(5-a+a^2).
\end{aligned}$$

Hence, it follows in the first approximation in a that

$$W^\mu = W_{\rm SR}\,\frac{4}{3}\,\xi^2\left\{\frac{1+\zeta\zeta'}{2}\left[\left(1+\frac{49}{12}a\right)\frac{\sin^2\nu}{2}+\frac{5-a}{12}\cos^2\nu\right]+\right.$$

$$\left.+\frac{1-\zeta\zeta'}{2}\left[\left(1+\frac{49}{12}a\right)\frac{1+\cos^2\nu}{2}+\frac{5-a}{12}\sin^2\nu+\frac{35\sqrt{3}}{64}\left(1+\frac{40}{9}a\right)\zeta\cos\nu\right]\right\}.$$

This formula coincides with the power of μ-radiation in Eq. (4.89) if averaging over the azimuthal angle λ is carried out there according to Eq. (4.90).

In the case of $a \gg 1$ we have

$$W^\mu = W_{\rm SR}\,\frac{a^6}{9}\,\xi^2\left[\frac{1+\zeta\zeta'}{2}\sin^2\nu+\frac{1-\zeta\zeta'}{2}(1+\zeta\cos\nu)^2\right].$$

It is easy to see that this expression coincides with Eq. (4.97) if one takes into account that

$$W_{\rm SR}\,\frac{4}{9}\,a^6\xi^2=\frac{128}{3}\,\frac{(\mu_0 a)^6H^4}{c^3\hbar^4}\,\gamma^4=W^\mu_{\uparrow\downarrow}$$

and

$$\frac{1+\zeta\zeta'}{2}\sin^2\nu+\frac{1-\zeta\zeta'}{2}(1+\zeta\cos\nu)^2=(1+\zeta\cos\nu)(1-\zeta'\sin\nu).$$

Let us consider in more detail the correspondence principle for the 'true' magnetic moment radiation. In the classical field theory the magnetic moment radiation at a relativistic point is specified by the expression[60]

$$\widetilde{H}^{\alpha\beta}=\frac{1}{c}\,\frac{\partial}{\partial\tilde{t}}\,\widetilde{A}^{[\alpha}n^{\beta]}-\frac{1}{c^2}\,\frac{\partial^2}{\partial\tilde{t}^2}\,\widetilde{Q}^{[\alpha\rho}n_\rho n^{\beta]}, \tag{4.98}$$

where

$$\widetilde{A}^\alpha=-e\,\frac{v^\alpha}{\tilde{r}_\rho v^\rho},\quad \widetilde{Q}^{\alpha\rho}=-c\mu\,\frac{\Pi^{\alpha\rho}}{\tilde{r}_\rho v^\rho}.$$

Here, as usual, $\Pi^{\alpha\rho}$ is the dimensionless antisymmetric spin tensor, $\tilde{t}$ is the time of fixation of radiation at the observation point, $\tilde{r}^{\rho} = (\tilde{r}, \tilde{\boldsymbol{r}})$, and the square brackets in the superscripts imply antisymmetry with respect to these superscripts.

The Fourier transform of the electromagnetic field tensor of Eq. (4.98) has the form

$$\widetilde{H}^{\alpha\beta}_{\tilde{\omega}} = \frac{1}{\tilde{r}c^2} \int\limits_{-\infty}^{\infty} \left(-ie\tilde{\omega}u^{[\alpha}n^{\beta]} + \mu\tilde{\omega}^2\pi^{[\alpha\rho}n_{\rho}n^{\beta]}\right) e^{i\tilde{\omega}\left[t-\frac{1}{c}(\boldsymbol{nr})\right]}dt,$$

where $u^{\alpha} = c(1, \boldsymbol{\beta})$, $\pi^{\alpha\rho} = \Pi^{\alpha\rho}/\gamma$, $n^{\alpha} = (1, \boldsymbol{n})$.

In more detail we have

$$u^{[\alpha}n^{\beta]} = c(\boldsymbol{\beta} - \boldsymbol{n},\ [\boldsymbol{\beta n}]), \quad \pi^{[\alpha\rho}n_{\rho}n^{\beta]} = (\boldsymbol{\kappa},\ [\boldsymbol{\kappa n}]),$$

where

$$\boldsymbol{\kappa} = [\boldsymbol{n\pi}] + [\boldsymbol{n}[\boldsymbol{n}[\boldsymbol{\beta\pi}]]].$$

We proceed from the vector $\boldsymbol{\pi}$ to the spin vector in the rest frame in the last expression for $\boldsymbol{\kappa}$, according to the Lorentz transformation

$$\boldsymbol{\pi} = \frac{\boldsymbol{\zeta}}{\gamma} - \frac{\gamma}{\gamma+1}\,[\boldsymbol{\beta}[\boldsymbol{\beta\zeta}]].$$

Then

$$\boldsymbol{\kappa} = -[\boldsymbol{\zeta}(\boldsymbol{n} - \boldsymbol{\beta})] + \frac{\gamma}{\gamma+1}\,(\boldsymbol{\zeta\beta})[\boldsymbol{\beta n}].$$

Separating from Eq. (4.98) the polarization components of the electric field and taking into account that $(\boldsymbol{n n}_s) = 0$, we have

$$\widetilde{E}_{s\tilde{\omega}} = i\,\frac{e\tilde{\omega}}{\tilde{r}c} \int\limits_{-\infty}^{\infty} \left(\boldsymbol{\beta}^{\text{eff}}_{\text{cl}}\boldsymbol{n}_s\right) e^{i\tilde{\omega}\left[t-\frac{1}{c}(\boldsymbol{nr})\right]}dt,$$

$$\boldsymbol{\beta}^{\text{eff}}_{\text{cl}} = \boldsymbol{\beta} - i\frac{\mu\tilde{\omega}}{ec}\left\{[\boldsymbol{\zeta}(\boldsymbol{n} - \boldsymbol{\beta})] - \frac{\gamma}{\gamma+1}(\boldsymbol{\zeta\beta})[\boldsymbol{\beta n}]\right\} = \boldsymbol{\beta} + \boldsymbol{\beta}^{L}.$$

We see that the latter expressions are different from Eqs (4.78) and (4.80) as a result of the substitution $\boldsymbol{\sigma} \to \boldsymbol{\zeta}$. Thus, for μL-radiation the correspondence principle is absolutely realized.

We can now readily follow all the problems that arise in classical and semiclassical calculations. The classical expression for the radiation power averaged over the period of spin precession has the form ($\nu = \pi/2$, $\zeta_{\perp}^2 = \zeta^2 = 1$)

$$\overline{W^{\mu}} = \frac{2}{3}\,\frac{\mu^2}{c^3}\,\omega_{\mu}^4\gamma^4. \tag{4.99}$$

In the nonrelativistic approximation, this formula is the well-known expression for the power of the magnetic dipole radiation.

In a similar form, the corresponding semiclassical expression is, according to Eq. (4.97),

$$W^{\mu} = \frac{8}{3}\,\frac{\mu^2}{c^3}\,\omega_{\mu}^4\gamma^4\,\frac{(1+\zeta\cos\nu)(1-\zeta'\cos\nu)}{4}.$$

At $\nu = 0$ we have the result of the quantum theory for radiation from a neutron[59] and at $\nu = \pi/2$ the classical Eq. (4.99).

In spite of the fact that the classical theory does not operate on the terms of spin-flip transitions which explain the Sokolov-Ternov effect, it is possible to obtain a correct expression for the neutron polarization time by means of Eq. (4.99). In order to perform this, one has to represent μ-radiation in the classical theory as a nonstationary process

$$\zeta_{\|}(t)\Big|_{t=0} = 0, \quad \zeta_{\perp}(t)|_{t=0} = \zeta = 1. \tag{4.100}$$

The corresponding energy balance in the rest frame of an electron has the form (the magnetic moment $\mu < 0$ is specified in the rest frame)

$$\frac{d}{d\tau}\,(\boldsymbol{\mu}\boldsymbol{H}_0) = -W_0^{\mu}.$$

Hence we find, with account of Eq. (4.100), that

$$\frac{d\zeta_{\|}}{d\tau} = \frac{1}{2T_0}\,(1-\zeta_{\|}^2), \tag{4.101}$$

where

$$T_0 = \frac{3}{64}\,\frac{\hbar^4 c^3}{|\,\mu\,|^5\,H_0^3}.$$

The solution of Eq. (4.101) has the form

$$\zeta_{\|}(\tau) = \frac{e^{-\frac{\tau}{T_0}}-1}{e^{-\frac{\tau}{T_0}}+1}.$$

Thus, T_0 is the spin relaxation time. At $\tau \to \infty$ the spin of a neutron will be oriented opposite to the magnetic field direction $\zeta_{\|}(\infty) = -1$.

To transform T_0 to the laboratory frame of reference, it should be taken into account that

$$T = \gamma T_0, \quad H_0 = \gamma H\sqrt{1-\beta^2\cos^2\alpha}.$$

Then we obtain

$$T = \frac{3}{64}\,\frac{\hbar^4 c^3}{|\,\mu\,|^5\,H_0^3\gamma^2}\left(1-\beta^2\cos^2\alpha\right)^{-3/2}.$$

The same expression was obtained in the quantum theory (see, for example, Refs[36,61]).

Note that unlike the electron the neutron can have 100 % self-polarization, but the polarization time for the neutron is very long. In the most effective case of orthogonal motion ($\alpha = \pi/2$)

$$T_n = 3\,\frac{(H^*/H)^3}{\alpha\gamma^2\omega_Z}\left|\frac{\mu_n}{\mu_0}\right|^5, \tag{4.102}$$

where $|\,\mu_n/\mu_0\,| \approx 1.04\cdot 10^{-3}$ is the neutron-electron magnetic moment ratio [the rest of the notations are the same as in Eq. (4.94)], a marked result can be obtained only for high energies of neutrons in strong magnetic fields. For the energy 10^{13} *GeV* and $H = 4\cdot 10^6$ *Oe* we have an effective length of self-polarization $uT \approx cT \approx 1\ m$.

Note that as applied to an electron ($|\,\mu_n/\mu_0\,| = 1$), Eq. (4.102) is only different from Eq. (4.94) by the numerical coefficient.

We have considered a particular case when the 'true' magnetic moment (a neutron) moves orthogonal to the magnetic field, but one can also show that the correspondence principle works in a similar way for any other orientation of the velocity vector relative to the magnetic field vector. Moreover, the classical[60] and the quantum[62,63,64] theories of radiation of a relativistic particle with the 'true' magnetic moment are in full accord for an arbitrary configuration of constant uniform electromagnetic fields (see Sec. 4.3).

It should be noted in conclusion that these simple considerations may seem trivial from the point of view of contemporary theoretical physics but these issues will have to be resolved sooner or later since a number of topics in electrodynamics associated with the magnetic moment and interaction of the charge and the spin magnetic moment of particles have not yet been considered. There are some single papers in this line (Refs[1–5] and others); for more complete list see Ref.[18], but note that the authors do not consider how their results correlate with the quantum theory. This area of electrodynamics has been *terra incognita* without the correspondence principle.

Bibliography

1. J. Frenkel, *Z. Phys.* **37** (1926) 243; see also Frenkel' Ya.I. *Collection of Selected Works, Vol.1 Electrodynamics* (Moscow, Leningrad, Izd. Akad. Nauk SSSR, 1956) [in Russian].
2. H.J. Bhabha, H.C. Corben, *Proc. Roy. Soc.* **A178** (1941) 273.
3. A. Bialas, *Acta Phys. Pol.* **22** (1962) 349.
4. M. Kolsrud, E. Leer, *Phys. Nor.* **2** (1967) 181.
5. J. Cohn, H. Wiebe, *J. Math. Phys.* **17** (1976) 1946.
6. S.R. de Groot, L.G. Suttorp, *Foundation of Electrodynamics* (North-Holland, Amsterdam, 1972).
7. I.F. Farkas, *Acta. Phys. Hung.* **15** (1961) 131.
8. J.I. Horváth, *Acta. Phys. Hung.* **3** (1954) 171.
9. A.A. Sokolov, I.M. Ternov, *Relativistic Electron* (Moscow, Nauka, 1983).
10. I.M. Ternov, V.A. Bordovitsyn, *Vestn. Mosk. Univ., III* **28** 2 (1987) 21 [in Russian].

11. V.A. Bordovitsyn, I.M. Ternov, V.G. Bagrov, *Physics-Uspekhi* **38** 1995 1037.
12. V.A. Bordovitsyn, N.N. Bysov, G.K. Razina, V.Ya. Epp, *Izv. Vuz. Fiz.* 5 (1978) 12 [in Russian].
13. V.A. Bordovitsyn, G.K. Razina, N.N. Bysov, *Izv. Vuz. Fiz.* 10 (1980) 33 [in Russian].
14. V.A. Bordovitsyn, G.K. Razina, *Izv. Vuz. Fiz.* 4 (1981) 120 [in Russian].
15. I.M. Ternov, V.A. Bordovitsyn, *Vestn. Mosk. Univ., III* 21 3 (1980) 8 [*Moscow Univ. Phys. Bull.* **35** 3 (1980) 7].
16. L.D. Landau, E.M. Lifshitz, *The Classical Theory of Fields* (Oxford, Pergamon Press, 1980).
17. I.M. Ternov, *Physics-Uspekhi* **38** 4 (1995) 409.
18. V.A. Bordovitsyn, The Doctoral Dissertation in Phys.-Math. Sciences (Tomsk, Moscow, MGU, 1983).
19. S.A. Belomestnykh, A.E. Bondar, M.N. Yegorychev, et al. *Nucl. Instrum. Methods Phys. Res.* **227** (1984) 173.
20. V.N. Korchuganov, G.N. Kulipanov, N.A. Mezentsev, et al., Preprint INP 77-83 (Novosibirsk, Institute of Nuclear Physics, 1977).
21. A.E. Bondar, E.L. Saldin, *Nucl. Instru. Methods* **195** (1982) 577.
22. A.A. Sokolov, I.M. Ternov, *Dokl. Akad. Nauk SSSR* **153** (1963) 1052 [*Sov. Phys. Docl.* **8** (1964) 1203].
23. W. Pauli. *Papers in Quantum Mechanics. 1928-1958* translated into Russian (Moscow, Nauka, 1977).
24. N.F. Mott, H.S.W. Massey, *The theory of Atomic Collisions* 3rd edition (Oxford, Oxford University Press, 1965).
25. I. Kessler, *Polarized Electrons* translated into Russian (Moscow, Mir, 1969).
26. S.V. Vonsovskii, *Magnetism of Microparticles* (Moscow, Nauka, 1973) [in Russian].
27. A.A. Sokolov, I.M. Ternov, *Synchrotron Radiation* (New York, Pergamon Press, 1968).
28. V.G. Bagrov, O.F. Dorofeev, A.A. Sokolov, *Dokl. Akad. Nauk SSSR* **221** 2 (1975) 312 [*Sov. Phys. Docl.* **20** (1975) 198].
29. I.M. Ternov, *Fiz. Elem. Chastits At. Yadra* **17** (1986) 884 [*Sov. J. Part. Nucl.* **17** (1986) 389].
30. I.M. Ternov, V.G. Bagrov, R.A. Rzaev, *Zh. Eksp. Teor. Fiz.* **46** (1964) 374 [*Sov. Phys. JETP* **19** (1964) 255].
31. J. Schwinger, *Proc. Acad. Nat. Sci. USA* **40** (1954) 132.
32. V.N. Baier, *Usp. Fiz. Nauk* **105** (1971) 441 [*Sov. Phys. Usp.* **14** (1972) 695].
33. V.N. Baier, V.M. Katkov, V.S. Fadin, *Radiation of Relativistic Electrons* (Moscow, Atomizdat, 1973) [in Russian].
34. V.B. Berestetskii, E.M. Lifshitz, L.P. Pitaevskii, *Quantum Electrodynamics* (Moscow, Nauka, 1980) [in Russian].
35. V.G. Bagrov, V.V. Belov, A.Yu. Trifonov, *J. Phys.* **A26** (1993) 6431.
36. J.D. Jackson, *Rev. Mod. Phys.* **48** (1976) 417.
37. I.M. Ternov, V.A. Bordovitsyn, V.Ya. Epp, *Izv. Vuz. Fiz.* **33** 5 (1990) 49 [*Sov.*

Phys. J. **33** (1990) 420].
38. I.M. Ternov, V.A. Bordovitsyn, V.Ya. Epp, *Izv. Vuz. Fiz.* **33** 6 (1990) 22 [*Sov. Phys. J.* **33** (1990) 478].
39. I.M. Ternov, V.A. Bordovitsyn, V.Ya. Epp, *Izv. Vuz. Fiz.* **33** 7 (1990) 103 [in Russian].
40. Ya.S. Derbenev, A.M. Kondratenko, *Zh. Eksp. Teor. Fiz.* **64** (1973) 1918 [*Sov. Phys. JETP* **37** (1973) 968].
41. D.P. Barber, S.R. Mane, *Phys. Rev.* **A37** (1988) 456.
42. A.A. Sokolov, I.M. Ternov, *Zh. Eksp. Teor. Fiz.* **24** (1953) 249 [in Russian].
43. Wu-Yang Tsai, Asim Yildiz, *Phys. Rev.* **D8** (1979) 3446.
44. V.G. Bagrov, *Izv. Vuz. Fiz.* 5 (1965) 121 [in Russian].
45. I.M. Ternov, V.A. Bordovitsyn, *Usp. Fiz. Nauk* **132** (1980) 345 [*Sov. Phys. Usp.* **23** (1980) 679].
46. V.A. Bordovitsyn, S.V. Sorokin, *Izv. Vuz. Fiz.* **26** 8 (1983) 125 [in Russian].
47. I.E. Tamm. *Dokl. Akad. Nauk SSSR* **29** (1940) 551; C.r. Acad. Sci. USSR, **29** (1940) 551; Phys. Rev. **58** (1940) 952, see also I.E. Tamm *Collection of Scientific Works, Vol.2* (Moscow, Nauka, 1975) [in Russian].
48. V.L. Ginzburg, in *Topics in Theoretical Physics. To the Memory of I.E. Tamm* (Moscow, Nauka, 1972); *Izv. Vuz. Radiophys.* **27** (1984) 852; *Trudy Fiz. Inst. Akad. Nauk (FIAN)* **176** (1986) 3 [in Russian].
49. I.M. Ternov, V.A. Bordovitsyn, *Vestn. Mosk. Univ., III* **23** 6 (1982) 72 [*Moscow Univ. Phys. Bull.* **37** 6 (1982) 84].
50. I.M. Ternov, V.G. Bagrov, V.A. Bordovitsyn, O.F. Dorofeev, *Zh. Eksp. Teor. Fiz.* **55** (1968) 2273 [*Sov. Phys. JETP* **28** (1969) 1206].
51. V.A. Bordovitsyn, *Izv. Vuz. Fiz.* **36** 11 (1993) 39 [*Russ. Phys. J.* **36** (1993) 1046].
52. I.M. Ternov, V.G. Bagrov, V.Ch. Zhukovskii, *Vestn. Mosk. Univ., III* (1) (1966) 30 [in Russian].
53. V.A. Bordovitsyn, V.S. Gushchina, I.M. Ternov, *Nucl. Instrum. Methods Phys. Res.* **A359** (1995) 34.
54. V.G. Bagrov, Candidate Dissertation in Phys.-Math. Sciences (Moscow, MGU, 1964).
55. A.A. Sokolov, A.N. Matveev, I.M. Ternov, *Dokl. Akad. Nauk SSSR* **102** 1 (1955) 65 [in Russian].
56. J. Schwinger, Wu-Yang Tsai, *Ann. Phys. (USA)* **110** (1978) 63.
57. Wu-Yang Tsai, *Phys. Rev.* **D8** (1979) 3460.
58. A.N. Matveev, *Zh. Eksp. Teor. Fiz.* **31** (1956) 479 [*Sov. Phys. JETP* **4** (1957) 409].
59. I.M. Ternov, V.G. Bagrov, A.M. Khapaev, *Zh. Eksp. Teor. Fiz.* **48** (1965) 921 [*Sov. Phys. JETP* **21** (1965) 613].
60. V.A. Bordovitsyn, V.S. Gushchina, *Izv. Vuz. Fiz.* **36** 2 (1993) 60; **36** 3 (1993) 73; **37** 1 (1994) 53; **38** 2 (1995) 63; **38** 3 (1995) 83 [*Sov. Phys. J.* **36** (1993); **36** (1993) 247; **37** (1994) 49; **38** (1995) 155; **38** (1995) 293].
61. V.L. Lyuboshits, *Yad. Fiz.* **4** (1966) 269 [*Sov. J. Nucl. Phys.* **4** (1966) 195].

62. V.G. Bagrov, V.A. Bordovitsyn, *Izv. Vuz. Fiz.* **23** 2 (1980) 67 [*Sov. Phys. J.* **23** (1980) 128].
63. I.M. Ternov, V.G. Bagrov, V.M. Kruzhkov, A.M. Khapaev, *Izv. Vuz. Fiz.* 4 (1967) 41 [in Russian].
64. V.G. Bagrov, P.V. Bozrikov, D.M. Gitman et al. *Izv. Vuz. Fiz.* **17** 6 (1974) 150 [in Russian].
65. V.L. Ginzburg, *On the Relativity Theory* (Moscow, Nauka, 1989) [in Russian].
66. V.A. Bordovitsyn, *Izv. Vuz. Fiz.* **40** (1997) 40 [in Russian].

Chapter 5

Undulator Radiation

V.Ya. Epp
Tomsk State Pedagogical University 634041 Tomsk, Russia[a]

5.1. General Theory

5.1.1. The Undulator Parameter

An undulator, in a broad sense is a device assigned to production of electromagnetic radiation by means of relativistic charged particles. The particles execute periodic oscillations and drift with relativistic velocity along the undulator axis. As a rule the particle is engaged in transverse oscillation, though the undulators with longitudinal oscillations are possible too. The emitted radiation is confined in a narrow cone of a width about γ^{-1}, where γ is the usual shorthand for the relativistic factor $1/\sqrt{1-\beta^2}$ and $\boldsymbol{\beta} = \boldsymbol{v}/c$ is the speed of a particle in units of speed of light. This feature defines many properties of radiation, because the spectrum sufficiently depends on the ratio between the velocity deviation angle α and γ^{-1}. In case of relatively weak magnetic field ($\alpha\gamma \ll 1$) the radiation cone is oriented in the vicinity of the drift direction and the radiation within the cone is formed by the whole trajectory. The radiation spectrum in this case sufficiently depends on the structure of the magnetic field in the undulator. In the opposite case of a strong field ($\alpha\gamma \gg 1$) radiation in the given direction is emitted within a short interval of time from a short segment of the path which can be replaced by an arc of circle. The undulator radiation spectrum in this case is close to synchrotron radiation one and slightly depends on the magnetic field structure. An undulator with a strong magnetic field is referred to as a wiggler. The relation between velocity deviation angle and the opening of radiation cone is characterized by the so-called undulator parameter.

Let us consider the equations of motion of a charged particle in the undulator

[a]E-mail: epp@tspu.edu.ru

field in order to define the undulator parameter. Assume that the particle has a considerably high energy, i.e. the relativistic parameter γ is much greater than unity. Let also the electric field in the undulator be equal to zero. Then the equations of motion in the magnetic field $\boldsymbol{H}(\boldsymbol{r})$ become

$$\frac{d^2\boldsymbol{r}}{dt^2} = \frac{e}{mc\gamma}[\boldsymbol{v}\boldsymbol{H}] \tag{5.1}$$

where e and m are the charge and the mass of the particle. Solution of Eq. (5.1) can be simplified under some reasonable assumptions. First, the high energy of the particle causes the speed vector $\boldsymbol{v}$ to retain close to the direction of the initial velocity. Second, we suppose the initial velocity to be parallel to the undulator axis. And finally, let the longitudinal component of the magnetic field be of an order or less than the transversal one. Hence, we impose the following conditions:

$$\gamma \gg 1,\ \beta_\perp \ll 1,\ \beta_\perp H_\| \ll H_\perp \tag{5.2}$$

Here $\beta_\perp$ and $H_\perp$ are the orthogonal with respect to the undulator axis components of the velocity and magnetic field respectively, $H_\|$ is the longitudinal component. The last condition allows us to neglect the longitudinal component of the magnetic field in the vector product in Eq. (5.1).

Let the direction of the undulator axis be x-axis. Integrating Eq. (5.1), we find the transferal components of the particle velocity

$$\beta_y = -\frac{e}{mc^2\gamma}\int H_z(x)dx,\quad \beta_z = \frac{e}{mc^2\gamma}\int H_y(x)dx, \tag{5.3}$$

We took into account that $v_x \approx c$ and replaced dt with dx/c. As shown below, we have to know the longitudinal component of the velocity up to γ^{-2} in order to calculate the radiation spectrum. Caring out the integration of x-component in Eq. (5.1) we get

$$\beta_x = \beta_\| + \frac{e}{mc^2\gamma}\int[\beta_y H_z(x) - \beta_z H_y(x)]dx, \tag{5.4}$$

where β_y and β_z are the known functions of x given by Eqs (5.3). The constant $\beta_\|$ is a mean velocity of the particle $\beta_\| = \langle\beta_x\rangle$. Respectively, the second term in Eq. (5.4) is an oscillating function of x with a mean value equal to zero. We still keep here $dt = dx/c$ though Eq. (5.4) is accurate up to the second order of γ^{-1}. The point is that the integrand is a small quantity of the first order. Hence, taking into account the difference between v_x and c we would get corrections of the third order.

In approximation $\gamma \gg 1$, the velocity deviation angle α is equal to the orthogonal component of $\boldsymbol{\beta}$:

$$\alpha = \beta_\perp = \sqrt{\beta_y^2 + \beta_z^2} \tag{5.5}$$

As discussed above, the character of radiation is determined by the relation of α and γ^{-1}. Thus, we can choose as a parameter specifying the undulator operating mode the product of γ by the mean square of $\beta_\perp$

$$k = \gamma\sqrt{\langle\beta_\perp^2\rangle} \tag{5.6}$$

This equation defines[b] the undulator parameter. In accordance with the above discussion undulator with $k \gg 1$ is referred to as a wiggler. It follows from the definition expressed by Eq. (5.6) and Eq. (5.3) that the undulator parameter is proportional to the transverse component of the magnetic field. Taking into account that the integrals in Eqs (5.3) vary periodically with the space period λ_0 we find that k is of the order

$$k \sim \frac{e\lambda_0 H_\perp}{mc^2}. \tag{5.7}$$

One can see from this relation that the parameter k does not depend on the particle energy but is defined exclusively by the magnitude of the magnetic field and the undulator period. Knowing the undulator parameter one can easily find the mean speed of the particle

$$\beta_\parallel = \langle\sqrt{\beta^2 - \beta_\perp^2}\rangle \approx \langle 1 - \frac{\gamma^{-2} + \beta_\perp^2}{2}\rangle = 1 - \frac{1+k^2}{2\gamma^2}. \tag{5.8}$$

Let us consider for example two special cases of the magnetic field structure.

1. *Sinusoidal magnetic field.* Let the coordinate-dependence of the magnetic field be as follows: $\boldsymbol{H}(0, 0, H\sin 2\pi x/\lambda_0)$. This is a good approximation for the on-axis field in conventional linearly polarized undulators like the one shown in Fig. 5.1. The longitudinal component of the magnetic field can be neglected because the conditions of Eq. (5.2) are typically fulfilled.

Exact solutions of equations of motion can be expressed in terms of elliptic functions. We can easily find the approximate solutions from Eqs (5.3) and (5.4)

$$\beta_y = \frac{e\lambda_0 H}{2\pi mc^2\gamma}\cos\omega_0 t, \quad \beta_x = \beta_\parallel - \left(\frac{e\lambda_0 H}{4\pi mc^2\gamma}\right)^2 \cos 2\omega_0 t. \tag{5.9}$$

The frequency of particle oscillations is given by $\omega_0 = 2\pi v_\parallel/\lambda_0$. Averaging β_y^2 we obtain from Eqs (5.6)

[b] A definition $k = \gamma\sqrt{2\langle\beta_\perp^2\rangle}$ is of frequent use too.

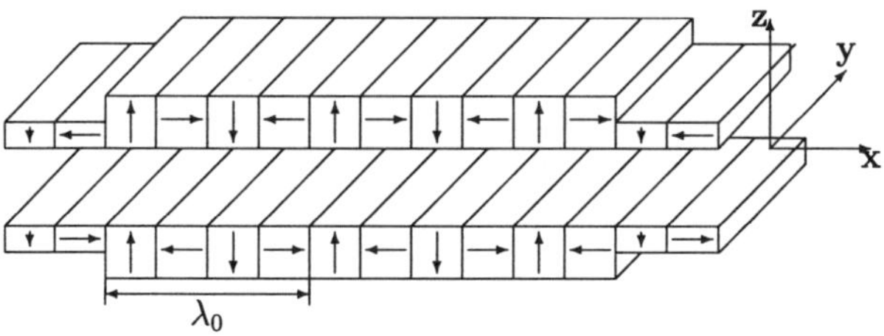

Fig. 5.1. Schematic picture of a plane undulator. Arrows indicate the polarization of permanent magnets.

$$k = \frac{e\lambda_0 H}{2\sqrt{2}\pi mc^2}, \tag{5.10}$$

To complete the discussion we integrate Eqs (5.9) once more and obtain the solution of the equation of motion

$$x = v_{\|}t - \frac{k^2\lambda_0}{8\pi\gamma^2}\sin 2\omega_0 t, \quad y = \frac{k\lambda_0}{\sqrt{2}\pi\gamma}\sin\omega_0 t \tag{5.11}$$

with $v_{\|}$ defined by Eqs (5.8) and (5.10).

2. *Helical magnetic field.* In order to produce radiation with circular polarization the magnetic field of the structure $\boldsymbol{H}(x) = (0, H_y \sin 2\pi x/\lambda_0, H_z \cos 2\pi x/\lambda_0)$ is widely used. The last expression describes in a good approximation the on-axis field in undulators like the one shown in Fig. 5.2.

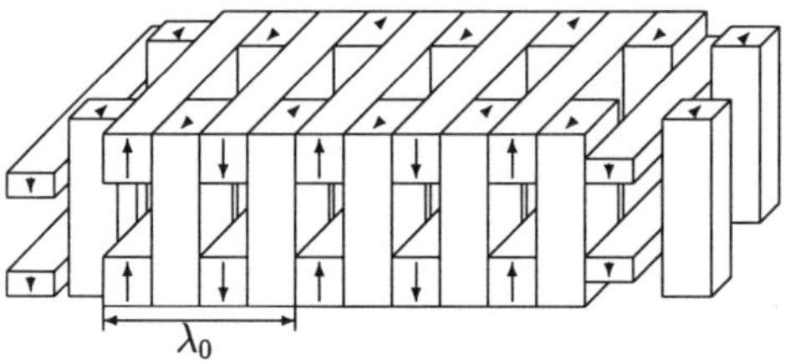

Fig. 5.2. Schematic picture of helical undulator.

Repeating the calculations performed in the previous example we get

$$\begin{aligned}
\beta_y &= -\frac{2\pi}{\lambda_0}a_y \sin\omega_0 t, \quad \beta_z = -\frac{2\pi}{\lambda_0}a_z\cos\omega_0 t, \\
\beta_x &= \beta_{\|} + \frac{4\pi}{\lambda_0}a_x\cos 2\omega_0 t.
\end{aligned} \tag{5.12}$$

where

$$a_x = \frac{e^2\lambda_0^3(H_y^2 - H_z^2)}{64\pi^3 m^2 c^4 \gamma^2}, \quad a_{y,z} = \frac{e\lambda_0^2 H_{z,y}}{4\pi^2 mc^2\gamma}. \tag{5.13}$$

The undulator parameter is equal to

$$k = \frac{e\lambda_0}{2\sqrt{2}\pi mc^2}\sqrt{H_y^2 + H_z^2}. \tag{5.14}$$

And finally, for the solution of the equation of motion we have

$$\begin{aligned} x &= v_{\parallel}t + a_x \sin 2\omega_0 t, \\ y &= a_y \cos\omega_0 t, \\ z &= -a_z \sin\omega_0 t. \end{aligned} \tag{5.15}$$

In the case of $H_y = H_z$ the longitudinal oscillations vanish and the trajectory becomes a helical one.

5.1.2. Spectrum and Angular Distribution

Consider a relativistic charged particle moving in an undulator of N periods. We can represent the radius vector in terms of a drift with a constant velocity $\boldsymbol{v}_{\parallel}$ and periodic oscillations in the following way

$$\begin{aligned} \boldsymbol{r}(t) &= \boldsymbol{v}_{\parallel}t + \boldsymbol{r}_1(t), \quad 0 < t < NT; \\ \boldsymbol{r}(t) &= \boldsymbol{v}_0 t, \qquad\qquad t < 0, \ t > NT. \end{aligned} \tag{5.16}$$

Here $\boldsymbol{r}_1(t)$ is a periodic function with a period T and $\boldsymbol{v}_0$ is the constant velocity of the particle outside the undulator. The last expression is an approximate description of the particle motion. In real undulators the first and the last magnet sections differ from the main ones. Thus, the particle motion inside the undulator is not a periodic one. The electric field of radiation at the time moment t' is given by expression [1,2]

$$\boldsymbol{E}(t') = \frac{e}{cR_e}\frac{[\boldsymbol{n}[(\boldsymbol{n} - \boldsymbol{\beta})\dot{\boldsymbol{\beta}}]]}{(1 - \boldsymbol{\beta n})^3}, \tag{5.17}$$

where the right part of the equation is a function of the retarded time $t = t' - R_e/c$, $\boldsymbol{R}_e$ is the vector from the position of charge at the retarded time to the point where the electric field is defined and $\boldsymbol{n} = \boldsymbol{R}_e/R_e$ is a unit vector. It is generally assumed that the distance between the undulator and the observation point is much greater then the undulator length. Calculation under this assumption is usually referred to as a far field approximation. Suppose this condition is met. Denoting by $\boldsymbol{R}$ the vector from coordinate origin placed somewhere inside the undulator to the observation point we compute

$$R_e = \sqrt{(\boldsymbol{R}-\boldsymbol{r})^2} \approx R - \boldsymbol{n}\boldsymbol{r} \tag{5.18}$$

with $\boldsymbol{n} = \boldsymbol{R}/R$. Thus, we have for the retarded time

$$t = t' - \frac{R}{c} + \frac{\boldsymbol{n}\boldsymbol{r}(t)}{c}. \tag{5.19}$$

Let us find the Fourier transform of the field of radiation

$$\boldsymbol{E}(\omega) = \frac{e}{cR} e^{ikR} \int\limits_0^{NT} \frac{[\boldsymbol{n}[(\boldsymbol{n}-\boldsymbol{\beta})\dot{\boldsymbol{\beta}}]]}{(1-\boldsymbol{\beta}\boldsymbol{n})^2} e^{i(\omega t - \boldsymbol{k}\boldsymbol{r})} dt. \tag{5.20}$$

Here ω is the frequency of radiation and $\boldsymbol{k} = \omega \boldsymbol{n}/c$ is the propagation vector. The integral in the latter expression can be represented as a sum of N integrals over each period. Integrating by parts in Eq. (5.20) and taking into account the sum of geometric progression

$$\sum_{n=1}^{N} e^{2ian} = \frac{\sin Na}{\sin a} e^{i(N+1)a} \tag{5.21}$$

we find

$$\begin{aligned} \boldsymbol{E}(\omega) &= \frac{2ie}{cR(1-\boldsymbol{\beta}_{\|}\boldsymbol{n})} e^{ikR + i\pi\nu(N-1)} \left\{ \frac{1-\boldsymbol{\beta}_{\|}\boldsymbol{n}}{1-\boldsymbol{\beta}_0\boldsymbol{n}} [\boldsymbol{n}[\boldsymbol{n}\boldsymbol{\beta}_0]] e^{i(\pi\nu - \boldsymbol{k}\boldsymbol{r}_0)} \sin \pi\nu N - \right. \\ & \left. \pi\nu \frac{\sin \pi\nu N}{\sin \pi\nu} \frac{1}{T} \int\limits_0^T [\boldsymbol{n}[\boldsymbol{n}\boldsymbol{\beta}]] e^{i(\omega t - \boldsymbol{k}\boldsymbol{r})} dt \right\}. \end{aligned} \tag{5.22}$$

Here

$$\nu = \frac{\omega}{\omega_0}(1 - \boldsymbol{\beta}_{\|}\boldsymbol{n}) \tag{5.23}$$

is the reduced frequency of radiation, $\omega_0 = 2\pi/T$ is the frequency of particle motion, $\boldsymbol{r}_0 = \boldsymbol{r}(0)$ and $\boldsymbol{\beta}_0 = \boldsymbol{\beta}(0) = \boldsymbol{\beta}(NT)$ are the coordinates and speed of the particle at the point of entry into the undulator.

The first term in Eq. (5.22) depends only on initial and final coordinates and velocity, hence it describes the radiation field emitted at the entrance and outlet of the undulator linked to the transition from motion with a uniform velocity to periodic oscillations and back. The second term is modulated by the function $\sin\pi\nu N/\sin\pi\nu$. This function peaks at frequencies $\nu = (m+1/2)/n$ and $\nu = n$, where m and n are the integers. The maximum of the absolute value at the integral ν is proportional to the number of periods N. Thus, at great N and integral ν the second term overrides the first one. Moreover, the real motion of the particle at undulator edges is not included in Eq. (5.16) because it is of a peculiar type for each device. Therefore, the first term in Eq. (5.22) does not carry any useful information. We will neglect this term on the assumption of great N and ν being close to integer (approximation of a long undulator). It is easy to show that the width of peaks is equal to $2/N$. Thus, from now on we assume that

$$N \gg 1, \qquad |\nu - n| < \frac{1}{N}. \tag{5.24}$$

The energy emitted per unit solid angle and unit frequency bandwidth is given by

$$\frac{d\mathcal{E}}{d\Omega d\omega} = \frac{cR^2}{4\pi^2}\left|\boldsymbol{E}(\omega)\right|^2. \tag{5.25}$$

Taking the square of module of Eq. (5.22) and keeping terms proportional to N^2 only, we find

$$\frac{d\mathcal{E}}{d\Omega d\omega} = \frac{e^2\omega_0^2\nu^2}{4\pi^2 c(1-\boldsymbol{\beta}_{\|}\boldsymbol{n})^2} I(\nu, N)\left|\int_0^T [\boldsymbol{n}\boldsymbol{\beta}]e^{i(\omega t - \boldsymbol{k}\boldsymbol{r})}dt\right|^2, \tag{5.26}$$

where

$$I(\nu, N) = \frac{\sin^2 \pi\nu N}{\sin^2 \pi\nu}. \tag{5.27}$$

The intensity formula of Eq. (5.26) factors into the term which describes the radiation from one period and the interference factor $I(\nu, N)$. A sample of radiation spectrum in a fixed direction is shown in Fig. 5.3.

It consists of sharp maxima at integral ν. The frequency of maximum of nth harmonic can be found from Eq. (5.23) by substitution $\nu = n$

$$\omega_n = \frac{n\omega_0}{1-\boldsymbol{\beta}_{\|}\boldsymbol{n}}. \tag{5.28}$$

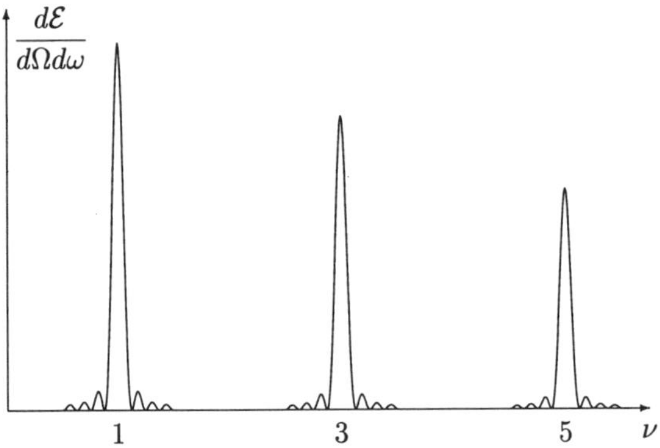

Fig. 5.3. A typical spectrum of undulator radiation in a fixed direction.

It is a multiple of the frequency of charge wiggle compressed by the Doppler effect. Each harmonic is of the width $2/N$ (between the adjacent minima) and of the magnitude proportional to N^2. The central lobe is surrounded by secondary maxima which decrease rapidly. Note that the secondary peaks are beyond the accuracy $1/N$ because the conditions of Eq. (5.24) do not stand. The distribution of radiated energy among the harmonics depends on the value of k.

The right hand side of Eq. (5.26) can be expressed as a sum of harmonics. Having this in mind, we expand $\sin \pi\nu$ in a power series around $\nu = n$

$$\frac{1}{\sin^2 \pi\nu} = \frac{1}{\pi^2(n-\nu)^2}\left[1 + \frac{\pi^2}{3}(n-\nu)^2 + \ldots\right].$$

The last term $\sim (n-\nu)^2 \sim 1/N^2$ can be neglected within the adopted accuracy. Then the profile of nth harmonic takes the form

$$I_n(\nu, N) = \frac{\sin^2 \pi\nu N}{\pi^2(n-\nu)^2}. \tag{5.29}$$

Multiplying the numerator and denominator by $(n+\nu)^2 \approx 4n^2$, we find another form of $I_n(\nu, N)$ in the vicinity of integral ν

$$I_n(\nu, N) = \frac{4n^2}{\pi^2}\frac{\sin^2 \pi\nu N}{(n^2-\nu^2)^2}. \tag{5.30}$$

In order to represent the radiated power as a sum of separate harmonics we take into account the fact that the righthand side of Eq. (5.26) is a product of the rapidly oscillating function $I(\nu, N)$ and a function slowly varying with frequency. The last multiplier can be taken approximately constant in a narrow band $|\nu - n| \sim 1/N$ with an accuracy up to $1/N$. Thus, we can replace Eq. (5.26) with

$$\frac{d\mathcal{E}}{d\Omega d\omega} = \frac{e^2}{c(1-\boldsymbol{\beta}_{\parallel}\boldsymbol{n})^2} \sum_{n=1}^{\infty} n^2 I_n(\nu, N) \left| \frac{1}{T} \int_0^T [\boldsymbol{n}\boldsymbol{\beta}] e^{i(\omega_n t - \boldsymbol{k}_n \boldsymbol{r})} dt \right|^2 , \tag{5.31}$$

where $\boldsymbol{k}_n = \boldsymbol{n}\omega_n/c$ with the frequency ω_n of nth harmonic defined by Eq. (5.28).

Note that the last equation describes a continuous spectral and angular distribution though the right side is represented as a sum of harmonics. It is apparent that in order to calculate the spectral density of radiation at any frequency ν close to integral m, one should take only one term of the sum in Eq. (5.31), namely, the term with $n = m$.

Spectral distribution of the emitted energy within a separate harmonic has a standard form according to the function $I_n(\nu, N)$ and does not depend on a specific undulator. But the distribution of energy among harmonics strongly depends on the structure and magnitude of the magnetic field in the undulator. The angular distribution of energy can be found by integration of Eq. (5.31) over the reduced frequency ν. It is easily seen that

$$\int_0^{\infty} I_n(\nu, N)\, d\nu = N. \tag{5.32}$$

Next we represent the angular distribution as a sum over harmonics

$$\frac{d\mathcal{E}}{d\Omega} = \sum_{n=1}^{\infty} \frac{d\mathcal{E}_n}{d\Omega}, \tag{5.33}$$

where the nth term has a meaning of angular distribution of the energy radiated at nth harmonic

$$\frac{d\mathcal{E}_n}{d\Omega} = \frac{e^2 n^2 \omega_0 N}{c(1-\boldsymbol{\beta}_{\parallel}\boldsymbol{n})^3} \left| \frac{1}{T} \int_0^T [\boldsymbol{n}\boldsymbol{\beta}] e^{i(\omega_n t - \boldsymbol{k}_n \boldsymbol{r})} dt \right|^2 . \tag{5.34}$$

It follows from Eqs (5.31) and (5.34) that the detailed spectral and angular distribution and the angular distribution of nth harmonic are related by a simple equation

$$\frac{d\mathcal{E}_n}{d\Omega d\omega} = \frac{\lambda_0(1-\boldsymbol{\beta}_{\parallel}\boldsymbol{n})}{2\pi c \beta_{\parallel} N} I_n(\nu, N) \frac{d\mathcal{E}_n}{d\Omega}. \tag{5.35}$$

Now we examine the spectrum as a function of direction to an observer. Consider the angular distribution of radiation at a fixed frequency. The main properties of this distribution are defined by the function $I(\nu, N)$. Denote by θ the angle between the vectors $\boldsymbol{n}$ and $\boldsymbol{\beta}_{\parallel}$, the latter aligned with the undulator axis. If we observe the radiation emitted at a fixed frequency ω and at an angle θ, we will detect the 'harmonic number' $\nu = \omega(1 - \boldsymbol{\beta}_{\parallel}\cos\theta)/\omega_0$. In particular, on the undulator axis $\nu = \omega(1-\boldsymbol{\beta}_{\parallel})/\omega_0$. If this continuous harmonic number is close to integral n, we see a bright spot in the center of the radiation pattern related to nth harmonic. As the angle θ increases, the harmonic number increases too and at some angle θ_{n+1} one can see a ring of a harmonic number $n+1$ and so on. Thus, the radiation pattern observed at a fixed frequency consists of a set of concentric rings, each representing the corresponding harmonic. It is easy to find from Eq. (5.28) that the maximum of nth harmonic is emitted at the angle

$$\cos\theta_n = \frac{1}{\beta_{\parallel}}(1 - \frac{n\omega_0}{\omega}). \tag{5.36}$$

The azimuthal distribution of radiated energy within the ring significantly depends on the harmonic number and the particle dynamics.

In case of the ultrarelativistic particle, the angle θ is a small quantity of the order γ^{-1}. Expanding Eqs (5.23) and (5.28) for small γ^{-1} and θ, we can rewrite the formulae for the harmonic number and the frequency of nth harmonic as

$$\nu = \frac{\omega}{2\omega_0\gamma^2}(1 + k^2 + \psi^2), \quad \omega_n = \frac{2\omega_0\gamma^2 n}{1 + k^2 + \psi^2}, \tag{5.37}$$

where $\psi = \gamma\theta$ is the reduced angle. In the same approximation the angle θ at which we see the maximum of nth harmonic (see Eq. (5.36)) is equal to

$$\theta_n = \gamma^{-1}\sqrt{\frac{2n\omega_0\gamma^2}{\omega} - 1 - k^2}. \tag{5.38}$$

We have defined the undulator parameter k through the root-mean-square value of the transversal velocity or the transversal magnetic field in the undulator. But due to magnetic field errors the calculated value of k differs from the 'real' one. That is, the real maximum of nth harmonic is shifted with respect to ω_n which is defined by Eq. (5.37). In this sense Eq. (5.37) can be taken as an empirical definition of the undulator parameter.

5.1.3. Polarization

The Fourier transform of the electric field of Eq. (5.20) can be written as follows

$$\boldsymbol{E}(\omega) = \boldsymbol{n}_\sigma E_\sigma(\omega) + \boldsymbol{n}_\pi E_\pi(\omega), \tag{5.39}$$

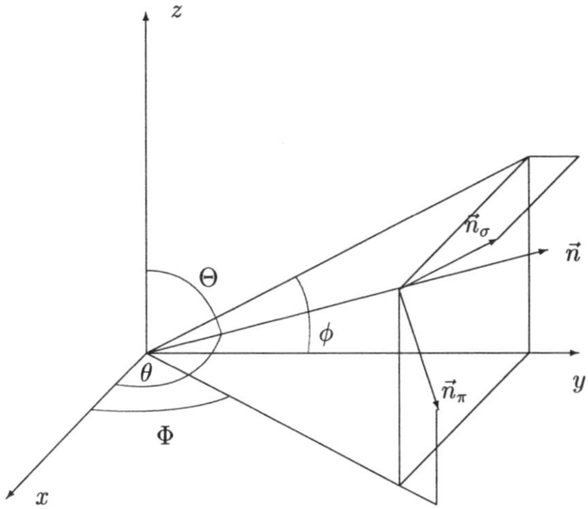

Fig. 5.4. The coordinate system and polarization vectors.

where $\boldsymbol{n}_\sigma$ and $\boldsymbol{n}_\pi$ are the unit polarization vectors shown in Fig. 5.4.

The particle drift is assumed to be in the x direction. Let us define the propagation direction $\boldsymbol{n}$ by angles Θ and Φ (see Fig. 5.4). Then the polarization vectors have coordinates

$$\begin{aligned} \boldsymbol{n}_\sigma &= (-\sin\Phi,\ \cos\Phi,\ 0), \\ \boldsymbol{n}_\pi &= (\cos\Theta\cos\Phi,\ \cos\Theta\sin\Phi,\ -\sin\Theta). \end{aligned} \tag{5.40}$$

In case of an ultrarelativistic particle, the angle between the drift direction and the direction of radiation is a small quantity of the order γ^{-1}. Then another set of angular coordinates is more convenient — θ and ϕ (see Fig. 5.4). The two sets of angles are related by equation

$$\begin{aligned} \boldsymbol{n} &= (\sin\Theta\cos\Phi,\ \sin\Theta\sin\Phi,\ \cos\Theta), \\ &= (\cos\theta,\ \sin\theta\cos\phi,\ \sin\theta\sin\phi), \end{aligned} \tag{5.41}$$

which results in

$$\boldsymbol{n}_\sigma = \frac{(-\sin\theta\cos\phi,\ \cos\theta,\ 0)}{\sqrt{1-\sin^2\theta\sin^2\phi}},$$

$$\boldsymbol{n}_\pi = \frac{(\cos\theta\sin\theta\sin\phi, \sin^2\theta\sin\phi\cos\phi,\ \sin^2\theta\sin^2\phi - 1)}{\sqrt{1-\sin^2\theta\sin^2\phi}}. \tag{5.42}$$

Expanding for small $\theta \sim \gamma^{-1}$ up to γ^{-2} we can rewrite the coordinates of orthogonal set of unit vectors in the form

$$\begin{aligned} \boldsymbol{n} &= (1 - \frac{1}{2}\theta^2,\ \theta\cos\phi,\ \theta\sin\phi), \\ \boldsymbol{n}_\pi &= (\theta\sin\phi,\ \theta^2\sin\phi\cos\phi,\ \frac{1}{2}\theta^2\sin^2\phi - 1), \\ \boldsymbol{n}_\sigma &= (-\theta\cos\phi,\ 1 - \frac{1}{2}\theta^2\cos^2\phi,\ 0). \end{aligned} \tag{5.43}$$

The polarization components of $\boldsymbol{E}(\omega)$ can be derived directly from Eq. (5.22). Restricting our consideration to the approximation of a long undulator $N \gg 1$ we find

$$E_j(\omega) = \frac{2\pi\omega i e}{cR\omega_0} e^{ikR + i\pi\nu(N-1)} \frac{\sin\pi\nu N}{\sin\pi\nu} B_j(\omega), \quad j = \sigma, \pi, \tag{5.44}$$

where

$$B_j(\omega) = \frac{1}{T}\int_0^T \boldsymbol{n}_j \boldsymbol{\beta}(t) e^{i(\omega t - \boldsymbol{kr})} dt. \tag{5.45}$$

The function $B_j(\omega)$ is complex and caries information on the phase difference between the polarization components. The radiation is linearly polarized in case B_σ and B_π have the same phase, or phases differing by an integral multiple of π. In the event of the arbitrary phase difference the simple harmonic wave is elliptically polarized. Polarization of the monochromatic component of radiation can be defined by one complex parameter

$$P(\omega) = \frac{E_\pi(\omega)}{E_\omega(\omega)} = p(\omega) e^{i\Delta(\omega)}, \tag{5.46}$$

or by two real parameters — the ratio of the amplitudes $p(\omega)$ and the phase difference $\Delta(\omega)$. There is more general formalism for describing polarization properties of fully or partly polarized light involving Stokes parameters.

It is convenient sometimes to use the circular polarization components E_+ and E_- instead of the linear ones E_σ and E_π. We define these components as follows

$$E_\pm = (E_\pi \pm iE_\sigma)/\sqrt{2} \tag{5.47}$$

The components E_+ and E_- are chosen in such a way that the intensity of the right- or lefthand polarized wave is proportional to $|E_+|^2$ or $|E_-|^2$, respective.

The spectral and angular distribution of emitted energy is given by

$$\frac{d\mathcal{E}_j}{d\Omega d\omega} = \frac{cR^2}{4\pi^2} \left|\boldsymbol{E}_j(\omega)\right|^2 \tag{5.48}$$

where the subscript j denotes a kind of polarization: σ and π for the linear components or '+' and '-' for the components of circular polarization. The last equation can be rewritten in the form of Eq. (5.31)

$$\frac{d\mathcal{E}}{d\Omega d\omega} = \frac{e^2}{c(1-\boldsymbol{\beta}_{\|}\boldsymbol{n})^2} \sum_{n=1}^{\infty} n^2 I_n(\nu, N) \left|B_{jn}\right|^2, \tag{5.49}$$

where

$$B_{jn} = \frac{1}{T} \int_0^T [\boldsymbol{n}\boldsymbol{\beta}]_j e^{i(\omega_n t - \boldsymbol{k}_n \boldsymbol{r})} dt \tag{5.50}$$

In order to obtain detailed information on polarization as a function of the direction $\boldsymbol{n}$ and frequency ω one has to perform the calculations in Eqs (5.49) and (5.50) for a particular trajectory of a charged particle. There are nevertheless some ways to estimate a type of polarization without going into detail.

For instance, a linear polarized radiation occurs in an undulator in which the particle describes a flat path satisfying the following condition: there is a point on the trajectory with respect to which the radius vector $\boldsymbol{r}(t)$ of the particle is an odd function of time [3]. In order to prove this, let us take the time origin at this point so that $\boldsymbol{r}(t) = -\boldsymbol{r}(-t)$. Then according to Eq. (5.44), the real part of the Fourier transform $E_j(\omega)$ is equal to zero for both the components. Actually,

$$\mathrm{Im}[B_j(\omega)] = \frac{1}{T} \int_{-T/2}^{T/2} \boldsymbol{n}_j \boldsymbol{\beta}(t) \sin(\omega t - \boldsymbol{k}\boldsymbol{r}) dt = 0, \tag{5.51}$$

because the function $\boldsymbol{\beta}(t) = \boldsymbol{\beta}_{\|} + d\boldsymbol{r}_1/dt$ is even. Hence, the phase difference between the polarization components vanishes.

The type of polarization can be estimated by the use of the following property of radiation [4]: the electric field produced by a nonrelativistic charge in the radiation zone follows the direction of the projection of the charge acceleration onto the plane orthogonal to the wave vector $\boldsymbol{k}$. If the particle follows the helical trajectory, the on-axis radiation is circularly polarized and it has elliptical polarization in the off-axis direction. Correspondingly, for the charge confined to a plane orbit, the radiation emitted in to the plane of the orbit is linearly polarized.

And finally, one can make a qualitative analysis of polarization knowing the particle trajectory and the instant angular distribution of the polarization components. Let us dwell on this characteristic because it is quite useful.

The instant value of the components of the electric field can be found directly from Eq. (5.17). Restricting our consideration to the case of orthogonal velocity and acceleration we obtain

$$\begin{aligned} E_\sigma &= (\boldsymbol{E}\boldsymbol{n}_\sigma) = \frac{e\dot\beta}{c^2R}\frac{n_x - \beta(1-n_z^2)}{\sqrt{1-n_z^2}\,(1-\beta n_x)^3}, \\ E_\pi &= (\boldsymbol{E}\boldsymbol{n}_\pi) = \frac{e\dot\beta}{c^2R}\frac{n_y n_z}{\sqrt{1-n_z^2}\,(1-\beta n_x)^3}. \end{aligned}$$

The instant power of the energy loss is given by

$$\frac{dW}{d\Omega} = \frac{cR^2}{4\pi}(1-\boldsymbol{\beta}_{\|}\boldsymbol{n})E^2(t'). \tag{5.52}$$

In terms of angles θ and ϕ the angular distribution takes the form

$$\begin{aligned} \frac{dW}{d\Omega} &= \frac{e^2\dot\beta^2}{4\pi c}(\rho_\sigma+\rho_\pi); \\ \rho_\sigma &= \frac{[\beta(1-\sin^2\theta\sin^2\phi)-\cos\theta]^2}{(1-\sin^2\theta\sin^2\phi)(1-\beta\cos\theta)^5}; \\ \rho_\pi &= \frac{\sin^4\theta\cos^2\phi\sin^2\phi}{(1-\sin^2\theta\sin^2\phi)(1-\beta\cos\theta)^5}; \\ \rho_\sigma+\rho_\pi &= \frac{(1-\beta\cos\theta)^2-\sin^2\theta\cos^2\phi(1-\beta)^2}{(1-\beta\cos\theta)^5}. \end{aligned} \tag{5.53}$$

The same functions are given by Eq. (1.41) where $\rho^H_{\theta,\phi}$ correspond to $\rho_{\pi,\sigma}$ of this chapter. The geometrical figures $\rho_\sigma(\theta,\phi)$ and $\rho_\pi(\theta,\phi)$ are referred to as indicatrix of radiation. They are shown in figures of Sec. 1.4. In case of the ultrarelativistic particle, we can expand the functions ρ_σ and ρ_π with respect to small γ^{-1} and θ

$$\rho_\sigma = \frac{8\gamma^6(1-\psi^2\cos 2\phi)^2}{(1-\psi^2)^5}, \quad \rho_\pi = \frac{8\gamma^6\psi^4\sin^2 2\phi}{(1-\psi^2)^5} \tag{5.54}$$

More detailed information see in Sec. 1.4 or in Ref.[5]

5.1.4. Radiation of an Infinite Undulator

It is useful sometimes to consider a long undulator as an infinite one. Clearly, the infinite undulator is a model. Moreover, we cannot use the standard formulae like

Eq. (5.26) to calculate the radiation of an infinite undulator because the condition that the distance between the undulator and an observer must be much greater than the undulator length (far field approximation) cannot be fulfilled, no matter how great the distance may be. In other words, approximation of Eq. (5.20) is not longer correct. The correct way of using the infinite undulator approximation in order to circumvent difficulties like those arising in the theory of radiation from a charge with collinear velocity and acceleration [6], is to find a desired function (a spectrum for instance, or angular distribution) for a finite undulator and then turn the number of periods N to infinity. In doing so the total energy $d\mathcal{E}$ emitted from an undulator into a solid angle $d\Omega$ and in a frequency band $d\omega$ tends to infinity, but the ratio of the energy to the time $\Delta t' = NT(1 - \boldsymbol{\beta}_{\|}\boldsymbol{n})$ it takes an observer to receive the radiation impulse from a particle is finite and presents the radiation intensity dI. Accordingly, the ratio of the energy $d\mathcal{E}$ to the time $\Delta t = NT$ it takes the particle to pass trough the undulator is finite as well and is referred to as the energy loss or radiated power dW.

$$\frac{dI}{d\Omega d\omega} = \frac{\omega_0}{2\pi(1 - \boldsymbol{\beta}_{\|}\boldsymbol{n})} \lim_{N\to\infty} \frac{d\mathcal{E}}{N d\Omega d\omega}, \tag{5.55}$$

$$\frac{dW}{d\Omega d\omega} = (1 - \boldsymbol{\beta}_{\|}\boldsymbol{n}) \frac{dI}{d\Omega d\omega}. \tag{5.56}$$

Next turning to the limit $N \to \infty$ in Eq. (5.31) and taking into account that

$$\lim_{N\to\infty} \frac{1}{N} I_n(\nu, N) = \delta(n - \nu),$$

where $\delta(n - \nu)$ is the Dirac delta function, we obtain

$$\frac{dI}{d\Omega d\omega} = \frac{e^2 \omega_0}{2\pi c(1 - \boldsymbol{\beta}_{\|}\boldsymbol{n})^3} \sum_{n=1}^{\infty} n^2 \left| \frac{1}{T} \int_0^T [\boldsymbol{n}\boldsymbol{\beta}] e^{i(\omega_n t - \boldsymbol{k}_n \boldsymbol{r})} dt \right|^2 \delta(n - \nu). \tag{5.57}$$

This equation shows that the radiation spectrum of an infinite undulator is a discrete one. The corresponding distribution for the radiated power can be easily found with Eq. (5.56).

The angular distribution is obtained by integrating over frequencies in the last equation. This yields

$$\frac{dI}{d\Omega} = \frac{e^2 \omega_0^2}{2\pi c(1 - \boldsymbol{\beta}_{\|}\boldsymbol{n})^4} \sum_{n=1}^{\infty} n^2 \left| \frac{1}{T} \int_0^T [\boldsymbol{n}\boldsymbol{\beta}] e^{i(\omega_n t - \boldsymbol{k}_n \boldsymbol{r})} dt \right|^2, \tag{5.58}$$

Although this expression is written as the angular distribution, it represents the radiation spectrum as well because each term of the sum refers to the corresponding harmonic. Comparing with Eqs (5.33) and (5.33) we obtain a natural result:

$$\frac{dI}{d\Omega} = \frac{\omega_0}{2\pi N(1 - \boldsymbol{\beta}_{\parallel}\boldsymbol{n})} \frac{d\mathcal{E}}{d\Omega} \tag{5.59}$$

One can find the angular distribution of the radiated power multiplying the last expression by the retardation factor $1 - \boldsymbol{\beta}_{\parallel}\boldsymbol{n}$.

Let us discuss a practical implementation of the derived formulae. The question arises: what is measured on registration of the undulator radiation, intensity of radiation or the radiated power (i.e. energy loss)? As is seen from Eq. (5.56) these quantities differ by the factor of $1 - \boldsymbol{\beta}_{\parallel}\boldsymbol{n}$. If the particle is moving within a limited region, the drift velocity $\boldsymbol{\beta}_{\parallel}$ is zero and the two quantities are equal. This is true, for instance, in the case of synchrotron radiation. But the model we used for the particle motion in an undulator implies an infinite motion. Of course, if we deal with an infinite undulator, we will measure the intensity of radiation. But the real undulator is a finite one. Moreover, one has usually to do with a beam of electrons.

Let a chain of ultrarelativistic electrons with the spacing between them l pass through the finite device of the length L. Then the impulse of radiation produced by each electron has the length $\Delta l = L(1 - \boldsymbol{\beta}_{\parallel}\boldsymbol{n}) \sim \gamma^{-2}L$ with the spacing between impulses equal to l. The measurement of the radiation intensity within such a short impulse requires corresponding time-resolving instruments. In practice, the time of measurement is much greater than the duration of a single impulse of radiation. The measured intensity in this situation is the intensity averaged over the beam of particles.[c] In our case, the averaging is equal to multiplying the intensity by the factor $\Delta l/l$). This yields

$$\langle dI \rangle = dI(1 - \boldsymbol{\beta}_{\parallel}\boldsymbol{n})\frac{L}{l} \tag{5.60}$$

One can readily see that the mean intensity of radiation from the beam of particles is equal to the mean power of radiation because the factor $L/\Delta l$ in Eq. (5.60) has the meaning of the number of particles located within the undulator simultaneously.

5.2. Approximate Methods

5.2.1. Radiation in a Small k Limit

One can obtain much more information about radiation properties in case of a relatively weak magnetic field in the undulator, i.e. in case $k \ll 1$. We begin with a proof of the statement that the particle moving in a magnetic field which satisfies the condition $k \ll 1$ has a nonrelativistic velocity in the reference frame which moves

[c]We assume that the radiation of the beam is incoherent.

with a constant speed $\beta_{\|}$. We consider here only an ultrarelativistic particle because in a nonrelativistic case it is obvious. It follows from $k \ll 1$ that $\beta_{\perp} \ll \gamma^{-1}$. The longitudinal velocity differs from the total one by the value of the order of $\beta_{\perp}^2$:

$$\beta_x = \sqrt{\beta^2 - \beta_{\perp}^2} = \beta + O(\beta_{\perp}^2). \tag{5.61}$$

The mean longitudinal velocity is given by

$$\beta_{\|} = \langle \beta_x \rangle = \beta + O(\beta_{\perp}^2). \tag{5.62}$$

The transversal velocity $\beta'_{\perp}$ in the moving reference frame is given by the Lorentz transformation

$$\beta'_{\perp} = \frac{\beta_{\perp}\sqrt{1-\beta_{\|}^2}}{1-\beta_x\beta_{\|}} = \beta_{\perp}\gamma[1+O(k^2)] \ll 1. \tag{5.63}$$

Correspondingly, the longitudinal velocity in the moving frame is equal to

$$\beta'_x = \frac{\beta_x - \beta_{\|}}{1-\beta_x\beta_{\|}} = O(k^2) \ll 1. \tag{5.64}$$

We see that in the moving reference frame the particle is, in fact, a nonrelativistic one. Hence, the radiation emitted by the particle is of the dipole nature.

The Fourier transform of the electric field of radiation in this case is given by [1]

$$\boldsymbol{E}(\omega) = \frac{e}{cR}\frac{e^{ikR}}{(1-\boldsymbol{\beta n})^2}\int_{-\infty}^{\infty}[\boldsymbol{n}[(\boldsymbol{n}-\boldsymbol{\beta})\dot{\boldsymbol{\beta}}(t)]]e^{i\omega t(1-\boldsymbol{\beta}_{\|}\boldsymbol{n})}dt, \tag{5.65}$$

where the velocity $\boldsymbol{\beta}$ is assumed to be constant.

Note, that we expand, with respect to small k, only the amplitude of the electric field keeping the exact expression for the phase. The reason is that even a small variation of k can produce a significant shift in frequency. Actually, we see from Eq. (5.37) that the frequency shift $\Delta\nu_k$ caused by a nonzero k is equal to $\Delta\nu_k = \omega k^2/2\omega_0\gamma^2 \sim \nu k^2$ while the natural width of the spectral line is equal to $\Delta\nu_N = 1/N$. Hence, we may neglect k only if

$$k^2 \ll 1/N\nu. \tag{5.66}$$

Splitting the integral in Eq. (5.65) into the sum of integrals over each period and taking into account Eq. (5.21) we can write the previous equation in the form

$$\boldsymbol{E}(\omega) = \frac{2\pi e}{cR\omega_0} e^{ikR+i(N-1)\pi\nu} \frac{\sin \pi\nu N}{\sin \pi\nu} \frac{[\boldsymbol{n}[(\boldsymbol{n}-\boldsymbol{\beta})\dot{\boldsymbol{\beta}}(\nu)]]}{(1-\boldsymbol{\beta n})^2} \tag{5.67}$$

with

$$\dot{\boldsymbol{\beta}}(\nu) = \frac{1}{T}\int_0^T \dot{\boldsymbol{\beta}}(t) e^{i\omega_0 \nu t} dt. \tag{5.68}$$

Thus, the Fourier transform of the electric field is represented in terms of the Fourier transform of acceleration. Acceleration itself depends on the magnetic field in the undulator. Dividing the velocity and the magnetic field into longitudinal and transversal components we write the equation of motion

$$\gamma m\ddot{\boldsymbol{r}} = e\left\{[\boldsymbol{\beta}_x \boldsymbol{H}_\perp] + [\boldsymbol{\beta}_\perp \boldsymbol{H}_\perp] + [\boldsymbol{\beta}_\perp \boldsymbol{H}_{||}]\right\}.$$

Obviously, the second term is much smaller than the first one, and so it can be neglected. The third term is small if the condition

$$H_{||} \ll \gamma H_\perp$$

is held. Suppose it is. Then we have

$$\gamma m\ddot{\boldsymbol{r}} = e[\boldsymbol{\beta}_{||}\boldsymbol{H}]. \tag{5.69}$$

Here $\boldsymbol{\beta}_{||} = \boldsymbol{\beta} = const$ is the initial velocity in accordance with Eqs ((5.61)) and (5.62), $\boldsymbol{H}$ stands for $\boldsymbol{H}_\perp$ because only the orthogonal component is valid. Now we can use in Eqs (5.67) either the Fourier transform of acceleration or that of the magnetic field according to

$$\dot{\boldsymbol{\beta}}(\nu) = \frac{e}{mc\gamma}[\boldsymbol{\beta}_{||}\boldsymbol{H}(\nu)], \tag{5.70}$$

where

$$\boldsymbol{H}(\nu) = \frac{1}{\lambda_0}\int_0^{\lambda_0} \boldsymbol{H}(x) \exp i\frac{2\pi\nu x}{\lambda_0}\, dx. \tag{5.71}$$

Here $\boldsymbol{H}(x)$ is the magnetic field along the particle trajectory which can be thought as a straight line, λ_0 is the length of the undulator period.

In order to calculate the spectrum and the angular distribution of the radiated energy, we substitute Eq. (5.67) in Eq. (5.25). This gives

$$\frac{d\mathcal{E}}{d\Omega d\omega} = \frac{e^2}{\omega_0^2(1-\boldsymbol{\beta n})^2} I(\nu, N) \left|\left[\boldsymbol{n}[(\boldsymbol{n}-\boldsymbol{\beta})\dot{\boldsymbol{\beta}}(\nu)]\right]\right|^2 . \tag{5.72}$$

We see that in case of small k the spectrum is completely defined by the Fourier transform of the acceleration or that of the magnetic field on a straight line close to the trajectory [3,7]. Moreover, this allows to calculate the radiation spectrum knowing only magnetization of the permanent magnets in undulator [8].

Let us proceed to the most frequently used case of an undulator that forms a flat trajectory of the particle. It is clear from Eq. (5.69) that the acceleration retains its direction. Let us denote this direction by a unit vector $\boldsymbol{e}_a$, so that $\dot{\boldsymbol{\beta}} = \dot{\beta}\boldsymbol{e}_a$. Then the triple product in Eq. (5.72) takes the form of $[\boldsymbol{n}[(\boldsymbol{n}-\boldsymbol{\beta})\dot{\boldsymbol{\beta}}(\nu)]] = [\boldsymbol{n}[(\boldsymbol{n}-\boldsymbol{\beta})\boldsymbol{e}_a]]\dot{\beta}(\nu)$. Dividing this vector into polarization components and substituting them into Eq. (5.48) we obtain

$$\frac{d\mathcal{E}}{d\Omega d\omega} = \frac{e^2(1-\boldsymbol{\beta}_{||}\boldsymbol{n})}{c\omega_0^2} \frac{\sin^2 \pi\nu N}{\sin^2 \pi\nu} (\rho_\sigma + \rho_\pi)|\dot{\boldsymbol{\beta}}(\nu)|^2, \tag{5.73}$$

where the angular functions ρ_σ and ρ_π are given by Eq. (5.53) or Eq. (5.54). Splitting the continuous function $I(\nu, N)$ into the sum over harmonic numbers as we did on deriving Eq. (5.31), we find an equivalent form of Eq. (5.73)

$$\frac{d\mathcal{E}}{d\Omega d\omega} = \frac{e^2(1-\boldsymbol{\beta}_{||}\boldsymbol{n})}{c\omega_0^2} (\rho_\sigma + \rho_\pi) \sum_{n=1}^{\infty} I_n(\nu, N)|\dot{\boldsymbol{\beta}}(n)|^2. \tag{5.74}$$

One can see that at least the angular distribution of radiation in case of the arbitrary flat trajectory is given in an explicit form. This is due to the fact that the deviation of the velocity and acceleration vectors is much less than the radiation cone. As a result the angular distribution of radiation is equal to the instant one. Let us find the radiation spectrum in the approximation of ultrarelativistic particle. Integration over the angle ϕ in Eq. (5.74) and substitution of Eq. (5.70) gives

$$\frac{d\mathcal{E}_{\sigma,\pi}}{d\omega} = \frac{2e^4\pi\gamma}{m^2c^3\omega_0^2} \sum_{n=1}^{\infty} \int_0^{\infty} |H(\nu)|^2 I_n(\nu, N) \left\{ \begin{array}{c} 2+\psi^4 \\ \psi^4 \end{array} \right\} \frac{d\psi^2}{(1+\psi^2)^4}. \tag{5.75}$$

From now on the first line of equations corresponds to the σ- and the second one to the π-component. The upper limit in the integral over θ is π but the bulk of the energy output is emitted within an angle $\theta \sim \gamma^{-1}$ which is much less then π. Hence, we put the upper limit equal to infinity. Define a new variable u in the following way (see Eq. (5.37))

$$u = \frac{\omega}{2\omega_0\gamma^2}(1 + k^2 + \psi^2), \quad d\psi^2 = \frac{2\omega_0\gamma^2}{\omega}du. \tag{5.76}$$

This gives

$$\frac{d\mathcal{E}_{\sigma,\pi}}{d\omega} = \frac{2e^4\xi}{\pi c^3 m^2 \omega_0^2} \sum_{n=1}^{\infty} \int_{\xi}^{\infty} \frac{\sin^2 \pi u N}{(n-u)^2} |H(u)|^2 \left\{ \begin{array}{c} 3\xi^2 - 2\xi u + u^2 \\ (u-\xi)^2 \end{array} \right\} \frac{du}{u^4}, \tag{5.77}$$

where

$$\xi = \frac{\omega(1 + k^2)}{2\omega_0\gamma^2}. \tag{5.78}$$

In approximation of a long undulator ($N \ll 1$) we can put $u = n$ everywhere except the fast oscillating factor $\sin^2 \pi\nu N/(n-u)^2$. Thus, the spectrum takes the form close to that given in Ref.[1] for a nonrelativistic particle

$$\frac{d\mathcal{E}_{\sigma,\pi}}{d\omega} = \frac{2e^4\pi N}{m^2 c^3 \omega_0^2} \sum_{n=1}^{\infty} \frac{\xi_n}{n} |H(n)|^2 \left\{ \begin{array}{c} 3\xi_n^2 - 2\xi_n + 1 \\ (1-\xi_n)^2 \end{array} \right\} \Theta(N, 1-\xi_n). \tag{5.79}$$

Here

$$\Theta(N, 1-\xi_n) = 1 + \frac{x \,\mathrm{si}\, 2x - \sin^2 x}{\pi x}, \quad \xi_n = \xi/n, \tag{5.80}$$

$\mathrm{si}\, x$ is the sine integral [9]

The function $\Theta(N, 1-\xi_n)$ is close to unity at large N in the region $\xi_n < 1$ and then drops steeply at $\xi_n \approx 1$. In the limit of $N \to \infty$ it transforms into the step-function $\Theta(1-\xi_n)$. Thus, the spectrum of the n-th harmonic extends to $\xi_n \approx 1$ or in terms of frequency to $\omega_n = 2\gamma^2\omega_0 n/(1+k^2)$. The shape of the spectral curve in terms of ξ_n is universal for any harmonic and is shown in Fig. 5.5 for several values of the product nN. The width of the region where the spectrum curve drops is of the order $\Delta\xi_n \sim 1/nN$. That is, the longer the undulator is or/and the higher the harmonic number is, the more steeply the curve slopes.

The parameter k appears in Eq. (5.79) as a factor k^2 in $|H(n)|^2$ and in the argument of the function $\Theta(N, 1-\xi_n)$ in the form

$$x = \pi N n \left[1 - \frac{\omega(1+k^2)}{2\omega_0\gamma^2 n} \right]. \tag{5.81}$$

The small parameter k can be neglected here if the difference $|\Delta x|$ caused by non-zero k is small enough

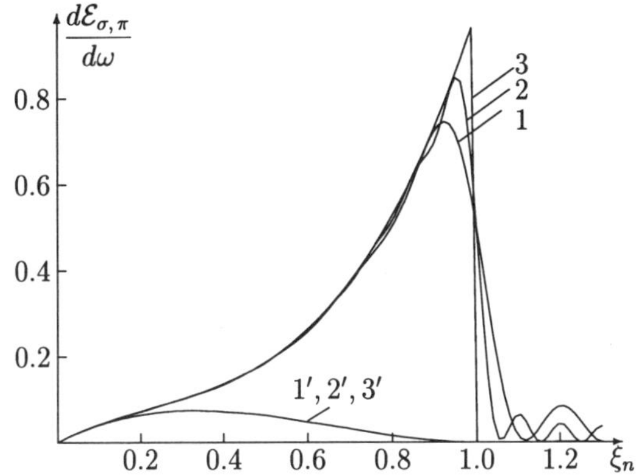

Fig. 5.5. The spectral distribution of total energy emitted in any harmonic at small k for different number of periods. 1 — 3: σ-component. 1: $Nn = 5$, 2: $Nn = 10$, 3: $Nn \to \infty$. $1' - 3'$: π-component (dependence on N is indistinguishable).

$$|\Delta x| = \frac{\pi N \omega k^2}{2\omega_0 \gamma^2} \ll \pi, \tag{5.82}$$

or

$$k^2 \ll \frac{2\omega_0 \gamma^2 N\omega}{N\omega}, \tag{5.83}$$

which is consistent with Eq. (5.66).

The angular distribution of radiated energy can be obtained by integration of Eq. (5.73) over the frequencies. Taking into account that $|H(n)|^2 \sin^2 \pi\nu N / \sin^2 \pi\nu$ has the meaning of the Fourier transform of the magnetic field and using the Parseval's theorem

$$\int_0^\infty |f(\omega)|^2 d\omega = \pi \int_{-\infty}^{\infty} |f(t)|^2 dt \tag{5.84}$$

we have

$$\frac{4\pi^2}{\omega_0^2} \int_0^\infty \frac{\sin^2 \pi\nu N}{\sin^2 \pi\nu} |H(\nu)|^2 d\omega = \frac{\pi}{1 - \boldsymbol{\beta}_\| \boldsymbol{n}} \int_0^{NT} [H(t)]^2 dt. \tag{5.85}$$

Thus, the angular distribution

$$\frac{d\mathcal{E}}{d\Omega} = \frac{e^4 N \lambda_0 \langle h^2 \rangle}{4\pi c^4 m^2 \gamma^2 \beta} (\rho_\sigma + \rho_\pi) \tag{5.86}$$

is proportional to the mean square value of the magnetic field

$$\langle H^2 \rangle = \frac{1}{\lambda_0} \int_0^{\lambda_0} H^2(x)\, dx. \tag{5.87}$$

As can be expected, the angular distribution of radiated energy is of the same form as the instant distribution of emitted power.

The total emitted energy can be derived by integrating either Eq. (5.86) over the angles or Eq. (5.79) over the frequencies. Any way, we obtain

$$\mathcal{E} = \frac{2e^4 \beta \gamma^2 \lambda_0 N \langle h^2 \rangle}{3c^4 m^2}. \tag{5.88}$$

5.2.2. Radiation in a Large k Limit

An undulator with a relatively strong field with $k \gg 1$ is referred to as a wiggler. Usually, it is supposed that the radiation in this case is similar to synchrotron radiation. Indeed, the main part of the radiation in a given direction is emitted from a short section of the trajectory which can be replaced by a circular arc [2]. Nevertheless, wiggler radiation differs from synchrotron one. First, the presence of several periods modulates the spectral and angular distribution at a fixed frequency. Second, the limited range of the velocity directions caused the angular distribution of the undulator radiation to be nonuniform in the trajectory plane. The intensity has its maximum in the on-axis direction and decreases to zero in the directions of maximum deviation of the velocity vector. Other peculiarities are considered in Sec. 5.3.4.

In order to calculate the wiggler radiation, we use the technique developed for investigation of synchrotron radiation of relativistic charged particles [10]. A similar procedure has been followed in studies of the undulator radiation [11,12] in case of great k. This is not surprising because it follows from $k \gg 1$ that the particle is a relativistic one in the reference frame moving with the velocity $\boldsymbol{\beta}_{\|}$. Actually, the particle energy in this frame is given by Lorentz transformation

$$mc^2 \gamma' = mc^2 \gamma \frac{1 - \beta_{\|} \beta_x}{\sqrt{1 - \beta_{\|}^2}}. \tag{5.89}$$

The longitudinal components of velocity involved here are of the order

$$\begin{aligned}\beta_x &= \sqrt{1-\gamma^{-2}(1+\gamma^2\beta_\perp^2)} \approx 1-\frac{1}{2}\gamma^{-2}k^2,\\ \beta_{\|} &= \langle\beta_x\rangle \approx 1-\frac{1}{2}\gamma^{-2}k^2.\end{aligned} \tag{5.90}$$

Substitution in Eq. (5.89) gives

$$\gamma' \sim k \gg 1,$$

which proves the statement.

Further we follow Ref.[11]. Let the particle motion satisfy the following conditions: i) $\gamma \gg 1$; ii) the trajectory is a flat one and the radius of curvature changes its sign twice within the period; iii) the velocity and acceleration of the particle vary smoothly enough, so that the derivatives of the radius vector $d^n\boldsymbol{r}/dt^n$ are less or of the order of λ_0/T, where λ_0 and T are the length and the duration of the period of motion. We study the radiation emitted within a plane tangent to the particle path at some point and orthogonal to the plane of the trajectory. In accordance with above discussion, there are two planes of this kind within each period of the trajectory. Denote the points of trajectory at which the planes tangent to it by numbers 1 and 2 as shown in Fig. (5.6)

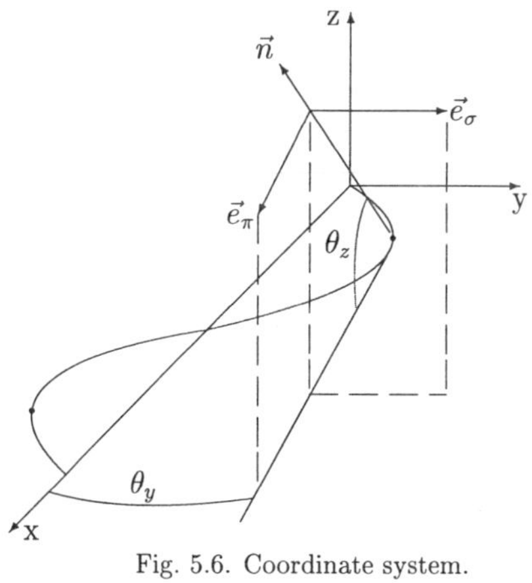

Fig. 5.6. Coordinate system.

Let the coordinates of points 1 and 2 be $\boldsymbol{r}_1^{(0)}$ and $\boldsymbol{r}_2^{(0)}$ and the time when the particle reaches these points t_1 and t_2, respectively. In case of great k, the radiation in a given direction is emitted from short arcs of the trajectory in the vicinity of points 1 and 2. This allows expanding the radius vector in a power series with respect to small $\tau_i = \omega_0(t_i - t)$

$$\boldsymbol{r}(t) = \boldsymbol{r}_i^{(0)} + \boldsymbol{r}_i^{(1)}\tau_i + \frac{1}{2}\boldsymbol{r}_i^{(2)}\tau_i^2 + \frac{1}{6}\boldsymbol{r}_i^{(3)}\tau_i^3 + \cdots, \tag{5.91}$$

where

$$\boldsymbol{r}_i^{(k)} = \left.\frac{d^k\boldsymbol{r}}{d\tau^k}\right|_{t=t_i}.$$

We demand the variable τ to be small enough to neglect the higher terms of expansion. That is, the duration of radiation is much less than the period of motion. Hence, the component of acceleration $r_\perp^{(2)}$ orthogonal to velocity must be large enough to provide the deviation of the velocity vector by the angle γ^{-1} in a time much less than the period. Namely[d]

$$r_\perp^{(2)} \gg \lambda_0\gamma^{-1} \tag{5.92}$$

which is another form of the condition $k \gg 1$.

The energy radiated per unit solid angle and unit band of frequency is given by standard formula

$$\frac{d\mathcal{E}}{d\Omega d\omega} = \frac{e^2\omega^2}{4\pi^2 c}\left|\int_{-\infty}^{\infty}[\boldsymbol{n}[\boldsymbol{n}\boldsymbol{\beta}]]e^{i(\omega t - \boldsymbol{k}\boldsymbol{r})}dt\right|^2, \tag{5.93}$$

but now we divide the integral into $2N$ parts, two parts in each period. The limits of each integral can be extended to infinity because the period is much longer than the duration of radiation. Denote the angle between the trajectory axis and the tangent to the trajectory at points 1 and 2 by θ_y and the angle between the vector $\boldsymbol{n}$ and the trajectory plane by θ_z as shown in Fig. (5.6). The unit polarization vectors $\boldsymbol{n}_\sigma$ and $\boldsymbol{n}_\pi$ are defined so that the vector $\boldsymbol{n}_\sigma$ is parallel to the trajectory plane. Expansion of the vector product in Eq. (5.93) with respect to small θ_y, θ_z and γ^{-1} gives

$$[\boldsymbol{n}[\boldsymbol{n}\boldsymbol{\beta}]]_\sigma = \frac{r_\perp^{(2)}}{\lambda_0}\tau, \quad [\boldsymbol{n}[\boldsymbol{n}\boldsymbol{\beta}]]_\pi = \theta_z.$$

The phase in Eq. (5.93) must be expanded to the third order of γ^{-1} since the second order term can be equal to zero or negligibly small

$$\begin{aligned}\omega t - \boldsymbol{k}\boldsymbol{r} &= \frac{\omega}{\omega_0}\left[-\frac{\omega_0}{c}(\boldsymbol{n}\boldsymbol{r}^{(0)}) + \frac{1}{2}(\theta_z^2 + \gamma^{-2})\tau - \frac{\omega_0}{2c}\frac{(\boldsymbol{r}^{(2)}\boldsymbol{\beta})}{\beta}\tau^2 - \right.\\ &\left.\frac{\omega_0}{6c}\frac{(\boldsymbol{r}^{(3)}\boldsymbol{\beta})}{\beta}\tau^3\right].\end{aligned} \tag{5.94}$$

[d]Here and further the subscript i is dropped for simplicity.

Further calculation significantly depends on the relation between the terms in this equation. The first term is constant and linked to the initial phase. The second one is of the order γ^{-3}. The third term and the last one depend on the factors $\omega_0(\boldsymbol{r}^{(k)}\boldsymbol{\beta})/v$ which are of the order of unity or less according to the assumption adopted above. Thus, the last term is of the order γ^{-3} or less while the term proportional to τ^2 can be much greater than the others. We see that the value of longitudinal acceleration is critical to further calculations. Let α_a be the angle between the acceleration of the particle and its velocity, and α_f be the angle between the velocity and the force vector causing the acceleration. Then the directions of the acceleration and force are related by the equation [1]

$$\tan\alpha_f = \gamma^{-2}\tan\alpha_a \tag{5.95}$$

The square in τ term is of the order

$$\frac{(\boldsymbol{r}^{(2)}\boldsymbol{\beta})}{\beta}\tau^2 = r_\perp^{(2)}\tau^2\cot\alpha_a \sim \frac{c}{\omega_0}\gamma^{-4}\cot\alpha_f.$$

Now we classify the types of particle dynamics into three categories:

1. $\alpha_f \sim 1$ — the square term is of the order γ^{-4} and can be neglected in comparison with the others;

2. $\alpha_f \sim \gamma^{-1}$ — the square term is comparable with the others;

3. $\alpha_f \sim \gamma^{-2}$ — the square term is much greater then the others.

We shall restrict our consideration to the first case which includes motion in a magnetic field in the absence of an electric one. It follows from Eq. (5.95) that in this case the velocity is orthogonal to acceleration and $\alpha_a \approx \pi/2$. Taking the derivative from the identity $(\dot{\boldsymbol{r}}\ddot{\boldsymbol{r}}) = 0$ we obtain $(\boldsymbol{v}\boldsymbol{r}^{(3)}) = -\omega_0(r^{(2)})^2$. Hence, Eq. (5.94) can be written in the form

$$\omega t - \boldsymbol{k}\boldsymbol{r} = \frac{3}{2}\omega'(x + \frac{1}{3}x^3) - (\boldsymbol{k}\boldsymbol{r}^{(0)}),$$

where

$$x = \frac{\omega_0\tau r^{(2)}}{c\varepsilon},\quad \omega' = \frac{c\varepsilon^3\omega}{3r^{(2)}\omega_0^2},\quad \varepsilon = \sqrt{\gamma^{-2} + \theta_z^2}.$$

Finally, substituting the last result in Eq. (5.93) and summing over all $2N$ points along the trajectory we obtain

$$\begin{aligned}\frac{d\mathcal{E}_\sigma}{d\Omega d\omega} &= \frac{e^2\omega^2 c\sin^2\pi\nu N}{3\pi^2\omega_0^4\ \sin^2\pi\nu}\left\{\frac{\varepsilon_1^4}{r_1^{(2)^2}}K^2_{2/3}(\omega_1') + \frac{\varepsilon_2^4}{r_2^{(2)^2}}K^2_{2/3}(\omega_2') + \right.\\ &+ \left. 2\frac{\varepsilon_1^2\varepsilon_2^2}{r_1^{(2)}r_2^{(2)}}\cos 2\eta K_{2/3}(\omega_1')K_{2/3}(\omega_2')\right\}, \qquad (5.96)\\ \frac{d\mathcal{E}_\pi}{d\Omega d\omega} &= \frac{e^2\omega^2 c\,\theta_z^2\sin^2\pi\nu N}{3\pi^2\omega_0^4\ \sin^2\pi\nu}\left\{\frac{\varepsilon_1^2}{r_1^{(2)^2}}K^2_{1/3}(\omega_1') + \frac{\varepsilon_2^2}{r_2^{(2)^2}}K^2_{1/3}(\omega_2') + \right.\\ &+ \left. 2\frac{\varepsilon_1\varepsilon_2}{r_1^{(2)}r_2^{(2)}}\cos 2\eta K_{1/3}(\omega_1')K_{1/3}(\omega_2')\right\}, \qquad (5.97)\end{aligned}$$

where $2\eta = \omega(t_2 - t_1 - (\boldsymbol{n}\boldsymbol{r}_{12})/c)$, $\boldsymbol{r}_{12} = \boldsymbol{r}_2^{(0)} - \boldsymbol{r}_1^{(0)}$, $K_{1/3}(\omega')$ and $K_{2/3}(\omega')$ are the modified Bessel functions. If the particle does not change its energy between points 1 and 2, then $\epsilon_1 = \epsilon_2$. The angular distribution depends on the horizontal angle θ_y through the parameters $r_i^{(2)} = r_i^{(2)}(\theta_y)$ while the dependence on the vertical angle θ_z is given in an explicit form. At the limits of $\theta_y \to \pm\theta_{y\,max}$, where $\theta_{y\,max}$ is the absolute value of the angle between the trajectory axis and the tangent to the points of inflection of trajectory, the radiation vanishes because of acceleration and, consequently, parameters $r_i^{(2)}$ at those points are equal to zero. The argument of the Bessel functions tends to infinity, hence, the Bessel functions themselves fall exponentially. Thus, all the energy is emitted in an interval $-\theta_{y\,max} < \theta_y < \theta_{y\,max}$. Strictly speaking, the cone of radiation extends to angles $\theta_y \approx \pm(\theta_{y\,max} + \gamma^{-1})$ as it will be shown in Sec. 5.3.4

Radiation at any fixed frequency is elliptically polarized since the phase difference between the σ- and the π-component of an electric field has a nonzero value if the acceleration at points 1 and 2 is not equal. This feature is used in the design of the so-called asymmetric undulators intended to produce elliptically polarized radiation [13,14].

Of special interest is the generally employed undulator with the particle trajectories symmetric with respect to the points of inflection. In this case, we have $r_1^{(2)} = r_2^{(2)}$ and $\varepsilon_1 = \varepsilon_2$. Eqs (5.96) and (5.97) reduce to very simple ones

$$\begin{aligned}\frac{d\mathcal{E}_\sigma}{d\Omega d\omega} &= \mathcal{E}_0\varepsilon^2 K^2_{2/3}(\omega')\sin^2\eta,\\ \frac{d\mathcal{E}_\pi}{d\Omega d\omega} &= \mathcal{E}_0\theta_z^2 K^2_{1/3}(\omega')\cos^2\eta, \qquad (5.98)\end{aligned}$$

where

$$\mathcal{E}_0 = \frac{4e^2\omega^2 c\varepsilon^2\sin^2\pi\nu N}{3\pi^2\omega_0^4 r^{(2)^2}\sin^2\pi\nu}.$$

Let us discuss the general properties of radiation given by Eqs (5.96) and (5.97). The slowly varying functions $K_{1/3}(\omega')$ and $K_{2/3}(\omega')$ are split into harmonics by the factor $\sin^2 \pi\nu N/\sin^2 \pi\nu$. The maximum of the spectral distribution lies in the vicinity of the frequency ω_{max} where the argument of the Bessel function is of the order of unity. The equality $\omega' = 1$ gives the estimation[e]

$$\omega_{max} \sim \frac{r^{(2)}\gamma^3\omega_0^2}{c}. \tag{5.99}$$

The inequality of Eq. (5.92) shows that this frequency far exceeds the frequency of the fundamental harmonic $\omega_1 \sim \omega_0\gamma^2 k^{-2}$ given by Eq. (5.37). The product $r^{(2)}\omega_0^2$ represents the centripetal acceleration of the particle at points 1 and 2

$$|\ddot{\boldsymbol{r}}| \approx \frac{eH_\perp}{m\gamma}. \tag{5.100}$$

Hence, the frequency ω_{max} can be rewritten as follows

$$\omega_{max} \sim \omega_0\gamma^2 k \sim \omega_1 k^3. \tag{5.101}$$

Thus, the maximum of radiated energy is emitted at the harmonics of numbers

$$n_{max} \sim k^3. \tag{5.102}$$

According to Eq. (5.99) the frequency ω_{max} is of the same order as the critical frequency of synchrotron radiation of the particle in a field equal to the local field at points 1 and 2. But the distance between harmonics in the spectrum of wiggler radiation sufficiently differs from that of synchrotron radiation. In order to prove this, we consider the trajectory within half a period as an arc of the angle δ. If the particle moves along a circle, it emits impulses of radiation with a period, say, T_c. The period of motion in a wiggler T_w' is much shorter, namely $T_w' = T_c\delta/\pi$. Moreover this time is compressed by the retardation factor $1 - \boldsymbol{\beta}_{\|}\boldsymbol{n} \approx \gamma^{-2}(\psi_y^2 + \psi_z^2 + k^2)/2$. Thus, the impulses of wiggler radiation recur with a period of $T_w \sim T_c\delta\gamma^{-2}k^2 = T_c\delta\beta_\perp$. Taking into account that $\beta_\perp \sim \delta$, we have $T_w/T_c \sim \delta^3$. Correspondingly, the frequency interval between the harmonics of wiggler radiation is δ^{-3} times as large as that of synchrotron radiation.

The radiation pattern of symmetric wiggler given by Eq. (5.98) differs from that of synchrotron radiation by factors which describe the interference of radiation emitted from each half the period of the trajectory.

In the case of nonuniform beams of particles the oscillations in the spectra–angular distribution under discussion are smoothed. In particular, the spread of the directions of motion causes the broadening of spectral lines comparable or greater than the

[e]We assume that ω_1' and ω_2' are of the same order.

distance between the harmonics at large k. The broadening of harmonics can appear also as a result of other known reasons. When the broadening is sufficiently large we can average over the fast oscillating functions. For this let us take the identity

$$\frac{\sin^2 aN}{\sin^2 a} = e^{ia(N+1)} \sum_{n=1}^{N} e^{2ian} \tag{5.103}$$

and square the module of both the sides:

$$\frac{\sin^2 aN}{\sin^2 a} = N + 2\sum_{n=1}^{N}\sum_{j=1}^{n} \cos 2ja \tag{5.104}$$

Then we obtain an average

$$\langle \sin^2 \eta \frac{\sin^2 \pi\nu N}{\sin^2 \pi\nu} \rangle = \frac{1}{2}(N + \sum_{k=1}^{N}\sum_{j=1}^{n} \delta_{\pi\nu j,\eta x}) = \frac{N}{2} \tag{5.105}$$

Here we have used the inequality $\pi/2 > \eta/\nu$, and hence $\pi\nu j \neq \eta$. In a similar way, for a factor in a π-component:

$$\langle \cos^2 \eta \frac{\sin^2 \pi\nu N}{\sin^2 \pi\nu} \rangle = \frac{N}{2} \tag{5.106}$$

Substitution in Eq. (5.98) gives an apparent result

$$\frac{d\mathcal{E}_{wiggler}}{d\Omega d\omega} = 2N \frac{d\mathcal{E}_{SR}}{d\Omega d\omega} \tag{5.107}$$

where the subscript SR stands for synchrotron radiation.

It is interesting to note that the assumption of flat trajectory has been only used to compute the vector product in Eq. (5.93). Hence, the energy distribution given by Eqs (5.96) and (5.97) is valid in the case of a three-dimensional trajectory if the planes tangent to it at points 1 and 2 are parallel and the coordinate plane xy is defined to be parallel to these planes. But the problem lies in the fact that the orientation of the tangent planes depends on the direction of vector $\boldsymbol{n}$. As a result, the angular coordinate system (θ_y, θ_z) can be rather complicated.

5.2.3. Paraxial Approximation

Solutions of equations of motion of a charged particle in the undulator magnetic field used in Sec. 5.1.1 rest on the so-called paraxial approximation. In this section we discuss the basis of this approximation in more detail.

It was assumed at the beginning of Sec. 5.1.1 that the magnetic field in the vicinity of the trajectory varies slowly enough to be taken as a uniform one in the directions orthogonal to the undulator axis. Besides, the longitudinal component of the magnetic field has been neglected. It has been shown recently that the longitudinal component of the magnetic field causes slow betatron oscillations orthogonal to the orbit plane [15 - 16], with a space period of the order of [15,17] $k^{-1}\gamma\lambda_0$. One can neglect the longitudinal field and the betatron motion if its oscillations period is large enough with respect to the undulator length, i.e.

$$k\gamma^{-1}N \ll 1 \tag{5.108}$$

with the number of undulator periods N. The exact solution of the equation of motion in a two-dimensional transversal magnetic field of an arbitrary configuration has been derived in Ref.[20]. The plane charge trajectory in a one-dimensional magnetic field is found in Refs[15,17] as a particular case in the absence of a longitudinal field component.

Next we study the conditions that must be kept for the employment of the paraxial method.

Let a relativistic charged particle cross a magnetic field which extends over a finite area. Let the x-axis be directed along the initial velocity $\boldsymbol{v}_0$. Denote by L the characteristic size of the area occupied by the magnetic field along the x-axis. Let H_m be the maximum value of the magnetic field on this axis and $\boldsymbol{h}(\boldsymbol{r})$ be a dimensionless function that defines the geometry of the field according to $\boldsymbol{H}(\boldsymbol{r}) = H_m\boldsymbol{h}(\boldsymbol{r})$. Let us solve the equations of motion using the perturbation method. As a small perturbation parameter we take

$$\epsilon = \frac{eH_mL}{mcv_0\gamma}\,, \tag{5.109}$$

where ϵ is the final deviation angle of the velocity vector if the magnetic field is uniform within the distance L. The condition $\epsilon \ll 1$ applied to an undulator transforms into that of Eq. (5.108). In the spirit of the perturbation theory we present the particle radius-vector as an expansion

$$\boldsymbol{r}(t) = \boldsymbol{r}_0(t) + \epsilon\boldsymbol{r}_1(t) + \epsilon^2\boldsymbol{r}_2(t) + \ldots, \tag{5.110}$$

where $\boldsymbol{r}_0(t) = [x(t)\;, y_0,\; z_0]$ are the particle coordinates in the absence of the magnetic field. The line $y = y_0$, $z = z_0$ is referred to as the trajectory axis. The magnetic field can be expanded in the neighborhood of the trajectory axis as follows

$$\boldsymbol{h}(r) = \boldsymbol{h}_0 + (\boldsymbol{r}_\perp\nabla_\perp)\boldsymbol{h}_0 + \ldots, \tag{5.111}$$

with $(\boldsymbol{r}_\perp \nabla_\perp) = (y - y_0)\partial/\partial y + (z - z_0)\partial/\partial z$. Subscript 0 denotes the field and its derivative on the trajectory axis. Substituting Eqs (5.110) and (5.111) into the equation of motion and equating coefficients of the like power in ϵ on both the sides of the resulting equation we find the velocity:

$$\dot{\boldsymbol{r}}(t) = v_0[1 - \frac{1}{2}\epsilon^2 a^2(v_0 t), \ -\epsilon a_y(v_0 t), \ -\epsilon a_z(v_0 t)], \tag{5.112}$$

and the trajectory

$$\begin{aligned} x(t) &= v_0 t - \frac{\epsilon^2}{2}\int_{-\infty}^{v_0 t} a^2(x)dx, \\ y(t) &= y_0 - \epsilon \int_{-\infty}^{v_0 t} a_y(x)dx, \\ z(t) &= z_0 - \epsilon \int_{-\infty}^{v_0 t} a_z(x)dx \end{aligned} \tag{5.113}$$

where

$$\begin{aligned} a_y(x) &= \frac{1}{L}\int_{-\infty}^{x} h_{0z}(x, y_0, z_0)dx, \\ a_z(x) &= -\frac{1}{L}\int_{-\infty}^{x} h_{0y}(x, y_0, z_0)dx \\ a^2(x) &= a_y^2(x) + a_z^2(x) \end{aligned} \tag{5.114}$$

Functions $a_i(x)$ are proportional to the vector potential of the magnetic field on the trajectory axis. The solutions given by Eqs (5.113) are the generalization of the well known expressions derived by several authors for various particular fields. Some special applications for undulators or wigglers can be found in Refs[16,21].

Let us find the conditions of validity of the solutions given by Eq. (5.113), i.e. the conditions of smallness of the second order terms. The second term in the x-component in Eq. (5.112) is much smaller then the first one if $\epsilon a(x) \ll 1$. If the magnetic field does not oscillate, i.e. the characteristic length of the field variation is comparable to L, then the functions $a_i(x)$ in Eqs (5.114) have the order of about unity and, therefore, the approximate Eq. (5.112) is valid if $\epsilon \ll 1$. To estimate the next terms in y and z coordinates of $\dot{\boldsymbol{r}}(t)$ we take the second order of the equations of motion

$$\ddot{y}_2 = \frac{v_0}{L}\left[-\frac{1}{\epsilon}v_0(\boldsymbol{r}_\perp\nabla_\perp)h_{0z} + \beta_{1z}h_{0x}\right], \quad \ddot{z}_2 = \frac{v_0}{L}\left[\frac{1}{\epsilon}v_0(\boldsymbol{r}_\perp\nabla_\perp)h_{0y} - \beta_{1y}h_{0x}\right].$$

These terms can be neglected if

$$|(\boldsymbol{r}_\perp\nabla_\perp)\boldsymbol{h}_{0\perp}| \ll |\boldsymbol{h}_{0\perp}|, \tag{5.115}$$

$$|\epsilon\boldsymbol{v}_{1\perp}h_{0x}| \ll v_0\,|\boldsymbol{h}_{0\perp}| \tag{5.116}$$

The relation of Eq. (5.116) holds true when $\epsilon \ll 1$ because it results from Eq. (5.112) that $|\boldsymbol{v}_{1\perp}| \sim v_0\,|\boldsymbol{h}_{0\perp}|$, and h_{0x} has an order of unity. As to inequality of Eq. (5.115) it means that the magnetic field is uniform in the directions orthogonal to the trajectory axis within the amplitude of transversal motion of the particle. If these conditions are satisfied, the solutions of Eqs (5.113) can be expressed in terms of the vector potential on the trajectory axis.

Eqs (5.112) and (5.113) allow expressing the spectral and angular distribution of the charge radiation in terms of the on-axis magnetic field. For this sake, we substitute the coordinates of Eq. (5.113) and the velocity of Eq. (5.112) in the Fourier expansion of the radiation field and take a square of the moduli. It gives

$$\frac{d\mathcal{E}_j}{d\Omega d\omega} = \frac{e^2\omega^2}{4\pi^2 c v_0^2}\left|\int_{-\infty}^{\infty} \beta_j(x)e^{i\omega f(x)/c}dx\right|^2. \tag{5.117}$$

where the subscript j denotes the polarization components and

$$f(x) = \frac{1}{\beta_0}\left\{x(1-\beta_0 n_x) + \epsilon\int_{-\infty}^{x}\left[\frac{\epsilon}{2}a^2(x) + \beta_0(\boldsymbol{n}\boldsymbol{a}(x))\right]dx\right\}. \tag{5.118}$$

The latter formula can be reduced for the case of an ultrarelativistic particle. Assume that the particle energy is large compared to the rest energy. If $\gamma \gg 1$, then the angle θ defined by Eq. (5.43) is of the magnitude γ^{-1}. Hence, we can expand Eqs (5.117) and (5.118) for small γ^{-1} and θ. Then, to the first order, we get

$$\frac{d\mathcal{E}_{\sigma,\pi}}{d\Omega d\omega} = \frac{e^2\omega^2}{4\pi^2 c^3\gamma^2}\left|\int_{-\infty}^{\infty}[\boldsymbol{p} + \epsilon\boldsymbol{a}(x)]_{y,z}\,e^{i\omega f(x)/c}dx\right|^2. \tag{5.119}$$

with

$$f(x) = \frac{1}{2}\gamma^{-2}\int_{-\infty}^{x}\left\{1 + [\boldsymbol{p} + \epsilon\boldsymbol{a}(x)]^2\right\}dx,$$

$\boldsymbol{p} = [0,\ \psi\cos\phi,\ \psi\sin\phi]$, and $\psi = \theta\gamma$. Eqs (5.117)–(5.119) express the spectral-angular distribution of radiated energy in terms of $\boldsymbol{a}(x)$, the reduced potential of the magnetic field on the trajectory axis.

5.2.4. Quantum Effects in Wiggler Radiation

In Sec. 4.4 a simple quasiclassical method was proposed to compute quantum corrections to the synchrotron radiation of ultrarelativistic electrons. This method can also be used in computation of quantum corrections to the electron radiation in an undulator with a magnetic field of a large magnitude. The quantum effects in undulator radiation have been studied in Refs[7,22,23]. In those papers, a number of general results were obtained, with a detailed study of the radiation in the dipole approximation, when the ratio of the transverse momentum $p_\perp$ to the longitudinal momentum $p_\parallel$ is much smaller than γ^{-1}. Concerning the other limiting case of $p_\perp/p_\parallel \gg \gamma^{-1}$, one usually notes that in this approximation, the radiation in a wiggler coincides with the synchrotron radiation. Strictly speaking, this is true only for a single half-period wiggler. The presence of several periodicity elements leads to interference in the probability amplitude of photon radiation, which explains why some properties of the wiggler radiation differ from those of synchrotron radiation[24,25]. In this section, we use the method described in Sec. 5.2.2 to study the radiation of an ultrarelativistic electron in a wiggler taking into consideration quantum corrections. We analyze the characteristic features of this radiation caused by the presence of several periods of the magnetic field.

Let the magnetic field of a wiggler and the particle motion satisfy the conditions accepted in Sec. 5.2.2. We use the semiclassical method discussed in Sec. 4.4. According to this method, the Fourier transform of the electric field of radiation with respect to the quantum corrections can be calculated as follows:

$$\boldsymbol{E}_j(\omega) = \frac{e\tilde{\omega}}{Rc}\int\limits_{-\infty}^{\infty} \langle f|\boldsymbol{\alpha n}_j|i\rangle \exp i\tilde{\omega}\left(t - \frac{\boldsymbol{nr}}{c}\right)dt, \tag{5.120}$$

where $|i\rangle$ and $|f\rangle$ are the free electron state vectors before and after the emission of a photon, $\tilde{\omega} = \omega/(1 - \hbar\omega/\varepsilon)$, ω is the photon frequency, ε is the energy of the electron, and $\boldsymbol{\alpha}$ are the Dirac matrices. The states $|i\rangle$ and $|f\rangle$ are related by the conservation laws

$$\varepsilon' = \varepsilon - \hbar\omega,\ \boldsymbol{p}\,' = \boldsymbol{p} - \frac{\hbar\omega}{c}\boldsymbol{n}, \tag{5.121}$$

where ε and $\boldsymbol{p}$ are the energy and momentum of the electron before, and ε' and $\boldsymbol{p}\,'$ after the emission of the photon.

In accordance with Sec. 5.2.2 we perform integration in Eq. (5.120) in the vicinity of points 1 and 2 where the trajectory slope is parallel to the projection of vector

$\boldsymbol{n}$ onto the trajectory plane. Let the assumptions on the particle dynamics accepted in the previous section be valid here too. We also assume that at point 1 the initial direction of spin is parallel to that of the magnetic field.

Thus, performing the calculations as it is described in Sec. 5.2.2 for the classical part and in Secs 4.4 and 4.7 for the quantum corrections we obtain the radiation field for the σ- and π-components of polarization:

$$\begin{aligned} E_\sigma(\omega) &= E_0 R^{-1} i \Big\{(2+\delta)u_+\eta\left[\rho_1 K_{2/3}(\omega_1') + e^{2i\chi}\rho_2 K_{2/3}(\omega_2')\right] - \\ &\quad \delta\left(\zeta u_+ + \psi u_-\right)\left[\rho_1 K_{1/3}(\omega_1') - e^{2i\chi}\rho_2 K_{1/3}(\omega_2')\right]\Big\}, \\ E_\pi(\omega) &= E_0 R^{-1}\Big\{u_-\delta\eta\left[\rho_1 K_{2/3}(\omega_1') + e^{2i\chi}\rho_2 K_{2/3}(\omega_2')\right] - \\ &\quad \left[(2+\delta)u_+\psi - u_-\delta\zeta\right]\left[\rho_1 K_{1/3}(\omega_1') - e^{2i\chi}\rho_2 K_{1/3}(\omega_2')\right]\Big\}. \end{aligned} \tag{5.122}$$

Here

$$\begin{aligned} E_0 &= \frac{e\tilde{\omega}\eta\sin\pi\nu N}{c^2\sqrt{3}\gamma^2\sin\pi\nu}, \quad \omega_i' = \frac{|\rho_i|\tilde{\omega}\eta^3}{3c\gamma^3}, \quad u_\pm = \frac{1\pm\zeta\zeta'}{2} \\ \nu &= \frac{\tilde{\omega}}{\omega_0}(1-\beta_{||}n_x), \quad \chi = \tilde{\omega}\left(t_{12} - \frac{\boldsymbol{n}\boldsymbol{r}_{12}}{c}\right), \quad \eta = \sqrt{1+\psi^2}, \end{aligned} \tag{5.123}$$

$v_{||} = c\beta_{||}$ is the average velocity of motion in the direction of the wiggler axis; ω_0 is the frequency of motion; $\psi = \gamma\theta_z$, $\delta = \hbar\omega/\varepsilon'$; $\zeta, \zeta' = \pm 1$ correspond to the initial and final projection of spin on the z-axis; $\rho_i = \varepsilon/eH_i$ is the local radius of curvature of the trajectory; H_i is the magnetic field at points 1 or 2.

Eqs (5.122) contain two kinds of terms: classical ones without δ and quantum corrections proportional to δ. One can see that σ- and π-components of the radiation have different phases, which suggests that the radiation is elliptically polarized.

Further, we divide our consideration into two parts. First we study the radiation properties of an arbitrary wiggler with $\rho_1 \neq \rho_2$. This kind of wiggler or undulator is referred to as an asymmetric one and is used for production of the circularly polarized radiation [13,14]. In this case, we restrict our consideration to the first order of quantum terms. Next we compute the spectrum and the angular distribution of the radiation for the special case of a symmetric wiggler with $\rho_1 = \rho_2$ taking into account the terms of the order of $\hbar^2$.

5.2.4.1. Asymmetric Wiggler Radiation

Taking the square of module of Eqs (5.122) and keeping terms linear to δ we obtain

$$\frac{d\mathcal{E}_\sigma}{d\Omega d\omega} = \frac{E_0^2 c\eta u_+}{\pi^2}\Big\{(1+\delta)\eta\left[\rho_1^2 K_{2/3}^2(\omega_1') + 2\rho_1\rho_2\cos 2\chi K_{2/3}(\omega_1')K_{2/3}(\omega_2') + \right.$$

$$\rho_2^2 K_{2/3}^2(\omega_2')\Big] - \delta\zeta\Big[\rho_1^2 K_{1/3}(\omega_1')K_{2/3}(\omega_1') - \rho_2^2 K_{1/3}(\omega_2')K_{2/3}(\omega_1') +$$
$$\rho_1\rho_2\cos 2\chi\left(K_{2/3}(\omega_2')K_{1/3}(\omega_1') - K_{2/3}(\omega_1')K_{1/3}(\omega_2')\right)\Big]\Big\},$$
$$\frac{d\mathcal{E}_\pi}{d\Omega d\omega} = \frac{E_0^2 c\psi^2 u_+(1+\delta)}{\pi^2}\Big[\rho_1^2 K_{1/3}^2(\omega_1') + \rho_2^2 K_{1/3}^2(\omega_2') -$$
$$2\rho_1\rho_2\cos 2\chi K_{1/3}(\omega_1')K_{1/3}(\omega_2')\Big]. \qquad (5.124)$$

Note that Eqs (5.124) do not contain terms proportional to factor $u_- = (1-\zeta\zeta')/2$ which describes the spin-flip radiation. Thus, there is no need to keep the factor u_+ any more. The presence of the oscillatory factors $\sin^2\kappa$ and $\cos^2\kappa$ results in a strong dependence of the ratio between the polarization components of radiation on the frequency and the observation angle. Let us make an estimation of the gap $\Delta\omega$ between the maxima of $\sin^2\kappa$ and $\cos^2\kappa$. It follows from Eq. (5.123) that it is of the same order of magnitude as the distance between the neighboring harmonics in the radiation spectrum. Setting $\Delta\nu = 1$ in Eq. (5.123), we find

$$\Delta\omega \sim \frac{\omega_0}{1-\beta_{||}} \approx \frac{2\omega_0}{1-\beta_{||}^2} \sim \frac{\omega_0}{\beta_\perp^2},$$

where $c\beta_\perp$ is the typical value of the transverse velocity. The main part of the radiation corresponds to the frequencies

$$\omega_c \sim \frac{c}{\rho}\gamma^3 \sim \omega_0\beta_\perp\gamma^3.$$

(We have taken into account that $\rho/\lambda \sim \beta_{||}/\beta_\perp$.) Therefore, the relative distance between the maxima in the radiation spectrum has the order of magnitude:

$$\frac{\Delta\omega}{\omega_c} \sim \frac{1}{(\beta_\perp\gamma)^3} \ll 1.$$

Averaging the fast oscillating factors by the use of Eqs (5.105) and (5.106) we find

$$\frac{d\mathcal{E}_\sigma}{d\Omega d\omega} = \frac{e^2\tilde{\omega}^2 N\eta^3}{3c^3\gamma^4\pi^2}\Big\{(1+\delta)\eta\Big[\rho_1^2 K_{2/3}^2(\omega_1') + \rho_2^2 K_{2/3}^2(\omega_2')\Big] -$$
$$\delta\zeta\Big[\rho_1^2 K_{1/3}(\omega_1')K_{2/3}(\omega_1') - \rho_2^2 K_{1/3}(\omega_2')K_{2/3}(\omega_1')\Big]\Big\}, \qquad (5.125)$$
$$\frac{d\mathcal{E}_\pi}{d\Omega d\omega} = \frac{e^2\tilde{\omega}^2 N\eta^2\psi^2(1+\delta)}{3c^3\gamma^4\pi^2}\Big[\rho_1^2 K_{1/3}^2(\omega_1') + \rho_2^2 K_{1/3}^2(\omega_2')\Big]. \qquad (5.126)$$

The sign of the quantum correction in the σ-component depends on the initial polarization ζ of the electron. Changing the electron polarization or the direction of the wiggler magnetic field one can find experimentally the value of the quantum term in Eq. (5.125). The relative magnitude of this term is defined by the ratio

$$D_\sigma = \delta \frac{\omega_1'^2 K_{1/3}(\omega_1') K_{2/3}(\omega_1') - \omega_2'^2 K_{1/3}(\omega_2') K_{2/3}(\omega_2')}{\omega_1'^2 K_{2/3}^2(\omega_1') + \omega_2'^2 K_{2/3}^2(\omega_2')}, \tag{5.127}$$

which depends on the frequency and the direction of the radiation.

Let us find the angular distribution. Integrating Eqs (5.125) and (5.126) over the frequencies we obtain

$$\begin{aligned} \frac{d\mathcal{E}_\sigma}{d\Omega} &= \frac{e^2 N \gamma^5}{4\eta^5} \sum_{i=1}^{2} \frac{1}{\rho_i} \left\{ \frac{7}{4} - \frac{160\sqrt{3}}{9\pi\eta^3} \xi_i + \frac{35}{12\eta^4} (-1)^i \zeta \xi_i \right\} \\ \frac{d\mathcal{E}_\pi}{d\Omega} &= \frac{e^2 N \gamma^5 \psi^2}{4\eta^7} \sum_{i=1}^{2} \frac{1}{\rho_i} \left\{ \frac{5}{4} - \frac{2^7\sqrt{3}}{9\pi\eta^3} \xi_i \right\}. \end{aligned} \tag{5.128}$$

Here ξ_i is a small quantum parameter

$$\xi_i = \frac{3\hbar\gamma^2}{2mc|\rho_i|}.$$

Next we integrate Eqs (5.125) and (5.126) over the angle θ_z in order to find the spectrum of the radiation emitted in the vertical aperture of the angular width $d\phi$. It gives

$$\begin{aligned} \frac{d\mathcal{E}_\sigma}{d\phi d\omega} &= A \sum_{i=1}^{2} \frac{\rho_i^2}{y_i} \left\{ (1+\delta) \left[\int_{y_i}^{\infty} K_{5/3}(x)dx + K_{2/3}(y_i) \right] - 2(-1)^i \delta \zeta K_{1/3}(y_i) \right\}, \\ \frac{d\mathcal{E}_\pi}{d\phi d\omega} &= A(1+\delta) \sum_{i=1}^{2} \frac{\rho_i^2}{y_i} \left[\int_{y_i}^{\infty} K_{5/3}(x)dx - K_{2/3}(y_i) \right], \end{aligned} \tag{5.129}$$

with

$$A = \frac{e^2\omega^2 N}{3\sqrt{3}c^3\gamma^5\pi}, \quad y_i = \frac{2}{3} \frac{|\rho_i|\omega}{\gamma^3 c\,(1 - \hbar\omega/mc^2\gamma)}.$$

Integrating over the frequency we find the total energy radiated per solid angle $d\phi$:

$$\begin{aligned} \frac{d\mathcal{E}_\sigma}{d\phi} &= \frac{2e^2\gamma^4 N}{3} \sum_{i=1}^{2} \frac{1}{|\rho_i|} \left[\frac{7}{8} - \frac{25\sqrt{3}}{12} \xi_i - (-1)^i \zeta \xi_i \right], \\ \frac{d\mathcal{E}_\pi}{d\phi} &= \frac{2e^2\gamma^4 N}{3} \sum_{i=1}^{2} \frac{1}{|\rho_i|} \left[\frac{1}{8} - \frac{5\sqrt{3}}{24} \xi_i \right]. \end{aligned} \tag{5.130}$$

The same result can be found by integrating of Eqs (5.128) over the angle θ_z. Finally, we find the total energy emitted by electron during a single pass through the wiggler. On integrating Eqs (5.130) over the ϕ we take into account that

$$d\phi = \frac{dx}{\rho_i}, \quad \rho_i = \frac{mc^2\gamma}{eH_i}. \tag{5.131}$$

Integration over the x should be performed over the wiggler period. Let point 1 belong to the interval (x_1, x_2), where x_1 and x_2 are the coordinates of the adjacent points in which the magnetic field is equal to zero. When integrating from x_1 to x_2 we take in Eqs (5.130) $i = 1$, but in the interval $(x_2, x_1 + \lambda)$ we substitute $i = 2$. As a result, we find

$$\mathcal{E}_\sigma = \frac{2e^4\gamma^2 N\lambda\overline{H^2}}{3m^2c^4}\left[\frac{7}{8} - \frac{25\sqrt{3}}{12}\bar{\xi} - \zeta\xi_{12}\right], \tag{5.132}$$

$$\mathcal{E}_\pi = \frac{2e^4\gamma^2 N\lambda\overline{H^2}}{3m^2c^4}\left[\frac{1}{8} - \frac{5\sqrt{3}}{12}\bar{\xi}\right]. \tag{5.133}$$

The following notations are introduced here

$$\bar{\xi} = \frac{3e\hbar\overline{H^3}}{2m^2c^3\overline{H^2}}, \quad \xi_{12} = \frac{3e\hbar(\overline{H_1^3} - \overline{H_2^3})}{2m^2c^3\overline{H^2}}, \quad \overline{H^n} = \frac{1}{\lambda}\int_{x_1}^{x_1+\lambda} |H^n(x, y_0)|dx,$$

$$\overline{H_1^3} = \frac{1}{\lambda}\int_{x_1}^{x_2} |H^3(x, y_0)|dx, \quad \overline{H_2^3} = \frac{1}{\lambda}\int_{x_2}^{x_1+\lambda} |H^3(x, y_0)|dx.$$

It is supposed that the particle moves in the plane $z = 0$ in the vicinity of the line $y = y_0$, parallel to the wiggler axis. We assume as well that within the trajectory region the dependence of the magnetic field on the transverse coordinate y can be neglected.

Up to now we studied the linear polarization. But the advantage of the asymmetric wiggler is the circular polarization of radiation even from a flat trajectory. Thus, it is of interest to calculate the quantum corrections in the components of circular polarization. Let us define as usual the right E_+ and left E_- components of an electric field of radiation according to

$$E_\pm = \frac{E_\sigma \pm iE_\pi}{\sqrt{2}}.$$

Substituting Eqs (5.122) in the latter equation and taking the square of module, we get spectral and angular distribution of the circularly polarized components

$$\begin{aligned}\frac{d\mathcal{E}_\pm}{d\Omega d\omega} = {} & \frac{e^2\omega^2\eta^2 N}{6\pi^2c^3\gamma^4}\frac{1+\zeta\zeta'}{2}\Big\{(1+\delta)\eta^2\left[\rho_1^2K_{2/3}^2(\omega_1') + \rho_2^2K_{2/3}^2(\omega_2')\right] + \\ & \psi[\psi(1+\delta) \pm \zeta\delta]\left[\rho_1^2K_{1/3}^2(\omega_1') + \rho_2^2K_{1/3}^2(\omega_2')\right] + \eta[\pm 2\psi(1+\delta) \\ & +\zeta\delta]\left[\rho_1^2K_{2/3}(\omega_1')K_{1/3}(\omega_1') - \rho_2^2K_{2/3}(\omega_2')K_{1/3}(\omega_2')\right]\Big\}.\end{aligned} \tag{5.134}$$

One can see that the radiation intensity depends on the spin polarization of the electron. Inversion of either the spin polarization or the direction of the wiggler magnetic field changes the radiation intensity. This phenomenon can be used for detecting electron beam polarization.

The degree of polarization defined as follows

$$P_\pm = \left[\frac{d\mathcal{E}_+}{d\Omega d\omega} - \frac{d\mathcal{E}_-}{d\Omega d\omega}\right] / \left[\frac{d\mathcal{E}_+}{d\Omega d\omega} + \frac{d\mathcal{E}_-}{d\Omega d\omega}\right]$$

is equal to

$$\begin{aligned} P_\pm &= \Big\{\psi\zeta\delta\left[\rho_1^2 K_{1/3}^2(\omega_1') + \rho_2^2 K_{1/3}^2(\omega_2')\right] + \\ &\quad 2\psi\eta(1+\delta)\left[\rho_1^2 K_{2/3}(\omega_1')K_{1/3}(\omega_1') - \rho_2^2 K_{2/3}(\omega_2')K_{1/3}(\omega_2')\right]\Big\} / \\ &\quad \Big\{(1+\delta)\eta^2\left[\rho_1^2 K_{2/3}^2(\omega_1') + \rho_2^2 K_{2/3}^2(\omega_2')\right] + \\ &\quad \psi^2(1+\delta)\left[\rho_1^2 K_{1/3}^2(\omega_1') + \rho_2^2 K_{1/3}^2(\omega_2')\right] + \\ &\quad \eta\zeta\delta\left[\rho_1^2 K_{2/3}(\omega_1')K_{1/3}(\omega_1') - \rho_2^2 K_{2/3}(\omega_2')K_{1/3}(\omega_2')\right]\Big\} \end{aligned} \tag{5.135}$$

Finally, let us discuss a 3-pole snake which is a kind of asymmetric wiggler. The first measurement of the quantum part of radiation related to the electron magnetic moment has been performed at this kind of wiggler [26]. The magnetic field of the snake as a function of longitudinal coordinate has three extremums. The transversal coordinate of the electron is calculated by double integration of the magnetic field function and has, therefore, only one extremum. Thus, the radiation in any given direction is emitted only from one point of the trajectory. Hence, the radiation of the 3-pole snake is equal to synchrotron radiation which has been studied well enough. Corresponding expressions can be easily derived from the equations of this section taking only one term in the sums over i, for example the term with $i = 1$

5.2.4.2. Symmetric Wiggler Radiation

Let us consider a particular case of symmetric wiggler. We put in Eqs (5.122) $\rho_1 = -\rho_2 = \rho$, $\omega_1' = \omega_2' = \omega'$ and obtain

$$\begin{aligned} E_\sigma(\omega) &= E'\left\{-u_+(2+\delta)\eta\sin\chi K_{2/3}(\omega') + i\delta\cos\chi K_{1/3}(\omega')(u_-\psi - u_+\zeta)\right\}, \\ E_\pi(\omega) &= E'\left\{u_-\delta i\eta\sin\chi K_{2/3}(\omega') + \cos\chi K_{1/3}(\omega')[u_+(2+\delta)\psi + u_-\delta\zeta]\right\} \end{aligned} \tag{5.136}$$

where

$$E' = \frac{2\rho\tilde{\omega}}{\omega} e^{i\chi} E_0$$

The classical terms in Eq. (5.136) have the same phase, which suggests that the main part of the radiation is linearly polarized. As to the quantum corrections for the

σ- and π-components, their phases are different, which means that the polarization is not a linear one due to the recoil and spin effects. In order to find the spectral and angular distribution of the polarization components of radiation, let us take a square of Eqs (5.136) (keeping terms of the order δ^2). We obtain

$$\begin{aligned}\frac{d\mathcal{E}_\sigma}{d\Omega d\omega} &= \frac{e^2\tilde{\omega}^2\rho^2\eta^2)\sin^2\pi\nu N}{3c^3\gamma^4\pi^2\sin^2\pi\nu}\left\{u_+(2+\delta)^2\eta^2\sin^2\chi\, K^2_{2/3}(\omega')+\right.\\ &\quad \left.\delta^2\cos^2\chi\, K^2_{1/3}(\omega')[u_+ + u_-\psi^2]\right\},\\ \frac{d\mathcal{E}_\pi}{d\Omega d\omega} &= \frac{e^2\tilde{\omega}^2\rho^2\eta^2\sin^2\pi\nu N}{3c^3\gamma^4\pi^2\sin^2\pi\nu}\left\{u_+(2+\delta)^2\psi^2)\cos^2\chi K^2_{1/3}(\omega')+\right.\\ &\quad \left.u_-\delta^2[\eta^2\sin^2\chi K^2_{2/3}(\omega')+\cos^2\chi K^2_{1/3}(\omega')]\right\}. \qquad (5.137)\end{aligned}$$

In the classical limit $\hbar \to 0$, Eqs (5.137) become equal to Eqs (5.98). Averaging over the fast-oscillating factors by the use of Eqs (5.105) and (5.106) we find

$$\begin{aligned}\frac{d\mathcal{E}_\sigma}{d\Omega d\omega} &= \frac{e^2\omega^2\rho^2 N}{6\pi^2c^3\gamma^4}\eta^2\left\{u_+(2+\delta)^2\eta^2K^2_{2/3}(\omega')+\delta^2(u_+ + \psi^2u_-)K^2_{1/3}(\omega')\right\},\\ \frac{d\mathcal{E}_\pi}{d\Omega d\omega} &= \frac{e^2\omega^2\rho^2 N}{6\pi^2c^3\gamma^4}\eta^2\left\{u_-\delta^2\eta^2K^2_{2/3}(\omega')+[u_+(2+\delta)^2\psi^2+u_-\delta^2]K^2_{1/3}(\omega')\right\}.\end{aligned}$$

The obtained expressions differ from the corresponding ones for synchrotron radiation in the absence of terms linear in the initial electron spin.

Instead of Eqs (5.129) we obtain

$$\begin{aligned}\frac{d\mathcal{E}_\sigma}{d\phi d\omega} &= \frac{\sqrt{3}e^2\gamma N y}{8\pi c(1+\delta)^2}\left\{u_+\left[(2+\delta)^2\left(\int_y^\infty K_{5/3}(x)dx + K_{2/3}(y)\right)+\right.\right.\\ &\quad \left.\left. 2\delta^2y^2\int_y^\infty K_{1/3}(x)dx\right] + u_-\delta^2\left[\int_y^\infty K_{5/3}(x)dx - K_{2/3}(y)\right]\right\},\\ \frac{d\mathcal{E}_\pi}{d\phi d\omega} &= \frac{\sqrt{3}e^2\gamma N y}{8\pi c(1+\delta)^2}\left\{u_+(2+\delta)^2\left[\int_y^\infty K_{5/3}(x)dx-\right.\right.\\ &\quad \left.\left. K_{2/3}(y)\right] + u_-\delta^2\left[\int_y^\infty K_{5/3}(x)dx + K_{2/3}(y) + 2\int_y^\infty K_{1/3}(x)dx\right]\right\}.\end{aligned}$$

Finally, integrating over the frequency, we find the total energy emitted in the solid angle $d\phi$:

$$\frac{d\mathcal{E}_\sigma}{d\phi} = \frac{4e^2\gamma^4 N}{3\rho}\left\{u_+\left[\frac{7}{8}-\frac{25\sqrt{3}}{12}\xi+\frac{325}{18}\xi^2+\frac{5}{9}\xi^2\right]+u_-\frac{1}{18}\xi^2\right\},$$

$$\frac{d\mathcal{E}_\pi}{d\phi} = \frac{4e^2\gamma^4 N}{3\rho}\left\{u_+\left[\frac{1}{8}-\frac{5\sqrt{3}}{24}\xi+\frac{25}{18}\xi^2\right]+u_-\frac{23}{18}\xi^2\right\}. \tag{5.138}$$

Let us look at the structure of Eqs (5.138) in more detail. It is useful to compare Eqs (5.138) with Eqs (3.94) and (4.91). The first term in the σ- and π-components corresponds to classical synchrotron radiation. The quantum corrections proportional to u_+ (radiation without a spin-flip) describe the recoil effects in radiation (except for the $(5/9)\xi^2$ term in the σ-component (see Eqs (4.91)). The $(5/9)\xi^2$ term and the terms corresponding to the radiation with a spin-flip, represent the radiation of the proper magnetic moment of the electron. These corrections are also present in synchrotron radiation; they were obtained by Sokolov, Klepikov and Ternov [27] and by Schwinger [28]. However, unlike synchrotron radiation, Eqs (5.138) do not contain terms linear in the electron spin, which reflects the fact that due to the assumed symmetry of $H(x)$, the wiggler field does not define any axial vector in space. The independence of the obtained expressions of the initial state of transverse polarization of the electron suggests that the effect of radiative self-polarization typical for synchrotron radiation does not appear in the symmetric wiggler.

5.3. An Undulator with the Sinusoidal Magnetic Field

5.3.1. A Double-Frequency Undulator

We begin our study of radiation from the undulator with a sinusoidal magnetic field with general calculations which can be applied not only to this undulator but to a variety of systems realizing similar oscillations of a charged particle.

Let us consider a charged particle moving according to equations [29] (with τ being a parameter)

$$\begin{aligned} \boldsymbol{r} &= \boldsymbol{v}_{\|}\frac{\tau}{\omega_0} + \boldsymbol{A}\sin\tau + \boldsymbol{B}\sin 2\tau, \\ ct &= c\frac{\tau}{\omega_0} + A_0\sin\tau + B_0\sin 2\tau. \end{aligned} \tag{5.139}$$

The particle path looks as if the particle traces out figure '8' with a simultaneous drift in an arbitrary direction. Such oscillations take place, for example, in the sinelike field of a magnetic undulator, in a plane linearly polarized electromagnetic wave [1], in the linearly increasing orthogonal electric and magnetic fields [30], etc. Taking, for example, $A_x = A_z = A_0 = B_y = B_z = B_0 = 0$, one gets a trajectory in the sinelike magnetic field of an undulator. In combined magnetic and electric quadrupoles fields are realized that depend linearly on the coordinates. The motion in such a system is described by Eqs (5.139) with $A_x = A_z = A_0 = B_y = B_z = B_y = B_0 = 0$. Turning A_0, B_0, $\boldsymbol{B}$ to zero, one gets a linear harmonic oscillator. Therefore, Eqs (5.139) unite many particular cases of practical importance.

As in the previous sections, the particle is assumed to accomplish N oscillations with a period $T = 2\pi/\omega_0$. When $t < 0$ and $t > NT$, the particle moves uniformly along a straight line. Suppose also that $N >> 1$. The distribution of energy emitted from the particle is given by Eqs (5.49) and (5.50). The phase of the exponent in Eq. (5.50) can be rearranged as follows

$$\omega_n t - \boldsymbol{k}_n \boldsymbol{r} = n(\tau - p\cos\tau - q\sin 2\tau),$$

where

$$\begin{aligned} p &= \frac{\boldsymbol{n}\boldsymbol{a} - a_0}{1 - \boldsymbol{\beta}_{\|}\boldsymbol{n}}, \quad q = \frac{\boldsymbol{n}\boldsymbol{a} - a_0}{1 - \boldsymbol{\beta}_{\|}\boldsymbol{n}}, \\ \boldsymbol{a} &= \frac{\omega_0}{c}\boldsymbol{A}, \quad \boldsymbol{b} = \frac{\omega_0}{c}\boldsymbol{B}, \quad a_0 = \frac{\omega_0}{c}A_0, \quad b_0 = \frac{\omega_0}{c}B_0. \end{aligned} \tag{5.140}$$

We define a generalized Anger function $\mathbf{J}_\nu(x, y)$ by an integral formula [31]

$$\mathbf{J}_\nu(x, y) = \frac{1}{2\pi}\int_{-\pi}^{\pi} e^{i(\nu\tau - x\sin\tau - y\sin 2\tau)} d\tau. \tag{5.141}$$

In fact, Eqs (5.49) and (5.50) are written in terms of the integral ν. But if we use the initial Eq. (5.44) then we have to deal with a nonintegral index. We discuss here the properties of the generalized Anger function for generality. In the case of ν integer, this function is referred to as a generalized Bessel function which has been introduced in Refs[32 – 33]. Functions of this type have been used in the description of the relativistic scattering problems in Refs[32,37,38]. The generalized Bessel function of two and more variables have been studied thoroughly in a number of papers beginning with Refs[35,36]. We derive here some necessary properties which are immediately obtained from the definition of Eq. (5.141).

1. *Recurrence relations*

$$\begin{aligned} \pi\nu\mathbf{J}_\nu(x, y) &= \sin\pi\nu + \frac{1}{2}\pi x[\mathbf{J}_{\nu+1}(x, y) + \mathbf{J}_{\nu-1}(x, y)] + \\ &\quad \pi y[\mathbf{J}_{\nu+2}(x, y) + \mathbf{J}_{\nu-2}(x, y)], \end{aligned} \tag{5.142}$$

$$\frac{\partial \mathbf{J}_\nu(x, y)}{\partial x} = \frac{1}{2}[\mathbf{J}_{\nu-1}(x, y) - \mathbf{J}_{\nu+1}(x, y)], \tag{5.143}$$

$$\frac{\partial \mathbf{J}_\nu(x, y)}{\partial y} = \frac{1}{2}[\mathbf{J}_{\nu-2}(x, y) - \mathbf{J}_{\nu+2}(x, y)]. \tag{5.144}$$

2. *Particular values*

$$\mathbf{J}_\nu(x, 0) = \mathbf{J}_\nu(x), \tag{5.145}$$

$$\mathbf{J}_\nu(0, y) = \cos\frac{\pi\nu}{2}\mathbf{J}_{\frac{\nu}{2}}(-y). \tag{5.146}$$

3. *Series expansions*

Making a substitution

$$e^{-iy\sin 2\tau} = \sum_{k=-\infty}^{\infty} e^{-2ik\tau} J_k(y)$$

in Eq. (5.141) we obtain

$$\mathbf{J}_\nu(x,y) = \sum_{k=-\infty}^{\infty} (-1)^k J_k(y)\mathbf{J}_{\nu+2k}(x) = \sum_{k=-\infty}^{\infty} J_k(y)\mathbf{J}_{\nu-2k}(x). \quad (5.147)$$

Here $J_k(x)$ and $\mathbf{J}_k(x)$ are the Bessel and the Anger functions. In order to find the Taylor-series expansion for the generalized Bessel function of the integral index we substitute in Eq. (5.141)

$$e^{-ix\sin\tau} = \sum_{l=0}^{\infty} \frac{(-ix\sin\tau)^l}{l!}, \quad e^{-iy\sin 2\tau} = \sum_{m=0}^{\infty} \frac{(-iy\sin 2\tau)^m}{m!}.$$

The result of integration is

$$J_n(x,y) = \sum_{l=0}^{\infty}\sum_{m=0}^{\infty} (-1)^{l+m} \frac{[1+(-1)^{n+l}]i^{l+n}(l+m)!}{2^{l+m+1}l!m!} \times \frac{F\left(-m, \frac{n-l}{2} - m; \frac{l+n}{2}+1; -1\right)}{\Gamma\left(\frac{l-n}{2}+m+1\right)\Gamma\left(\frac{l+n}{2}+1\right)} x^l y^m. \quad (5.148)$$

Here $F(a,b;c;z)$ is the hypergeometric function [9]. In order to exclude the zero terms of the sum with $l+n$ odd and $l < n-2m$, we rename the indices as follows: $l = 2j+n-2m$. Then the expansion takes the form [21] (see also Ref.[36]

$$J_n(x,y) = x^n \sum_{j=0}^{\infty} \frac{(-1)^j}{j!} \sum_{m=0}^{[j+n/2]} (x^2)^{j-m} y^m \frac{(2j+n-m)!F(-m,-j;j+n-m+1;-1)}{2^{n+2j-m}m!(j+n-m)!(n+2j-2m)!}, \quad (5.149)$$

where $[j+n/2] = j+n/2$ if n is an even number and $[j+n/2] = j+(n-1)/2$ in the case of odd n.

4. *Symmetry relations*

$$\begin{aligned} J_n(-x,y) &= (-1)^n J_n(x,y), \\ J_{-n}(x,y) &= J_n(-x,-y) = (-1)^n J_n(x,-y). \end{aligned} \quad (5.150)$$

Here n is an integer.

Let us return to the radiation of the double-frequency oscillator. Functions B_{nj} of Eq.(5.50) take the form

$$\begin{aligned} B_{nj} &= (\beta_{\|j} + \frac{b_j}{q})J_n(np,nq) + \frac{1}{2}(a_j - \frac{p}{q}b_j)[J_{n+1}(np,nq) + \\ & J_{n-1}(np,nq)]. \end{aligned} \tag{5.151}$$

The quantities $B_{n\sigma}$ and $B_{n\pi}$ are real, hence, the radiation is linearly polarized. Substituting B_{jn} into Eq. (5.49), one gets a spectral-angular distribution of the polarization components.

$$\begin{aligned} \frac{d\mathcal{E}_j}{d\Omega d\omega} &= \frac{e^2}{4cq^2(1-\boldsymbol{\beta}_{\|}\boldsymbol{n})^2}\sum_{n=1}^{\infty} n^2 I_n(\nu,N)\left\{2(q\beta_{\|j}+b_j)J_n(np,nq)+\right. \\ &+ \left.(qa_j - pb_j)[J_{n+1}(np,nq)+J_{n-1}(np,nq)]\right\}^2. \end{aligned} \tag{5.152}$$

Next we discuss the radiation properties in the limit of small amplitudes of oscillations. Assume that the arguments p and q of the generalized Bessel function are much less than unity. Consider the ultrarelativistic particle. The denominator in Eq. (5.140) is of the order of γ^{-2}. So that the parameters p and q are small we suppose that $(\boldsymbol{na}), a_0, (\boldsymbol{nb}), b_0 \ll \gamma^{-1}$. The scalar product can be written in the form of $(\boldsymbol{na}) = (\boldsymbol{n}_\perp \boldsymbol{a}_\perp) + (\boldsymbol{n}_\| \boldsymbol{a}_\|)$, where the subscripts '$\perp$' and '$\|$' denote the orthogonal and parallel projections of vectors with respect to the direction of the mean velocity $\boldsymbol{\beta}_\|$. Taking into account that $n_\perp \sim \gamma^{-1}$ and $n_\| \sim 1$ we obtain the following conditions of smallness of the parameters p and q

$$|\boldsymbol{a}_\perp|, |\boldsymbol{b}_\perp| \ll \gamma^{-1}, \quad a_0, b_0 \ll \gamma^{-2}. \tag{5.153}$$

The requirement for smallness of a_0 and b_0 provide the linear dependence of time on the parameter τ in Eq. (5.139). The conditions under discussion allow retaining only the first terms in the expansion of the function $J_n(np,nq)$ into a series of small p and q. Eq. (5.148) gives

$$J_0(x,y) = 1,\ J_1(x,y) = x/2,\ J_2(x,y) = y/2,\ J_3(x,y) = xy/4. \tag{5.154}$$

Substituting the latter equation in Eq. (5.152) we obtain in the first approximation

$$\begin{aligned} \frac{d\mathcal{E}_{1j}}{d\Omega d\omega} &= \frac{e^2}{4\pi^2 c(1-\boldsymbol{\beta}_\|\boldsymbol{n})^2}(p\beta_{\|j}+a_j)^2\frac{\sin^2\pi\nu N}{(1-\nu)^2}, \\ \frac{d\mathcal{E}_{2j}}{d\Omega d\omega} &= \frac{4e^2}{\pi^2 c(1-\boldsymbol{\beta}_\|\boldsymbol{n})^2}(q\beta_{\|j}+b_j)^2\frac{\sin^2\pi\nu N}{(2-\nu)^2}. \end{aligned} \tag{5.155}$$

As it should be expected, the main contribution to the radiation energy in the dipole approximation is due to the first and second harmonics only. Moreover, the radiation energy of the first harmonic depends only on the amplitude of oscillations with the frequency ω_0, and that of the second harmonic only on oscillations with the frequency $2\omega_0$. We rewrite the last equation in terms of the angular coordinates θ and ϕ. Expanding Eq. (5.155) with respect to small γ^{-1} and θ one gets for the first and the second harmonics (with $\psi = \gamma\theta$)

$$\begin{aligned} \frac{d\mathcal{E}_{1\sigma,\pi}}{d\Omega d\omega} &= \frac{e^2\gamma^4}{\pi^2 c(1+\psi^2)^4}\frac{\sin^2\pi\nu N}{(1-\nu)^2}\left[2\gamma\psi\sin\phi(a_x-a_0)+\right. \\ &\left. a_{z,y}\psi^2\sin 2\phi - a_{y,z}(1\mp\psi^2\cos 2\phi)\right]^2, \\ \frac{d\mathcal{E}_{2\sigma,\pi}}{d\Omega d\omega} &= \frac{16e^2\gamma^4}{\pi^2 c(1+\psi^2)^4}\frac{\sin^2\pi\nu N}{(2-\nu)^2}\left[2\gamma\psi\sin\phi(b_x-b_0)+\right. \\ &\left. b_{z,y}\psi^2\sin 2\phi - b_{y,z}(1\mp\psi^2\cos 2\phi)\right]^2. \end{aligned} \tag{5.156}$$

The angular distribution can be easily obtained by the use of equality

$$\int_0^\infty \frac{\sin^2\pi\nu N}{(n-\nu)^2}d\omega = \frac{2\pi^2\omega_0\gamma^2 N}{1+\psi^2}. \tag{5.157}$$

Integrating over the angles in Eq. (5.157) one gets the radiation spectrum

$$\frac{d\mathcal{E}_j}{d\omega} = \frac{d\mathcal{E}_{1j}}{d\omega} + \frac{d\mathcal{E}_{2j}}{d\omega}, \tag{5.158}$$

where

$$\begin{aligned} \frac{d\mathcal{E}_{1\sigma,\pi}}{d\omega} &= \frac{e^2\pi N\gamma^2\xi}{2c}\left[4\gamma^2(a_x-a_0)^2\xi(1-\xi)+a_{z,y}^2(\xi^2-2\xi+1)+\right. \\ &+ \left. a_{y,z}^2(3\xi^2-2\xi+1)\right]\Theta(N,1-\xi), \\ \frac{d\mathcal{E}_{2\sigma,\pi}}{d\omega} &= \frac{e^2\pi N\gamma^2\xi}{4c}\left[4\gamma^2(b_x-b_0)^2\xi(2-\xi)+b_{z,y}^2(\xi^2-4\xi+4)+\right. \\ &+ \left. b_{y,z}^2(3\xi^2-4\xi+4)\right]\Theta(N,2-\xi). \end{aligned} \tag{5.159}$$

Here $\xi = \omega/2\omega_0\gamma^2$ is the reduced frequency and the function $\Theta(N, n-x)$ is defined by Eq. (5.80). As can be seen from Eqs (5.158) and (5.159) in the region of frequencies $\xi \le 1$ the spectral density of the radiation is a superposition of the first and the second harmonics and depends on a_i and b_i, whereas in the region $\xi > 1$ the spectrum is determined by the amplitudes b_i only.

Finally, let us obtain the total radiated energy. When integrated over the frequencies in the $1/N$-approximation the function $U_n(x)$ can be replaced by a step-function. Then we get

$$\mathcal{E}_{1\sigma,\pi} = \frac{2\pi e^2 N\omega_0\gamma^4}{3c}\left[\frac{1}{2}\gamma^2(a_x - a_0)^2 + \frac{1}{8}a_{z,y}^2 + \frac{7}{8}a_{y,z}^2\right],$$
$$\mathcal{E}_{2\sigma,\pi} = \frac{16\pi e^2 N\omega_0\gamma^4}{3c}\left[\frac{1}{2}\gamma^2(b_x - b_0)^2 + \frac{1}{8}b_{z,y}^2 + \frac{7}{8}b_{y,z}^2\right]. \tag{5.160}$$

It can be seen from Eqs (5.159) and (5.160) that in the dipole approximation the spectral and angular distributions break up in the sums of the corresponding characteristics for single oscillators performing oscillations in the direction of the x-, y- or z-axis. Particularly, the radiation spectrum of the first harmonic is composed of the spectrum of the longitudinal oscillator which is described by the function $\xi^2(1-\xi)$ and the spectrum of the transverse oscillator of the well known shape $\xi(2\xi^2 - 2\xi + 1)$ (here the summation is performed over the polarization components). Note that the degree of polarization, for example, for the first harmonic,

$$P = \frac{|\mathcal{E}_{1\sigma} - \mathcal{E}_{1\pi}|}{\mathcal{E}_{1\sigma} + \mathcal{E}_{1\pi}} = \frac{3}{4}\frac{|a_z^2 - a_y^2|}{\gamma^2(a_x - a_0)^2 + a_y^2 + a_z^2} \tag{5.161}$$

depends on the correlation between the oscillation amplitudes. It is maximal and equal to the degree of polarization of synchrotron radiation ($P = 3/4$), when $a_x - a_0 = 0$ and a_y or a_z vanishes. One can always make the components a_y or a_z be equal to zero by a corresponding turn of the coordinate system, i.e. by the choice of the polarization axes. Thus, only the dependence on $a_x - a_0$ is essential. Similarly, the polarization of the second harmonic depends on the amplitudes b_i.

5.3.2. An Undulator with the Sinusoidal Magnetic Field

One of the important special cases of the above particle motion is the motion in the magnetic field $\boldsymbol{H}(\boldsymbol{r}) = (0, 0, H\sin(2\pi x/\lambda_0))$. This field is a good approximation of the magnetic field realized in the midplane of the linearly polarized undulator shown in Fig. 5.1. The solution of equations of motion is given by Eq. (5.11). Comparing with Eq. (5.139) we obtain for the reduced amplitudes

$$a_x = a_z = a_0 = b_y = b_z = b_0 = 0, \quad a_y = \sqrt{2}\beta_{\parallel}k\gamma^{-1}, \quad b_x = -\beta_{\parallel}k^2\gamma^{-2}/4.$$

We write the spectral and angular distribution in the approximation of high energy of the particle. Using Eqs (5.155), (5.43) and (5.152) we obtain ($\psi = \gamma\theta$)

$$\frac{d\mathcal{E}_\sigma}{d\Omega d\omega} = \frac{2e^2\gamma^2}{c(1+k^2+\psi^2)^2}\sum_{n=1}^{\infty} n^2 I_n(\nu,N)\left\{\sqrt{2}\psi\cos\phi J_n(np,nq) - k[J_{n+1}(np,nq)+J_{n-1}(np,nq)]\right\}^2, \quad (5.162)$$

$$\frac{d\mathcal{E}_\pi}{d\Omega d\omega} = \frac{4e^2\gamma^2\psi^2\sin^2\phi}{c(1+k^2+\psi^2)^2}\sum_{n=1}^{\infty} n^2 I_n(\nu,N) J_n^2(np,nq), \quad (5.163)$$

where

$$p = \frac{2\sqrt{2}k\psi\cos\phi}{1+k^2+\psi^2}, \quad q = -\frac{k^2}{2(1+k^2+\psi^2)}. \quad (5.164)$$

Let us consider the on-axis brightness of radiation. Putting $\psi = 0$ in Eqs (5.162) and (5.163) we find that the π-component is equal to zero and the σ-component takes the form

$$\frac{d\mathcal{E}_\sigma}{d\Omega d\omega} = \frac{e^2\gamma^2}{\pi^2 c}\sum_{n=1}^{\infty}[1-(-1)^n]\frac{\sin^2\pi\nu N}{(n-\nu)^2} I_n(\kappa), \quad (5.165)$$

where

$$I_n(\kappa) = \frac{k^2 n^2}{(1+k^2)^2}\left[J_{\frac{n-1}{2}}(\frac{1}{2}n\kappa^2) - J_{\frac{n+1}{2}}(\frac{1}{2}n\kappa^2)\right]^2,$$

$\kappa = k/\sqrt{1+k^2}$. We see that in this direction the σ-component contains only odd harmonics. The energy of radiation in the maximum of nth harmonic is given by the expression

$$\frac{d\mathcal{E}_\sigma}{d\Omega d\omega} = \frac{2e^2\gamma^2 N^2}{c} I_n(\kappa), \quad (5.166)$$

The function $I_n(\kappa)$ for the first three harmonics is plotted in Fig. 5.7.

One can see that at small k only the first harmonic is represented. The other harmonics appear as k increases. Each harmonic has its maximum at certain k. With a further increase of k the brightness drops because of an increase of the solid angle in which the energy is spread. Though the total energy emitted in a definite harmonic increases, the energy per unit solid angle decreases with k and tends to zero when $k \to \infty$

In order to obtain the angular distribution of polarization components of the radiation, we integrate over time the instant power of radiation given by Eq. (5.56)

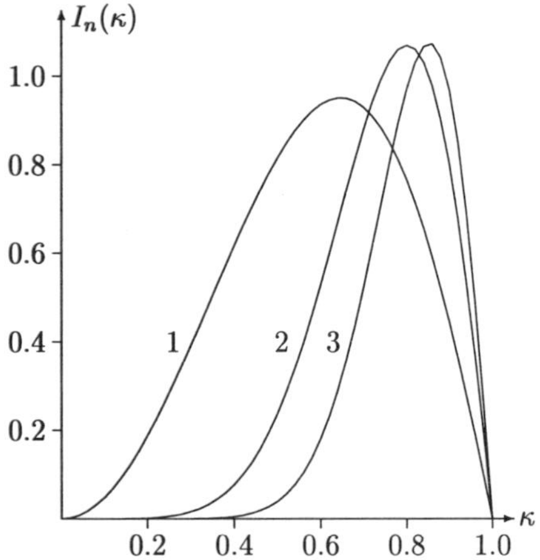

Fig. 5.7. Spectral brightness of the on–axis radiation versus κ for the first (1), third (2) and fifth (3) harmonics.

$$\frac{d\mathcal{E}_i}{d\Omega} = \frac{e^2}{4\pi c}\int_0^{NT} \frac{\left[(\boldsymbol{e}_i\boldsymbol{\beta})(\boldsymbol{n}\dot{\boldsymbol{\beta}}) + (\boldsymbol{e}_i\dot{\boldsymbol{\beta}})(1-\boldsymbol{\beta}\boldsymbol{n})\right]^2}{(1-\boldsymbol{\beta}\boldsymbol{n})^5}dt. \tag{5.167}$$

The direction of the vector $\boldsymbol{n}$ is given by the angles θ_z and θ_y between the vector and the planes xy and xz, respectively. For the ultra-relativistic charge the angles θ_z and θ_y are small. Expanding Eq. (5.167) in a power series in θ_z, θ_y and γ^{-1} and using the following designations $\psi_y = \gamma\theta_y$, $\psi_z = \gamma\theta_z$, we obtain:

$$\frac{d\mathcal{E}_\sigma}{d\Omega} = 4\mathcal{E}_0 k^2(I_3 - 4\eta^2 I_4 + 4\eta^4 I_5), \quad \frac{d\mathcal{E}_\pi}{d\Omega} = 16\mathcal{E}_0 k^2\psi_z^2(I_4 - \eta^2 I_5), \tag{5.168}$$

where $\eta^2 = 1 + \psi_z^2$,

$$I_n = \frac{1}{\pi}\int_0^{\pi} \frac{\sin^2 x\, dx}{[(k\cos x - \psi_y)^2 + \eta^2]^n}, \qquad \mathcal{E}_0 = \frac{e^2\gamma^4\omega_0 N}{c}.$$

The integrals I_n can be calculated by differentiating the equality

$$\frac{1}{\pi}\int_0^{\pi} \frac{dx}{q\cos^2 x + r\cos x + s} = \frac{\tan\mu}{\left(r^2 - 2qr\tan\frac{\mu}{2}\right)^{1/2}}, \quad \sin\mu = \frac{r}{q+s}$$

with respect to q or s.

Omitting simple but cumbersome details of integration, we can write down the result in the following form:

$$\begin{aligned} I_3 &= \frac{a}{4g^3}[(3b-1)g+2u], \\ I_4 &= \frac{a}{8g^5}[(5b^2-4b-1)g^2+2(3b-1)ug+4u^2], \\ I_5 &= \frac{a}{64g^7}[(35b^3-45b^2-3b+5)g^3+4(15b^2-12b-5)ug^2+ \\ & \quad 20(3b-1)u^2g+40u^3], \end{aligned} \tag{5.169}$$

where

$$u = 1+\psi^2+k^2, \quad g=\sqrt{(1+\psi^2-k^2)^2+4\eta^2k^2},$$
$$\psi^2=\psi_y^2+\psi_z^2, \quad a=\frac{b^{3/2}}{u+g}, \quad b=\left(1-\frac{2k^2}{u+g}\right)^{-1}.$$

The graphical display of the radiation pattern was considered in Ref.[12], where the numerical integrations of Eq. (5.167) have been made and the angular distribution of the total emitted mean power has been plotted. One can find more detailed information on the angular distribution in Ref.[39].

5.3.3. *Radiation in a Small k Limit*

In the case of small k, we can expand Eqs (5.162) — (5.164) in a power series using Eqs (5.149). One should take into account that $p \sim k$ but $q \sim k^2$. Let us rewrite the dependence on k in an explicit form. In doing so we introduce $p = kP$ and $q = k^2Q/2$. Keeping only the lowest degree of k we obtain from Eq. (5.149)

$$\begin{aligned} J_n(pn,qn) &= \left(\frac{k}{2}\right)^n \sum_{m=0}^{[n/2]} \frac{(nP)^{n-2m}(nQ)^m}{(n-2m)!m!}, \\ J_{n+1}(pn,qn)+J_{n-1}(pn,qn) &= \left(\frac{k}{2}\right)^{n-1} \sum_{m=0}^{n/2} \frac{(P_n)^{n-2m-1}(Qn)^m}{(n-2m-1)!m!}. \end{aligned}$$

Substituting these expressions in Eqs (5.162) and (5.163) we obtain the angular and spectral distribution of radiation in nth harmonic in the first approximation.

$$\left(\frac{d\mathcal{E}_{\sigma,\pi}}{d\Omega d\omega}\right)_n = \frac{4e^2\gamma^2\psi^2}{c(1+\psi^2)^2} I_n(\nu,N) a_{\sigma,\pi}^2 k^{2n}, \tag{5.170}$$

where

$$a_\sigma = \cos\phi \sum_{m=0}^{[(n+1)/2]} (n+2m+1)b_{mn},$$

$$a_\pi = \sin\phi \sum_{m=0}^{[n/2]} (n - 2m + 1) b_{mn},$$

$$b_{mn} = \frac{(-1)^m n^{n-m+1} (\sqrt{2}\,\psi\cos\phi)^{n-2m}}{2^{2m}(1+\psi^2)^{n-m} m!(n-2m+1)!}.$$

Let us study in detail the angular distribution of radiation in the first three harmonics [40,41]. Setting in Eq. (5.170) $n = 1, 2, 3$ and using the formula of Eq. (5.35) we arrive at

$$\frac{d\mathcal{E}_{jn}}{d\Omega d\omega} = \frac{\sin^2 \pi\nu N}{(n-\nu)^2} \frac{(1+\psi^2)\lambda_0}{4\pi^3\gamma^2 Nc} \frac{d\mathcal{E}_{jn}}{d\Omega}, \quad j = \sigma, \pi, \tag{5.171}$$

where $d\mathcal{E}_{jn}/d\Omega$ is the angular distribution of nth harmonic:

$$\frac{d\mathcal{E}_{\sigma 1}}{d\Omega} = K_1 (1 - \psi^2 \cos 2\phi)^2, \tag{5.172}$$

$$\frac{d\mathcal{E}_{\pi 1}}{d\Omega} = K_1 \psi^4 \sin^2 2\phi, \tag{5.173}$$

$$\frac{d\mathcal{E}_{\sigma 2}}{d\Omega} = 2K_2 \psi^2 \cos^2\phi\,[8\psi^2\cos^2\phi - 5(1+\psi^2)]^2, \tag{5.174}$$

$$\frac{d\mathcal{E}_{\pi 2}}{d\Omega} = 2K_2 \psi^2 \sin^2\phi\,[8\psi^2\cos^2\phi - 1 - \psi^2]^2; \tag{5.175}$$

$$\frac{d\mathcal{E}_{\sigma 3}}{d\Omega} = \frac{3^4}{2^4} K_3 \left[24\psi^4\cos^4\phi - 18\psi^2\cos^2\phi\,(1+\psi^2) + \right. \tag{5.176}$$

$$\left. +(1+\psi^2)^2\right]^2, \tag{5.177}$$

$$\frac{d\mathcal{E}_{\pi 3}}{d\Omega} = \frac{3^6}{2^4} K_3 \psi^4 \sin^2 2\phi\,[4\psi^2\cos^2\phi - 1 - \psi^2]^2, \tag{5.178}$$

with

$$K_n = \frac{8\pi e^2 \gamma^4 k^{2n} N}{\lambda_0 (1+\psi^2)^{2n+3}}. \tag{5.179}$$

The expression for K_n shows that the main energy is emitted in the first harmonic, and in the case of small k the higher harmonics are negligibly small. But the radiation of different harmonics is separated in frequencies. Thus, one can observe, for example, the third harmonic at a frequency $\omega_3 = 3\omega_1$ without interference from the first or second harmonic.

5.3.4. Radiation in a Large k Limit

Radiation in an undulator with sinusoidal magnetic field in a large k limit has been discussed in Refs[42,12,43].

Let $k \gg 1$ but the transverse velocity of the charged particle is small: $\beta_\perp \ll 1$. Combination of these conditions gives an inequality

$$1 \ll k \ll \gamma.$$

The spectral and angular distribution of polarization components in this case are given by Eq. (5.98) where the reduced acceleration $r^{(2)} = |\ddot{\vec{r}}|/\omega_0^2$ can be calculated by the use of Eq. (5.11). The acceleration $|\ddot{\vec{r}}|$ is to be calculated at point 1 or 2 according to Fig. 5.6. Thus, it is a function of the angle θ_y. As far as θ is small we have

$$\theta_y = \frac{dy}{dx} = -\sqrt{2}k\gamma^{-1}\sin\omega_0 t. \tag{5.180}$$

Obviously, $y_1 = -y_2$, hence $\cos\omega_0 t_1 = -\cos\omega_0 t_2$, $\sin\omega_0 t_1 = \sin\omega_0 t_2$ and thus θ_y is related with t_1 and t_2 as follows:

$$\sin\omega_0 t_{1,2} = -\alpha, \quad \cos\omega_0 t_{1,2} = \mp\sqrt{1-\alpha^2}, \tag{5.181}$$

where

$$\alpha = \frac{\theta_y\gamma}{\sqrt{2}k}. \tag{5.182}$$

We find the acceleration at points 1 and 2 to be

$$\ddot{r} = \sqrt{2}ck\omega_0\gamma^{-1}\sqrt{1-\alpha^2}. \tag{5.183}$$

Substituting it into Eq. (5.98) for the spectra-angular distribution of the wiggler radiation energy, we obtain

$$\begin{aligned} \frac{d\mathcal{E}_\sigma}{d\Omega d\omega} &= \mathcal{E}_0(1+\psi_z^2)\sin^2\chi K_{2/3}^2(\omega'), \\ \frac{d\mathcal{E}_\pi}{d\Omega d\omega} &= \mathcal{E}_0\psi_z^2\cos^2\chi K_{1/3}^2(\omega'), \end{aligned} \tag{5.184}$$

where

$$\begin{aligned} \mathcal{E}_0 &= \frac{2e^2\omega^2(1+\psi_z^2)}{3\pi^2 c\gamma^2 k^2\omega_0^2(1-\alpha^2)}\frac{\sin^2\pi\nu N}{\sin^2\pi\nu}, \\ \omega' &= \frac{\omega(1+\psi_z^2)^{3/2}}{3\sqrt{2}\gamma^2 k\omega_0\sqrt{1-\alpha^2}}, \\ \chi &= \frac{\omega}{2}(t_2 - t_1 - \frac{\vec{n}\vec{l}}{c}), \quad \nu = \frac{\omega}{\omega_0}(1-\beta_\| n_x). \end{aligned} \tag{5.185}$$

Expressions of Eq. (5.184) contain products of factors $(\sin^2 \pi\nu N)/(\sin^2 \pi\nu)$, $\sin^2 \chi$, $\cos^2 \chi$ which vary fast, and slowly varying functions of θ_y, θ_z and ω. Let us first treat a slowly varying part of Eq. (5.184)

The argument of the Bessel functions $K_{1/3}(z)$, $K_{2/3}(z)$ shows that in the trajectory plane the angular distribution is limited by directions $\alpha = \pm 1$, i.e. $\theta_y = \pm k\gamma^{-1}\sqrt{2}$. In accordance with Eq. (5.181) these directions coincide with the maximum deviations of the velocity vector from the axis of the trajectory. In the xz plane the main part of the energy is emitted within the limits of the angle of the order of γ^{-1} in magnitude. The frequency at which the radiation spectrum has its maximum is of the order of

$$\omega \sim \frac{\gamma^2 k\sqrt{1-\alpha^2}}{(1+\psi_z^2)^{3/2}}\omega_0 \tag{5.186}$$

and varies from zero for directions $\alpha = \pm 1$ up to $\omega_m \sim \gamma^2 k\omega_0$ for the x axis direction. For the radiation along the wiggler axis the Eqs (5.184) yield

$$\frac{d\mathcal{E}_\sigma}{d\Omega d\omega} = \frac{12e^2\gamma^2}{\pi^2 c}\frac{\sin^2 \pi\nu N \sin^2 \frac{\pi\nu}{2}}{\sin^2 \pi\nu}\omega'^2 K_{2/3}^2(\omega'), \qquad \frac{d\mathcal{E}_\pi}{d\Omega d\omega} = 0. \tag{5.187}$$

We see that in this direction the radiation spectrum includes only odd harmonics. A maximum of spectrum corresponds to $\omega' \approx 0.42$, or $\omega_m \approx 1.8\gamma^2 k\omega_0$.

Let us investigate the properties of radiation which are due to the factors

$$\frac{\sin^2 \pi\nu N}{\sin^2 \pi\nu}\left\{\begin{array}{c}\sin^2\chi \\ \cos^2\chi\end{array}\right\}. \tag{5.188}$$

Expanding Eqs (5.185) in powers of small γ^{-1}, k^{-1}, $k\gamma^{-1}$, we have

$$\nu = \frac{\omega k^2}{2\omega_0\gamma^2}(1+2\alpha^2). \tag{5.189}$$

Using Eqs (5.181) 5 we obtain

$$\begin{aligned}\chi &= \frac{\omega k^2}{2\omega_0\gamma^2}\Phi(\alpha),\\ \Phi(\alpha) &= (1+2\alpha^2)\arccos\alpha - 3\alpha\sqrt{1-\alpha^2}.\end{aligned} \tag{5.190}$$

As it was mentioned above, the maximum of the radiation spectrum lies in the region $\omega \sim \gamma^2 k\omega_0$, hence the numbers ν of the harmonics, as well as the parameter χ, are of the order of k^3 (except for the boundaries of the radiation cone at which $1-\alpha^2 \ll 1$). Note that we have neglected the terms in the expressions of Eqs (5.189) and (5.190) which are small in comparison with the remaining ones, but much greater than unity. For example, in order to calculate $\sin \pi\nu$ or $\sin \chi$ we should use the expressions of Eqs (5.185) which are more correct than Eqs (5.189) and (5.190). However, we can use Eqs (5.189) and (5.190) to analyze ν and χ versus angles, frequencies and the parameter k.

We turn now to the spectrum of radiation. The factor $\sin^2 \pi\nu N/\sin^2 \pi\nu$ cuts out harmonics of the widths $\Delta\nu \sim 1/N$ from the spectrum. In the main part of the spectrum the relative width of the harmonic is $\Delta\omega/\omega = \Delta\nu/\nu \sim (Nk^3)^{-1}$ and the relative distance between the harmonics is $\Delta\omega/\omega \sim k^{-3}$. The functions $\sin^2 \chi$ and $\cos^2 \chi$ modulate the spectrum with a period of the same order. Thus, the factors $\sin^2 \pi\nu N/\sin^2 \pi\nu$, $\sin^2 \chi$ and $\cos^2 \chi$ vary fast with respect to the functions $K_{1/3}(\omega')$ and $K_{2/3}(\omega')$. Note that the maxima and minima of the σ-component and those of the π-component alternate with each other in the spectral distribution. This means that in any given direction we have neighboring frequencies at which either the σ- or the π-component is emitted.

The effect of the factor $\sin^2 \pi\nu N/\sin^2 \pi\nu$ on the structure of the angular distribution has been investigated in detail in Sec. 5.1.2 and we shall not discuss it.

The direction dependence of factors $\sin^2 \chi$ and $\cos^2 \chi$ is governed by the function $\Phi(a)$ which is defined in Eq. (5.190). The function monotonously decreases as α increases, having the values 3π and 0 for $\alpha = -1$ and $\alpha = +1$, respectively. Directions for which $\sin^2 \chi$ or $\cos^2 \chi$ turn out to be zero or maximum are defined by the equation $\chi = \pi l$, where $l = 0, 1, 2, \ldots, l_m$. The number l_m is the greatest integer satisfying the condition

$$l_m \leq \frac{3\omega k^2}{2\omega_0 \gamma^2} \sim k^3. \tag{5.191}$$

The χ does not depend on θ_z, hence these directions in this approximation actually form the planes which include the axis z.

We see from Eq. (5.190) that the spectral-angular distribution is not symmetric with respect to the xz plane. This asymmetry is related with the positions of points 1 and 2 which are different for $a_y < 0$ and for $a_y > 0$ in a finite wiggler. However, when $N \to \infty$ the asymmetry vanishes. Indeed, we can rewrite Eq. (5.190) in the following equivalent form

$$\chi = \nu \left(\frac{\pi}{2} - \arcsin \alpha - \frac{3\alpha\sqrt{1-\alpha^2}}{1+2\alpha^2} \right). \tag{5.192}$$

When $N \to \infty$ we obtain the radiation emitted at frequencies corresponding to the integral ν.

$$\lim_{N\to\infty} \frac{\sin^2 \pi\nu N}{N \sin^2 \pi\nu} = \sum_{n=-\infty}^{\infty} \delta(\nu - n). \tag{5.193}$$

As follows from Eq. (5.192) the functions $\sin^2\chi$ and $\cos^2\chi$, in the case of the integral ν are even with respect to α. Of course, the radiation with integral ν does not depend on a sign of α at any N, but the point is that with a small N the radiation at nonintegral ν makes a significant contribution to the total spectrum.

Moreover, the maximum of the harmonics are not centered at the integral ν.

We can average the spectrum over short intervals containing many harmonics since the spacing between harmonics is much less than the whole radiation band. Using Eqs (5.105) and (5.106) we finally obtain the spectral-angular distribution

$$\begin{aligned}
\frac{d\mathcal{E}_\sigma}{d\Omega d\omega} &= \mathcal{E}'(1+\psi_z^2)K_{2/3}^2(\omega'),\\
\frac{d\mathcal{E}_\pi}{d\Omega d\omega} &= \mathcal{E}'\psi_z^2)K_{1/3}^2(\omega'),\\
\mathcal{E}' &= \frac{e^2\omega^2(1+\psi_z^2)N}{3\pi^2 c\gamma^2 k^2\omega_0^2(1-\alpha^2)}
\end{aligned} \tag{5.194}$$

which differs from synchrotron radiation distribution only in the dependence on the angle θ_y. If we replace the uniform magnetic field by a local $H_\alpha = H_0\sqrt{1-\alpha^2}$ in the vicinity of a radiating point in synchrotron radiation formulae, then the latter are transformed into Eq. (5.194).

The formulae of Eq. (5.194) are asymptotic expressions for the spectral-angular distribution in the case of $k \gg 1$. The relative error is about k^{-2}.

In order to find the angular distribution, we integrate Eq. (5.194) over the frequency, using the formulae

$$\begin{aligned}
\int_0^\infty \nu^2 K_{1/3}^2[\frac{\nu}{3}(1+\psi_z^2)^{3/2}]d\nu &= \frac{15}{16}\frac{\pi^2}{(1+\psi_z^2)^{9/2}},\\
\int_0^\infty \nu^2 K_{2/3}^2[\frac{\nu}{3}(1+\psi_z^2)^{3/2}]d\nu &= \frac{21}{16}\frac{\pi^2}{(1+\psi_z^2)^{9/2}}.
\end{aligned} \tag{5.195}$$

After the integration we have

$$\frac{d\mathcal{E}_\sigma}{d\Omega} = \frac{21\gamma^2\mathcal{E}\sqrt{1-\alpha^2}}{16\sqrt{2}k(1+\psi_z^2)^{5/2}}, \quad \frac{d\mathcal{E}_\pi}{d\Omega} = \frac{15\gamma^2\mathcal{E}\sqrt{1-\alpha^2}}{16\sqrt{2}k(1+\psi_z^2)^{7/2}}, \tag{5.196}$$

where

$$\mathcal{E} = \frac{4e^2\gamma^2k^2\omega_0 N}{3c} \tag{5.197}$$

is the total energy emitted by a particle during one passage through the wiggler.

The dependence of the radiation power on the angle ψ_z is the same as that for synchrotron radiation. In the orbital plane the radiation is bounded by directions $\theta_y = \pm\sqrt{2}\gamma^{-1}k$.

Strictly speaking the boundaries of the radiation cone at $\alpha = \pm 1$ cannot be so abrupt as it follows from Eq. (5.196). According to the nature of the angular distribution, it rather tails off within an angle of the order of γ^{-1}. This is because the method used here gives only the first term of asymptotic expansion, which is proportional to k as indicated by Eq. (5.196). A higher approximation can be obtained from the exact formula of Eq. (5.168) in the limit of $k \to \infty$. The structure of Eq. (5.168) allows two different limiting cases:

(1) the range of θ_y lies between the boundary directions $\theta_y = \pm k\gamma^{-1}$, $|\theta_y| \sim k\gamma^{-1}$, $k\gamma^{-1} - |\theta_y| \gg \gamma^{-1}$;

(2) angle θ_y varies in the vicinity of the boundary direction: $k\gamma^{-1} - |\theta_y| \sim \pm\gamma^{-1}$.

Let us begin with case (1). Turning to the limit $k \to \infty$ at a fixed $\psi_y/k \le 1$ in Eqs (5.168), we obtain Eqs (5.196). In case (2) with $|\psi_y| \sim k$ we denote $\delta = |\psi_y| - k$ and turn to the limit $k \to \infty$, with δ and $\psi_z \sim 1$. Then from Eq. (5.168) we obtain:

$$\begin{aligned}
&\frac{d\mathcal{E}_\sigma}{d\Omega} = K[1 + 3q - (1 + \psi_z^2)Q], \qquad \frac{d\mathcal{E}_\pi}{d\Omega} = K\psi_z^2 Q, \\
&Q = \frac{2}{\mu^2}(1 + 3q + 5q^2) - \frac{5(1 + \psi_z^2)}{4\mu^4}(1 + 3q + 6q^2 + 7q^3), \\
&q = \frac{\mu}{2(\mu + \delta)}, \quad \mu = \sqrt{1 + \psi_z^2 + \delta^2}, \\
&K = \frac{e^2\gamma^4 N\omega_0}{4c}\sqrt{\frac{q^3 k}{2\mu}}.
\end{aligned} \tag{5.198}$$

Eqs (5.198) make it possible to describe the radiation in the neighborhood of directions $|\psi_y| = k$, where it increases as the square root of k. More detailed analysis can be found in Ref.[39].

In order to calculate the total output spectrum, we integrate Eq. (5.194) over the angles. The integral over ψ_z is transformed into the integral of $K_{5/3}$ in the same way as in the case of synchrotron radiation:

$$\frac{d\mathcal{E}}{d\omega} = \frac{\sqrt{6}e^2 Nk}{\pi c}\int_{-1}^{1} d\alpha\, y \int_y^\infty K_{5/3}(x)dx, \tag{5.199}$$

where

$$y = \frac{\xi}{\sqrt{1 - \alpha^2}}, \quad \xi = \frac{\sqrt{2}\omega}{3\omega_0\gamma^2 k}. \tag{5.200}$$

Inverting the order of integration we have

$$\frac{d\mathcal{E}}{d\omega} = \frac{4\sqrt{2}\pi e^2 kN}{9c} F(\xi), \tag{5.201}$$

where

$$F(\xi) = \frac{9\sqrt{3}}{2\pi^2}\xi \int_\xi^\infty K_{5/3}(x) \arccos \frac{\xi}{x} dx. \tag{5.202}$$

The function $F(\xi)$ is normalized to unity:

$$\int_0^\infty F(\xi) d\xi = 1. \tag{5.203}$$

The change of variables in the second integral in Eq. (5.199) leads to an equivalent expression for $F(\xi)$:

$$F(\xi) = \frac{9\sqrt{3}}{4\pi^2}\xi \int_\xi^\infty K_{5/6}^2(\frac{x}{2}) dx. \tag{5.204}$$

The function $F(\xi)$ is plotted in Fig. 5.8 Its maximum occurs at $x = 0.20$.

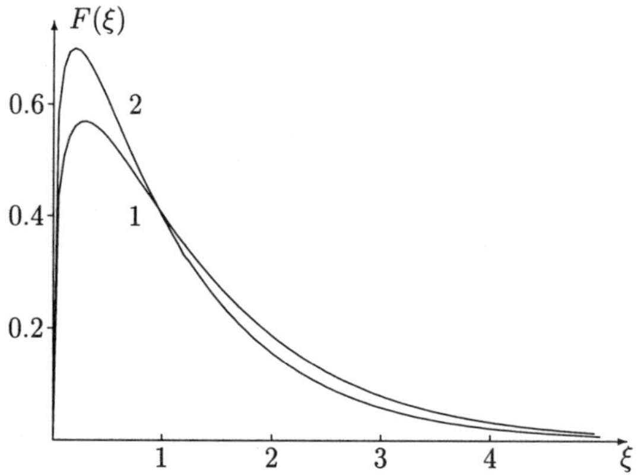

Fig. 5.8. Total output spectra of synchrotron (1) and wiggler (2) radiation.

As it follows from Eqs (5.200), the frequency corresponding to the maximum of radiation is proportional both to the magnetic field in the wiggler and to the particle energy. For comparison the synchrotron radiation spectrums is plotted as a function

$$F_s(\xi) = \frac{9\sqrt{3}}{8\pi}\xi \int_{\xi}^{\infty} K_{5/3}(x)dx, \quad \int_{0}^{\infty} F_s(\xi)d\xi = 1. \tag{5.205}$$

Since both of the spectra are normalized to the same value, this means that the synchrotron spectrum corresponds to the radiation in a homogeneous magnetic field H_s equal to the mean-square of the wiggler field H along the trajectory: $H_s = \sqrt{\langle H^2 \rangle}$.

Now we shall investigate both the low- and high-frequency asymptotic of the wiggler spectrum. In order to calculate the integral of Eq. (5.204) for $\xi \ll 1$, we transform it as follows:

$$\int_{\xi}^{\infty} K_{5/6}^2(\frac{x}{2})dx = -\frac{3}{5}\int_{\xi}^{\infty} x[K_{5/6}(\frac{x}{2})K_{1/6}(\frac{x}{2}) + K_{5/6}(\frac{x}{2})K'_{5/6}(\frac{x}{2})]dx. \tag{5.206}$$

When $\xi \to 0$ the integral of the first term is finite but that of the second term diverges. Neglecting the first term and integrating by parts the second one we find

$$F(\xi) = \frac{3^3\sqrt{3}}{2^3\pi^2}\xi^2 K_{5/6}^2(\frac{\xi}{2}). \tag{5.207}$$

We can use approximate expressions for the modified Bessel functions in the region of small arguments [9]

$$K_\nu(x) = \frac{2^{\nu-1}}{x^\nu}\Gamma(\nu), \tag{5.208}$$

then we have for $\xi \ll 1$

$$F(\xi) = \frac{3^4\sqrt{3}}{(2\pi)^3}\left[\Gamma(\frac{2}{3})\right]^4 \xi^{1/3}. \tag{5.209}$$

Thus, in the low-frequency range of both the synchrotron and the wiggler spectrum we have the same law $\omega^{1/3}$ of variation of the intensity, but with a greater factor in the later case. The reduced frequency ξ cannot be as small as desired for the reasons that the numbers ν of the harmonics are large as it was assumed above. Taking into consideration that $\xi \sim \nu/k^3$ we find the region of the validity of Eq. (5.209): $k^{-3} \ll \xi \ll 1$.

In order to obtain the high-frequency asymptotics of the spectrum, we use the approximate expression [9]

$$K_\nu(x) = \sqrt{\frac{\pi}{2x}} e^{-x}, \quad x \gg 1. \tag{5.210}$$

After integration in Eq. (5.204) in case $\xi \gg 1$ we have

$$F(\xi) = \frac{9\sqrt{3}}{4\pi} e^{-\xi} \tag{5.211}$$

5.4. Radiation of a Harmonic Oscillator

5.4.1. General Theory

The particle motion in an undulator is defined by the electric and magnetic fields. There is a variety of fields in which a charged particle executes harmonic oscillations with a drift of a constant velocity. This kind of motion can be realized in such fields as an uniform magnetic field, uniform crossed electric and magnetic fields [1], specific kinds of nonuniform static fields [44] and others. Besides, in many cases, it is possible to express an approximate solution of the equation of motion in terms of trigonometric functions. Radiation of a particle engaged in harmonic oscillations has been investigated in some specific cases. There are many books and papers where the radiation of a helically moving charge is described in detail (see, for instance [45,46,47,48,21] and Chapt. 1 of this book). Radiation of a one- or two-dimensional harmonic oscillator has been studied at the beginning of the century (see Ref.[49,50]). In this section we derive general expressions for spectral and angular distribution of the radiation of an arbitrary harmonic oscillator moving with a constant drift velocity. Then some particular cases are studied.

The most general case of harmonic oscillations in a four-dimensional form can be described by equation

$$r^\mu = \frac{v_\parallel^\mu}{\omega_0}\tau + A^\mu \sin(\tau + \psi^\mu), \quad \mu = 0, 1, 2, 3, \tag{5.212}$$

where r^μ is the four-vector of the particle position; ω_0 is the frequency of oscillation; τ is the parameter; $v_\parallel^\mu = (c, \boldsymbol{v}_\parallel)$, $a^\mu = (a_t, \boldsymbol{a})$ and $\psi^\mu = (\psi_t, \psi_x, \psi_y, \psi_z)$ are four-component quantities; vector $\boldsymbol{v}_\parallel = c\boldsymbol{\beta}_\parallel$ stands for the constant drift velocity. Let x-axis of a coordinate system be aligned with the vector $\boldsymbol{v}_\parallel$ and the directions of y- and z-axes be chosen so that $\psi_y = \pi/2 - \psi_z$. Moreover, we can put $\psi_z = 0$ by properly choice of zero-point of the parameter τ. Thus, instead of Eq. (5.212) we write

$$\begin{aligned} x &= \frac{v_{\parallel}}{\omega_0}\tau + A_x \sin(\tau + \psi_x), \\ y &= A_y \cos\tau, \\ z &= A_z \sin\tau, \\ ct &= \frac{c}{\omega_0}\tau + A_t \sin(\tau + \psi_t). \end{aligned} \tag{5.213}$$

One can find the energy radiated per unit solid angle and unit band of frequency by the use of Eq. (5.49). In doing so we replace the variable t in Eq. (5.50) by τ. It results in

$$B_{jn} = \frac{1}{cT}\int_0^{2\pi} (\boldsymbol{e}_j \frac{d\boldsymbol{r}}{d\tau} e^{i(\omega_n t - \boldsymbol{k}_n \boldsymbol{r})} d\tau, \quad j = \sigma, \pi. \tag{5.214}$$

The phase of the exponent can be transformed into

$$\omega_n t - \boldsymbol{k}_n \boldsymbol{r} = n[\tau - \kappa \sin(\tau - \alpha)], \tag{5.215}$$

where

$$\kappa = \frac{A}{1 - \boldsymbol{\beta}_{\parallel}\boldsymbol{n}}. \tag{5.216}$$

The constants A and α are defined by equations

$$\begin{aligned} A\sin\alpha &= a_t \sin\psi_t - a_x n_x \sin\psi_x - a_y n_y, \\ A\cos\alpha &= -a_t \cos\psi_t + a_x n_x \cos\psi_x + a_z n_z, \end{aligned} \tag{5.217}$$

where n_i are the components of the unit vector $\boldsymbol{n}$ and $a_i = A_i\omega_0/c$. The constants A and $\sin\alpha$ are defined by Eqs (5.217) with an accuracy to a sign, but the sign of the product $A\sin\alpha$ is uniquely determined. In other words, Eqs (5.217) are invariant relative to transformations

$$\kappa \to -\kappa, \quad \alpha \to \alpha + \pi. \tag{5.218}$$

We can assume that $\kappa \geq 0$. Then the sign of $\sin\alpha$ is unambiguously defined by Eqs (5.217). The derivatives of the components $r_\sigma = (\boldsymbol{r}\boldsymbol{n}_\sigma)$ and $r_\pi = (\boldsymbol{r}\boldsymbol{n}_\pi)$ with respect to τ yield

$$\frac{2\pi}{cT}\frac{dr_j}{d\tau} = P_j + Q_j \cos(\tau - \alpha) + R_j \sin(\tau - \alpha).$$

Here P_j, Q_j and R_j are the functions of the angular coordinates:

$$\begin{aligned}
P_\pi &= \beta_{||}n_{\pi x}, \quad P_\sigma = \beta_{||}n_{\sigma x}, \\
Q_\pi &= a_x n_{\pi x}\cos(\alpha+\psi_x) + a_z n_{\pi z}\cos\alpha - a_y n_{\pi y}\sin\alpha, \\
Q_\sigma &= a_x n_{\sigma x}\cos(\alpha+\psi_x) - a_y n_{\sigma y}\sin\alpha, \\
R_\pi &= -a_x n_{\pi x}\sin(\alpha+\psi_x) - a_z n_{\pi z}\sin\alpha - a_y n_{\pi y} cos\alpha, \\
R_\sigma &= -a_x n_{\sigma x}\sin(\alpha+\psi_x) - a_y n_{\sigma y} cos\alpha.
\end{aligned} \tag{5.219}$$

The components of polarization vectors are to be substituted from Eqs (5.40) or (5.42) according to which a set of angles (Θ, Φ) or (θ, ϕ) is accepted. The integration of Eq. (5.214) can be easily performed using the integral representation of the Bessel functions [9], which gives

$$\frac{1}{2\pi}\int_{-\pi}^{\pi}\cos\tau\, e^{in(\tau - z\sin\tau)}d\tau = \frac{1}{z}J_n(zn), \quad \frac{1}{2\pi}\int_{-\pi}^{\pi}\sin\tau\, e^{in(\tau - z\sin\tau)}d\tau = iJ_n'(zn).$$

The prime denotes the derivative with respect to the argument zn. Putting the pieces together we obtain

$$B_{jn} = e^{in\alpha}\left[(P_j + \frac{1}{\kappa}Q_j)J_n(\kappa n) + iR_jJ_n'(\kappa n)\right]. \tag{5.220}$$

This equation shows that the radiation in each harmonic is, in general, elliptically polarized because the phase difference between the σ- and π-components has a nonzero value. Note that Eq. (5.220) is invariant with respect to the transformations given by Eq. (5.218).

The angular and spectral distribution of polarization components in accordance with Eq. (5.49) are equal to

$$\begin{aligned}
\frac{d\mathcal{E}}{d\Omega d\omega} &= \frac{e^2}{c(1-\boldsymbol{\beta}_{||}\boldsymbol{n})^2}\sum_{n=1}^{\infty} I_n(\nu, N)\left\{(P_j + \frac{1}{\kappa}Q_j)^2 J_n^2(\kappa n) + \right. \\
&+ \left. R_j^2 {J_n'}^2(\kappa n)\right\}.
\end{aligned} \tag{5.221}$$

In order to find the angular distribution, we integrate this equation over the frequencies. Using Eq. (5.32) and identities ??? [51]

$$\begin{aligned}
\sum_{n=1}^{\infty} n^2 J_n^2(\kappa n) &= \frac{\kappa^2(4+\kappa^2)}{16(1-\kappa^2)^{7/2}}, \\
\sum_{n=1}^{\infty} n^2 {J_n'}^2(\kappa n) &= \frac{4+3\kappa^2}{16(1-\kappa^2)^{5/2}}
\end{aligned} \tag{5.222}$$

we easily obtain

$$\frac{d\mathcal{E}_j}{d\Omega} = \frac{e^2\omega_0 N}{16c(1-\boldsymbol{\beta}_{\parallel}\boldsymbol{n})^3}\left\{(\kappa P_j + Q_j)^2\frac{4+\kappa^2}{(1-\kappa^2)^{7/2}} + R_j^2\frac{4+3\kappa^2}{(1-\kappa^2)^{5/2}}\right\} \tag{5.223}$$

The denominator in this expression cannot be equal to zero because $\kappa < 1$. Actually, it follows from Eq. (5.215) that

$$t - \frac{\boldsymbol{n}\boldsymbol{r}}{c} = \frac{1-\boldsymbol{\beta}_{\parallel}\boldsymbol{n}}{\omega_0}[\tau - \kappa\sin(\tau-\alpha)]. \tag{5.224}$$

Taking the derivative with respect to time we obtain

$$1 - \boldsymbol{\beta}\boldsymbol{n} = \frac{1-\boldsymbol{\beta}_{\parallel}\boldsymbol{n}}{\omega_0}\left(\frac{dt}{d\tau}\right)^{-1}[1-\kappa\cos(\tau-\alpha)]. \tag{5.225}$$

It is obvious that $1-\boldsymbol{\beta}_{\parallel}\boldsymbol{n} > 0$ and $1-\boldsymbol{\beta}\boldsymbol{n} > 0$. The derivative $dt/d\tau$ is positive too because the time function $t(\tau)$ is a monotone increasing one. Hence, $\kappa > 1$.

Now let us study a special case of oscillations with small amplitudes, namely $a_i \ll \gamma^{-1}$. This kind of motion is associated with the case of $k \ll 1$ studied in Sec. 5.2.1. But we cannot use here the results of this section because they are valid only for motion in a pure magnetic field while the motion given by Eq. (5.212) implies that the particle experiences an electric field. Alternatively, the condition $k \ll 1$ constrains the transversal component of acceleration which is not the only one in our case. We restrict our consideration to the type of motion with $\psi_x = \psi_t = \psi$. We also assume that the particle moves with ultrarelativistic speed. In this case, the angular coordinates defined by Eq. (5.43) are more convenient. We calculate first

$$(1-\boldsymbol{\beta}_{\parallel}\boldsymbol{n}) \approx \frac{1}{2}(\gamma^{-2} + \beta_{\perp}^2 + \theta^2).$$

It follows from Eq. (5.213) that

$$\beta_{\perp}^2 \approx a_y^2\sin^2\tau + a_z^2\cos^2\tau \ll \gamma^{-2}.$$

Thus,

$$(1-\boldsymbol{\beta}_{\parallel}\boldsymbol{n}) \approx \frac{1}{2}(\gamma^{-2} + \theta^2).$$

In the approximation of small γ^{-1} Eqs (5.217) yield

$$\begin{aligned}\kappa^2 &= \frac{4}{(\gamma^{-2}+\theta^2)^2}\left\{(a_x - a_t)^2 + 2\theta(a_x - a_t)(a_z\sin\phi\cos\psi + \right.\\ &+ \left. a_y\cos\phi\sin\psi) + a_y^2\theta^2\cos^2\phi + a_z^2\theta^2\sin^2\phi\right\}.\end{aligned} \tag{5.226}$$

At this point we assume that a_x is close to a_t so that $|a_x - a_t| \ll \gamma^{-2}$. The whole set of the adopted assumptions looks like these

$$\gamma \gg 1, \quad a_i \ll \gamma^{-1}, \quad |a_x - a_t| \ll \gamma^{-2}. \tag{5.227}$$

These conditions provide the smallness of the parameter κ.

After expansion of Eq. (5.221) up to the first order with respect to small κ only the basic harmonic remains

$$\frac{d\mathcal{E}_{j1}}{d\Omega d\omega} = \frac{e^2}{4\pi^2 c(1-\boldsymbol{\beta}\|\boldsymbol{n})^2} \frac{\sin^2 \pi\nu N}{(1-\nu)^2} \left\{(\kappa P_j + Q_j)^2 + R_j^2\right\} \tag{5.228}$$

Note, that we keep κ in the product κP_j because $P_j \gg Q_j$ as one can see from Eq. (5.219).

The angular distribution can be easily obtained from Eq. (5.223)

$$\frac{d\mathcal{E}_j}{d\Omega} = \frac{e^2\omega_0 N}{4c(1-\boldsymbol{\beta}_{\|}\boldsymbol{n})^3} \left\{(\kappa P_j + Q_j)^2 + R_j^2\right\}. \tag{5.229}$$

The spectrum of radiation can be found by integration over the entire sphere in Eq. (5.228)

$$\frac{d\mathcal{E}_j}{d\Omega} = \frac{4e^2\gamma^2}{\pi^2 c} \int\limits_0^\infty \frac{\psi d\psi}{(1+\psi^2)^2} \frac{\sin^2 \pi\nu N}{(1-\nu^2)^2} \int\limits_0^{2\pi} [(\kappa P_j + Q_j)^2 + R_j^2] d\phi. \tag{5.230}$$

Following the calculations which resulted in Eq. (5.79) we obtain

$$\begin{aligned} \frac{d\mathcal{E}_{\sigma,\pi}}{d\omega} &= \frac{\pi e^2\gamma^2\omega_0^2 N}{2c^3}\xi \left\{4\gamma^2(A_x - \beta_{\|}A_t)^2(1-\xi)\xi + \right. \\ &+ \left. \begin{pmatrix} A_z^2 \\ A_y^2 \end{pmatrix} (1-\xi)^2 + \begin{pmatrix} A_y^2 \\ A_z^2 \end{pmatrix} (1-2\xi+3\xi^2) \right\} \Theta(N, 1-\xi). \end{aligned} \tag{5.231}$$

Here the usual notation $\xi = \omega/2\gamma^2\omega_0$ is adopted. The sum over the polarization components gives the total spectrum

$$\begin{aligned} \frac{d\mathcal{E}}{d\omega} &= \frac{\pi e^2\gamma^2\omega_0^2 N}{c^3}\xi \left\{4\gamma^2(A_x - \beta_{\|}A_t)^2(1-\xi)\xi + \right. \\ &+ \left. (A_y^2 + A_z^2)(1-2\xi+2\xi^2)\right\} \Theta(N, 1-\xi). \end{aligned} \tag{5.232}$$

One can see that it is represented in the form of a sum of spectral distribution of the radiation of three independent oscillators vibrating in mutually orthogonal directions.

Finally, we obtain the total radiated energy

$$\mathcal{E} = \frac{2e^2\pi\omega_0^3\gamma^4 N}{3c^3}[(A_x - \beta_{\|}A_t)^2\gamma^2 + A_y^2 + A_z^2]. \tag{5.233}$$

In the next three sections we apply the obtained results to the description of the radiation emitted by a one-dimensional oscillator and a helically moving charged particle.

5.4.2. A Transversal Harmonic Oscillator

Consider a charged particle performing one-dimensional oscillations orthogonal to the direction of the drift velocity. The coordinates of the particle are governed by equations

$$x = v_{\|}t, \quad y = A_y \cos\omega_0 t, \quad z = 0 \tag{5.234}$$

which follow from Eq. (5.213) in the case of $A_x = A_z = A_t = 0$. The fields realizing this kind of motion have been obtained in Ref.[44]. Besides, the particle motion obeys the same law in an undulator with transversal fields which vary according to the equation $F = F_0 \sin 2\pi x/\lambda_0$ where F stands for an electric or a magnetic field. This oscillator has been used as the first model of the undulator [52]. Some of the properties of radiation have been investigated as early as 1912 by Schott [50]. A detailed analysis and the history of the problem can be found in Ref.[53].

We obtain the radiation properties from the general expressions of the previous section. Eqs (5.217) and (5.219) result in

$$\begin{aligned} \kappa &= \frac{a_y \sin\Theta\sin\Phi}{1 - \boldsymbol{\beta}_{\|}\boldsymbol{n}}, \quad 1 - \boldsymbol{\beta}_{\|}\boldsymbol{n} = 1 - \beta_{\|}\sin\Theta\cos\Phi, \\ Q_\pi &= a_y\cos\Theta\sin\Phi, \quad Q_\sigma = a_y\cos\Phi, \quad R_\pi = R_\sigma = 0. \end{aligned} \tag{5.235}$$

It follows herefrom that the radiation is linearly polarized because the amplitudes B_π and B_σ of Eq. (5.220) have the same phase. Substituting the latter equations in Eq. (5.221) we obtain

$$\begin{aligned} \frac{d\mathcal{E}_\sigma}{d\Omega d\omega} &= \frac{e^2}{c(1 - \boldsymbol{\beta}_{\|}\boldsymbol{n})^2}\sum_{n=1}^{\infty} n^2 I_n(\nu, N)\frac{(\cos\Phi - \beta_{\|}\sin\Theta)^2}{\sin^2\Theta\sin^2\Phi}J_n^2(\kappa n), \\ \frac{d\mathcal{E}_\pi}{d\Omega d\omega} &= \frac{e^2}{c(1 - \boldsymbol{\beta}_{\|}\boldsymbol{n})^2}\sum_{n=1}^{\infty} n^2 I_n(\nu, N)\cot^2\Theta\, J_n^2(\kappa n). \end{aligned} \tag{5.236}$$

We write the distribution of radiation summarized over the polarization components in the form invariant with respect to the system of angular coordinates

$$\frac{d\mathcal{E}}{d\Omega d\omega} = \frac{e^2}{c}\sum_{n=1}^{\infty} n^2 I_n(\nu, N)\frac{(1-\boldsymbol{\beta}_{\|}\boldsymbol{n})^2 - (1-\beta_{\|}^2)n_y^2}{n_y^2(1-\boldsymbol{\beta}_{\|}\boldsymbol{n})^2} J_n^2(\kappa, n). \tag{5.237}$$

The angular distribution follows from Eq. (5.223)

$$\frac{d\mathcal{E}_{\sigma,\pi}}{d\Omega} = \frac{e^2 A^2 \omega_0^2 N(4+\kappa^2)}{16c^3(1-\boldsymbol{\beta}_{\|}\boldsymbol{n})^5(1-\kappa^2)^{7/2}} \left\{ \begin{array}{l} (\cos\Phi - \beta_{\|}\sin\Theta)^2 \\ \cos^2\Theta\sin^2\Phi \end{array} \right\}. \tag{5.238}$$

The total energy emitted by the particle can be obtained by integration over the angles

$$\mathcal{E} = \frac{e^2\omega_0 \pi N a_y'^2 (4-3a_y'^2)}{6c(1-a_y'^2)^{3/2}(1-\beta_{\|}^2)}. \tag{5.239}$$

Here $a' = a/\sqrt{1-\beta_{\|}^2}$ is the value of a in the reference frame moving at a velocity $v_{\|}$ in the direction of x-axis.

Let us turn to the radiation of the particle oscillating with a small amplitude. Suppose the conditions of Eq. (5.227) are fulfilled. Then the angular and spectral distribution follow from Eq. (5.228)

$$\frac{d\mathcal{E}_{\sigma,\pi}}{d\Omega d\omega} = \frac{4e^2 A^2 \omega_0^2 \gamma^4}{\pi^2 c^3 (1+\psi^2)^4} \frac{\sin^2 \pi\nu N}{(1-\nu^2)^2} \left\{ \begin{array}{l} (1-\psi^2\cos 2\phi)^2 \\ \psi^4 \sin^2 2\phi \end{array} \right\}. \tag{5.240}$$

The angular distribution is equal to

$$\frac{d\mathcal{E}_{\sigma,\pi}}{d\Omega} = \frac{e^2 a^2 \omega_0^3 \gamma^4 N}{c^3(1+\psi^2)^5} \left\{ \begin{array}{l} (1-\psi^2\cos 2\phi)^2 \\ \psi^4 \sin^2 2\phi \end{array} \right\}. \tag{5.241}$$

One can see that the angular distribution of polarization components is the same as the instant distribution given by Eq. (5.54). The reason is that in the approximation $a \ll 1$ the velocity of the particle can be considered to be a constant. The acceleration is constant in its direction but varies in its value. Thus the radiation pattern keeps its geometry varying in amplitude.

The spectrum of radiation is given by Eq. (5.231) which, in this, case is written in the form

$$\frac{d\mathcal{E}_{\sigma,\pi}}{d\omega} = \frac{\pi e^2 \gamma^2 \omega_0^2 a^2 N}{2c^3} \left\{ \begin{array}{l} (1-2\xi+3\xi^2)\xi \\ (1-\xi)^2\xi \end{array} \right\} \Theta(N, 1-\xi), \tag{5.242}$$

and the total energy emitted by Eq. (5.239) with $a'_y \ll 1$.

By the way, in the case under consideration, we can introduce the undulator parameter k according to Eq. (5.6) because the relativistic factor γ can be taken as being constant $\gamma \approx 1 - \beta_{\|}^2$ and acceleration is orthogonal to the velocity vector. It follows from Eq. (5.6) that $k = \gamma a/\sqrt{2} \ll 1$. Thus we can deduce Eqs (5.240), (5.241) and (5.242) from the corresponding equations of Sec. refs6.2.1.

Now we consider the radiation in the limit of large k. Namely, we suppose that the particle has relativistic velocity in the reference frame moving at a speed $\boldsymbol{v}_{\|}$. One can find from Eq. (5.234) that the maximum value of the particle velocity in the moving frame is $v'_{max} = A\omega_0\sqrt{1-\beta_{\|}}$ or divided by the speed of light $\beta'_{max} = a/\sqrt{1-\beta_{\|}} = a'$. In line with the last assumption $a' \approx 1$. Hence, the parameter κ in Eq. (5.235) can be close to unity. In this case, the Bessel function $J_n(\kappa n)$ can be represented in terms of modified Bessel functions as follows [9]

$$J_n(\kappa n) \approx \frac{sqrt1 - \kappa^2}{\pi\sqrt{3}} K_{1/3}\left[\frac{n}{3}(1-\kappa^2)^{3/2}\right]. \tag{5.243}$$

Eqs (5.236) and (5.237) are still valid but because of large n it is convenient to replace the discrete n with continuous ν and $\sum_n I_n(\nu, N)$ with $I(\nu, n)$. It is much the same as to return from Eq. (5.31) to Eq. (5.26).

The spectrum of radiation is derived by integration over the angles. In the approximation of the infinite undulator it gives [53]

$$\frac{dW}{dy} = \frac{e^2\omega^2}{12c(1-b'^2_y)^{3/2}}\phi(y); \quad y = \frac{\omega}{3\omega_0}(1-b'^2_y)^{3/2}, \tag{5.244}$$

where

$$\phi(y) = \frac{24y}{\pi^2}[K_{1/3}(y)K_{2/3}(y) + 3yK^2_{1/3}(y) - 3yK^2_{2/3}(y)], \tag{5.245}$$

with

$$\int_0^\infty \phi(y)dy = 1.$$

It has been shown in Ref.[53] that in spite of large n or ν the maximum of spectral distribution falls on the first harmonic because $\phi(y)$ is a monotone decreasing function.

5.4.3. A Longitudinal Harmonic Oscillator

By a longitudinal harmonic oscillator we mean a particle that executes harmonic oscillations

$$x' = A' \sin \omega' t', \quad y' = z' = 0 \tag{5.246}$$

in the reference frame which moves at a constant velocity $v_{||}$ along the direction of oscillations. In the laboratory reference frame the particle moves according to equations

$$\begin{aligned} x &= \frac{v_{||}}{\omega_0}\tau + A \sin \tau, \\ y &= z = 0, \\ ct &= \frac{c}{\omega_0}\tau + \beta_{||} A \sin \tau, \end{aligned} \tag{5.247}$$

where

$$A = A'/\sqrt{1-\beta_{||}^2}, \quad \omega_0 = \omega_0'\sqrt{1-\beta_{||}^2}, \quad \tau = \omega_0' t'.$$

The particle velocity cannot exceed the velocity of light, hence

$$a = \frac{A'\omega'}{c} = \frac{A\omega}{c} < 1.$$

The radiation generated by this kind of oscillator has been studied in Ref.[54]. We derive the corresponding equations from the results of Sec. 5.4.1. Comparing Eqs (5.247) and (5.213) we find that

$$A_y = A_z = \psi_x = \psi_t = 0, \quad A_x = A, \quad A_t = \beta_{||} A.$$

The angular distribution is evidently axial symmetric with respect to the collinear vectors of velocity $\boldsymbol{v}$ and acceleration $\dot{\boldsymbol{v}}$ because they distinguish the only physically motivated direction. Thus, the most natural angular coordinates are θ and ϕ of Eq. (5.43). The directions of y- and z-axes do not matter in studying the net angular or spectral distribution, but the distribution of the polarization components essentially depends on the choice of the coordinate system.

It follows from Eqs (5.216) and (5.217) that

$$\kappa = a\frac{\cos\theta - \beta_{||}}{1 - \beta_{||}\cos\theta}. \tag{5.248}$$

We have to replace the components of vectors $\boldsymbol{e}_i$ in Eqs (5.219) by proper functions of θ and ϕ. For instance, we calculate P_π

$$P_\pi = \beta_{||}\frac{\cos\theta\sin\theta\sin\phi}{\sqrt{1-\sin^2\theta\sin^2\phi}}.$$

After some geroic algebra we obtain from Eq. (5.221)

$$\frac{d\mathcal{E}_{\sigma,\pi}}{d\Omega d\omega} = \frac{e^2 \sin^2\theta (1-\beta_{\|}^2)^2 \sum\limits_{n=1}^{\infty} I_n(\nu, N) J_n^2(\kappa n)}{c(1-\boldsymbol{\beta}_{\|}\boldsymbol{n})^4(\cos\theta - \beta_{\|})^2(1-\sin^2\theta\sin^2\phi)} \left\{ \begin{array}{c} \cos^2\phi \\ \cos^2\theta\sin^2\phi \end{array} \right\} \tag{5.249}$$

Qualitative evaluation of polarization can be easily specified by the use of the ultimate expression of Eq. (5.17). It is clear that the electric vector of the wave is aligned with the vector $[\boldsymbol{n}[\boldsymbol{n}\dot{\boldsymbol{\beta}}]]$ since $[\boldsymbol{n}[\boldsymbol{\beta}\dot{\boldsymbol{\beta}}]] = 0$. But the direction of acceleration does not vary. Hence, the radiation is linearly polarized and the plane of polarization goes through the vectors $\boldsymbol{n}$ and $\boldsymbol{\beta}$.

The angular distribution of Eq. (5.223) becomes

$$\frac{d\mathcal{E}_{\sigma,\pi}}{d\Omega} = \frac{e^2 a^2 \omega_0 (1-\beta_{\|}^2)^2 \sin^2\theta\,(4+\kappa^2)}{16c(1-\beta_{\|}\cos\theta)^5(1-\kappa^2)^{7/2}(1-\sin^2\theta\sin^2\phi)} \left\{ \begin{array}{c} \cos^2\phi \\ \cos^2\theta\sin^2\phi \end{array} \right\}. \tag{5.250}$$

In the case of ultrarelativistic particle, we expand Eqs (5.249) and (5.250) with respect to small γ^{-1} and θ. It is easy to rewrite Eqs (5.249) and (5.250) in an ultrarelativistic form. We present here only the last one

$$\frac{d\mathcal{E}}{d\Omega} = \frac{2e^2a^2\omega_0 N}{c(1-\beta_{\|}^2)^2(1+\kappa^2)^2} \frac{\psi^2}{(1+\psi^2)^5} \frac{4+\kappa^2}{(1-\kappa^2)^{7/2}} \left\{ \begin{array}{c} \cos^2\phi \\ \sin^2\phi \end{array} \right\}, \tag{5.251}$$

with

$$\kappa = a\frac{1-\psi^2}{1+\psi^2}, \qquad \psi = \gamma\theta.$$

It can be seen that the σ- and π-components of polarization differ only in a turn through $\pi/2$ around the x-axis. The sum over the polarization components depends naturally only on the angle θ. The total emitted energy is obtained either by integration of Eq. (5.251) or directly from Eq. (5.239) which is valid in the case of longitudinal oscillator as well. The point is that in the reference frame moving at a velocity $v_{\|}$ along the x-axis both the oscillators are equivalent with an accuracy to orientation which does not matter in this case. Thus, all we have to do is to replace a' in Eq. (5.239) by the corresponding value for the longitudinal oscillator. It follows from Eqs (5.246) and (5.247) that $a' = A'\omega'/c = a$. Consequently, we write

$$\mathcal{E} = \frac{e^2\omega_0\pi N a^2(4-3a^2)}{6c(1-a^2)^{3/2}(1-\beta_{\|}^2)}. \tag{5.252}$$

In the case of small-amplitude oscillations we present only the angular and the spectral distribution of an ultrarelativistic particle

$$\frac{d\mathcal{E}_{\sigma,\pi}}{d\Omega d\omega} = \frac{16e^2a^2\gamma^2}{c\pi^2} \frac{\sin^2 \pi\nu N}{(1-\nu^2)^2} \frac{\psi^2}{(1+\psi^2)^4} \left\{ \begin{array}{c} \cos^2\phi \\ \sin^2\phi \end{array} \right\}. \tag{5.253}$$

The angular distribution can be readily found from Eqs (5.250) and (5.251) in the limit $\kappa \to 0$ and the spectrum from Eq. (5.232). Note that the maximum of the spectral distribution is located at $\xi = 2/3$ but not at $\xi = 1$ as in the case of a transversal oscillator. The reason is that no radiation is produced in the on-axis direction $\psi = 0$ which is linked to the high frequency radiation according to the Doppler effect.

5.4.4. A Helical Undulator

Radiation from a particle confined to a helical orbit has been widely discussed in scientific periodicals [55,47,45,48] as well as in books [46,21]. See also Chapter 1. We present this material here as a reference source and for the sake of completeness.

There is a variety of electromagnetic field structures which can provide a helical motion of a charged particle. The simplest of them are the uniform magnetic field and a rotating field introduced at the end of Sec.5.1.1. The drawback of the uniform field is that only the component orthogonal to the velocity vector is used. The rotating undulator field $\boldsymbol{H} = (0, H_0 \sin 2\pi x/\lambda, H_0 \cos 2\pi x/\lambda)$ is more preferable.

Let the charged particle execute helical motion

$$x = v_{\|}t; \quad y = R\cos\omega_0 t; \quad z = R\sin\omega_0 t. \tag{5.254}$$

We put in equations of Sec. 5.4.1 $a_x = a_t = 0$, $a_y = a_z = R\omega_0/c = \beta_\perp$, where $\beta_\perp$ is the orthogonal component of velocity. In the approximation of a long undulator the angular distribution of radiation may be considered as axially symmetric one. Hence, it would suffice to describe the radiation properties in any plane confining the helix axis. Let, for instance, $\phi = 0$. Then Eq. (5.220) becomes

$$\begin{aligned} B_{\pi n} &= e^{i(n\alpha - \pi/2)}\beta_\perp J'_n(\kappa n), \\ B_{\sigma n} &= -e^{in\alpha}\frac{\cos\theta - \beta_{\|}}{\sin\theta} J_n(\kappa n), \end{aligned} \tag{5.255}$$

with

$$\kappa = \frac{\beta_\perp \sin\theta}{(1-\boldsymbol{\beta}_{\|}\boldsymbol{n})}, \quad 1 - \boldsymbol{\beta}_{\|}\boldsymbol{n} = 1 - \beta_{\|}\cos\theta.$$

We see that the σ- and π-components are $\pi/2$ out of phase. This means that the radiation is elliptically polarized. The exception is the cone with the opening angle $\cos\theta = \beta_{\|}$, where the radiation has linear polarization with the vector $\boldsymbol{E}$ orthogonal to the vectors $\boldsymbol{n}$ and $\boldsymbol{v}_{\|}$.

A helical undulator is used with the aim to produce a circularly polarized radiation. Thus, we place a particular emphasis upon the degree of circular polarization. Combining Eqs (5.255), and (5.47) we obtain

$$B_{\pm} = \frac{1}{\sqrt{2}} e^{in\alpha} \left(\frac{\cos\theta - \beta_{\|}}{\sin\theta} J_n(\kappa n) \pm \beta_{\perp} J'_n(\kappa n) \right).$$

It follows from Eq. (5.48) that

$$\frac{d\mathcal{E}_{\pm}}{d\Omega d\omega} = \frac{e^2}{2c(1-\boldsymbol{\beta}_{\|}\boldsymbol{n})^2} \sum_{n=1}^{\infty} n^2 I_n(\nu, N) \left[\frac{\cos\theta - \beta_{\|}}{\sin\theta} J_n(\kappa n) \pm \beta_{\perp} J'_n(\kappa n) \right]^2. \quad (5.256)$$

The first term in the square brackets is positive inside the cone $\theta = \arccos\beta_{\|}$ and negative outside. Hence, within the cone the value of the expression in the brackets is greater if the plus sign is applied than in the case of minus sign. It causes a right circularly polarized radiation inside the cone $\theta = \arccos\beta_{\|}$. Note that according to Eq. (5.254), the trajectory represents a right-hand helix too. This agreement becomes more apparent if we consider the problem in a reference frame moving at a velocity $\boldsymbol{v}_{\|}$. The particle path in this frame makes a circle. The synchrotron radiation emitted in the directions confined to the plane of the circle is linearly polarized. These directions form a cone $\theta = \arccos\beta_{\|}$ in the laboratory reference frame. The radiation observed at a greater angle has been launched backward with respect to the moving reference frame. Consequently, the electric field of the wave emitted in any direction rotates in the same fashion as the 'visible' position of the particle. The on-axis radiation contains only the first harmonic of the right circular polarization.

The angular distribution can be found by the use of Eqs (5.222) and (5.32)

$$\frac{d\mathcal{E}}{d\Omega} = \frac{e^2\omega_0 N \beta_{\perp}^2}{16c(1-\boldsymbol{\beta}_{\|}\boldsymbol{n})^3} \left\{ \left(\frac{\cos\theta - \beta_{\|}}{1 - \beta_{\|}\cos\theta} \right)^2 \frac{4+\kappa^2}{(1-\kappa^2)^{7/2}} + \frac{4+3\kappa^2}{(1-\kappa^2)^{5/2}} \right\}. \quad (5.257)$$

Corresponding expressions for an ultrarelativistic charge can be easily obtained by substitutions

$$\cos\theta - \beta_{\|} = \frac{1}{2\gamma^2}(1 + k^2 - \psi^2), \quad 1 - \beta_{\|}\cos\theta = \frac{1}{2\gamma^2}(1 + k^2 + \psi^2),$$
$$\kappa = \frac{2k\psi}{1 + k^2 + \psi^2}, \quad \psi = \gamma\theta, \quad (5.258)$$

where $k = \gamma\beta_{\perp}$ is the undulator parameter.

Radiation in the forward direction scales with the undulator parameter as

$$\frac{d\mathcal{E}}{d\Omega} \sim \frac{k^2}{(1+k^2)^3}.$$

That is, it grows with k at low k having the maximum at $k = 1/\sqrt{2}$ and then falls abruptly. In the limit of high k the major part of radiation is emitted in the direction tangent to the trajectory, i.e. at an angle $\theta \approx \beta_\perp$ or $\psi \approx k$.

The total energy emitted by a particle results from Eq. (5.233)

$$\mathcal{E} = \frac{4\pi e^2 \omega_0 \gamma^2 k^2 N}{3c} \tag{5.259}$$

which coincides of course with that of synchrotron radiation.

Integrating Eq. (5.256) over the angles we obtain the spectrum for an arbitrary velocity

$$\begin{aligned} \frac{d\mathcal{E}}{d\omega} &= \frac{2\pi e^2 \omega N}{c\omega_0 \beta_\parallel \gamma_\parallel^2} \sum_{n=1}^{\infty} \left\{ \frac{(2\xi - n)^2 J_n^2(\kappa_n)}{4\xi n - n^2 \gamma_\parallel^{-2} - 4\xi^2} + \beta_\perp^2 \gamma_\parallel^2 J'^2_n(\kappa_n) \right\} \times \\ & \left\{ \Theta(N, n - \frac{\omega}{\omega_0}(1 - \beta_\parallel)) - \Theta(N, n - \frac{\omega}{\omega_0}(1 + \beta_\parallel)) \right\}, \end{aligned} \tag{5.260}$$

where the Θ-function is defined by Eq. (5.80), $\gamma_\parallel = (1 - \beta_\parallel^2)^{-1/2}$ and

$$\kappa_n = \beta_\perp n \left[\left(\frac{\omega}{n\omega_0} \right)^2 - \frac{1}{\beta_\parallel^2} \left(\frac{\omega}{n\omega_0} - 1 \right)^2 \right]^{1/2}, \quad \xi = \frac{\omega}{2\omega_0 \gamma_\parallel^2}.$$

The braces with Θ-function restrict the spectral band of nth harmonic to a range of

$$\frac{n\omega_0}{1 + \beta_\parallel} < \omega < \frac{n\omega_0}{1 - \beta_\parallel}$$

with smeared-out boundaries.

Let us turn to the radiation of ultrarelativistic particle in a small k limit. It follows from Eq. (5.256) that the first harmonic only remains

$$\frac{d\mathcal{E}_\pm}{d\Omega d\omega} = \frac{2e^2 \gamma^4 \beta_\perp^2}{\pi^2 c (1 + \psi^2)^2} \frac{\sin^2 \pi\nu N}{(1 - \nu^2)^2} \left[\frac{1 - \psi^2}{1 + \psi^2} \pm 1 \right]^2. \tag{5.261}$$

Applying the identity of Eq. (5.32) one can obtain angular distribution

$$\frac{d\mathcal{E}}{d\Omega} = \frac{4e^2 \gamma^6 \beta_\perp^2 \omega_0 N (1 + \psi^4)}{c (1 + \psi^2)^5}. \tag{5.262}$$

The spectrum of radiation in this approximation follows from Eq. (5.232)

$$\frac{d\mathcal{E}}{d\omega} = \frac{2\pi e^2 \gamma^2 \omega_0^2 R^2 N}{c^3} \xi (1 - 2\xi + 2\xi^2) \Theta(N, 1 - \xi). \tag{5.263}$$

To study the radiation properties in the high k limit we apply the results of Sec. 5.2.2. Let $k \gg 1$ but $\beta_\perp \ll 1$. The plane tangent to the trajectory makes an angle $\theta_0 = \beta_\perp$ with the helix axis. We place vector $\boldsymbol{n}$ in the plane which is orthogonal to the tangent plane in accordance with section Sec. 5.2.2. Denote the angle between $\boldsymbol{n}$ and the tangent plane by ξ. In other words $\theta = \theta_0 + \xi$. The angle ξ is of order of γ^{-1}. The radiation in any given direction is emitted once a period. Hence, we have to keep in Eq. (5.96) the terms with subscript 1 (or 2) only. Substituting $R^{(2)} = v_\perp/\omega_0$ we obtain

$$\frac{d\mathcal{E}}{d\Omega d\omega} = \frac{e^2\omega^2(1+x^2)}{3\pi^2\omega_0^2\gamma^4 c\beta_\perp^2} I(\nu, N)\left\{(1+x^2)K_{2/3}^2(\eta) + x^2 K_{1/3}^2(\eta)\right\}, \tag{5.264}$$

where

$$x = \gamma\chi, \quad \eta = \frac{\omega}{3\beta_\perp\gamma^3\omega_0}(1+x^2)^{3/2}. \tag{5.265}$$

The main part of the energy is emitted in the vicinity of the angle $\theta_0 = \beta_\perp$ with an angular width $\Delta x \sim 1$ or $\Delta\xi \sim \gamma^{-1}$, that is, the radiation is confined between two cones of the angles $\sim \theta_0 - \gamma^{-1}$ and $\sim \theta_0 + \gamma^{-1}$. As to the spectrum, functions $K_{1/3}(\eta)$ and $K_{2/3}(\eta)$ are valuable at $\eta \sim 1$. Therefore, most of the energy is emitted at the frequencies $\omega = \omega_0 k\gamma^2$. It follows from Eq. (5.37) that these frequencies correspond to harmonic numbers $\nu \sim k^3$ as it was indicated in Eq. (5.102). The oscillating factor $I(\nu, N)$ can be averaged over frequency.

In order to obtain the spectrum, we integrate Eq. (5.264) over the solid angle $d\Omega = 2\pi\beta_\perp d\xi$. It gives

$$\frac{d\mathcal{E}}{d\omega} = \frac{2e^2\omega^2 N}{3\pi c\omega_0^2 k\gamma^4}\int_{-\infty}^{\infty}\left\{(1+x^2)K_{2/3}^2(\eta) + x^2K_{1/3}^2(\eta)\right\}(1+x^2)dx. \tag{5.266}$$

Or using the identities [56]

$$\begin{aligned} \int_{-\infty}^{\infty}(1+x^2)^2K_{2/3}^2(\eta)dx &= \frac{\pi}{y\sqrt{3}}\left[\int_y^{\infty}K_{5/3}(x)dx + K_{2/3}(y)\right], \\ \int_{-\infty}^{\infty}x^2(1+x^2)K_{1/3}^2(\eta)dx &= \frac{\pi}{y\sqrt{3}}\left[\int_y^{\infty}K_{5/3}(x)dx - K_{2/3}(y)\right]. \end{aligned} \tag{5.267}$$

with

$$y = \frac{2\omega}{3\beta_\perp\gamma^3\omega_0}$$

we rewrite Eq. (5.266) in the form

$$\frac{d\mathcal{E}}{d\omega} = \frac{2e^2\omega N}{c\omega_0\gamma^2\sqrt{3}} f(y), \quad f(y) = y\int\limits_y^\infty K_{5/3}(x)dx. \tag{5.268}$$

The shape of the spectrum is determined by the function $f(y)$ which is just the same as in the case of synchrotron radiation. The difference occurs in the dependence $y(\omega)$. Consequently, the wiggler spectrum is scaled with respect to synchrotron one.

Angular distribution is obtained by integration of Eq. (5.264) over the frequencies

$$\frac{d\mathcal{E}}{d\Omega} = \frac{e^2\omega_0\gamma^4 kN}{16c}\frac{7+12x^2}{(1+x^2)^{7/2}}. \tag{5.269}$$

The same result follows from the exact formula of Eq. (5.257) in the assumption that $k \gg 1$ and the substitution $\theta = \theta_0 + \xi$ or $\psi = k + x$ made in Eq. (5.258).

The total energy of radiation in the approximation $k \gg 1$ is given by the exact formula of Eq. (5.259).

Bibliography

1. L.D. Landau and E.M. Lifshitz, *The Classical Theory of Fields* (Pergamon, New York, 4th edition, 1975).
2. J.D. Jackson, *Classical Electrodynamics* (2nd ed. Wiley, New York, 1975).
3. V.G. Bagrov, D.M. Gitman, A.A. Sokolov, et al., *Sov. Phys. Techn. Phys.* **20** (1226) 1975.
4. R.P. Feynman, R.B. Leighton and M. Sands, *The Feynman Lectures on Physics* (Addison–Wesley, Reading MA, Commemorative Issue, 1989), vol. 2.
5. V.G. Bagrov, V.A. Bordovitsyn, G.F. Kopytov and V.Ya. Epp, *Sov. Phys. J.* **17** (1974) 46.
6. W. Pauli, *Theory of Relativity* (Pergamon Press, 1958).
7. V.N. Baier, V.M. Katkov and V.M. Strachovenko, *Sov. Phys. – JETP* **36** (1973) 1120.
8. N.V. Smolyakov, *Zh. Tekh. Fiz.* **58** (1988) 489; 2174.
9. M. Abramowitz and I.A. Stegun, eds, *Handbook on Mathematical Functions* (Dover, New York, 1970).
10. J. Schwinger, *Phys. Rev.* **75** (1949) 1992.
11. M.B. Moiseev, M.M. Nikitin and V.Ya. Epp, *Sov. Phys. J.* **24** (1981) 870.
12. K.J. Kim, in *Physics of Particle Accelerators, AIP Conf. Proc. 1989*, vol. 1, p. 565.
13. J.Goulon et al., *Nucl. Instr. Meth.* **A254** (1987) 192.
14. J.Pfluger, *Rev. Sci. Instr.* **63** (1992) 295.
15. A. Luccio and S. Krinsky, in *Free-Electron Generators of Coherent Radiation* (Addison-Wesley, New York, 1982), p. 181.

16. E. E. Koch, ed. *Handbook of Synchrotron Radiation* (North Holland, Amsterdam, 1983).
17. P. Torggler and C. Leubner, *Phys. Rev.* **A 39** (1989) 1989.
18. T.J. Karr, *Phys. Rev.* **A 35** (1697) 1987.
19. M.N. Rosenbluth, *IEEE J. Quantum Electron.* **QE-21** (1985) 966.
20. M.B. Moiseev, M.M. Nikitin, S.V. Sorokin and N.I. Fedosov, *Sov. Phys. J.* **27** (1984) 634.
21. M. M. Nikitin and V.Ya. Epp, *Undulator Radiation* (Energoatomizdat, Moskow, in Russian, 1988).
22. D. F. Alferov, Yu. A. Bashmakov and E. G. Bessonov, in *Proceedings of the Lebedev Physics Institute* (Consulting Bureau, N.Y., 1976), vol. 80, p. 97.
23. I.M. Ternov, V.R. Khalilov, V.G. Bagrov and M.M. Nikitin, *Sov. Phys. J.* **23** (1980) 79.
24. V.A. Bordovitsyn, I.M. Ternov and V.Ya. Epp, *Sov. Phys. J.* **34** (1991) 391.
25. V.A. Bordovitsyn and V.Ya. Epp, *Nucl. Instr. and Meth.* **A405** (1998) 220.
26. S.A. Belomestnych, A.E. Bondar, M.N. Yegorychev et al., *Nucl. Instr. and Meth.* **A227** (1984) 173.
27. A.A Sokolov, N.P. Klepikov and I.M. Ternov, *Zh. Eksp. Teor. Fiz.* [Sov. Phys. JETP] **23** (1952) 632.
28. J. Schwinger, *Proc. Nat. Acad. Sci. USA* **40** (1954) 132.
29. M.M. Nikitin, V.Yu. Muravyatkin and V.Ya. Epp, *Nucl. Instr. and Meth.* **A282** (1989) 481.
30. V.A. Bordovitsyn, *Izv. Vuz. Fiz.* [Sov. Phys. J] **10** (1972) 88.
31. A.N. Didenko, A.V. Kozhevnikov, A.F. Medvedev, M.M. Nikitin and V.Ya. Epp, *Sov. Phys. JETP* **49** (1979) 973.
32. H.R. Reiss, *Phys. Rev.* **A22** (1980) 1786.
33. G. Dattoli, A. Renieri and A. Torre, *Lectures on the Free Electron Laser Theory and Related Topics* (World Scientific, 1993).
34. C. Leubner, *Phys. Rev.* **A23** (1981) 2877.
35. G. Dattoli, L. Gianessi, L. Mezi and A. Torre, *Nuovo Cimento* **105B** (1990) 327.
36. G. Dattoli, A. Torre, S. Lorenzutta, G. Maino and C. Chiccoli, *Nuovo Cimento* **106B** (1991) 21.
37. A.I. Nikishov and V.I. Ritus, *Zh. Eksp. Teor. Fiz.* **46** (1964) 776.
38. L.S. Brown and T.W.B. Kibble, *Phys. Rev.* **133** (1964) 705.
39. V.Ya. Epp, V.F. Zalmezh and Yu.M. Nikitina, *Nucl. Instr. and Meth.* **A368** (1996) 543.
40. M.M. Nikitin, A.F. Medvedev, M.B. Moiseev and V.Ya. Epp, *Sov. Phys. Tech. Phys.* **26** (1981) 350.
41. M.M. Nikitin, A.F. Medvedev, M.B. Moiseev, V.Ya. Epp and G.K. Razina, *Sov. Phys. Tech. Phys.* **26** (1981) 355.
42. K-J. Kim, *Nucl. Instr. Meth.* **A246** (1986) 67.
43. V.Ya. Epp and G.K. Razina, *Nucl. Instr. and Meth.* **A307** (1991) 562.
44. V.Ya. Epp and V.A. Churicov, *Sov. Phys. J.* **27** (1984) 121.

45. D.F. Alferov, Yu.A. Bashmakov and E.G. Bessonov, *Sov. Phys. Techn. Phys.* **21** (1976) 1408.
46. I.M. Ternov, V.V. Mikhailin and V.R. Khalilov, *Synchrotron Radiation and its Application* (Harwood Academic Publishers, 1986).
47. T. Takakura, *Publ. Astron. Soc. Jap.* **12** (1960) 352.
48. B.M. Kincaid, *J. Appl. Phys.* **48** (1977) 2684.
49. M. Abraham, *Ann. d. Phys.* **14** (1904) 236.
50. G.A. Schott, *Electromagnetic Radiation* (Cambridge University Press, 1912).
51. G.N. Watson, *A Treatise on the Theory of Bessel Functions* (Cambridge University Press, London, 1958).
52. V.L. Ginzburg, *Izv. Akad. Nauk SSSR* **11** (1947) 165.
53. V.G. Bagrov, L.G. Glazov, Yu.N. Kargin and G.K. Razina, Preprint HCEI No 35 (High Current Electronics Institute of Russian Academy of Science, Tomsk, 1986), [in Russian].
54. A.B. Kukanov and El K. Suleiman, *Izv. Vuz. Fiz.* **11** (1975) 7.
55. A.A Sokolov, D.V. Galtsov and V.Ch. Zhukovski, *Sov. Phys. Techn. Phys.* **18** (1973) 430.
56. K. Westfold, *Astrophys. J.* **130** (1959) 241.

Chapter 6

Radiation in Special Electrodynamical Systems

V.G. Bagrov and V.Ya. Epp
Tomsk State University 634050 Tomsk, Russia[a]
Tomsk State Pedagogical University 634041 Tomsk, Russia

6.1. Introduction

Synchrotron and undulator radiation of a charged particle is an important phenomenon from the point of view of theory as well as practice. But a variety of systems emitting electromagnetic radiation is not restricted to accelerators and undulators. One of the peculiarities of these devices is that the radiation spectrum is bounded below. Radiation frequency can not be less than that of the fundamental harmonic. That is why a long–wave radiation can not be produced by installations of this kind. There are nevertheless different radiophysical devices which can produce electromagnetic radiation in any spectral region. But it is of theoretical and practical interest to investigate the possibility of generation of wide spectrum by means of one charged point particle moving in an external electromagnetic field. Of theoretical interest are two kinds of electrodynamic systems — a particle moving along a finite arc of a circle, and systems of a "short magnet" type. Both of them are discussed in this chapter.

There are tens of papers where theoretical treatment of radiation in the devices of this kind is given. In early studies [1–17] the basic properties of radiation have been investigated. The particle motion providing emission of radiation with the given properties has been found in Refs[18,19]. Further studying discovered new interesting peculiarities of radiation in the systems under discussion [20–26].

[a]E-mail: bagrov@phys.tsu.tomsk.su

6.2. Radiation of Electrons Moving in an Arc of a Circle

Classical and quantum theories of synchrotron radiation are based on the assumption that an electron makes infinitely many revolutions in a circle and radiation is observed from the whole circular trajectory. In reality, as a rule, one observes radiation from a small part of the trajectory and the number of electron revolutions is finite under any conditions. And if we consider, for example, synchrotron radiation of a charged particle in the space environment (synchrotron component of cosmic radiation), the particle perhaps does not perform a complete revolution. The same phenomenon takes place on the terrestrial scale, for example, as a beam of particles propagates in a bending magnet. There are certain convincing qualitative physical estimates (see e.g. Ref.[27]) showing that the main properties of radiation generated by a charged particle moving at a relativistic speed along an arc of angular dimension $2\alpha > mc^2/\epsilon = \gamma^{-1}$ (ϵ is a total particle energy) coincide with the properties of synchrotron radiation. However, these qualitative estimates can be made more exact to reveal some peculiarities of radiation in the case of trajectory having a form of an arc of arbitrary length. Such an investigation may be useful for applications to radiation of a particle moving in systems of a 'short magnet' type and alike [28–31].

Here we give an analysis of radiation properties of an electron moving with a constant speed $v = c\beta$ in an arc of the angular dimension 2α. To examine this, methods of classical electrodynamics are used. Let us consider an electron trajectory lying in the plane $z = 0$. We place the coordinate origin in the center of a circle and suppose the trajectory to be symmetric with respect to the plane yz.

6.2.1. General Expressions

Let the radius of a circle be R, then the total time of an electron in an arc of the circle is $T = l/c\beta = 2\alpha/\omega_0$, where $\omega_0 = c\beta/R$ and $l = 2\alpha R$ is the length of the arc. If, for example, an electron moves in an external magnetic field $\boldsymbol{H}$ then

$$R = \beta\epsilon/eH, \quad \omega_0 = ceH/\epsilon. \tag{6.1}$$

An electric vector $\boldsymbol{E}$ of an electron radiation field in the wave zone at the point with coordinates $\boldsymbol{r}$ at the moment t is well known [27]

$$\begin{aligned} \boldsymbol{E}(\boldsymbol{r},t) &= (e\beta/2\pi cr)\int_{-\infty}^{\infty} \boldsymbol{f}(\omega,\boldsymbol{n})\exp[i\omega(t-r/c)]d\omega, \\ \boldsymbol{f}(\omega,\boldsymbol{n}) &= \beta^{-1}\int_{-\infty}^{\infty}[\boldsymbol{n}[(\boldsymbol{n}-\boldsymbol{\beta})\dot{\boldsymbol{\beta}}]][1-\boldsymbol{n\beta}]^{-2}\exp\{-i\omega[t-\boldsymbol{na}/c]\}dt, \end{aligned} \tag{6.2}$$

where

$$\boldsymbol{\beta} = \frac{\boldsymbol{v}}{c}, \quad \dot{\boldsymbol{\beta}} = \frac{d\boldsymbol{\beta}}{dt}, \quad \boldsymbol{n} = \frac{\boldsymbol{r}}{r} = (\sin\theta\cos\phi, \sin\theta\sin\phi, \cos\theta)$$

and $\boldsymbol{a} = \boldsymbol{a}(t')$ is a radius vector of a moving charge at the time t'. From Eq. (6.2) we get the spectral-angular distribution of the total radiated energy W

$$\begin{aligned} dW &= (e^2\beta^2/4\pi^2 c)|\boldsymbol{f}(\omega,\boldsymbol{n})|^2 d\omega d\Omega, \\ d\Omega &= \sin\theta d\theta d\phi, \quad \omega > 0. \end{aligned} \tag{6.3}$$

By introducing unit vectors of polarization $\boldsymbol{n}_\sigma$ and $\boldsymbol{n}_\pi$ we obtain from Eq. (6.2)

$$\boldsymbol{f} = f_\sigma \boldsymbol{n}_\sigma + f_\pi \boldsymbol{n}_\pi, \quad |\boldsymbol{f}|^2 = |f_\sigma|^2 + |f_\pi|^2,$$

where the quantities $f_\sigma = (\boldsymbol{f}\boldsymbol{n}_\sigma)$, $f_\pi = (\boldsymbol{f}\boldsymbol{n}_\pi)$ are characteristics for σ- and π-components of linear polarization of radiation of an electron. In the case of motion along an arc of a circle the following expressions, accurate to an unessential phase factor, are easily obtained:

$$\begin{aligned} f_\sigma &= \int\limits_{\phi-\alpha}^{\phi+\alpha} (\cos x - \mu) p^{-2}(x) e^{-iq\psi} dx, \\ f_\pi &= \cos\theta \int\limits_{\phi-\alpha}^{\phi+\alpha} p^{-2}(x) e^{-iq\psi} \sin x dx, \\ p(x) &= 1 - \mu\cos x, \quad \psi = \psi(x) = x - \mu\sin x, \quad \mu = \beta\sin\theta, \quad q = \omega/\omega_0. \end{aligned} \tag{6.4}$$

These equations are the starting point for the analysis of radiation properties.

Taking Eq. (6.2) into account, we integrate Eq. (6.3) over the frequencies and angles and obtain the total radiated energy W as follows:

$$W = W_0 T, \quad W_0 = (2ce^2/3R^2)\beta^4\gamma^{-4}, \tag{6.5}$$

where W_0 is the total power of synchrotron radiation of an electron moving in a circle of a radius R. Thus, we have obtained physically evident results, i.e. the total radiative energy is equal to the power of synchrotron radiation multiplied by the total time of motion in an arc of a circle. This simple fact shows that it is convenient to represent the spectral-angular distribution of Eq. (6.3) in the following form:

$$\begin{aligned} dW = W_0 T F(\beta,\alpha,q;\theta,\phi) dq d\Omega, \quad F = F_\sigma + F_\pi, \\ F_{\sigma,\pi}(\beta,\alpha,q;\theta,\phi) = 3(1-\beta^2)^2(16\pi^2\alpha)^{-1}|f_{\sigma,\pi}|^2. \end{aligned} \tag{6.6}$$

Here F is the distribution function of radiation normalized as follows

$$\int\limits_0^\infty dq \oint d\Omega F(\beta,\alpha,q;\theta,\phi) = 1.$$

Integration over the angles and spectrum of polarization components, each taken separately, yields the result

$$p_{\sigma,\pi} = \int_0^\infty dq \oint d\Omega F_{\sigma,\pi}(\beta, \alpha, q; \theta, \phi),$$
$$p_\sigma = (6 + \beta^2)/8, \quad p_\pi = (2 - \beta^2)/8, \tag{6.7}$$

which is well known in the theory of synchrotron radiation. Thus, the distribution in the σ- and π-components of linear polarization is independent of α. This results from the general properties of radiation proved in Ref.[32]. It is easy to show (see Eq. (6.4)) that $f_{\sigma,\pi}$ decreases as q^{-1} when $q \to \infty$. Thus, for $\omega \to \infty$ spectral distribution decreases as $\sim \omega^{-2}$, i.e., much slower than for synchrotron radiation, which is decreasing exponentially with the increase of frequency. The reason for this discrepancy is that for a finite l and for a chosen character of motion the acceleration $\dot{\boldsymbol{\beta}}$ is different from zero at $t \in T$ and turns to zero step-wise at t beyond T, i.e. acceleration is a discrete function of time. Then it follows from Eq. (6.2) and from general properties of Fourier transform that when $\omega \to \infty$, Fourier images for the piecewise continuous function decrease as ω^{-1}. If acceleration is continuous, as is the case for any real physical conditions, then the Fourier images of Eq. (6.2) decrease with $\omega \to \infty$ not slower than ω^{-2} and hence, the spectrum decreases not slower than ω^{-4}. These conclusions will be confirmed by particular examples in Sec. 6.3.

6.2.2. Angular Distribution of Radiation

Integration over the frequency in Eq. (6.6) can be carried out exactly and after simple but cumbersome calculations we obtain

$$\begin{aligned} F_{\sigma,\pi}(\beta, \alpha, \theta, \phi) &= \int_0^\infty F_{\sigma,\pi}(\beta, \alpha, q; \theta, \phi) dq \\ &= \Phi^0_{\sigma,\pi}(\beta, \theta) + \Phi^1_{\sigma,\pi}(\phi + \alpha) - \Phi^1_{\sigma,\pi}(\phi - \alpha), \end{aligned} \tag{6.8}$$

where Φ^0 defines an angular distribution of synchrotron radiation averaged over one revolution

$$\Phi^0_\sigma(\beta, \theta) = \frac{3(1 - \beta^2)^2(4 + 3\mu^2)}{64\pi(1 - \mu^2)^{5/2}},$$
$$\Phi^0_\pi(\beta, \theta) = \frac{3(1 - \beta^2)^2(4 + \mu^2)\cos^2\theta}{64\pi(1 - \mu^2)^{7/2}} \tag{6.9}$$

and functions $\Phi^1_{\sigma,\pi}$ are defined as follows:

$$\begin{aligned}
\Phi^1_\sigma(x) &= Q(1-\mu^2)\left\{\frac{6\mu p^4(x)(4+3\mu^2)}{\sqrt{1-\mu^2}}\arctan\frac{\mu\sin x}{p(x)+\sqrt{1-\mu^2}}\right. \\
&\quad +[(23\mu^2-2)p^3(x)+(9\mu^2-2)(1-\mu^2)p^2(x) \\
&\quad \left. -2(1-\mu^2)^2p(x)+6(1-\mu^2)^3]\sin x\right\}, \\
\Phi^1_\pi(x) &= Q\cos^2\theta\left\{\frac{6\mu p^4(x)(4+\mu^2)}{\sqrt{1-\mu^2}}\arctan\frac{\mu\sin x}{p(x)+\sqrt{1-\mu^2}}\right. \\
&\quad +[(2+13\mu^2)p^3(x)+(3\mu^2+2)(1-\mu^2)p^2(x) \\
&\quad \left. +2(1-\mu^2)^2p(x)-6(1-\mu^2)^3]\sin x\right\}, \\
Q &= (1-\beta^2)^2[128\pi\mu\alpha(1-\mu^2)^3p^4(x)]^{-1}. \qquad (6.10)
\end{aligned}$$

Let us analyze angular distribution of Eqs (6.8) – (6.10). If we integrate in Eq. (6.8) over ϕ, or put $\alpha = \pi N$, where $N = 1, 2, \ldots$ (i.e. an electron makes N revolutions in a circle), we obtain

$$\frac{1}{2\pi}\int_0^{2\pi} F_{\sigma,\pi}(\beta,\alpha,\theta,\phi)d\phi = F_{\sigma,\pi}(\beta,\pi N;\theta,\phi) = \Phi^0_{\sigma,\pi}(\beta,\theta), \qquad (6.11)$$

that is, an angular distribution of synchrotron radiation averaged over one revolution. This property is also a consequence of the general statements proved in Ref.[32].

If $\alpha \to 0$ (or, more exactly, if $\alpha \ll \gamma^{-1}$), we have from Eqs (6.8)–(6.10)

$$\begin{aligned}
F_\sigma(\beta,0;\theta,\phi) &= 3(1-\beta^2)^2(\mu-\cos\phi)^2[8\pi p^5(\phi)]^{-1}, \\
F_\pi(\beta,0;\theta,\phi) &= 3(1-\beta^2)^2\cos^2\theta\sin^2\phi[8\pi p^5(\phi)]^{-1}. \qquad (6.12)
\end{aligned}$$

These are the instantaneous angular distributions of σ- and π-polarization components of synchrotron radiation (see Sec. 1.3).

Thus, when α changes from 0 to π, Eqs (6.8)–(6.11) describe a continuous transition from the instantaneous angular distribution to the one, averaged per revolution.

It is easy to see from Eqs (6.8)–(6.11) that $F_{\sigma,\pi}(\beta,\alpha;\theta,\phi) = F_{\sigma,\pi}(\beta,\alpha;\theta,-\phi)$, i.e., angular distribution is symmetric with respect to the plane $y = 0$ and consequently it is sufficient to consider only the region of angles $0 \le \phi \le \pi$. Radiation of a relativistic electron ($\gamma \gg 1$) is the most interesting one. All characteristics of the radiation of a nonrelativistic particle have rather simple analytical expressions and they will be considered in Sec. 6.2.5. At $\gamma \gg 1$ the main part of radiation is concentrated in a narrow interval of angles $\delta\theta \sim \gamma^{-1}$ in the vicinity of $\theta = \pi/2$.

Let us consider the function $x(\phi) = F(\beta,\alpha;\pi/2,\phi)$ which describes the angular distribution in the plane $\theta = \pi/2$. We restrict our consideration to $0 \le \alpha \le \pi$. The function $x(\phi)$ reaches an extremum at the points $\phi = 0$ and $\phi = \pi$ for all α and

β. This function can also have some extremums at the points $\phi = \phi_k$, $k = 1, 2, 3, 4$ (more exactly, the points ϕ_1 and ϕ_3 can be points of minimum, while ϕ_2 and ϕ_4 being points of maximum). We shall call these extremums secondary. The position of the points ϕ_k depends on α and β, i.e. $\phi_k = \phi_k(\alpha, \beta)$. At a fixed β all ϕ_k are monotonically increasing functions of α, while at a fixed α, the functions ϕ_1 and ϕ_2 being monotonically increasing functions of β, and the functions ϕ_3, and ϕ_4 being monotonically decreasing ones. The functions ϕ_k satisfy the inequalities

$$0 \leq \phi_1 \leq \phi_2 \leq \alpha \leq \phi_3 \leq \phi_4 \leq \pi.$$

At a fixed α we also have $\lim\limits_{\beta \to 1} \phi_k(\beta, \alpha) = \alpha$, while at $\gamma \gg 1$ we obtain the asymptotic expressions

$$\phi_{1,2}(\beta, \alpha) \approx \alpha - \gamma^{-1} \mp \gamma^{-4}(2\sqrt{2}\sin^3\alpha)^{-1}, \quad \phi_{3,4}(\beta, \alpha) \approx \alpha + \gamma^{-1} \mp \gamma^{-4}(2\sqrt{2}\sin^3\alpha)^{-1}.$$

Assuming that $\beta > \sqrt{2/5}$ we find two angles ψ_1 and ψ_2

$$0 < \psi_1 < \psi_2 < \pi/2, \quad \cos\psi_1 = \beta, \quad \cos\psi_2 = (5\beta^2 - 2)/3\beta. \tag{6.13}$$

For the ultrarelativistic case $\psi_1 \approx \gamma^{-1}$, $\psi_2 \approx \gamma^{-1}\sqrt{21}/3$. As for the functions $\phi_k(\beta, \alpha)$, we have the following additional information about the intervals in which the functions are defined and about the particular values

$$\begin{aligned}
0 \leq \phi_1(\beta, \alpha) \leq \pi - \psi_2, &\quad \psi_2 \leq \alpha \leq \pi, \\
\phi_1(\beta, \psi_2) = 0, &\quad \phi_1(\beta, \pi) = \pi - \psi_2, \\
0 \leq \phi_2(\beta, \alpha) \leq \pi - \psi_1, &\quad \psi_1 \leq \alpha \leq \pi, \\
\phi_2(\beta, \psi_1) = 0, &\quad \phi_2(\beta, \pi) = \pi - \psi_1, \\
\psi_1 \leq \phi_3(\beta, \alpha) \leq \pi, &\quad 0 \leq \alpha \leq \pi - \psi_1, \\
\phi_3(\beta, 0) = \psi_1, &\quad \phi_3(\beta, \pi - \psi_1) = \pi, \\
\psi_2 \leq \phi_4(\beta, \alpha) \leq \pi, &\quad 0 \leq \alpha \leq \pi - \psi_2, \\
\phi_4(\beta, 0) = \psi_2, &\quad \phi_4(\beta, \pi - \psi_2) = \pi.
\end{aligned} \tag{6.14}$$

The functions $\phi_k(\beta, \alpha)$ can be found as implicit functions from the equation

$$\begin{aligned}
&[\beta - \cos(\phi_k + \alpha)]^2[1 - \beta\cos(\phi_k - \alpha)]^5 - \\
&[\beta - \cos(\phi_k - \alpha)]^2[1 - \beta\cos(\phi_k + \alpha)]^5 = 0.
\end{aligned} \tag{6.15}$$

Now we can describe in details the structure of the angular distribution $x(\phi)$.

If $0 \leq \alpha \leq \psi_1$, then $x(\phi)$ reaches an absolute maximum at the point $\phi = 0$; at the point $\phi = \phi_3(\beta, \alpha)$ we have a secondary minimum; at the point $\phi = \phi_4(\beta, \alpha)$ there is a secondary maximum; and at $\phi = \pi$ we have a minimum again. Thus, $x(\phi)$ has four extremums in the interval $[0,\ \pi]$ at $0 \leq \alpha \leq \psi_1$.

For increasing α, when $\psi_1 < \alpha < \psi_2$, at the point $\phi_2(\beta, \alpha)$ an absolute maximum of $x(\phi)$ appears, while $\phi = 0$ becomes a point of a minimum for these α. Consequently, as $\psi_1 < \alpha < \psi_2$, the function $x(\phi)$ has five extremums on the interval $[0,\ \pi]$.

If $\psi_2 < \alpha < \pi - \psi_2$, then a new point of minimum $\phi = \phi_1(\beta, \alpha)$ appears and $\phi = 0$ becomes a maximum again. Hence, in this case $x(\phi)$ reaches its maximums at $\phi = 0,\ \phi_2,\ \phi_4$, and minimums at $\phi = \phi_1,\ \phi_3,\ \pi$. For α close to point ψ_2 this point is the point of absolute maximum, while for α exceeding a certain $\alpha > \psi_2$, an absolute maximum appears at $\phi = 0$.

Further increase in α leads to the coincidence of a maximum ϕ_4 with a minimum $\phi = \pi$ at $\alpha = \pi - \psi_1$ as it follows from Eq. (6.14), and the point $\phi = \pi$ at $\pi - \psi_2 < \alpha < \pi - \psi_1$ becomes a maximum, while the maximum ϕ_4 disappears. Thus, $x(\phi)$ has five extremums at $\pi - \psi_2 < \alpha < \pi - \psi_1$, an absolute maximum being at the point $\phi = 0$.

At $\alpha = \pi - \psi_1$ a secondary minimum ϕ_3 coinciding with the point $\phi = \pi$ disappears according to Eq. (6.14); at $\pi - \psi_1 < \alpha < \pi$ the point $\phi = \pi$ becomes a minimum of $x(\phi)$ again. The function $x(\phi)$ has two maximums at these α in the interval $[0,\ \pi]$, i.e. an absolute maximum at $\phi = 0$, a secondary one at $\phi = \phi_2$, and two minimums at $\phi = \phi_1,\ \pi$.

With increasing α absolute difference between all the extremums decreases and for $\alpha = \pi$ all the extremums coincide with each other, which in accordance with Eq. (6.11) indicates the independence of the function $x(\phi)$ of ϕ at $\alpha = \pi$.

6.2.3. Generation of the Low-Frequency Radiation

Simple qualitative considerations show that for a relativistic electron at $\alpha \gg \gamma^{-1}$ the radiation spectrum, at least in a frequency range of the main contribution to the radiation, must have a synchrotron character, in particular its maximum lies at the frequencies $\omega \sim \omega_0\gamma^3$. However, the character of the spectrum in the low-frequency region $\omega \leq \omega_0$ can not be deduced from the qualitative considerations but requires direct calculations.

Expanding Eqs (6.4) in powers of ω to an accuracy of terms of the order of ω and integrating over the angles we obtain

$$\begin{aligned}
dW &= W_0 T F(\beta, \alpha, q) dq, \\
F(\beta, \alpha, q) &= \oint F(\beta, \alpha, q; \theta, \phi) d\Omega, \\
F(\beta, \alpha, q) &= F^0(\beta, \alpha) + q^2 F^1(\beta, \alpha) + \cdots, \\
F^k(\beta, \alpha) &= 3(1-\beta^2)^2 (8\pi\alpha\beta^2)^{-1} f^k(\beta, \alpha), \\
f^0_\sigma(\beta, \alpha) &= 2\nu \ln(1+\nu)/(1-\nu), \\
f^0_\pi(\beta, \alpha) &= 2\nu^{-1} \ln(1+\nu)/(1-\nu) - 4,
\end{aligned}$$

$$\begin{aligned}
f_\sigma^1(\beta,\alpha) &= \left(1-\frac{\sin\alpha}{\alpha\sin\alpha}\right)^2\left(\beta\ln\frac{1+\beta}{1-\beta}-\nu\ln\frac{1+\nu}{1-\nu}\right)+\\
&\left(\frac{\sin^2\alpha}{\alpha^2}-1\right)\left(\beta\ln\frac{1+\beta}{1-\beta}+\ln(1-\beta^2)\right),\\
f_\pi^1(\beta,\alpha) &= \left(1-\frac{\sin\alpha}{\alpha\cos\alpha}\right)^2\left(\frac{1}{\beta}\ln\frac{1+\beta}{1-\beta}-\frac{1}{\nu}\ln\frac{1+\nu}{1-\nu}\right)+\\
&\left(\frac{\sin^2\alpha}{\alpha^2}-1\right)\left(2-\frac{1}{\beta}\ln\frac{1+\beta}{1-\beta}-\ln(1-\beta^2)\right),\\
\nu &= \beta\sin\alpha(1-\beta^2\cos^2\alpha)^{-1/2}.
\end{aligned} \tag{6.16}$$

The values $F^0(\beta,\alpha)$ are characteristics of a relative contribution of the narrow range of frequencies $d\omega/\omega_0$ in the vicinity of the point $\omega = 0$ to the total radiation. From Eq. (6.16) we obtain $F^0_{\sigma,\pi}(\beta,0) = F^0_{\sigma,\pi}(\beta,\pi) = 0$, while the functions $F^0_{\sigma,\pi}(\beta,\alpha)$ are positive in the interval $0 < \alpha < \pi$. Consequently, as $\alpha = \bar{\alpha}(\beta)$, the functions $F^0(\beta,\bar{\alpha}(\beta))$ reach their maximum values. Values of $\bar{\alpha}(\beta)$ and $F^0(\beta,\bar{\alpha}(\beta))$ depending on β have been calculated numerically and are summarized in Table 6.1.

Table 6.1. Functions $\bar{\alpha}(\beta)$ and $F^0(\beta,\bar{\alpha}(\beta))$.

β	$\bar{\alpha}_\sigma(\beta)$	$\bar{\alpha}_\pi(\beta)$	$\bar{\alpha}(\beta)$	F^0_σ	F^0_π	F^0
0.0	1.166	1.166	1.166	0.346	0.115	0.461
0.1	1.163	1.164	1.163	0.341	0.112	0.454
0.2	1.153	1.158	1.155	0.325	0.109	0.434
0.3	1.137	1.148	1.140	0.299	0.102	0.400
0.4	1.112	1.134	1.118	0.264	0.091	0.355
0.5	1.176	1.112	1.086	0.221	0.078	0.299
0.6	1.023	1.080	1.039	0.173	0.063	0.236
0.7	0.946	1.033	0.971	0.122	0.046	0.167
0.8	0.828	0.955	0.865	0.071	0.028	0.099
0.9	0.629	0.801	0.678	0.027	0.011	0.038
0.99	0.212	0.325	0.240	0.001	0.000	0.001

Functions $\bar{\alpha}(\beta)$ monotonically decrease with increasing β, all $\bar{\alpha}(\beta)$ being equal at $\beta = 0$ and $\beta = 1$ ($\bar{\alpha}(1) = 0$). In the interval $0 < \beta < 1$ they satisfy the inequalities $\bar{\alpha}_\sigma(\beta) < \bar{\alpha}(\beta) < \bar{\alpha}_\pi(\beta)$, and as $\beta \to 1$, the functions $\bar{\alpha}(\beta)$ have the following asymptotics:

$$\bar{\alpha}(\beta) \approx a\gamma^{-1},$$

where

$$a_\sigma \approx 1.509,\ a_\pi \approx 2.391,\ a \approx 1.722.$$

Functions $F^0(\beta,\bar{\alpha}(\beta))$ monotonically decrease to zero with increasing β.

We can see that with $\omega \to 0$, the spectral distribution does not tend to zero at $\alpha \neq 0, \pi$. The region of frequencies about $\omega = 0$ make the most important relative contribution to the total radiated energy at small β. The decrease of contribution of low frequencies with increasing β suggests that either the radiation spectrum effectively spreads to more and more high frequencies with increasing β or a maximum in the spectrum shifts to the high-frequency region or both these factors take place. As will be shown in Sec. 6.3, when $\alpha \ll \gamma^{-1}$, an absolute maximum is at the frequency $\omega = 0$. An effective spectrum width is about $\delta\omega \sim \omega_0\gamma^2/\alpha$. Consequently, the effective spectrum width increases in this case. If α is close to π, the spectrum actually coincides with the synchrotron radiation spectrum for which $\omega_{max} \sim \omega_0\gamma^3$, while the contribution to the low-frequency radiation is negligible. In this connection it is highly interesting to examine the term $F^1(\beta,\alpha)$. If $F^1(\beta,\alpha) > 0$, then $\omega = 0$ is the point of a minimum. If $F^1(\beta,\alpha) < 0$, then at $\omega = 0$ the spectrum has a maximum and alternatively, this maximum is an absolute one (as in the case of $\alpha \ll \gamma^{-1}$) or it lies in the high-frequency region. Thus, the equations $F^1(\beta,\alpha) = 0$ define $\alpha = \alpha(\beta)$ that delineate the above mentioned cases. Numerical calculations show that functions $\alpha(\beta)$ increase with β and $\alpha_\pi(\beta) < \alpha(\beta) < \alpha_\sigma(\beta)$ for $\beta > 0$. In the limiting points we have

$$\alpha_0 = \alpha_\pi(0) = \alpha_\sigma(0) = \alpha(0) \approx 1.303,$$

$$\alpha_\pi(1) \approx 1.467, \quad \alpha(1) \approx 1.953, \quad \alpha_\sigma(1) \approx 2.260.$$

Thus, if $\alpha < \alpha_0 \sim 75^\circ$ (i.e., if the angle of the arc is $< 150^\circ$), $\omega = 0$ is a point of a maximum for the spectrum (but this is not necessary an absolute maximum).

From the point of view of physics it is obvious that for $\alpha > \gamma^{-1}$ a radiation spectrum of a relativistic electron, in its main part, must be close to a synchrotron one, now it is also obvious that this statement is true only in the high-frequency region. As to the low-frequency region, $\omega < \omega_0$ at $\alpha < \alpha(\beta)$, the spectrum is analogous to the radiation spectrum for a "short" magnet (see Sec. 6.3) and only at $\alpha > \alpha(\beta)$ one may expect an analogy to the synchrotron radiation spectrum practically all over the frequencies significantly contributing to radiation.

It is easy to deduce from Eqs (6.2) and (6.4), that the radiation in the given direction θ, ϕ in the low-frequency region is completely linearly polarized and an electrical vector of the radiation field at a zero frequency is parallel to the vector $\boldsymbol{j}(\alpha)$

$$\boldsymbol{j}(\alpha) = (\cos\phi - \mu\cos\alpha)\boldsymbol{e}_\phi + \cos\theta\sin\phi\boldsymbol{e}_\theta. \tag{6.17}$$

Eqs (6.16) allow to find a criterion of smallness of α, i.e. α can be neglected when $\nu \sim 1$, which is equivalent to the inequality

$$\alpha \ll \sqrt{1-\beta^2} = mc^2/\epsilon. \tag{6.18}$$

6.2.4. High-frequency Asymptotics of Spectrum

One can easily derive a high-frequency asymptotics of spectrum. Let us twice perform integration by parts in Eq. (6.4). It results in the identity

$$\begin{aligned} f_{\sigma,\pi} &= \frac{i}{q}\left\{\frac{b_{\sigma,\pi}(x)}{p^3(x)}e^{-iq\psi(x)}\Big|_{\phi-\alpha}^{\phi+\alpha} - \frac{i}{q}\frac{e^{-iq\psi(x)}}{p(x)}\frac{d}{dx}\frac{b_{\sigma,\pi}(x)}{p^3(x)}\Big|_{\phi-\alpha}^{\phi+\alpha} + \right. \\ &+ \left. \frac{i}{q}\int\limits_{\phi-\alpha}^{\phi+\alpha} dx\, e^{-iq\psi(x)}\frac{d}{dx}\left[\frac{1}{p(x)}\frac{d}{dx}\frac{b_{\sigma,\pi}(x)}{p^3(x)}\right]\right\}, \\ b_\sigma(x) &= \cos x - \mu, \quad b_\pi(x) = \cos\theta\sin x. \end{aligned} \tag{6.19}$$

Keeping terms decreasing with $q \to \infty$ not slowly than q^{-3} we obtain

$$\begin{aligned} F_{\sigma,\pi} &= \frac{3(1-\beta^2)^2}{8\pi^2\alpha q^2}\int\limits_0^\pi \sin\theta d\theta \int\limits_0^{2\pi} d\phi \left[\frac{b^2_{\sigma,\pi}(\phi+\alpha)}{2p^6(\phi+\alpha)} + \frac{b^2_{\sigma,\pi}(\phi-\alpha)}{2p^6(\phi-\alpha)}\right. \\ &- \left.\frac{b_{\sigma,\pi}(\phi+\alpha)b_{\sigma,\pi}(\phi-\alpha)}{p^3(\phi+\alpha)p^3(\phi-\alpha)}\cos 2q(\alpha-\mu\cos\phi\sin\alpha)\right]. \end{aligned} \tag{6.20}$$

A set of simple calculations leads to the equations

$$\begin{aligned} &\int\limits_0^\pi \sin\theta d\theta \int\limits_0^{2\pi} d\phi \frac{b^2_{\sigma,\pi}(\phi\pm\alpha)}{p^6(\phi\pm\alpha)} = \frac{\pi l_{\sigma,\pi}}{15(1-\beta^2)^3}, \\ &l_\sigma = 30 + 25\beta^2 - 3\beta^4, \quad l_\pi = 10 - 9\beta^2 + 3\beta^4. \end{aligned} \tag{6.21}$$

The last integral in Eq. (6.20) decreases as q grows if the argument $2q(\alpha - \mu\cos\phi\sin\alpha)$ of cosine really depends on the variables θ and ϕ. But one can easily see that there is no such dependence only if $\beta\sin\alpha = 0$. Hence, either $\beta = 0$ — the electron is a nonrelativistic one, or $\alpha = \pi n$ (n is an integer) — the number of revolutions is integral. Both the cases will be considered below. Thus, the main term of the high frequency asymptotics of spectrum takes the form

$$\begin{aligned} F_\sigma(\beta,\alpha;q) &\approx \frac{30 + 25\beta^2 - 3\beta^4}{40\pi\alpha(1-\beta^2)q^2}, \\ F_\pi(\beta,\alpha;q) &\approx \frac{10 - 9\beta^2 + 3\beta^4}{40\pi\alpha(1-\beta^2)q^2}. \end{aligned} \tag{6.22}$$

If $\beta \neq 0$ but $\alpha = \pi n$, we obtain from Eq. (6.20)

$$
\begin{aligned}
F_\sigma(\beta, \pi n; q) &\approx \frac{30 + 25\beta^2 - 3\beta^4}{20n(1-\beta^2)} \left(\frac{\sin \pi n q}{\pi q} \right)^2 , \\
F_\pi(\beta, \pi n; q) &\approx \frac{10 - 9\beta^2 + 3\beta^4}{20n(1-\beta^2)} \left(\frac{\sin \pi n q}{\pi q} \right)^2 .
\end{aligned}
\tag{6.23}
$$

Note high degree of linear polarization of radiation in high-frequency limit. We see from Eq. (6.22) that at $\beta \approx 1$ the π-component takes only 1/14 of the total power of radiation.

6.2.5. *Radiation of a Nonrelativistic Electron*

Principal characteristics of radiation of a nonrelativistic electron can be obtained in a simple analytical form. From Eqs (6.2)–(6.4) and Eq. (6.6) at $\beta = 0$ we find the spectral-angular distribution

$$
\begin{aligned}
F_\sigma(0, \alpha, q; \theta, \phi) &= \frac{3}{16\pi^2 \alpha} \left[\frac{\sin^2 \alpha(q-1)}{(q-1)^2} + \frac{\sin^2 \alpha(q+1)}{(q+1)^2} + \right. \\
&\quad \left. 2 \frac{\sin \alpha(q-1) \sin \alpha(q+1)}{q^2 - 1} \cos 2\phi \right] , \\
F_\pi(0, \alpha, q; \theta, \phi) &= \frac{3 \cos^2 \theta}{16\pi^2 \alpha} \left[\frac{\sin^2 \alpha(q-1)}{(q-1)^2} + \frac{\sin^2 \alpha(q+1)}{(q+1)^2} - \right. \\
&\quad \left. 2 \frac{\sin \alpha(q-1) \sin \alpha(q+1)}{q^2 - 1} \cos 2\phi \right] .
\end{aligned}
\tag{6.24}
$$

After integration over the angles we obtain the spectral distribution of radiation

$$
\begin{aligned}
F_\sigma(0, \alpha, q) = \frac{3}{4} F, \quad & F_\pi(0, \alpha, q) = \frac{1}{4} F, \\
F = \frac{\sin^2 \alpha(q-1)}{\pi \alpha (q-1)^2} &+ \frac{\sin^2 \alpha(q+1)}{\pi \alpha (q+1)^2} .
\end{aligned}
\tag{6.25}
$$

If we integrate in Eq. (6.24) over the frequency we have the angular distribution of radiation for a nonrelativistic electron

$$
\begin{aligned}
F_\sigma(0, \alpha, \theta, \phi) &= \frac{3}{16\pi} \left[1 + \frac{\sin \alpha}{\alpha} \cos \alpha \cos 2\phi \right] , \\
F_\pi(0, \alpha, \theta, \phi) &= \frac{3 \cos^2 \theta}{16\pi} \left[1 - \frac{\sin \alpha}{\alpha} \cos \alpha \cos 2\phi \right] .
\end{aligned}
\tag{6.26}
$$

This results from Eq. (6.8) when $\beta = 0$. It is clear from Eq. (6.25), that the main contribution to radiation is given by the frequency region $\omega \sim \omega_0$ and the effective width of the spectrum is $\Delta\omega \sim \omega_0/\alpha$.

6.2.6. Radiation in Case of Integral Number of Revolutions

Now let an electron make a finite number N of revolutions ($\alpha = \pi N,\ N = 1, 2, \ldots$). It is convenient in this case to fix a point of inlet and outlet of an electron for example, on y-axis. Then Eq. (6.6) takes the form

$$
\begin{aligned}
dW &= W_0 T g_N(q)(\overline{F}_\sigma + \overline{F}_\pi) dq\, d\Omega, \quad q > 0, \\
\overline{F}_\sigma &= \frac{3(1-\beta^2)^2}{4\pi^3} \left| \left\{ q \int_0^\phi \sin x e^{-iq\psi(x)} dx - ie^{-iq\psi(\phi)} p^{-1}(\phi) \sin\phi + \pi q \boldsymbol{E}'_q(q\mu) \right\} \right. \\
&\quad \left. \cdot \sin \pi q + \pi q \boldsymbol{J}'_q(q\mu) \cos \pi q \right|^2, \\
\overline{F}_\pi &= \frac{3(1-\beta^2)^2 \cot^2\theta}{4\pi^3\beta^2} \left| \left\{ q \int_0^\phi e^{-iq\psi(x)} dx - ie^{-iq\psi(\phi)} p^{-1}(\phi) + i\pi q \boldsymbol{E}_q(q\mu) \right\} \right. \\
&\quad \left. \cdot \sin \pi q + i\pi q \boldsymbol{J}_q(q\mu) \cos \pi q \right|^2, \\
g_N(q) &= (\sin^2 \pi q N)(N \sin^2 \pi q)^{-1}.
\end{aligned}
\tag{6.27}
$$

Here $\boldsymbol{J}_q(x)$ and $\boldsymbol{E}_q(x)$ are the Anger and Weber functions, respectively; the primed symbols denote their derivatives. We can see from this formula that the number of revolutions occurs only in the factor $g_N(q)$ and the rest of the expressions in Eq. (6.27) is independent of N and coincides with a spectral-angular distribution of radiation of an electron making one revolution ($g_1(q) = 1$). Spectral-angular distribution depends on the angle ϕ. It is physically obvious that asymmetry of the angular distribution is associated with the presence of the point of inlet (outlet). Integration over the frequency in Eq. (6.27) leads to the formulae in Sec. 6.2.2 and eliminates the asymmetry in ϕ.

Eq. (6.27) allows to trace how the spectrum of synchrotron radiation forms while $N \to \infty$. It can be easily seen that

$$
\lim_{N\to\infty} g_N(q) = \sum_{n=-\infty}^{\infty} \delta(q - n). \tag{6.28}
$$

Besides, if $N \gg 1$, then the effective width of a spectral line at $q = n$ is $\Delta q \approx 1/N$. In the limit $N \to \infty$ we find from Eq. (6.27) that

$$
\lim_{N\to\infty} dW/T = (ce^2/2\pi R^2) \sum_{n=1}^{\infty} n^2 [\beta^2 J_n'^2(n\mu) + \cot^2\theta J_n^2(n\mu)] d\Omega. \tag{6.29}
$$

We have taken into account, that $\boldsymbol{J}_n(x) = J_n(x)$ where $J_n(x)$ is the Bessel function of an integral index. This is a well known Schott's formula, i.e., a classical expression for the spectral-angular distribution of synchrotron radiation.

As $N \to \infty$, the terms producing asymmetry of the field in Eq. (6.27) fall out, being smaller by a factor of $1/N$ than the terms independent of ϕ.

Thus, we have succeeded in revealing the character of formation of the synchrotron radiation spectrum in the case of an electron making a finite, but large number of revolutions in a circle.

6.2.7. Representation of Spectrum in Terms of Two-Dimensional Integrals

Integrating function $F(\beta, \alpha; q; \theta, \phi)$ of Eq. (6.6) over the angles, we obtain the spectral distribution

$$F(\beta, \alpha; q) = \int_0^{\pi} \sin\theta d\theta \int_0^{2\pi} d\phi F(\beta, \alpha; q; \theta, \phi), \tag{6.30}$$

which satisfies the obvious condition

$$\int_0^{\infty} F(\beta, \alpha; q) dq = 1. \tag{6.31}$$

Taking into account Eq. (6.4) one can find the following integral representation of spectral distribution

$$\begin{aligned}
F(\beta, \alpha; q) &= F_\sigma(\beta, \alpha; q) + F_\pi(\beta, \alpha; q), \quad F_{\sigma,\pi}(\beta, \alpha; q) = \frac{3(1-\beta^2)^2}{16\pi^2\alpha}\Phi_{\sigma,\pi}, \\
\Phi_\sigma &= \int_0^{\pi} \sin\theta d\theta \int_0^{2\pi} d\phi \int_{\phi-\alpha}^{\phi+\alpha} dx \int_{\phi-\alpha}^{\phi+\alpha} dy \frac{(\cos x - \mu)(\cos y - \mu)}{p^2(x) p^2(y)} \cos 2q\chi, \\
\Phi_\pi &= \cos^2\theta \int_0^{\pi} \sin\theta d\theta \int_0^{2\pi} d\phi \int_{\phi-\alpha}^{\phi+\alpha} dx \int_{\phi-\alpha}^{\phi+\alpha} dy \frac{\sin x \sin y}{p^2(x) p^2(y)} \cos 2q\chi, \\
2\chi &= \psi(y) - \psi(x),
\end{aligned} \tag{6.32}$$

which can be used as a basis of numerical analysis of the functions $F(\beta, \alpha; q)$. But the calculation of four-dimensional integrals at different values of β, α and q is too cumbersome and not very convenient.

It is possible to transform Eq. (6.32) into a form which allows to integrate it over two variables. Thus, the result can be written in terms of two-dimensional integrals.

Shifting the limits of integration with respect to x $(x \to x + \phi)$ and y $(y \to y + \phi)$, we rewrite Eqs (6.32) in the form

$$\Phi_{\sigma,\pi} = \int_0^{\pi} \sin\theta d\theta \int_0^{2\pi} d\phi \int_{-\alpha}^{\alpha} dx \int_{-\alpha}^{\alpha} dy r^{-2} a_{\sigma,\pi} \cos 2q\chi, \tag{6.33}$$

where

$$\begin{aligned}
a_\sigma &= [\cos(\phi + x) - \mu][\cos(\phi + y) - \mu] = \mu^2 - 2\mu\cos(\phi + \frac{x+y}{2})\cos\frac{x+y}{2} \\
&+ \cos(\phi + x)\cos(\phi + y), \\
a_\pi &= \cos^2\theta \sin(x + \phi)\sin(y + \phi), \\
\chi &= \frac{y-x}{2} + \mu\cos(\phi + \frac{x+y}{2})\sin\frac{x-y}{2}, \\
r &= p(x+\phi)p(y+\phi) = 1 - 2\mu\cos(\phi + \frac{x+y}{2})\cos\frac{x-y}{2} \\
&+ \mu^2\cos(\phi + x)\cos(\phi + y).
\end{aligned} \tag{6.34}$$

The integrals in Eq. (6.33) are periodical (with the period 2π) functions of ϕ and integration should be performed over the period. Hence, making a shift $\phi \to \phi - (x + y)/2$ we do not change the range of integration over ϕ and Eq. (6.33) keeps its form. But Eqs (6.34) transform into

$$\begin{aligned}
a_\sigma &= \mu^2 + \cos(\phi + \frac{x-y}{2})\cos(\phi - \frac{x-y}{2}) - 2\mu\cos\phi\cos\frac{x-y}{2}, \\
a_\pi &= \cos^2\theta\sin(\phi + \frac{x-y}{2})\sin(\phi - \frac{x-y}{2}), \\
\chi &= \frac{y-x}{2} + \mu\cos\phi\sin\frac{x-y}{2}, \\
r &= 1 - 2\mu\cos\phi\cos\frac{x-y}{2} + \mu^2\cos(\phi + \frac{x-y}{2})\cos(\phi - \frac{x-y}{2}).
\end{aligned} \tag{6.35}$$

Now let us make a shift $y \to y + x$ which gives

$$\Phi_{\sigma,\pi} = \int_0^{\pi} \sin\theta d\theta \int_0^{2\pi} d\phi \int_{-\alpha}^{\alpha} dx \int_{-\alpha-x}^{\alpha-x} dy\, r^{-2} a_{\sigma,\pi} \cos 2q\chi \tag{6.36}$$

where the integrands take the form

$$a_\sigma = \mu^2 + \cos(\phi + \frac{y}{2})\cos(\phi - \frac{y}{2}) - 2\mu\cos\phi\cos\frac{y}{2},$$

$$\begin{aligned} a_\pi &= \cos^2\theta\sin(\phi+\frac{y}{2})\sin(\phi-\frac{y}{2}), \\ \chi &= \frac{y}{2}-\mu\cos\phi\sin\frac{y}{2}, \\ r &= 1-2\mu\cos\phi\cos\frac{y}{2}+\mu^2\cos(\phi+\frac{y}{2})\cos(\phi-\frac{y}{2}). \end{aligned} \tag{6.37}$$

The integrand in Eq. (6.36) contains a variable x within the limits of integration only over y. Let us consider the integral

$$J(\alpha)=\int\limits_{-\alpha}^{\alpha}dx\int\limits_{-\alpha-x}^{\alpha-x}f(y)dy, \tag{6.38}$$

where $f(y)$ is an arbitrary function of y. Taking a derivative from Eq. (6.38) with respect to α we obtain

$$J'(\alpha)=\int\limits_{-2\alpha}^{0}f(y)dy+\int\limits_{0}^{2\alpha}f(y)dy+\int\limits_{-\alpha}^{\alpha}f(\alpha-y)dy+\int\limits_{-\alpha}^{\alpha}f(-\alpha-y)dy. \tag{6.39}$$

We change the variables as follows: in the first integral $y=-x$, in the third one $y=\alpha-x$, and in the fourth one $y=x-\alpha$. It results in

$$J'(\alpha)=2\int\limits_{0}^{2\alpha}[f(x)+f(-x)]dx. \tag{6.40}$$

Integrating Eq. (6.40) over α with regard to the initial condition $J(0)=0$ following from Eq. (6.38) we find

$$J(\alpha)=\int\limits_{0}^{2\alpha}(2\alpha-x)[f(x)+f(-x)]dx. \tag{6.41}$$

Thus, the two-dimensional integral in Eq. (6.38) is reduced to a simple one. Finally, replacing $x\to 2x$ and taking into account that $f(x)$ is even according to Eq. (6.37), we write Eq. (6.36) in the form

$$\Phi_{\sigma,\pi}=8\int\limits_{0}^{\pi}\sin\theta d\theta\int\limits_{0}^{2\pi}d\phi\int\limits_{0}^{\alpha}dx(\alpha-x)r^{-2}a_{\sigma,\pi}\cos 2q\chi, \tag{6.42}$$

where

$$\begin{aligned}
a_\sigma &= \beta^2\sin^2\theta - 2\beta\sin\theta\cos\phi\cos x + \cos^2\phi - \sin^2 x,\\
a_\pi &= \cos^2\theta(\cos^2 x - \cos^2\phi),\\
\chi &= x - \beta\sin\theta\cos\phi\sin x,\\
r &= (1-\beta\sin\theta\cos\phi\cos x)^2 - \beta^2\sin^2\theta\sin^2\phi\sin^2 x.
\end{aligned} \tag{6.43}$$

We have substituted μ from Eq. (6.4). Let us change the variables as follows:

$$\begin{gathered}
\sin\theta\cos\phi = \cos\xi, \quad 0\le\xi\le\pi, \quad 0\le\eta\le 2\pi,\\
\sin\theta\sin\phi = \sin\xi\sin\eta, \quad \sin\theta d\theta d\phi = \sin\xi d\xi d\eta,\\
\cos\theta = \sin\xi\cos\eta, \quad \sin^2\theta = A = \cos^2\xi + \sin^2\xi\sin^2\eta.
\end{gathered} \tag{6.44}$$

As a result we obtain from Eqs (6.42) and (6.43)

$$\begin{aligned}
\Phi_{\sigma,\pi} &= 8\int_0^\pi \sin\xi d\xi \int_0^{2\pi} d\eta \int_0^\alpha dx(\alpha - x) I_{\sigma,\pi}\cos[2q(x - \beta\cos\xi\sin x)],\\
I_\sigma &= \frac{\beta^2A^2 - A(2\beta\cos\xi\cos x + \sin^2 x) + \cos^2\xi}{A(B - A\beta^2\sin^2 x)^2},\\
I_\pi &= \frac{(1-A)(A\cos^2 x - \cos^2\xi)}{A(B - A\beta^2\sin^2 x)^2},\\
B &= 1 - 2\beta\cos\xi\cos x + \beta^2\cos^2\xi.
\end{aligned} \tag{6.45}$$

We see now that the variable η occurs only in A and one can perform the integration over ν exactly. Quantities $I_{\sigma,\pi}$ can be represented in the form

$$I_{\sigma,\pi} = \frac{C^{(1)}_{\sigma,\pi}}{A} + \frac{C^{(2)}_{\sigma,\pi}}{B - A\beta^2\sin^2 x} + \frac{C^{(3)}_{\sigma,\pi}}{(B - A\beta^2\sin^2 x)^2}. \tag{6.46}$$

Comparing Eqs (6.45) and (6.46) we obtain

$$\begin{aligned}
C^{(1)}_\sigma &= -C^{(1)}_\pi = \frac{\cos^2\xi}{B^2}, \quad C^{(2)}_\sigma = \frac{\beta^2\sin^4 x\cos^2\xi - B}{B^2\sin^2 x},\\
C^{(2)}_\pi &= \frac{B^2\cos^2 x - \beta^4\sin^4 x\cos^2\xi}{B^2\beta^2\sin^2 x},\\
C^{(3)}_\sigma &= [\beta^2\sin^4 x\cos^2\xi - B(2\beta\cos x + \sin^2 x)\sin^2 x + B^2](B\sin^2 x)^{-1},\\
C^{(3)}_\pi &= (\beta^2\sin^2 x - B)(B\cos^2 x - \beta^2\sin^2 x\cos^2\xi)(B\beta^2\sin^2 x)^{-1}.
\end{aligned} \tag{6.47}$$

Taking into account that

$$\int_0^{2\pi} \frac{d\eta}{1+\varepsilon\sin^2\eta} = \frac{2\pi}{\sqrt{1+\varepsilon}}, \quad \int_0^{2\pi} \frac{d\eta}{(1+\varepsilon\sin^2\eta)^2} = \frac{\pi(2+\varepsilon)}{(1+\varepsilon)^{3/2}}, \tag{6.48}$$

we calculate

$$\begin{aligned} \int_0^{2\pi} \frac{d\eta}{A} &= \frac{2\pi}{|\cos\xi|}, \\ \int_0^{2\pi} \frac{d\eta}{B - A\beta^2\sin^2 x} &= \frac{2\pi}{(1-\beta\cos\xi\cos x)\sqrt{B-\beta^2\sin^2 x}}, \\ \int_0^{2\pi} \frac{d\eta}{(B - A\beta^2\sin^2 x)^2} &= \frac{\pi[(1-\beta\cos\xi\cos x)^2 + B - \beta^2\sin^2 x]}{(1-\beta\cos\xi\cos x)^3(B-\beta^2\sin^2 x)^{3/2}}. \end{aligned} \tag{6.49}$$

This gives the result of integration over a variable β. Making one more change of a variable $y = \cos\xi$, finally we derive the following representation of spectral distribution

$$F(\beta,\alpha;q) = \frac{3(1-\beta^2)^2}{2\pi\alpha}\int_{-1}^{1} dy \int_0^{\alpha} dx(\alpha - x)I(\beta;x,y)\cos 2q(x - \beta y\sin x), \tag{6.50}$$

where for the radiation summarized over the polarization components one should set the following

$$\begin{aligned} I_0 &= I(\beta;x,y) = (B-\beta^2\sin^2 x)^{-3/2}\{2(1-\beta y\cos x) \\ &- \frac{(1+3\beta^2)y^2\cos^2 x - 2(1+\beta^2)y^2 - (3+\beta^2)\cos^2 x + 4}{1-\beta y\cos x} \\ &+ \left.\frac{\beta^2(1-y^2)[2-(1+\beta^2)y^2]\sin^4 x}{(1-\beta y\cos x)^3}\right\}, \\ B &= 1 - 2\beta y\cos x + \beta^2 y^2. \end{aligned} \tag{6.51}$$

For the σ- and π-components of linear polarization we derive

$$\begin{aligned} I_\pi &= -\frac{2|y|}{B^2} + \frac{1}{B^2(B-\beta^2\sin^2 x)^{1/2}}\left\{(1-\beta y\cos x)[\cos^2 x + y^2(1+\sin^2 x)]\right. \\ &- \left.\frac{(1+2\sin^2 x)(1-y^2)\beta^2 y^2\sin^2 x}{1-\beta y\cos x} - \frac{\beta^4 y^4(1-y^2)\sin^6 x}{(1-\beta y\cos x)^3}\right\}, \\ I_\sigma &= I_0 - I_\pi. \end{aligned} \tag{6.52}$$

Eqs (6.51) and (6.52) are the starting point for numerical calculations.

6.2.8. Numerical Analysis of Spectrum

Radiation spectrum at low frequencies has been analyzed in Sec. 6.2.3. In case of ultrarelativistic motion the location of maximum in terms of a variable q practically does not depend on α if $\alpha \gg \gamma^{-1}$ and belongs to q defined by the synchrotron radiation theory

$$q_{max} \approx a\gamma^3, \quad a = 0.4287, \quad a_\sigma = 0.5032, \quad a_\pi = 0.2158.$$

The main results of numerical calculations are represented in Figs 6.1– 6.13 where a set of functions $F(\beta, \alpha; q)$ is plotted for different values of parameters. The set of functions for a fixed β and different α is shown in Figs 6.1–6.7, while the functions for a fixed α and a varying β are plotted in Figs 6.8 – 6.13.

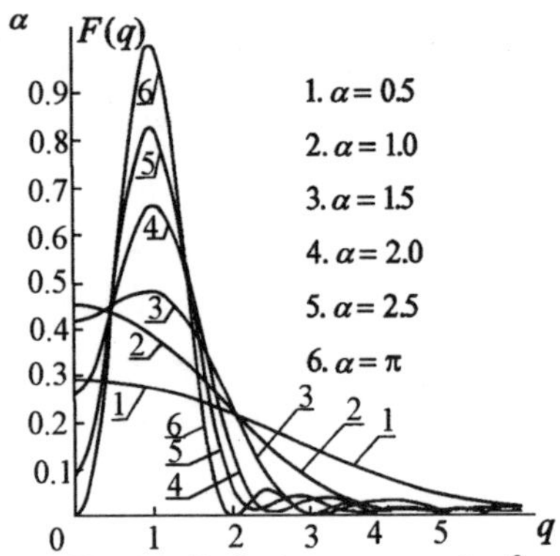

Fig. 6.1. Radiation spectrum for $\beta = 0$.

6.3. Radiation in a 'Short Magnet'

Let us consider electromagnetic radiation generated by an electron moving through a "short" magnet system. By a "short" magnet system we will understand the following: before entering a "short" magnet an electron travels uniformly and linearly with a velocity $v = c\beta$ (c is the velocity of light) but in a "short" magnet external electromagnetic fields change the direction of electron motion without visible change in its total energy, and after leaving a "short" magnet an electron moves uniformly and linearly again, with practically the same velocity, moreover, a deflection angle α (the angle between the initial and final electron momentum) is small and fulfills the condition

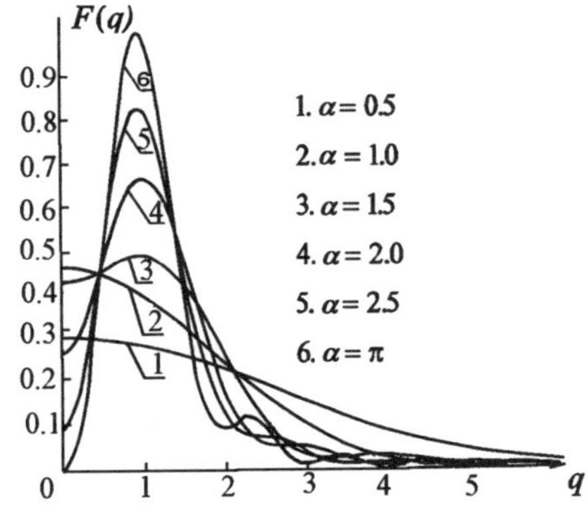

Fig. 6.2. Radiation spectrum for $\beta = 0.2$.

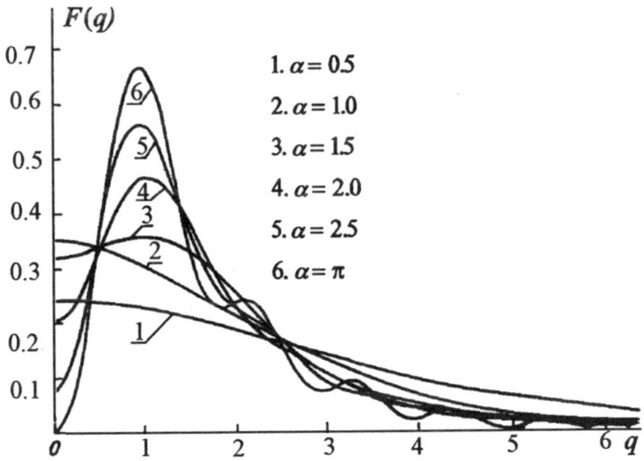

Fig. 6.3. Radiation spectrum for $\beta = 0.4$.

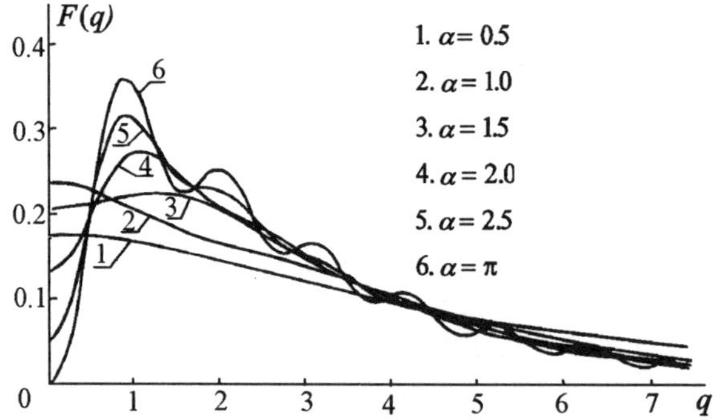

Fig. 6.4. Radiation spectrum for $\beta = 0.6$.

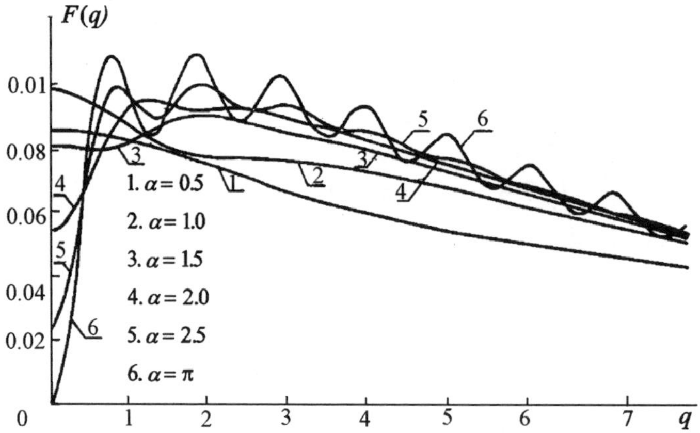

Fig. 6.5. Radiation spectrum for $\beta = 0.8$.

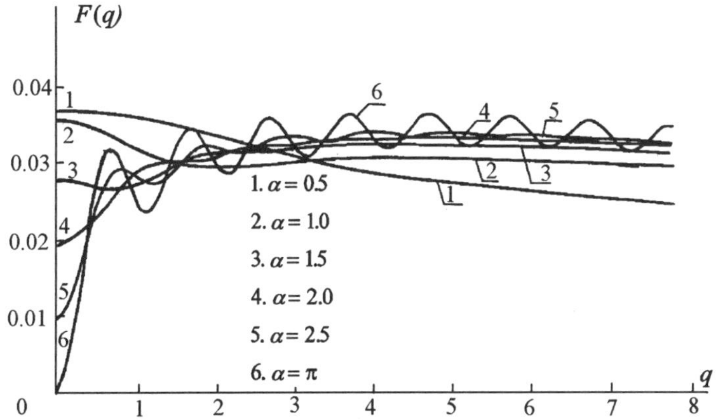

Fig. 6.6. Radiation spectrum for $\beta = 0.9$.

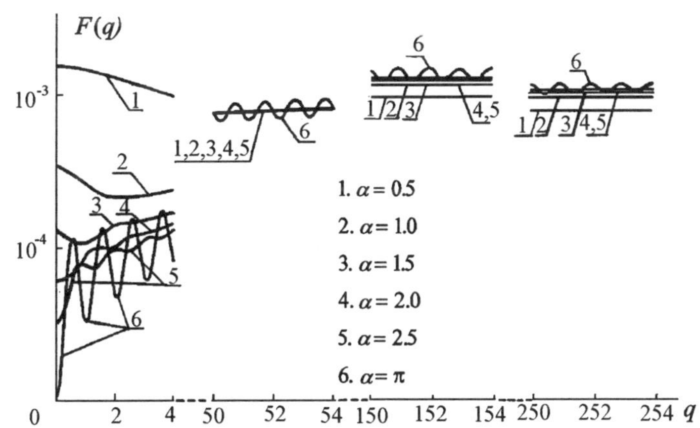

Fig. 6.7. Radiation spectrum for $\beta = 0.99$.

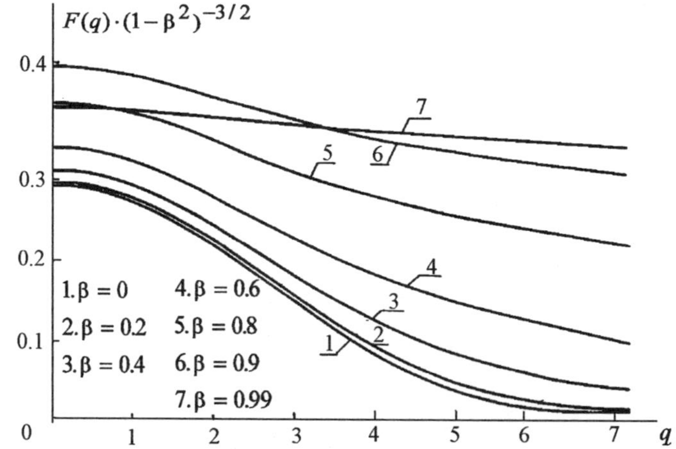

Fig. 6.8. Radiation spectrum for $\alpha = 0.5$

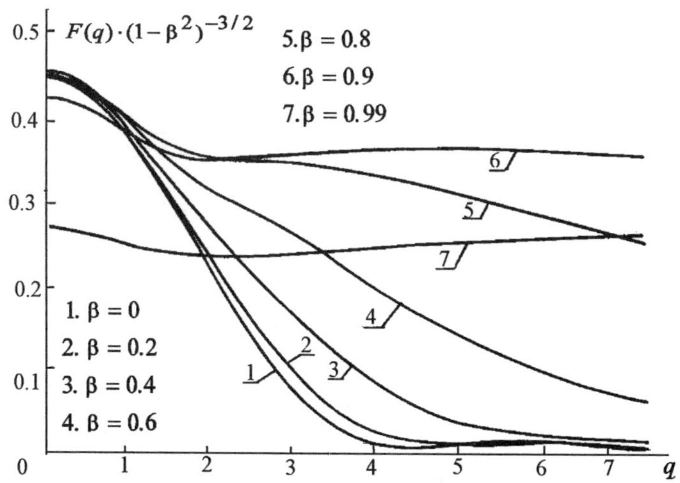

Fig. 6.9. Radiation spectrum for $\alpha = 1$

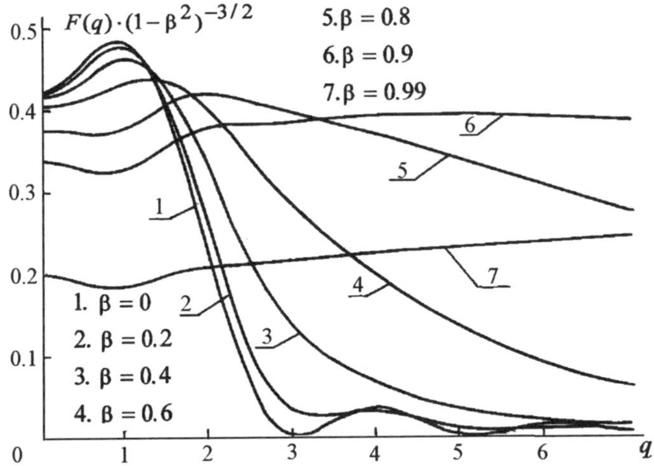

Fig. 6.10. Radiation spectrum for $\alpha = 1.5$

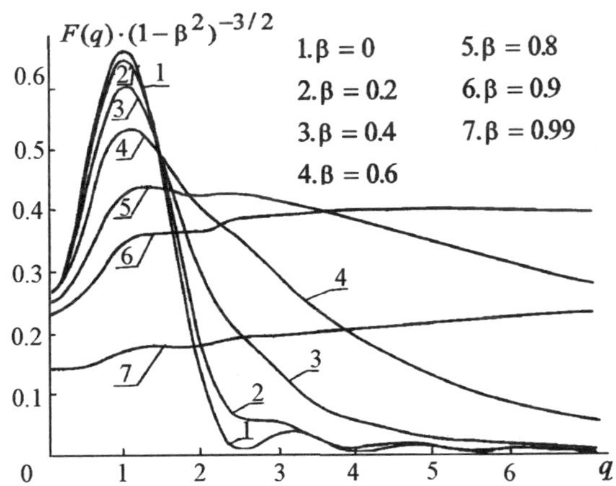

Fig. 6.11. Radiation spectrum for $\alpha = 2$

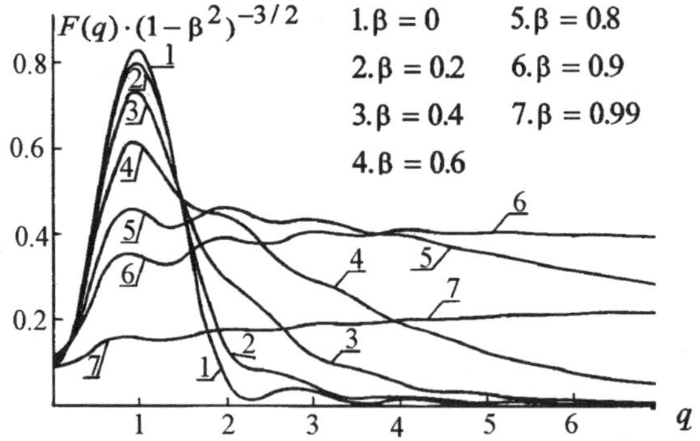

Fig. 6.12. Radiation spectrum for $\alpha = 2.5$

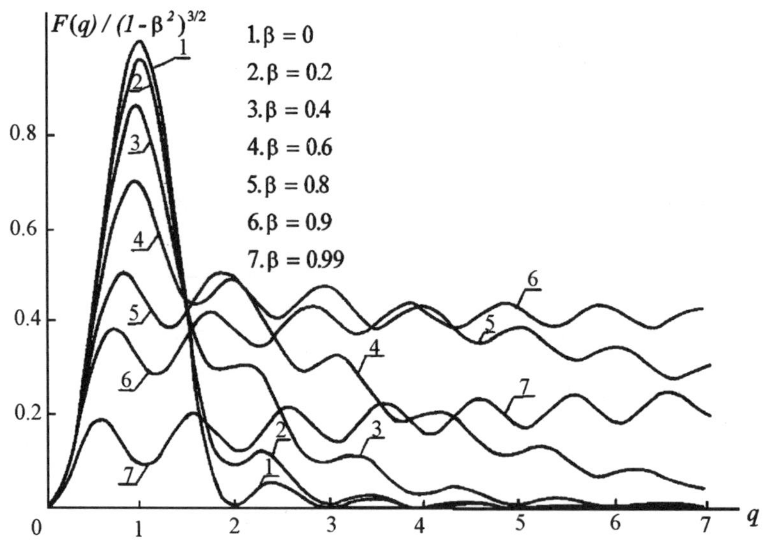

Fig. 6.13. Radiation spectrum for $\alpha = \pi$

$$\alpha \ll \gamma^{-1}, \quad \gamma^{-1} = \sqrt{1-\beta^2}, \tag{6.53}$$

where mc^2 is the electron rest energy. Thus, the motion in a "short" magnet is in fact quasi-elastic electron scattering on an external electromagnetic field at a small angle.

Some of the simplest properties of radiation in a "short" magnet have been treated in the handbooks on classical electrodynamics (for instance, some properties of this radiation have been discussed in the book [27]) and in papers [1 – 7,10,12,13,15,16,20,23,25]. Here, we will show that the main radiation characteristics in a "short" magnet can be obtained in closed analytical form by the methods of classical electrodynamics and are interesting from the physical point of view.

The system under discussion is a particular case of an undulator with a small undulator parameter k considered in Sec. 5.2.1. But, contrary to the undulator field we assume here that external electric and magnetic fields do not change their direction and sign within the interaction region.

Electron velocity components before entering and after escape from a "short" magnet are $v_x = c\beta\cos\alpha/2, v_y = \pm c\beta\sin\alpha/2$ and, applying condition of Eq. (6.53) we can write in the spirit of a dipole approximation [27]

$$v_x = c\beta, \quad v_y = v_z = 0. \tag{6.54}$$

In this case we obtain from the Lorentz equation $|\dot{\beta}_x| \ll |\dot{\beta}_y|$ and for $\dot{\beta}_y$ we have the expression

$$mc\gamma\dot{\beta}_y = e\beta(H_z - \beta^{-1}E_y), \quad \dot{\beta}_x = 0. \tag{6.55}$$

A dot denotes a time derivative; H_z, E_y are the corresponding electromagnetic field components. We introduce the effective field on the trajectory as follows:

$$H_{eff} = H_z - \beta^{-1}E_y = Hf(s/l), \tag{6.56}$$

where H is a constant, $f(s/l)$ is the field distribution function on the trajectory s. s is the coordinate along the trajectory and l is the characteristic length of a "short" magnet and will be defined below.

Let $f \geq 0$. We select a reference point s so that a maximum of the effective field is in the vicinity of $s = 0$. Let function f obey the condition

$$l^{-1}\int\limits_{-\infty}^{\infty} f^2(s/l)\,ds = \int\limits_{-\infty}^{\infty} f^2(u)\,du = 1. \tag{6.57}$$

Then H is a root-mean-square field amplitude. Let us introduce a mean radius of the trajectory in a "short" magnet by the relation $R = \beta mc^2\gamma/eH$ and the characteristic length by $l = \alpha R$. We also introduce an effective time T of the electron motion in the "short" magnet $T = l/c$.

Taking into account the above assumptions we find after simple calculations the expression for sn electric vector of the radiation field at a point $\boldsymbol{r}$ and at a time t

$$\boldsymbol{E}(\boldsymbol{r}, t) = \frac{\alpha e\beta \boldsymbol{j}}{2\pi c r p^2} \int_{-\infty}^{\infty} \psi(q) \exp[i\omega(t - r/c)]\, d\omega. \tag{6.58}$$

Here vector $\boldsymbol{j}$ does not depend on frequency ω and in terms of unit vectors $\boldsymbol{e}_\theta$, $\boldsymbol{e}_{phi}$ of the spherical coordinate system it has the form

$$\boldsymbol{j} = (\cos\phi - \beta\sin\theta)\boldsymbol{e}_\phi + \cos\theta\sin\phi\boldsymbol{e}_\theta. \tag{6.59}$$

Function $\psi(q)$ is the Fourier component of an effective field

$$\psi(q) = \int_{-\infty}^{\infty} f(u)\exp(-iqu)\, du$$
$$q = T\omega p, \quad p = 1 - \beta\sin\theta\cos\phi. \tag{6.60}$$

and fulfills, according to Eq. (6.57), the condition

$$\int_0^{\infty} |\psi(q)|^2\, dq = \pi. \tag{6.61}$$

It follows from Eq. (6.58) that the electron radiation in a "short" magnet in a given direction θ, ϕ is completely linearly polarized and the electric vector of the radiation field is parallel to the vector $\boldsymbol{j}$. For different directions θ, ϕ polarization vectors are different. The expansion of the vector $\boldsymbol{j}$ in terms of the unit vectors of the spherical coordinate system corresponds to the decomposition of a radiation field into σ- and π-components of linear polarizations.

From Eq. (6.58) we obtain the spectral angular distribution of the radiated energy $\mathcal{E}$, that can be represented in the following form

$$\frac{d\mathcal{E}}{d\Omega d\omega} = W_0 T G(\theta, \phi, q), \tag{6.62}$$

where

$$W_0 = \frac{2ce^2\beta^4\gamma^4}{3R^2}, \quad G(\theta, \phi, q) = (\rho_\sigma + \rho_\pi) p g(q), \quad g(q) = \frac{1}{\pi}|\psi(q)|^2. \tag{6.63}$$

Note that the obvious property follows from Eq. (6.61)

$$\int_0^\infty g(q)\, dq = 1, \tag{6.64}$$

where the angular functions ρ_σ and ρ_π are equal to the functions ρ_ϕ^H and ρ_θ^H defined by Eqs (1.41). In alternative coordinate system these functions are given in Eqs (5.53).

Integrating angular distribution of the radiated energy over ω in Eq. (6.62) and using Eq. (6.64) we obtain

$$\frac{d\mathcal{E}}{d\Omega} = W_0 T(\rho_\sigma + \rho_\pi) \tag{6.65}$$

and integrating in Eq. (6.65) over the angles we find the total radiated energy and the σ- and π-components of the linear polarization

$$\mathcal{E} = W_0 T, \quad \mathcal{E}_\sigma = \mathcal{E}(6+\beta^2)/8, \quad \mathcal{E}_\pi = \mathcal{E}(2-\beta^2)/8. \tag{6.66}$$

W_0 is the total power of synchrotron radiation from an electron moving in a circle of a radius R. Thus, the total radiated energy is the product of synchrotron radiation power and the effective time of the electron motion in the "short" magnet. The same follows from Eq. (5.88) if we replace $\lambda_0 N \to l$. The degree of linear polarization of radiation in terms of σ- and π-components coincides with the corresponding expressions for synchrotron radiation.

The angular distribution of radiation is completely characterized by the functions ρ_σ and ρ_π and coincides with the instantaneous angular distribution of synchrotron radiation. In particular, for relativistic ($\beta \sim 1$) electron the radiation is concentrated in a narrow cone $\Delta\Omega \sim \gamma^{-2}$ with the axis coinciding with the x-axis (the radiation maximum is reached at $\theta = \pi/2, \phi = 0$).

Spectral properties of radiation are defined by the function $g(q)$. While $f(u)$ noticeably differs from zero only for $|u| < 1$, the function $g(q)$ noticeably differs from zero at $q < 1$, and if $q > 1$ it decreases with increasing q. As follows from the general properties of Fourier integrals, if $f(u)$ is continuous, then the spectral distribution decreases with increasing ω not slower than ω^{-4}, but if $f(u)$ is discontinuous (obviously, this case is physically impossible), then $g(q)$ decreases not slower than ω^{-2}. At $\omega = 0$ we have

$$g(0) = |\psi(0)|^2\pi, \quad \psi(0) = f_0 = \int_{-\infty}^{\infty} f(u)\, du > 0. \tag{6.67}$$

Moreover, it is easy to find from Eq. (6.60) that for $\omega \to 0$ the following expansion is valid

$$|\psi(q)|^2 = f_0^2[1 - q^2(\overline{u^2} - \overline{u}^2) + \ldots], \quad \overline{u^n} = f_0^{-1} \int_{-\infty}^{\infty} u^n f(u)\, du. \tag{6.68}$$

As $f(u) \geq 0$, we have $\overline{u^2} > \overline{u}^2$ and the point $\omega = 0$ is the point of maximum of the function $g(q)$, moreover $g'(q = 0) = 0$. It follows from the above that the radiation spectrum of an electron in a "short" magnet is practically "white noise" spreading from zero up to the frequencies $\omega \sim \omega_c$, where the critical frequency is $\omega_c = (c\beta/l)\gamma^4$. This result refines the general statement that the main part of radiation is concentrated in the frequency region [27] $\omega \sim \omega_c$.

Let us give some simplest examples of the function $\psi(q)$ for various distributions of an effective field along an electron trajectory.

1. Step-function

$$f(u) = \begin{cases} 1, & |u| < 1/2, \\ 0, & |u| > 1/2, \end{cases}$$

$$\psi(q) = \frac{2}{q} \sin \frac{q}{2}.$$

2. Half period of cosine

$$f(u) = \begin{cases} \sqrt{2} \cos \pi u, & |u| < 1/2, \\ 0, & |u| > 1/2, \end{cases}$$

$$\psi(q) = \frac{2\sqrt{2}\,\pi \cos q/2}{\pi^2 - q^2}.$$

3. Half period of squared cosine

$$f(u) = \begin{cases} \sqrt{8/3}\, \cos^2 \pi u, & |u| < 1/2, \\ 0, & |u| > 1/2, \end{cases}$$

$$\psi(q) = \frac{8\sqrt{2}\,\pi^2 \sin q/2}{q\sqrt{3}(4\pi^2 - q^2)}.$$

4. Normal distribution

$$f(u) = (2\pi)^{-1/4} \exp(-u^2/4),$$

$$\psi(q) = (8\pi)^{1/4} \exp(-q^2).$$

For the critical frequency ω_c at $l \sim 10 - 100$ cm, $mc^2\gamma \sim 1$ GeV we obtain approximately $\omega \sim 10^{16} - 10^{15}$ s^{-1}. These are frequencies in the visible region of spectrum. Thus, the electron radiation in a "short" magnet with such parameters is a "white noise", overlapping a radio-high-frequency range and spreading up to visible light.

As one can see, the spectral properties of radiation considered here, are significantly different from those of synchrotron radiation.

6.4. Inverse Problem of Radiation

An inverse problem of electromagnetic radiation is interesting from different points of view. There are two aspects which should be discussed in this context:

(i) determination of the trajectory of a charged particle using the known radiation parameters;

(ii) design and construction of radiation sources with preassigned spectral and polarization properties.

For example, several ways to produce radiation with certain properties from undulators have been proposed [33–37]. For a limited range of frequencies the desired spectrum could be produced by choosing a suitable magnetic field along a trajectory of a charged particle. The general solution to this problem for the case of a single charged particle has been obtained in Refs[18,19].

The electric vector $\boldsymbol{E}(t)$ of a radiation field within a wave zone approximation lies in a plane orthogonal to the unit vector $\boldsymbol{n}$, directed from a particle to an observer. Therefore, we have only two independent scalar time functions for a given $\boldsymbol{E}(t)$. Evidently this is insufficient to determine three coordinates of a particle. Hence, the inversion problem of radiation theory contains one arbitrary scalar function. In order to define the unique trajectory one needs one more time function. Several ways to solve the reverse problem described in (i) and (ii) are possible.

It is assumed that one observes electromagnetic radiation emitted in a given direction $\boldsymbol{n}$ and emitted by a single moving charge at a large distance from an observer (in the wave zone). The problem is to calculate the motion of a charged particle if the electric vector $\boldsymbol{E}(t)$ of a radiation field is a given time function at the position of an observer.

Evidently, this problem is equivalent to the problem of producing a radiation with a given spectrum and a given polarization. An analytical solution to this problem can be obtained as follows. If an electric field $\boldsymbol{E}(t)$ of the radiation is known, then a magnetic field $\boldsymbol{H} = [\boldsymbol{nE}]$ of a wave can be found [27]. The relationship between $\boldsymbol{E}(t)$ and the parameters of a particle trajectory is expressed by means of the Liénard-Wiechert equation [27]

$$\boldsymbol{E}(t) = \frac{e[\boldsymbol{n}[(\boldsymbol{n}-\boldsymbol{\beta})\dot{\boldsymbol{\beta}}]]}{cR[1-(\boldsymbol{n\beta})]^3}, \quad \boldsymbol{\beta}(t') = \frac{d\boldsymbol{r}(t')}{cdt'},$$
$$\dot{\boldsymbol{\beta}}(t') = \frac{d\boldsymbol{\beta}(t')}{dt'}, \quad t = t' + \frac{R}{c} - \frac{(\boldsymbol{nr}(t'))}{c}, \quad \boldsymbol{n} = \frac{\boldsymbol{R}}{R}, \tag{6.69}$$

where $\boldsymbol{R}$ is the radius-vector, connecting the origin (which is placed somewhere near a radiating particle) with the observation point, $R = |\boldsymbol{R}|$. All functions in the right hand side of Eq. (6.69) depend only on t'. Therefore, Eq. (6.69) can be considered as a parametric form for $\boldsymbol{E}(t)$. The radius-vector $\boldsymbol{r}(t)$ of a radiating particle is an unknown function.

The electromagnetic potential $\boldsymbol{A}(t)$ is given by

$$\boldsymbol{A}(t) = \frac{e}{R}\boldsymbol{f}(t), \quad \boldsymbol{f}(t) = \frac{\boldsymbol{\beta} - \boldsymbol{n}(\boldsymbol{n\beta})}{1 - \boldsymbol{n\beta}}, \quad \boldsymbol{nf} = 0. \tag{6.70}$$

Taking into account relation

$$\frac{\partial t'}{\partial t} = [1 - \boldsymbol{n\beta}]^{-1}, \tag{6.71}$$

one can easily see that

$$\boldsymbol{E}(t) = -\frac{1}{c}\frac{\partial \boldsymbol{A}(t)}{\partial t} = -\frac{e}{c\boldsymbol{R}}\frac{\partial t'}{\partial t}\frac{\partial \boldsymbol{f}}{\partial t'}. \tag{6.72}$$

Since the function $\boldsymbol{E}(t)$ is known, $\boldsymbol{f}(t)$ is known as well

$$\boldsymbol{f}(t) = -\frac{c\boldsymbol{R}}{e}\int \boldsymbol{E}(t)dt. \tag{6.73}$$

The scalar product $\boldsymbol{n\beta}$ can be determined from Eq. (6.70)

$$\boldsymbol{n\beta} = (f^2 + \varphi)(1 + f^2)^{-1}, \quad \varphi = \varphi(t) = \pm[\beta^2 - f^2(1 - \beta^2)]^{1/2} \tag{6.74}$$

Note that the inequalities

$$f^2 \leq \beta^2(1 - \beta^2)^{-1}, \quad \varphi^2(t) < 1 \tag{6.75}$$

result from Eq. (6.74). Taking Eq. (6.71) into account, we find

$$\frac{\boldsymbol{n\beta}}{1 - \boldsymbol{n\beta}} = \frac{1}{c}\frac{\partial}{\partial t}(\boldsymbol{nr}(t')) \tag{6.76}$$

and from Eq. (6.74)

$$\frac{\boldsymbol{n\beta}}{1 - \boldsymbol{n\beta}} = \frac{\beta^2 + \varphi}{1 - \beta^2}. \tag{6.77}$$

Comparing Eqs (6.76) and (6.77) we get the relation

$$\frac{\partial}{\partial t}(\boldsymbol{n}\boldsymbol{r}(t')) = c\frac{f^2+\varphi}{1-\beta^2}. \tag{6.78}$$

Using Eqs (6.70), (6.71) and (6.76) one finds

$$\frac{\boldsymbol{\beta} - \boldsymbol{n}(\boldsymbol{n}\boldsymbol{\beta})}{1-\boldsymbol{n}\boldsymbol{\beta}} = \frac{1}{c}\frac{\partial}{\partial t}[\boldsymbol{r}(t') - \boldsymbol{n}(\boldsymbol{n}\boldsymbol{r}(t'))] = \boldsymbol{f}(t). \tag{6.79}$$

With Eqs (6.79) and (6.78) one obtains

$$\frac{\partial \boldsymbol{r}(t')}{\partial t} = c\boldsymbol{f}(t) + c\boldsymbol{n}\frac{\beta^2+\varphi}{1-\beta^2}. \tag{6.80}$$

The relation between t and t' can be established using Eqs (6.71) and (6.74)

$$\frac{\partial t'}{\partial t} = \frac{1+\varphi}{1-\beta^2}. \tag{6.81}$$

If β^2 were a known function of t, then Eqs (6.80) and (6.81) would be a solution to the problem in a parametric form. But one can easily see that for every arbitrary function $\beta^2(t)$ Eqs (6.80) and Eq. (6.81) are a also solutions to the problem.

Thus, one can define the trajectory using a known function $\boldsymbol{f}(t)$

$$\boldsymbol{r}(t') = c\int \boldsymbol{f}(t)dt + c\boldsymbol{n}\int\frac{\beta^2+\varphi}{1-\beta^2}dt, \quad t' = \int\frac{1+\varphi}{1-\beta^2}dt - \frac{R}{c}, \tag{6.82}$$

where β^2 is an arbitrary function of t (with $\beta^2(t) < 1$), and the $\varphi(t)$ function is connected with β^2 via Eq. (6.74). The solution given by Eq. (6.82) can be rewritten using Eq. (6.74)

$$\beta^2 = (f^2+\varphi^2)(1+f^2)^{-1}. \tag{6.83}$$

Now Eq. (6.82) can be expressed as follows

$$\boldsymbol{r}(t') = c\int \boldsymbol{f}(t)dt + c\boldsymbol{n}\int\frac{f^2+\varphi}{1-\varphi}dt, \quad t' = \int\frac{1+f^2}{1-\varphi}dt - \frac{R}{c}, \tag{6.84}$$

where $\varphi(t)$ is an arbitrary function, but it must satisfy Eq. (6.75). Finally, one may write the solution as follows

$$\boldsymbol{r}(t') = c\int \boldsymbol{f}(t)dt + c\boldsymbol{n}[\psi(t) - t], \quad t' = \psi(t) - \frac{R}{c}, \tag{6.85}$$

where $\psi(t)$ is an arbitrary, monotonously increasing function of t, satisfying the condition $2\psi'(t) > 1 + f^2$. $\varphi(t)$ is connected with $\psi(t)$

$$\varphi(t) = 1 - \frac{1 + f^2}{\psi'(t)}. \tag{6.86}$$

This can be easily proved by using Eqs (6.82), (6.84) or Eq. (6.85)

$$\boldsymbol{\beta}(t') = \frac{[\boldsymbol{f}(t)(1-\varphi) + \boldsymbol{n}(f^2 + \varphi)]}{1 + f^2} = \frac{\boldsymbol{f}(t) + \boldsymbol{n}(\psi' - 1)}{\psi'}. \tag{6.87}$$

Thus, Eqs (6.82), (6.84) or (6.85) represent the solution to the problem. As can be seen there are several solutions possible since one arbitrary function is left. We note here that this means that there are several solutions for the trajectory along $\boldsymbol{n}$ which produce the required wave field. This means that the relation between the time t' of radiation of the field and the time t for an observer is also arbitrary within the given limits.

The obvious question arises whether it is possible to exploit this ambiguity to define not only $\boldsymbol{E}(t)$ for a fixed $\boldsymbol{n}$ but also full $\boldsymbol{E}(\boldsymbol{n}, t)$ dependence. Generally speaking, the answer is negative since one arbitrary function of one variable ($\varphi(t)$, for example) is not sufficient to define two variables of $\boldsymbol{n}$ ($\boldsymbol{n}$ is a unit vector and only two of its components are independent).

In the following some special aspects of the solution obtained above are discussed. If $\boldsymbol{f}(t)$ is given by the $\boldsymbol{E}(t)$ field but with the uncertainty of an arbitrary constant vector, then $\boldsymbol{r}(t')$ is also defined up to arbitrary Lorentz-transformations made in a plane perpendicular to the unit vector $\boldsymbol{n}$. Hence, if $\boldsymbol{f}(t)$ is finite one can always choose the motion in such a way that the average velocity in the direction orthogonal to $\boldsymbol{n}$ is zero. This is impossible along $\boldsymbol{n}$ in general. However, if $f^2(t) < 1$ and assuming that $\varphi(t) = -f^2(t)$ we obtain an extremely simple result

$$\boldsymbol{r}(t') = c\int \boldsymbol{f}(t)dt, \quad t' = t - \frac{R}{c}. \tag{6.88}$$

This means that the radiation with the field having such a time dependence is produced by motion in a plane perpendicular to $\boldsymbol{n}$. However, in this case the amplitude of the vector potential of the radiation field is limited. If the function $f^2(t)$ is limited (but $f^2(t)$ may be greater than 1), than one can always choose $\psi(t) = at$ in Eq. (6.85), where a is a constant satisfying the condition $2a > 1 + f^2$. In this case the desired spectrum is generated by a complex motion in the plane orthogonal to $\boldsymbol{n}$ and by the translation along $\boldsymbol{n}$. The trajectory in the plane orthogonal to $\boldsymbol{n}$ is determined unambiguously by the function $\boldsymbol{f}(t)$. If a linear polarization of the radiation is required ($\boldsymbol{f}(t)$ has only one component which differs from zero) then the radiation should be generated by a charged particle moving along a trajectory in a plane (with an accuracy up to the Lorentz translation) parallel to $\boldsymbol{n}$. These properties of the radiation emitted by an electron moving in a plane magnetic undulator have been discussed in Sec. 5.1.3.

From our point of view the following conclusion is of importance: as long as $\boldsymbol{f}(t)$ is a finite function, a solution always exists which corresponds to the motion of a particle at a constant speed ($\beta^2 = const.$). Such motion can be realized, for example, in an external magnetic field. If $\boldsymbol{f}(t)$ is infinite, then the motion can be realized only for $\beta^2(t) \to 1$ (see Eq. (6.75)). This means that such motion can be produced only by an external electromagnetic field with a nonzero electric component.

The obtained theoretical results can be illustrated by the following applications. We assume that $\boldsymbol{n}$ is parallel to z-axis.

1) For example, if radiation should be monochromatic and circularly polarized, then

$$\boldsymbol{f}(t) = f_0(\cos\omega t,\, \sin\omega t,\, 0), \quad f_0 = \text{const}, \quad \omega = \text{const}. \tag{6.89}$$

Using Eq. (6.85) we find

$$\boldsymbol{r}(t') = \frac{c}{\omega}\{f_0 \sin\omega t,\, -f_0\cos\omega t,\, \omega\psi(t) - t\omega\}, \quad t' = \psi(t) - \frac{R}{c}. \tag{6.90}$$

The trajectory is a spiral, of a radius cf_0/ω (helical motion) and an arbitrary space period in the direction of an axis of a helix. Monochromatic radiation is emitted in the direction of the axis and is circularly polarized. An analogous result is known from the synchrotron radiation theory (β=const).

2) Linearly polarized and monochromatic radiation can be emitted by a charged particle moving with a constant velocity. Assuming that

$$\boldsymbol{f}(t) = \{\beta(1-\beta^2)^{-1/2}\cos\omega t, 0, 0\}, \tag{6.91}$$

and using Eq. (6.82), we find

$$\begin{aligned} \boldsymbol{r}(t') &= c\beta a\{\sqrt{1-\beta^2}\sin\omega t,\, 0,\, \beta\omega t \pm \cos\omega t\}, \\ t' &= a\omega t \pm a\beta\cos\omega t - \frac{R}{c}, \quad a^{-1} = \omega(1-\beta^2). \end{aligned} \tag{6.92}$$

Eq. (6.92) describes the motion in a plane magnetic undulator. In this case, the first harmonic of radiation can be observed on the axis of the undulator only.

3) Finally we treat the case where linear polarization and a broad spectrum in a given spectral range are required

$$\boldsymbol{f}(t) = \{\beta(1-\beta^2)^{-1/2}\tanh\omega t,\, 0,\, 0\}, \tag{6.93}$$

β and ω are constants. With this $\boldsymbol{f}(t)$ an electric field will vary as $\boldsymbol{E}(t) \sim (\cosh\omega t)^{-2}$. If the absolute value of particle velocity is constant and equal to $c\beta$ we obtain with Eq. (6.82)

$$\begin{gathered} \boldsymbol{r}(t') = c\beta a\{\sqrt{1-\beta^2}\ln\cosh\omega t,\, 0,\, \beta\omega t \pm \arctan\sinh\omega t\}, \\ t' = a\omega t \pm a\beta\arctan\sinh\omega t - \frac{R}{c}, \quad a^{-1} = \omega(1-\beta^2). \end{gathered} \tag{6.94}$$

Eqs (6.94) describes motion of a charged particle in a bending magnet and a special case of relativistic motion $1-\beta^2 \ll 1$ in a "short" magnet system, which have been studied in the previous section.

Now we consider the case when an electric field vector is known at two observation points at a large distance from a charged particle. The simplest way to solve the problem is to use Feynman's equation [38]

$$\boldsymbol{E}(t) = \frac{e}{c^2}\frac{d^2\boldsymbol{n}}{dt^2}. \tag{6.95}$$

The unit vector $\boldsymbol{n}$ is taken at a time t'. Though $\boldsymbol{n}$ is constant, its time derivative is not.

We assume that radiation with electric field vectors $\boldsymbol{E}_1(t_1)$ and $\boldsymbol{E}_2(t_2)$ is observed in directions $\boldsymbol{n}_1$ and $\boldsymbol{n}_2$ and at distances R_1 and R_2 from a charged particle, respectively; t_1 and t_2 are the times at two observation points. We introduced two Cartesian coordinate systems $O_1(x_1, y, z_1)$ and $O_2(x_2, y, z_2)$ having a common axis y. The x_1 and x_2 axes point in the direction of $\boldsymbol{n}_1$ and $\boldsymbol{n}_2$, respectively. We assume that $\boldsymbol{E}_1$, $\boldsymbol{E}_2$, $\boldsymbol{n}_1$ and $\boldsymbol{n}_2$ are not in the same plane. α is the angle between $\boldsymbol{n}_1$ and $\boldsymbol{n}_2$; x_i, y, z_i $(i = 1, 2)$ are the coordinates of the particle at the time t'. Integrating Eq. (6.95) we obtain

$$\boldsymbol{r}_{1\perp}(t') = \boldsymbol{F}_1(t_1), \quad \boldsymbol{r}_{2\perp}(t') = \boldsymbol{F}_2(t_2), \tag{6.96}$$

where

$$\boldsymbol{F}_i(t_i) = R_i\left[\frac{c^2}{e}\int_0^{t_i} dt^* \int_0^{t^*} \boldsymbol{E}_i(t_i)dt_i + \boldsymbol{b}_i t_i\right] + \boldsymbol{r}_{i0},$$

$\boldsymbol{r}_{1\perp}$ and $\boldsymbol{r}_{2\perp}$ are the projections of a radius-vector of a charged particle on the planes yz_1 and yz_2, respectively; $\boldsymbol{b}_i$, $\boldsymbol{r}_{i0}$ are constants. The equations

$$y_1(t') = F_{1y}(t_1), \quad z_1(t') = F_{1z}(t_2), \tag{6.97}$$
$$y_2(t') = F_{2y}(t_2), \quad z_2(t') = F_{2z}(t_2), \tag{6.98}$$

together with equations of retardation of electromagnetic waves

$$t_1 = t' + \frac{x_1(t')}{c}, \quad t_2 = t' + \frac{x_2(t')}{c}, \tag{6.99}$$

lead to the equation of the trajectory of a charged particle in a non-orthogonal coordinate system ($y = y_1 = y_2, z_1, z_2$) as a function of two parameters t_1 and t_2 (constant terms in Eqs (6.99) have been neglected, since the origin for the times t_1 and t_2 can be chosen arbitrarily). The times t_1 and t_2 are related via $F_{1y}(t_1) = F_{2y}(t_2)$, i.e. $t_2 = \xi(t_1)$. It is easy to express x_1 and x_2 in terms of z_1 and z_2:

$$x_1(t') = (F_{2z}[\xi(t_1)] - F_{1z}(t_1)\cos\alpha)(\sin\alpha)^{-1}, \tag{6.100}$$
$$x_2(t') = (F_{2z}[\xi(t_1)]\cos\alpha - F_{1c}(t_1))(\sin\alpha)^{-1}. \tag{6.101}$$

Eqs (6.100) and (6.99) lead to

$$t' = t_1 - (F_{2z}[\xi(t_1)] - F_{1z}(t_1)\cos\alpha)(c\sin\alpha)^{-1}. \tag{6.102}$$

Eqs (6.97), (6.100) and (6.102) define the trajectory of a charged particle in the coordinate system (x_1, y, z_1). In a similar way one can obtain the equivalent equations for a coordinate system $(x_2, y.z_2)$.

Note that Eqs (6.97)–(6.99) are overdetermined since we have four scalar functions for choosing three coordinates of a charge.

The derived expressions are valid under assumption that we observe radiation of a single charged particle or particles with such compact distribution that they can be regarded as a point source (the domain of distribution is much smaller than the radiation wavelength).

Bibliography

1. R. Coisson, *Phys. Rev.* **A20** (1979) 524.
2. R. Coisson, *Opt. Commun.* **22** (1977) 135.
3. R. Coisson, *Nucl. Instr. and Meth.* **143** (1977) 241.

4. I.M. Ternov, V.R. Khalilov, V.G. Bagrov and M.M. Nikitin, *Sov. Phys. J.* **23** (1980) 79.
5. V.G. Bagrov, M.B. Moiseev, M.M. Nikitin and N.I. Fedosov, *Izv. Vuz., Fiz.* [Sov. Phys. J.] **24** (1980) 26.
6. V.G. Bagrov, M.B. Moiseev, M.M. Nikitin and N.I. Fedosov, *Nucl. Instr. and Meth.* **195** (1982) 569.
7. V.G. Bagrov, M.M. Nikitin, I.M. Ternov and N.I. Fedosov, in *IV Symposium on High Current Electronics*, Part II (Tomsk, Russia, 1982), p. 160.
8. V.G. Bagrov, I.M. Ternov and N.I. Fedosov, *Zh. Teor. Eksp. Fiz.* [Sov. Phys. JETP] **82** (1982) 1442.
9. V.G. Bagrov, I.M. Ternov and N.I. Fedosov, *Dokl. Akad. Nauk.* [Sov. Phys. Dokl.] **263** (1982) 1339.
10. B.M. Bolotovskiy, in *Trudy FIAN* [Proceedings of the Lebedev Physics Institute] **140** (1982) 35.
11. V.G. Bagrov, M.B. Moiseev, M.M. Nikitin and N.I. Fedosov, *Sov. Phys. J.* **25** (1982) 648.
12. V.G. Bagrov, M.M. Nikitin, I.M. Ternov and N.I. Fedosov, *Izv. Vuz., Radiofiz.* **26** (1983) 253.
13. V.G. Bagrov, M.M. Nikitin, I.M. Ternov and N.I. Fedosov, *Nucl. Instr. and Meth.* **208** (1983) 167.
14. V.G. Bagrov, N.I. Fedosov and I.M. Ternov, *Phys. Rev.* **D28** (1983) 2464.
15. A.G. Bonch-Osmolovskiy, V.L. Luboshitz and M.I. Podgoretskiy, JINR Preprint P2-83-157 (Dubna, 1983).
16. E.G. Bessonov, *Zh. Tekh. Fiz.* [Sov. Phys. Techn. Phys.] **53** (1983) 1369.
17. N.V. Smolyakov, *Zh. Tekh. Fiz.* [Sov. Phys. Techn. Phys.] **58** (1988) 489.
18. V.G. Bagrov, M.M. Nikitin, N.I. Fedosov and V.Ya. Epp, *Izv. Vuz. Radiofiz.* **27** (1984) 1287.
19. V.G. Bagrov, M.M. Nikitin, N.I. Fedosov, V.F. Zalmezh and V.Ya. Epp, *Nucl. Instr. and Meth.* **A239** (1985) 579.
20. A.G. Bonch-Osmolovskiy and M.I. Podgoretskiy, JINR Preprint P2-84-136 (Dubna, 1984).
21. Yu.N. Kargin, G.K. Razina and N.I. Fedosov, *Izv. Vuz., Fiz.* [Sov. Phys. J.] **28** (1985) 58.
22. R. Gonzales Felipe and Yu.G. Pavlenko, *Vestn. Mosk. Univ., Fiz.* **26** (1985) 13.
23. A.G. Bonch-Osmolovskiy and M.I. Podgoretskiy, *Zh. Tekh. Fiz.* [Sov. Phys. Techn. Phys.] **55** (1985) 1500.
24. B.M. Bolotovskiy and S.N. Stolyarov, *Usp. Fiz. Nauk* [Phys. Usp.] **162** (1992) 195.
25. B.M. Bolotovskiy and A.V. Serov, *Zh. Teor. Eksp. Fiz.* [Sov. Phys. JETP] **102** (1992) 1506.
26. B.M. Bolotovskiy and A.V. Serov, *Zh. Teor. Eksp. Fiz.* [Sov. Phys. JETP] **105** (1994) 43.
27. L.D. Landau and E.M. Lifshitz, *The Classical Theory of Fields* (Pergamon, New

York, 1975).

28. M.M. Nikitin, A.F. Medvedev, M.B. Moiseev and V.Ya. Epp, *Sov. Phys. JETP* **52** (1980) 388.
29. E.G. Bessonov, *Sov. Phys. JETP* **53** (1981) 433.
30. V.G. Bagrov, M.M. Nikitin, N.N. Sirota and N.I. Fedosov, *Nucl. Instr. and Meth.* **A282** (1989) 402.
31. O.V. Chubar and N.V. Smoljakov, *J. Optics* **24** (1993) 117.
32. I.M. Ternov, V.G. Bagrov and V.A. Bordovitsyn, *Vestn. Mosk. Univ., Fiz., Astron.* **19** (1972) 248.
33. M.M. Nikitin and V.Ya. Epp, *Sov. Phys. Tech. Phys.* **21** (1976) 1404.
34. M.M. Nikitin and N.I. Fedosov, *Sov. Phys. Tech. Phys.* **22** (1977) 1438.
35. M.B. Moiseev, M.M. Nikitin and N.I. Fedosov, *Sov. Phys. J.* **21** (1978) 332.
36. E.G. Bessonov and A.V. Serov, Preprint FIAN No 62 (Lebedev Physics Institute, Moscow, 1982.
37. N.V. Smoljakov and A.I. Chechin, *Nucl. Instr. and Meth.* **A359** (1995) 97.
38. R.P. Feynman, R.B. Leighton and M. Sands *The Feynman Lectures on Physics*, vol. 2. (Addison-Wesley, Reading MA, Commemorative Issue, 1989).

Chapter 7

Synchrotron Radiation in a Strong Magnetic Field

O.F. Dorofeyev, A.V. Borisov, and V.Ch. Zhukovsky
Physics Faculty, Moscow State University,
119899 Moscow, Russia[a]

Professor I.M. Ternov was one of those who created the quantum theory of synchrotron radiation, which is a unique physical phenomenon widely used in various branches of science and technology. Prof. I.M. Ternov and Prof. A.A. Sokolov discovered two fundamental effects: quantum excitation of radial oscillations of electrons in cyclic accelerators (1953) and radiative polarization of electrons and positrons as a result of quantum fluctuations of SR (1963). I.M. Ternov made a substantial contribution to the development of the general theory of quantum processes in strong external fields basing upon the method of exact solutions of relativistic wave equations, that was also proposed by him. He is one of the founders of the modern high energy spin physics. In the present paper, which is dedicated to the memory of I.M. Ternov, presented is a review of the theory of SR and the processes in a strong magnetic field related to SR. It is with great gratitude that the authors remember the years of collaboration and fruitful discussions with I.M. Ternov, an outstanding scholar and a man.

7.1. Introduction

The investigation of the quantum properties of SR proved to be very interesting from the point of view of the development of the quantum theory of macroscopic motion. It led to the discovery of fundamental, at times unexpected effects of great

[a]E-mail: zhukovsk@th180.phys.msu.su

theoretical and practical significance [1,2]. The original ideas about the possible manifestation of quantum effects in SR were not obvious, SR had been well described in its basic features by the methods of the classical electrodynamics of Maxwell and Lorentz. Indeed, the usual criterion for the applicability of classical methods of describing the radiation of a relativistic charge is that the energy E_ϕ of the emitted photon must be small compared with the electron energy. At the same time, as was shown by Vladimirskii [3] and Schwinger [4] quantum effects can be manifested only if $E_\phi = (\hbar c/R)\,(E/mc^2)^3 \sim E$. Thus, the criterion for the possibility of a classical description is

$$E \ll E_{1/2} = mc^2 \left(\frac{mcR}{\hbar}\right)^{1/2} \tag{7.1}$$

(the estimate of the photon energy is taken at the maximum of the spectral curve). This criterion was also confirmed by arguments related to the invariance of the power of SR, by virtue of which the power depends only on the invariant parameter (we ignore the motion along the field: $cp_\perp \cong E$.)

$$\chi = \frac{1}{H^*}\left[-\left(F_{\mu\nu}\,p^\nu\right)^2\right]^{1/2} = \frac{H}{H^*}\frac{E}{mc^2} = \left(\frac{E}{E_{1/2}}\right)^2, \tag{7.2}$$

where H^* is Schwinger's magnetic field. However, it was found that the $E_{1/2}$ criterion did not cover all features associated with the discrete properties of SR, above all the influence of discreteness of radiation on the motion of a particle. Indeed, since the energy

$$E_\phi = (c\hbar/R)\left(E/\left(mc^2\right)\right)^3$$

of an emitted photon is fairly high, the number of such high-energy photons emitted during one gyration of an electron is finite and equal to

$$N_{\text{gyr}} \cong W/E_\phi = \frac{4\pi}{3}\left(e^2/c\hbar\right)\left(E/mc^2\right),$$

where

$$W = \frac{2}{3}\left(e^2 c/R^2\right)\left(E/mc^2\right)^4$$

is the power of the radiation. To get a clearer estimate of the influence of the discreteness of radiation, one can find the length L traversed by the electron without the emission of high-energy photons (the "mean free" path)

$$L = \frac{cE_\phi}{W} = \frac{3}{2}\frac{\hbar^2}{mc^2}\frac{H^*}{H} = 34.9 \cdot 10^4/H. \tag{7.3}$$

For ordinary field strengths for accelerators and storage rings, $H \sim 10^4$ G and $L \sim 30$ cm, i.e., on the average one photon is emitted per a path of length 30 cm.

The discreteness of radiation, which appears here as an important factor, can influence the particle trajectory, producing quantum fluctuations of it as a consequence of the electron recoil on photon emission. To estimate a critical value of the energy at which radiation begins to affect the particle trajectory, we note that

$$\langle R^2\rangle = \frac{n+s+1/2}{\gamma}, \qquad \langle \Delta R^2\rangle = \frac{s}{2\gamma}, \qquad \gamma = \frac{eH}{2c\hbar} \tag{7.4}$$

where $\langle \Delta R^2\rangle = \langle R^2\rangle - (\langle R\rangle)^2$ is the mean-square fluctuation of the radius, and $s = 0, 1, 2, ...$ is the radial quantum number, which characterizes the position of the orbit center of electron gyration. It is obvious that excitation of radial fluctuations commences if the photon energy is

$$E_\phi = \frac{\hbar c}{R}\left(\frac{E}{mc^2}\right)^3 = \left(\langle \Delta E^2\rangle\right)^{1/2} = \sqrt{e^2H^2\langle \Delta R^2\rangle} = \sqrt{eHc\hbar s}, \tag{7.5}$$

from which we obtain the expression for a critical value of the energy

$$E_c = E_{1/5} = mc^2\left(\frac{mcR}{\hbar}\right)^{1/5}. \tag{7.6}$$

The $E_{1/5}$ criterion, as the condition for excitations of radial fluctuations of the electron trajectory, was first found by Sokolov [5], who began the development of the quantum theory of SR.

Thus, the limits of applicability of the classical theory of synchrotron radiation are determined by three quantities.

1. $E_{1/2} = mc^2\left(mcR/\hbar\right)^{1/2}$, which is the limit of applicability of the classical theory of SR and it means physically that the energy of a radiated photon must be less than the energy of an electron: $\hbar\omega << E$.

2. $E_{1/5} = mc^2\left(mcR/\hbar\right)^{1/5}$, which is the limit of applicability of the equations of motion with allowance for radiative friction force. Physically, this means that the energy of an emitted photon must be less than the energy of quantum excitations of the radial degrees of freedom

$$\hbar\omega \ll \left(\langle \Delta E^2\rangle\right)^{1/2} = \sqrt{eHc\hbar s}.$$

Otherwise, when $E \sim E_{1/5}$, the radiation reaction force must be modified to take into account the discreteness of radiation, i.e. the effect of quantum fluctuations on an electron trajectory. Note that $E_{1/5}$ occurs only in the expression for the probability of quantum transitions and does not occur when one considers the power of SR. Thus, the number $E_{1/5}$ essentially distinguishes a "quasiquantum" range of energies: The classical description of the radiation is still valid, but the discrete nature of the radiation is already beginning to be manifested in the form of quantum fluctuations of the electron trajectory (quantum "broadening" of the orbit [6]).

3. The region of small quantum numbers, when the quantizing properties of the magnetic field begin to be manifested most strongly is of special interest. Under these conditions, the dynamical parameter

$$\chi = (H/H^*)\,(p_\perp/mc) = \sqrt{2n}\,(H/H^*)^{3/2}$$

takes an essentially discrete series of values, and although for $H/H^* << 1$ it remains small, a classical description of SR is impossible (see also Ref. [7] for an analysis of the criteria of applicability of the classical theory).

The quantum theory of SR was developed by Sokolov — Ternov group on the fundamental quantum description of the state of an electron in a homogeneous magnetic field (solution to the Dirac equation) and the rigorous method of quantum electrodynamics (Furry picture). This method, the method of exact solutions [2,8], did not lead to any restrictions on the field strength and made it possible to treat the problem even in the range of the fields above the critical Schwinger value $(H/H^* > 1)$.

This approach and choice of the model (a homogeneous magnetic field) opened up a possibility of predicting and establishing some essentially new physical effects: quantum fluctuations of the electron trajectory under the conditions of macroscopic motion (Sokolov and Ternov, 1953) [6], radiative self-polarization of electrons and positrons moving in storage rings (Sokolov and Ternov, 1963) [9], and a dynamical nature of the anomalous magnetic moment of the electron (Ternov, Bagrov, Bordovitsyn, and Dorofeev [10]). The method of exact solutions also made it possible to consider the exotic region of magnetic fields of the order of $H^* \cong 4 \cdot 10^{13}$ G which are reliably observed. As was shown by Klepikov in 1954 [11], under these conditions new reaction channels are opened: single-photon annihilation and single-photon production of electron-positron pairs.

In order to analyze spin effects, we have developed and applied a method of separating solutions of the Dirac equation

$$i\hbar\frac{\partial}{\partial t}\Psi = \left\{c\,(\boldsymbol{\alpha P}) + \rho_3 mc^2 + e\Phi\right\}\Psi, \tag{7.7}$$

with respect to spin states (see Refs [1,9]), in accordance with which the electron wave function satisfies not only the equation $\widehat{H}\Psi = E\Psi$ but also the subsidiary equation $\hat{s}\Psi = \zeta\Psi$, where $\hat{s}$ is the spin operator that commutes with the Hamiltonian. To describe longitudinal polarization (spin projection along the direction of motion of an electron), this operator can be taken to be

$$\hat{s}^0 = \frac{(\boldsymbol{\sigma}\widehat{\boldsymbol{P}})}{\left|\langle\widehat{\boldsymbol{P}}\rangle\right|}, \tag{7.8}$$

where $\boldsymbol{\sigma} = \rho_1\boldsymbol{\alpha}$ are the Dirac matrices, and $\widehat{\boldsymbol{P}}$ is the operator of the kinetic momentum: $\widehat{\boldsymbol{P}} = -i\hbar\boldsymbol{\nabla} - (e/c)\boldsymbol{A}$. Then the wave function will satisfy besides Eq. (7.7) the equation $\hat{s}^0\Psi = \zeta_{\parallel}\Psi$, where $\zeta_{\parallel} = \pm 1$ corresponds to two possible spin orientations along or against the direction of an electron motion. To investigate the projection of the spin onto the direction of the magnetic field, it is expedient to introduce the polarization operator (see Chapter 2)

$$\widehat{\boldsymbol{\Sigma}} = \rho_3\boldsymbol{\sigma} + \rho_1\,\frac{c\widehat{\boldsymbol{P}}}{E} - \rho_3\,\frac{mc^2\widehat{\boldsymbol{P}}\left(\boldsymbol{\sigma}\widehat{\boldsymbol{P}}\right)}{E\,(E+mc^2)}, \tag{7.9}$$

which in the electron rest frame becomes the Pauli spin operator. The projection of the spin onto the magnetic field is a constant of the motion [for the field $\boldsymbol{H} = (0, 0, H)$, it is $\widehat{\Sigma}_3$ component], and the wave function must satisfy the equation $\widehat{\Sigma}_3\Psi = \zeta_\perp\Psi$, where $\zeta_\perp = \pm 1$ corresponds to the two possible spin orientations: along and opposite to the external field $\boldsymbol{H}$.

7.2. Quantum Theory of Radiation at Small χ

Using general methods of the quantum theory of radiation[2], it is possible to obtain the following expression for the power of SR

$$W_i = \frac{e^2 c}{2\pi} \sum_{n',s',\zeta',k_3'} \int d\kappa^3\, \delta\left(\kappa - \kappa_{nn'}\right) \Phi_{i,}, \tag{7.10}$$

where $c\kappa_{nn'} = \left(E_n - E_{n'}\right)/\hbar$ is the radiation frequency; n, s, k_3, ζ are the quantum numbers that determine the state of an electron; and i corresponds to the two polarization components of a photon. The expressions for the function Φ_i ($i = \sigma$, $i = \pi$ are the two components of the linear polarization) have the form

$$\Phi_\sigma = \left|\langle\alpha_1\rangle\right|^2, \quad \Phi_\pi = \left|\langle\alpha_2\rangle \cos\vartheta - \langle\alpha_3\rangle \cos\vartheta\right|^2,$$

and the matrix elements of the Dirac matrices

$$\langle\alpha_i\rangle = \langle n', s', k_3', \zeta' \left|\alpha_i\right| n, s, k_3, \zeta\rangle$$

are related to the Laguerre functions $I_{nn'}(x)$ and $I_{ss'}(x)$:

$$\left.\begin{array}{c} -i\langle\alpha_1\rangle \\ \langle\alpha_2\rangle \end{array}\right\} = \left[AI_{n,n'-1}(x) \mp BI_{n-1,n'}(x)\right] I_{ss'}(x)\, \delta_{k_3',\,k_3-\kappa\cos\vartheta}, \tag{7.11}$$

$$\langle\alpha_3\rangle = \left[CI_{n-1,n'-1}(x) + DI_{nn'}(x)\right] I_{ss'}(x)\, \delta_{k_3',\,k_3-\kappa\cos\vartheta},$$

in which A, B, C, D are spin coefficients that depend on the choice of a polarization operator. Here, we are not able to dwell on the calculations in more detail (see Ref. [2]) and turn to the conclusions.

We first consider a semiclassical approximation.

7.2.1. Semiclassical Approximation

In this case $n \gg 1$, the energy spectrum is quasicontinuous, and the dynamical parameter $\chi = (H/H^*)(p_\perp/mc)$ is a continuous quantity. Under these conditions, the sum over n' can be replaced by an integral

$$\sum_{\nu=0}^{n} \longrightarrow \int_0^n d\nu \quad (\nu = n - n'),$$

and after the necessary change of an integration variable we obtain the expression for the radiation power

$$W = \frac{27}{16\pi}\frac{e^2c}{R^2\epsilon_0^{9/2}}\int_0^\infty dy\, y^2\, (1+\xi_0\, y)^{-6}\oint d\Omega \sum_i \Phi_i. \tag{7.12}$$

Note that this expression for power is valid only in the semiclassical approximation ($n \gg 1$). In derivating this expression a very important factor was the relation [12]

$$\sum_{s'=0}^{\infty} I_{ss'}^2\ (x) = 1, \tag{7.13}$$

by virtue of which the number $E_{1/5}$ disappears from the expression for radiation power. In the semiclassical region, we may approximate the Laguerre function by means of the Macdonald function [2,13]

$$I_{nn'}\ (x) \simeq \frac{1}{\pi\sqrt{3}}\ (1-x/x_0)^{1/2}\ K_{1/3}\left(\frac{2}{3}\ \left(x_0^2 nn'\right)^{1/4}\ (1-x/x_0)^{3/2}\right),$$

where $x_0 = \left(\sqrt{n}-\sqrt{n'}\right)^2$. Summing now over the polarization states of the electron and photon, we obtain

$$W = W^{\rm cl}\,\frac{9\sqrt{3}}{8\pi}\int_0^\infty \frac{ydy}{(1+\xi y)^3} \tag{7.14}$$

$$\times\left(\int_y^\infty K_{5/3}\ (x)\ dx\ +\ \frac{(\xi y)^2}{1+\xi y}\ K_{2/3}\ (y)\right).$$

Here

$$W^{\rm cl} = \frac{2}{3}\ \left(e^2c/R^2\right)\left(E/mc^2\right)^4$$

and

$$\xi = 3\chi/2.$$

Thus, radiation power depends only on one parameter χ. Integrating over the spectrum, we obtain

$$W = W^{\rm cl}\left[1-\frac{55\sqrt{3}}{16}\chi\ +\ 48\chi^2\ -\ldots\right],\qquad \chi\ll 1. \tag{7.15}$$

The correction in χ was first found by Sokolov, Klepikov, and Ternov [14] and was later confirmed for a spinless electron by Schwinger [15]. It is interesting to note that the quantum corrections have by now been experimentally confirmed. If in Eq. (7.14) we retain those terms that depend on the initial spin orientation, then we obtain[2]

$$W = W^{\rm cl}\,\frac{9\sqrt{3}}{8\pi}\int_0^\infty ydy\left(\int_y^\infty K_{5/3}\ (x)\ dx - \zeta\xi y K_{2/3}\ (y)\right). \tag{7.16}$$

Here we have retained only the terms linear in ξ. In 1977, a method was proposed at the Institute of Nuclear Physics of the Siberian Branch of the USSR Academy of Sciences for measuring SR dependence on a spin[16]. As it follows from Eq. (7.16), an additional power of SR associated with the orientation of a spin relative to the direction of the magnetic field ($\zeta_{\perp} = \zeta$) has the form

$$\frac{dW_{\varsigma}}{dy} = \frac{dW^{\rm cl}}{dy}\Delta, \quad \Delta = \frac{|\zeta|\ \xi y\, K_{2/3}\,(y)}{\int_y^{\infty} K_{5/3}\,(x)\,dx}. \tag{7.17}$$

The most favorable possibility for observation is opened up in the short-wavelength region of the spectrum: $y \gg 1$ (since $y \longrightarrow y^{\rm cl} = \frac{2}{3}\nu\,(mc^2/E)^3$). Then, remembering the asymptotic behavior $K_{\mu}\,(x) \simeq \sqrt{\pi/2x}\exp{(-x)}$ of the Macdonald function, we readily find that

$$\Delta = |\zeta|\ \xi y, \qquad \xi = \frac{3}{2}\chi = \frac{3}{2}\frac{\hbar}{mcR}\left(\frac{E}{mc^2}\right)^2.$$

In the case of observation of the electron beam radiation, $|\zeta|$ in this expression characterizes mean polarization the beam acquires in the radiation process. Therefore, observing the power of SR at a fixed frequency in the spectrum, one can determine polarization characteristics. It is particularly advantageous to carry out the experiment in a storage ring since radiative energy losses of electrons are compensated in the ring by the external sources of rf field. As a result, the particle energy remains on the average constant. Thus, the quantum correction to synchrotron power [see Eq. (7.16)] can be regarded as an experimentally established fact, and the actual method of determining the polarization properties of electron and positron particle beams.

7.2.2. Macroscopic Quantum Fluctuations of an Electron Trajectory

Thus, in the case when the energy of a radiated photon is sufficient for excitation of radial degrees of freedom of an electron

$$E_{\phi} = \left(\langle\Delta E^2\rangle\right)^{1/2} = eH\left(\langle R^2\rangle - (\langle R\rangle)^2\right)^{1/2},$$

i.e., $E \sim E_{1/5}$, radial transitions are excited, and the classical theory of the particle motion becomes invalid since it cannot explain the fluctuating nature of the radiation reaction.

Returning to the expression of Eq. (7.14) for the power of SR, we divide the integrand by the photon energy $\hbar\omega$ and find that the probability of quantum transitions has the form [2]

$$\frac{dw_{ss'}}{dy} = \frac{\sqrt{3}}{2\pi}\frac{e^2}{\hbar R}\frac{E}{mc^2}\frac{1}{(1+\xi y)^2} \tag{7.18}$$

$$\times\left(\int_y^{\infty} K_{5/3}\,(x)\,dx + \frac{(\xi y)^2}{(1+\xi y)} K_{2/3}\,(y)\right) I^2_{ss'}\,.$$

We have here restored the radial factor $I^2_{ss'}(x)$, which on summation over s' becomes unity and does not contribute to the synchrotron power. The radial factor $I^2_{ss'}$, reflects the stochastic nature of the excitation of the radial degrees of freedom of an electron. This becomes particularly clear if one assumes that in the initial state the orbit center of electron gyration is at the origin ($s = 0$). Then

$$I^2_{0s'}(x) = \frac{e^{-x}x^{s'}}{s'!}, \qquad x = \frac{\xi_1 y^2}{(1+\xi y)^2}, \qquad \xi_1 = \frac{9}{8}\frac{\hbar}{mcR}\left(\frac{E}{mcR}\right)^5$$

takes the form of an ordinary Poisson distribution.

In the approximation of the classical theory ($\hbar \to 0$), the argument in this formula also vanishes, and therefore in the limit

$$\lim_{x\to 0} I^2_{0,s'}(x) = \delta_{0,s'},$$

in other words, the radial number s is not subject to change in the classical theory. If the orbit at the initial time was circular, it remains such at subsequent times too. Therefore, from the point of view of the classical theory, radiation results only in a reduction in the orbit radius, and in accelerators and storage rings it is compensated by an external rf electric field, and fluctuations of the orbit radius do not occur

$$s = \frac{eH}{c\hbar}\left(\langle R^2\rangle - \langle R\rangle^2\right) = \text{const.}$$

The problem of the change in the squared fluctuation of the radius was solved for the first time by the methods of quantum theory by Sokolov and Ternov [6] on the basis of the method of exact solutions of the Dirac equation. The result was the effect of quantum "broadening" of the trajectory.

Using the Eq. (7.18), it is readily seen that the change in the radial quantum number s has the form

$$\frac{ds}{dt} = \int_0^\infty dy \sum_{s'} (s' - s)\, w_{ss'}.$$

Then, bearing in mind that $\sum_{s'=0}^{\infty} I^2_{ss'}(x)\,(s' - s) = x$, we obtain [6]

$$\frac{ds}{dt} = \frac{d}{dt}\left\{\frac{eH}{c\hbar}\left(\langle R^2\rangle - \langle R\rangle^2\right)\right\}. \tag{7.19}$$

In the case of semiclassical motion, when $\xi \ll 1$, it follows that

$$\frac{ds^{\text{scl}}}{dt} = \frac{55}{48\sqrt{3}}\frac{e^2}{mcR^2}\left(\frac{E}{mc^2}\right)^5 t. \tag{7.20}$$

Thus, for a constant mean energy of an electron the stochastic process of radiation leads to an expression characteristic of Brownian motion and reflecting the effect on the particle of random forces: $\langle x^2\rangle = 2Dt$, where D is the diffusion coefficient. It should be noted here that this effect is similar to the influence of a vacuum electromagnetic field on an electron. Vacuum fluctuations of the field, as was particularly clearly emphasized by Bogolyubov and Tyablikov [17], induce a certain smearing of an effective radius of an electron. At the same time, the electron executes a Brownian motion with a definite square of the displacement.

The change of the squared fluctuation of the radius can be regarded as a macroscopic manifestation of quantum fluctuations of the radiation — a consequence of the discrete nature of photon emission and the recoil effect which an electron is subjected to. Electrons move in the radial direction in accordance with the quantum laws, whereas the motion in the circle remains on the average classical ("macroscopic atom" [2]).

However, the situation is changed if radiative energy losses of an electron are not compensated. This occurs, for example, under astrophysical conditions and, it is interesting to emphasize that by virtue of two processes — radiative reduction in the radius of an electron orbit and the growth of squared fluctuation — a certain equilibrium is established. Since,

$$\langle R^2 \rangle = (2c\hbar/(eH))\,(n+s+1/2)\,,$$

we find that

$$\frac{d\langle R^2\rangle}{dt} = \frac{2c\hbar}{eH}\left[\frac{dn}{dt} + \frac{ds}{dt}\right], \tag{7.21}$$

where ds/dt is determined by Eq. (7.20), and the change of the principal quantum number n is related to the change in the energy

$$\frac{dn}{dt} = -\frac{H^*}{H}\frac{EW}{(mc^2)^2}, \qquad W = \frac{2}{3}\frac{e^2c}{R^2}\left(\frac{E}{mc^2}\right)^4.$$

Then it follows from Eq. (7.21) that $d\langle R^2\rangle/dt = 0$ under the condition of a minimum value of a radius

$$R_{\min} = \frac{55}{32\sqrt{3}}\frac{\hbar}{mc}\left(\frac{E}{mc^2}\right)^2.$$

This expression is analogous to the expression in the case of an electron motion in a focusing field of a storage ring[1,2], when, besides the quantum broadening of a particle trajectory, forces of radiative damping occur, and the amplitude of the electron fluctuation reaches a certain minimum value. It should be said that the quantum excitation of radial fluctuations is important in engineering practice in the construction of storage rings, determining the condition of focusing of beams. Quantum fluctuations were observed in the United States (Sands [18]) and in the USSR (Korolev [19]), and they are in complete agreement with the prediction of the theory (for more details, see Refs[1,12]). The $E_{1/5}$ criterion for the onset of their excitation has also been confirmed.

7.2.3. Self-Polarization of Electrons and Positrons in Storage Rings

Radiative polarization of electrons and positrons takes the form of a directed process of orientation of the particle spins under the influence of SR during prolonged circulation in storage rings. This effect was predicted by Ternov [20] in a collaboration with Sokolov [9,21] and was rigorously established on the basis of exact solutions of the Dirac equation separated with respect to polarization states.

Calculation of the probability of quantum transitions per unit time [1,2] accompanied by spin-flip (spin-flip transitions) leads to a characteristic dependence of the probability on the initial orientation of the spin ζ with respect to the direction of a magnetic field

$$\frac{dw^{\uparrow\downarrow}}{d\Omega dy} = \frac{3}{32\pi}\frac{e^2}{\hbar R}\frac{\epsilon}{(\epsilon_0)^3}\xi^2 y^3 \tag{7.22}$$

$$\times\left[\cos^2\vartheta\, K_{1/3}^2(z) + \left(\sqrt{\epsilon}K_{2/3}(z) + \zeta\sqrt{\epsilon_0}K_{1/3}(z)\right)^2\right],$$

where

$$\epsilon = 1-\beta^2\sin\vartheta,\ \epsilon_0 = 1-\beta^2,\ z = \frac{1}{2}y\ (\epsilon/\epsilon_0)^{3/2},$$

$$y = \frac{2}{3}\nu\ (\epsilon_0)^{3/2},\ \omega = \nu\omega_0.$$

An explicit dependence on the spin orientation also remains in the integrated probability (see Refs[2,9])

$$w_\zeta^{\uparrow\downarrow} = \frac{1}{2\tau}\left(1+\zeta\frac{8\sqrt{3}}{15}\right), \tag{7.23}$$

where the polarization time has the form

$$\tau = \frac{8\sqrt{3}}{15}\frac{\hbar^2}{mce^2}\left(\frac{mc^2}{E}\right)^2\left(\frac{H^*}{H}\right)^3 = \frac{15\sqrt{3}}{16}\frac{\hbar}{mcR}\left(\frac{E}{mc^2}\right)^2\frac{W}{E}\,\xi^{-1}.$$

Hence it follows that because of radiation the electron has a tendency to go over to states with the preferred spin orientation in the direction opposite to a magnetic field[9]. For positron, spin orientation is along a magnetic field direction. These states correspond to a minimum value of the potential energy of particles possessing magnetic moment $\mu = \zeta\,(e_0\hbar/\,(2mc))$ in a magnetic field

$$U = -\frac{e}{|e|}\mu H.$$

Here, we shall not dwell in detail on the kinetics of the polarization process (see Refs[9,21,22]) and, turning to an ensemble of electrons, we shall characterize the polarization of a beam of particles by the mean value $\zeta(t) = \langle\zeta\rangle$, bearing in mind that the interaction of an electron with an electromagnetic field leads to transition of the particle to a mixed state. We then find

$$\frac{d}{dt}\zeta(t) = \sum_{\zeta'} \left(\zeta' - \zeta\right) w_{\zeta}^{\uparrow\downarrow} = -2\sum_{\zeta} \zeta\, w_{\zeta}^{\uparrow\downarrow},$$

from which it follows that

$$\zeta(t) = -\frac{8\sqrt{3}}{15}\left(1 - \exp\left(-t/\tau\right)\right), \tag{7.24}$$

and the limiting degree of polarization (when $t \gg \tau$) reaches the value

$$P(\infty) = 8\sqrt{3}/15 = 0.924.$$

The estimate of the polarization time τ shows that for the magnetic field strength typical of accelerators ($H \sim 10^4$ G) the polarization effect is accessible to observation only in the case of prolonged circulation of particles in the magnetic field (of the order of 1 h). Such a possibility is realized in storage rings, and although under actual conditions of a storage ring there are phenomena that depolarize the particle beam, the effect of the radiative polarization does exist in them and provides a unique basis for creating polarized electrons and positrons of high energy. Radiative polarization has been observed in the USSR, France, Germany, the United States and Japan, and it is in good agreement with the Sokolov-Ternov theory; these questions are considered in detail in Ternov's review[23].

7.2.4. Ultraquantum Aspects of SR ("Magnetic Bremsstrahlung")

We now consider the limiting ultra quantum case when the dynamical invariant parameter $\xi = 3\chi/2$ takes large values: $\chi = (H/H^*)(p_\perp/mc) \gg 1$. This region is of great interest in connection with astrophysical problems associated with radiation of high-energy electrons in an ultrastrong magnetic field $H \sim H^* = 4 \cdot 10^{13}$ G.

We return to Eq. (7.14) for the power of SR. After integration by parts, this expression for the "magnetic bremsstrahlung" takes the form

$$W = W^{\mathrm{cl}}\,\frac{9\sqrt{3}}{8\pi}\int_0^\infty dy\left[\frac{1}{2}\,\frac{y^2}{(1+\xi y)^2}K_{5/3}(y) + \frac{\xi^2 y^3}{(1+\xi y)^4}\,K_{2/3}(y)\right]. \tag{7.25}$$

We have deliberately introduced the expression "magnetic bremsstrahlung" since in the case $\xi \gg 1$ the properties of this radiation are very different from the ordinary properties of SR.

Indeed, let us find the expression for W^{uq} in the ultraquantum region, when $\xi \gg 1$. We note that the functions $y^2 K_{5/3}(y)$and $y^3 K_{2/3}(y)$ reach a maximum value at $y \sim 1$, then they decrease exponentially. In the ultraquantum case, by virtue of the factor $(1+\xi y)^2$ in the denominator, the spectrum is truncated at $\xi \gg 1$. Then for the Macdonald functions $K_\mu(y)$ we can use the asymptotic expression $K_\mu(y) \cong 2^{\mu-1}\Gamma(\mu)/y^\mu$, and, integrating in Eq. (7.25) by means of the integral

$$\int_0^\infty dy y^{p-1}(1+\xi y)^{-q} = \frac{1}{\xi^p}\frac{\Gamma(p)\Gamma(q-p)}{\Gamma(q)},$$

we find [2,12]

$$W^{uq} = \frac{16}{27}\frac{\Gamma(2/3)}{(3\chi^4)^{1/3}} W^{\mathrm{cl}}. \tag{7.26}$$

Thus, in the ultraquantum case the power of "magnetic bremsstrahlung" is strongly suppressed ($\chi \gg 1$). In contrast to the semiclassical case ($\chi \ll 1$), when quantum effects yielded only small corrections to the classical expressions (see Eq. (7.15)), in Eq. (7.26) the main term is already a quantum one, and the transition to the classical approximation is here impossible.

To obtain a more transparent interpretation of our result, we consider the form of the radiation spectrum. By means of standard methods of the quantum theory, we can readily obtain an expression for the radiation frequency

$$\omega_{\max} = \frac{3}{2}\frac{c}{R}\frac{\beta y_{\max}}{\epsilon_0^{3/2}}\left(1+\frac{3}{2}\chi\beta y_{\max}\right)^{-1} \cong \frac{3}{2}\frac{c}{R}\frac{1}{\epsilon_0^{3/2}\left(1+\frac{3}{2}\chi\right)}, \tag{7.27}$$

where $\beta y_{\max}$ is of the order of unity. From this there follow two limiting cases

$$\omega_{\max} = \begin{cases} \frac{3}{2}\frac{c}{R}\left(\frac{E}{mc^2}\right)^3, & \chi \ll 1, \\ \frac{E}{\hbar}, & \chi \gg 1. \end{cases} \tag{7.28}$$

Thus, in the ultraquantum case the radiation spectrum terminates at frequency $\omega'_{\max} = E/\hbar$ which is much less than the characteristic frequency of SR: $\omega'_{\max} \ll \omega^{\mathrm{cl}}_{\max}$ (see Ref. [2]), since $\omega'_{\max} \cong \omega^{\mathrm{cl}}_{\max}/\chi$. In other words, an electron cannot radiate energy greater than its energy E. Fig. 7.1 shows the graph of the spectral distribution of the

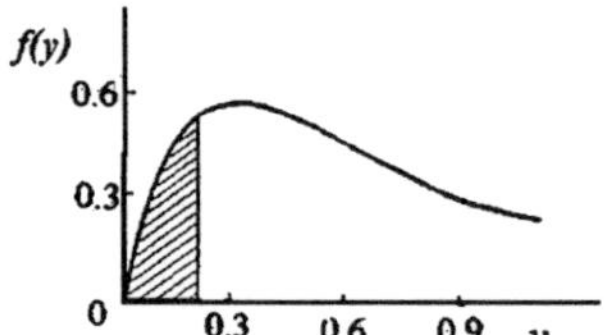

Fig. 7.1. Schematic spectral distribution of synchrotron power.

power of SR

$$W = W^{\mathrm{cl}}\int_0^\infty f\,(y)\,dy$$

in the classical approximation, when

$$f\ (y) = \frac{9\sqrt{3}}{8\pi}\, y \int_y^\infty K_{5/3}(x)\, dx.$$

Characteristic features of this function are determined by its asymptotic behavior

$$f\ (y) = \frac{9\sqrt{3}}{8\pi} \begin{cases} 2^{2/3}\Gamma(2/3)\, y^{1/3} & \text{for } y \ll 1, \\ \sqrt{\pi/2}\,\sqrt{y}\, \exp(-y) & \text{for } y \gg 1. \end{cases}$$

Since

$$f^{\,uq}(y) = \frac{9\sqrt{3}}{8\pi}\,\frac{\Gamma(2/3)}{2^{1/3}} \left\{ \frac{2}{3}\,\frac{y^{1/3}}{(1+\xi y)^2} + \frac{\xi^2 y^{7/3}}{(1+\xi y)^4} \right\}$$

[this follows when $y \ll 1$ from Eq. (7.25)], we can see that there is a similarity of these functions in the region of small values of y. For comparison, we find the energy of radiation in accordance with the classical theory, restricting the integral to the area of the curve up to a certain

$$y_{\max} = \frac{2}{3}\,(R/c)\,(E/\hbar)\,\epsilon_0^{3/2}.$$

Then we obtain

$$\begin{aligned} W^{\rm scl} &= \frac{3\sqrt{3}}{4\pi}\, 2^{2/3}\Gamma(2/3)\, \frac{e^2 c}{(R\epsilon_0)^2} \int_0^{y_{\max}} y^{1/3} dy = \\ & \frac{3\cdot 3^{7/6\cdot}}{8\pi}\,\Gamma(2/3)\,\chi^{-4/3} W^{\rm cl}. \end{aligned}$$

If we compare this expression with the exact expression of Eq. (7.26), the difference is small enough

$$\frac{W^{\rm scl}}{W^{uq}} = \frac{3\cdot 3^{7/6}\cdot 3^{1/3}\cdot 3^3}{8\pi\cdot 16} = 1.04.$$

Thus, in this model the decrease in the power of SR in the ultraquantum case acquires an admittedly formal but very transparent interpretation.

Thus, the radiation spectrum and its power are very different from those of ordinary SR. We now consider the angular distribution of the power and compare it with the corresponding characteristics of SR, which, as is well-known [2], possesses a "projector effect": The radiation is directed forward along the motion of an electron and is concentrated in a narrow cone with the opening angle $\Delta\vartheta \sim mc^2/E$.

However, in the ultraquantum region, with the parameter $\chi \leq 1$, the position is changed greatly. Indeed, let us return to Eq. (7.14) for the radiation power and restore in it the integration over the angle ϑ. Then for the spectral-angular distribution of the radiation power we obtain

$$W = \frac{9}{8\pi}\,\frac{e^2 c}{R^2}\,\epsilon_0^{-9/2}\,(J_1 + J_2)\,, \tag{7.29}$$

$$J_1 = \int_0^\infty \frac{dyy^2}{(1+\xi y)^3} \int_0^\pi \sin\vartheta d\vartheta \left[\epsilon^2 K_{2/3}^2(x) + \epsilon \cos^2\vartheta^2 K_{1/3}^2(x)\right],$$

$$J_2 = \int_0^\infty \frac{dyy^2}{(1+\xi y)^4} \int_0^\pi \sin\vartheta d\vartheta \left[\epsilon^2 K_{2/3}^2(x) + \left(\epsilon \cos^2\vartheta^2 + \epsilon\epsilon_0\right) K_{1/3}^2(x)\right].$$

Here $x = (y/2)(\epsilon/\epsilon_0)^{3/2}$, $\epsilon = 1 - \beta^2 \sin^2\vartheta$, $\epsilon_0 = 1-\beta^2$. To analyze the angular distribution of the radiation power, we introduce a new variable, setting $y = 2x(\epsilon/\epsilon_0)^{-3/2}$. Further, introducing the angle $\vartheta = \pi/2 - \sqrt{\epsilon_0}\Psi$, we have for small $\sqrt{\epsilon_0}\Psi$ $\epsilon = \epsilon_0(1+\Psi^2)$, $\cos^2\vartheta = \epsilon_0\Psi^2$, and we then find that

$$J_1 = 8\epsilon_0^5 \int_0^\infty dx x^2 \int_{-\infty}^\infty d\Psi \left[1 + \frac{2\xi x}{(1+\Psi^2)^{3/2}}\right]^{-1} \times \left[\frac{K_{2/3}^2(x)}{(1+\Psi^2)^{5/2}} + \frac{\Psi^2 K_{1/3}^2(x)}{(1+\Psi^2)^{7/2}}\right], \tag{7.30}$$

$$J_2 = 16\xi^2\epsilon_0^8 \int_0^\infty dx x^4 \int_{-\infty}^\infty d\Psi \left[1 + \frac{2\xi x}{(1+\Psi^2)^{3/2}}\right]^{-4} \frac{\left(K_{2/3}^2(x) + K_{1/3}^2(x)\right)}{(1+\Psi^2)^{11/2}}.$$

Since the Macdonald functions $K_{1/3}$ and $K_{2/3}$ are essentially different from zero only for an argument x of the order of unity, and for $x >> 1$ these functions decrease exponentially, the limits of integration over Ψ, which varies in the range from $-\beta/\epsilon_0$ to β/ϵ_0, can be extended to infinity. Unfortunately, exact integration over the spectrum is difficult, and therefore to qualitatively analyse the angular distribution of the radiation power we estimate the integrals, assuming that at the maximum of the integrand in Eq. (7.29), i.e., for $x \sim 1$, we can set $x = 1$ in the denominator, and then the integrals over the spectrum can be calculated. By means of the integral

$$\int_0^\infty x^{\mu-1} K_\rho^2(x)\, dx = \frac{2^{\mu-3}}{\Gamma(\mu)} \Gamma\left(\rho + \frac{\mu}{2}\right) \Gamma\left(\frac{\mu}{2} - \rho\right) \Gamma^2\left(\frac{\mu}{2}\right)$$

we find that

$$W = \frac{e^2 c}{16 R^2 \epsilon_0^2} \int_{-\infty}^\infty \Phi(\Psi)\, d\Psi, \tag{7.31}$$

where

$$\Phi(\Psi) = \frac{\left\{(7+12\Psi^2)(1+\Psi^2) + 35\xi^2\sqrt{1+\Psi^2}\left[(1+\Psi^2)^{3/2} + 2\xi\right]^{-1}\right\}}{\left[(1+\Psi^2)^{3/2} + 2\xi\right]^3}$$

In the classical approximation, when $\xi \to 0$, we obtain a well-known formula for the angular distribution of the power characterized by the "projector effect" with all the radiation concentrated in a narrow cone, and the deflection of the radiation cone from the direction of a particle velocity is very small: $\Psi \sim \sqrt{\epsilon_0} = mc^2/E$,

$$\Phi^{cl}(\Psi) = \left(7 + 12\Psi^2\right)\left(1+\Psi^2\right)^{-7/2}$$

(see Ref. [2]). However, in the quantum case the sharpness of the radiation peak is strongly smoothed,and for $\xi \sim 1$ the pronounced directionality of the radiation disappears altogether (see Fig. 7.2). Thus, under the conditions of the ultraquantum

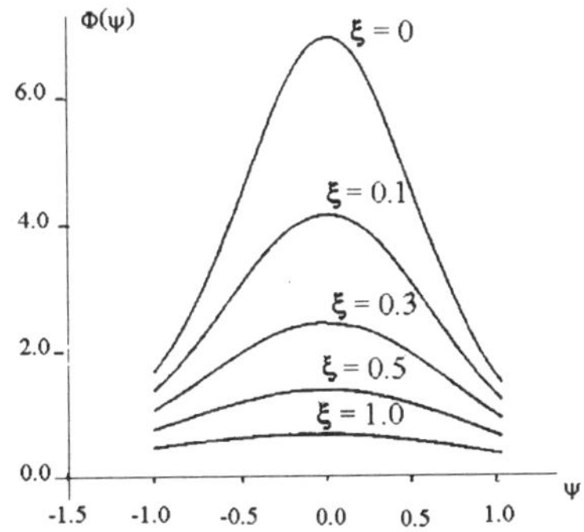

Fig. 7.2. Projector effect in SR and transition to the quantum case.

case the dependence of radiation on the angle ϑ becomes weakly expressed.

Further, we consider the polarization properties of the radiation and compare them with the case of semiclassical motion of an electron, when SR has a pronounced linear polarization. As is well-known (see Refs[24,2]), $(7/8)W^{\text{cl}}$ is the so-called σ component of the radiation, characterized by the fact that the electric vector of the radiation lies in the plane of the orbit of gyration of the particle and is directed along the radius to the center of the orbit, while $(1/8)W^{\text{cl}}$ corresponds to the π component, for which the electric vector is directed almost along the direction of an external magnetic field.

We now return to Eq. (7.29) for the spectral-angular distribution of the radiation power. After integration over the angle with allowance for the components of linear polarization[2], it takes the form

$$W_{\sigma,\pi} = W^{\text{cl}} \int_0^\infty dy\, f_{\sigma,\pi}(y)\,, \tag{7.32}$$

$$f_{\sigma,\pi}(y) = \frac{9\sqrt{3}}{16\pi} \frac{y}{(1+\xi y)^3} \left\{ \int_y^\infty K_{5/3}(x)\, dx \pm K_{2/3}(y) + \frac{\xi^2 y^2}{1+\xi y} K_{2/3}(y) \right\}.$$

In the approximation of the classical theory, $\xi \to 0$, we obtain a well-known result, according to which $W_\sigma = (7/8)W^{\text{cl}}$, $W_\pi = (1/8)W^{\text{cl}}$. We now consider this expression in the ultraquantum case, when $\xi = 3\chi/2 \gg 1$. Then the Macdonald functions K_μ must be taken in the asymptotic limit $y \ll 1$

$$W^{uq}_{\sigma,\pi} = W^{\text{cl}} \frac{9\sqrt{3}}{16\pi} \frac{\Gamma(2/3)}{2^{1/3}} \int_0^\infty dy \left[\frac{2}{3} \frac{y^{1/3}}{(1+\xi y)^2} \pm \frac{y^{1/3}}{(1+\xi y)^3} + \frac{\xi^2 y^{7/3}}{(1+\xi y)^4} \right].$$

After integration we obtain [25]

$$W_\sigma^{uq} = \frac{41}{64} W^{uq}, \qquad W_\pi^{uq} = \frac{23}{64} W^{uq}, \tag{7.33}$$

where

$$W^{uq} = 2^{8/3}\Gamma\left(2/3\right) W^{\mathrm{cl}} \left(9\xi^{4/3}\right)^{-1}.$$

Thus, the preferred polarization of the radiation is maintained, although the difference between the radiation powers of the σ and π components is somewhat smoothed out: $W_\sigma/W_\pi \simeq 5/3$ instead of $7/1$.

The case of the circular polarization is very interesting. One finds [25] that

$$W_l^{uq} = \frac{1}{2} W^{uq} \left(1 + \frac{11}{16}\, l\, \zeta\right), \tag{7.34}$$

where $l = \pm 1$ corresponds to the sign of the circular polarization of photons, and $\varsigma = \pm 1$ to the sign of the longitudinal polarization of an electron. We see that electrons with longitudinal spin orientation emit photons predominantly with the same circular polarization. For unpolarized electrons, circular polarization will be absent.

The radiation properties considered here require essentially the case of macroscopic motion, when the quantum number $n = \gamma R^2 = (E/mc^2)\, mcR/2\hbar$ is very large. In the case of semiclassical motion of electrons, the dynamical parameter $\chi = (H/H^*)\,(p_\perp/mc)$ takes continuous values, since the electron energy spectrum is quasicontinuous

$$\chi = (H/H^*)^{3/2} \sqrt{2n}, \qquad n \gg 1.$$

However, it is obvious that one also considers a different range of energy values in which the quantum effects will have a strong influence. This is the region of small quantum numbers $n = 0, 1, 2, \ldots$, i.e., electron states for which the magnetic field manifests its quantizing properties. If we assume an extremely strong magnetic field, $H \gtrsim H^*$, then we have the parameter $\chi > 1$ remaining discrete, and an electron has the relativistic spectrum

$$E = mc^2 \sqrt{1 + 2nH/H^*}$$

for any $n \neq 0$, and in this connection it is not valid to treat radiation as a continuous process. This confirms once more that the model of a crossed field can give a description of electron radiation only in the semiclassical approximation. The expression

$$W^{uq} = 2^{8/3}\Gamma\left(2/3\right) W^{\mathrm{cl}} \left(9\xi^{4/3}\right)^{-1}$$

for the radiation power obtained for the case of a purely magnetic field was also found by considering the radiation of an electron in a crossed field, but this method cannot bring out the quantizing properties of the field.

We consider some results relating to the calculation of the intensity of the radiation of an electron in a weakly excited state in a strong magnetic field. This problem was solved [26,27] by means of the method of exact solutions. We first of all consider the week-field case: $H << H^*$. Then for the radiation power we obtain the expression

$$W = \frac{2n-1}{2n} W^{\,\mathrm{cl}}, \qquad W^{\mathrm{cl}} = \frac{2}{3}\frac{eH^2}{m^2c^3}\frac{\beta^2}{1-\beta^2}, \tag{7.35}$$

which differs more strongly from the classical expression, the lower is the number n of the energy level.

It is not difficult to calculate the lifetime of the excited state. Indeed, since the quantum transitions in the considered case ($\beta << 1$) have a dipole nature, the emission process is accompanied by transitions to a neighboring level. For the total lifetime of an electron in the state n, we obtain

$$T(n) = \sum_{k=1}^{n} \frac{1}{w(k)},$$

where

$$w(k) = \frac{2}{3}\frac{2k-1}{T_0}\left(\frac{H}{H^*}\right)^2,$$

and

$$T_0 = \hbar^2/\left(mce^2\right) = 1.7 \cdot 10^{-19}\ \mathrm{s}.$$

Then

$$T\ (n) = \frac{3}{2}\ \frac{m^3c^5}{e^4H^2}\sum_{k=1}^{n}\frac{1}{2k-1} = \frac{3}{2}\ \frac{m^3c^5}{e^4H^2}\ \left[C/2 + \ln 2\sqrt{n} + \frac{1}{48n^2}\cdots\right],$$

where C is Euler's constant. Thus, the lifetime of an electron in the excited state does not differ strongly from the classical value

$$T^{\mathrm{cl}} = \frac{3}{2}\ \left(m^3c^5/e^4H^2\right).$$

This expression can be obtained by integrating the classical result

$$W^{\,\mathrm{cl}} = -\frac{\partial E}{\partial t} = \frac{2}{3}\frac{e^4H^2}{m^4c^7}\ \left(E^2 - m^2c^4\right), \qquad \frac{E}{mc^2} = \coth\left(\frac{t+t_0}{T^{\mathrm{cl}}}\right).$$

For the fields $H \sim 10^4$ G, the lifetime of the excited state is approximately 5 sec.

In the case of an ultrastrong field, $H > H^*$, an electron is relativistic already when it is in the first excited level, and therefore the energy of the emitted photon in the dipole transition $n = 1 \to 0$ is practically equal to the energy of the particle. Unfortunately, in the case of an ultrastrong field, $H > H^*$, the calculations can be made only numerically (see Refs[26,27,28]). The following expressions are obtained

$$W_\sigma = 0.742\,W, \qquad W_\pi = 0.258\,W, \qquad W = 0.302\ \frac{e^2m^2c^3}{\hbar^2}\ \left(\frac{E}{mc^2}\right)^2.$$

It is important to emphasize that these expressions agree neither with the classical expression of Eq. (7.35) nor with the ultraquantum case of Eq. (7.26). Thus, the radiation of electrons in the first excited states has important properties that can be described neither by the classical theory nor even by the semiclassical method. The results presented here may be of interest in astrophysics (see Chapter 8). In particular, in pulsar magnetospheres [29,30], (see Chapter 8) in contrast to storage rings, the radiative energy losses of electrons are not compensated. Therefore, the radiation of weakly excited charges in extremely strong magnetic fields can be particularly interesting.

7.2.5. SR of Relativistic Electrons and their Quantum Transitions to the Ground State

It was already mentioned above that in the last decades the development of astrophysics has led to the necessity of generalizing quantum theory of SR (SR) to the region of superstrong magnetic fields $H \gtrsim H^*$. It has been found that there are magnetic fields $H \sim 10^{12} \div 10^{14}$ G near the surface of pulsars, which are definitely determined to be neutron stars, and that one of the main mechanisms of radiation of pulsars is the synchrotron mechanism (see e.g. Ref. [30]). In superstrong magnetic fields $H \gtrsim H^*$ as opposed to the case of $H << H^*$, radiative transitions of high energy electrons to the low lying Landau levels turn out to be quite essential, as it has been for the first time shown in Ref. [31]. Such transitions serve as a source of hard gamma-quanta, which can be a basis for one of the possible mechanisms for observed since 1962 cosmic γ-outbursts, whose nature is so far not finally understood (see the review in Ref. [32]).

Thus in Ref. [33] consideration is presented of the magneto-Coulomb radiation, determined by the excitation of higher Landau levels of electrons in Coulomb scattering in magnetized plasma and by the subsequent radiation transition to the ground state, as well as the efficiency of this mechanism for the generation of gamma-quanta in conditions of cosmic astrophysical objects (e.g. pulsars), possessing strong magnetic fields, is shown. In Ref. [34] it was shown that the observed spectra of γ-outbursts can be explained by SR of electrons, that are excited by high energy photons in cold magnetosphere of a highly magnetized neutron star, under whose surface a thermonuclear explosion took place. Neutron stars as sources of γ-outbursts were considered in many papers (see e.g. Ref. [35]), it must however be mentioned that a number of alternative models of γ-outbursts have been proposed [32]. But anyhow, SR is an effective mechanism of production of high energy photons in neutron stars.

We now study the spectrum of SR in a magnetic field of arbitrary strength with a special attention paid to the case of superstrong magnetic fields $H \gtrsim H^*$ [36].

Making use of the energy spectrum of an electron in a constant magnetic field [2] we obtain the expression for the energy of a photon, emitted in electron transition from the state $|np_z\rangle$ to the state $|n'p'_z\rangle$

$$\hbar\omega = \hbar\omega_{nn'} = E - E' = \frac{E}{\sin^2\vartheta}\left(1 - \beta_z \cos\vartheta\right) \tag{7.36}$$

$$\times \left[1 - \left(1 - \frac{c^2\left(p_\perp^2 - (p'_\perp)^2\right)\sin^2\vartheta}{E^2\left(1 - \beta_z\cos\vartheta\right)^2}\right)^{1/2}\right].$$

Here ϑ is the angle between the photon momentum $\hbar\boldsymbol{k}$ and a magnetic field $\boldsymbol{H}$, $v_z = c\beta_z = cp_z/E$ is the longitudinal (along $\boldsymbol{H}$) electron velocity, $p_\perp = mc\,(n/a)^{1/2}$ is the transverse momentum and the following parameter is introduced

$$a = \frac{H^*}{2H}. \tag{7.37}$$

The probability of the radiative transition $w_{nn'}$ is expressed in terms of the Laguerre function (see Eqs (7.10), (7.11))

$$I_{nn'}(x) = \left(\frac{n'!}{n!}\right)^{1/2} \exp\left(-x/2\right) x^{\frac{n-n'}{2}} L_{n'}^{n-n'}(x), \tag{7.38}$$

where $L_{n'}^{n-n'}(x)$ is the Laguerre polynomial [37] of the invariant (with respect to the Lorentz boosts along the Oz axis) argument

$$x = a\left(\frac{\hbar\omega}{mc^2}\right)^2 \sin^2\vartheta = \frac{m}{1-\epsilon}\left[1 + \left(\epsilon + (1-\epsilon)\frac{n'}{n}\right)^{1/2}\right], \tag{7.39}$$

$\epsilon = 1 - \beta_0^2 \sin^2\vartheta_0$, ϑ_0 is the radiation angle of the photon in the reference frame, where the initial longitudinal electron momentum $p_z = 0$. The relation between the angles ϑ_0 and ϑ is as follows [2]

$$\sin\vartheta_0 = \frac{\left(1-\beta_z^2\right)^{1/2}}{1 - \beta_z\cos\vartheta}\sin\vartheta, \quad \cos\vartheta_0 = \frac{\cos\vartheta - \beta_z}{1 - \beta_z\cos\vartheta}.$$

The invariant velocity $v_0 = c\beta_0$ is introduced above

$$\beta_0 = \beta_\perp\left(1 - \beta_z^2\right)^{-1/2}, \tag{7.40}$$

where $c\beta_\perp = v_\perp = cp_\perp/E$ is the transverse velocity. Using Eq. (7.40), the transverse energy is defined

$$E_\perp = \left(m^2c^4 + c^2p_\perp^2\right)^{1/2} = mc^2\left(1 - \beta_0^2\right)^{-1/2}. \tag{7.41}$$

We observe that the case when relativistic electrons move at small angles to the field $\boldsymbol{H}$ direction (pitch-angles [38,39]) is of special interest for astrophysical applications.

For the relativistic transverse momentum ($p_\perp/mc >> 1$) we have $\beta_0 \simeq 1$, and the quantity ϵ in Eq. (7.39) is small

$$\epsilon \simeq 1 - \beta_0^2 + \cos^2\vartheta_0 \sim 1 - \beta_0^2 = \left(\frac{mc^2}{E_\perp}\right)^2 << 1. \tag{7.42}$$

The small pitch-angle ($p_\perp/p_z << 1$) means that the transverse velocity is nonrelativistic: $\beta_\perp << 1$, though the transverse energy in Eq. (7.41) is large in this case: $E_\perp \simeq cp_\perp >> mc^2$. Radiation is concentrated in the angular interval having the width

$$\Delta\vartheta \sim \left(1 - \beta_z^2\right)^{1/2}\left(\frac{mc}{p_\perp}\right) \simeq \frac{mc^2}{E}$$

with the center corresponding to the pitch-angle

$$\vartheta_H \simeq \frac{p_\perp}{p_z} \simeq \left(1 - \beta_z^2\right)^{1/2},$$

i.e. (see Fig. 7.3)

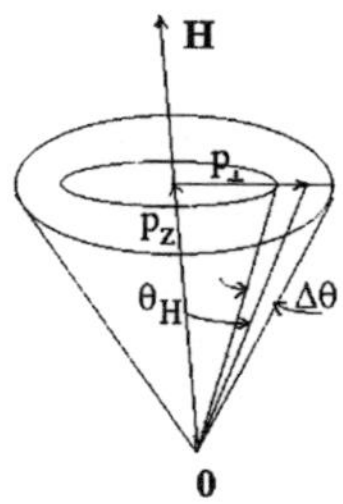

Fig. 7.3. Synchrotron radiation angular distribution of a relativistic electron for the case of a small pitch-angle.

$$\frac{\Delta\vartheta}{\vartheta_H} \sim \frac{mc}{p_\perp} << 1.$$

Consider radiative transition from the state with the principle quantum number $n >> 1$ to the state with $n' << n$. The energy of a photon emitted in this case according to Eqs (7.36) and (7.42) is equal to

$$\hbar\omega = E\left[1 - \left(\epsilon + \frac{n'}{n}\right)^{1/2} + \cos\vartheta_0 + O(\epsilon)\right], \tag{7.43}$$

and the argument of the Laguerre function in Eq. (7.38) is

$$x \simeq n - 2\left[n\left(n' + n\epsilon\right)\right]^{1/2}. \tag{7.44}$$

In the case in question ($n >> 1,\ n' << n$) the following asymptotics for the Laguerre function (valid up to $O\left(n^{-1/2}\right)$) is known [31,36]

$$I_{nn'}(x) \simeq \left[n'!\,(2\pi n)^{1/2}\right]^{-1/2} D_{n'}(t)\,. \qquad (7.45)$$

Here

$$D_{n'}(t) = 2^{-n'/2}\exp\left(-t^2/4\right) H_{n'}\left(t/\sqrt{2}\right)$$

is the parabolic cylinder function of the argument (see Eq. (7.44))

$$t = \frac{n-x}{\sqrt{n}} = 2\ (n'+n\epsilon)^{1/2} = \sqrt{n}\,\frac{E'}{E} + \frac{E}{E'}\,\frac{n'+a}{\sqrt{n}}.$$

We emphasize that the asymptotics of Eq. (7.45) considerably differs from the semi-classical asymptotics in terms of the Macdonald function $K_{1/3}$ (see the formula that follows Eq. (7.13)) obtained for $n \sim n' >> 1$ [2].

Basing upon Eq. (7.45) the differential in p'_z probability of SR is found in the form [36]

$$\frac{dw_{n'n}}{dp'_z} = \frac{\alpha c}{2\hbar}\,\frac{E_\perp^2}{EE'}\,\frac{t}{\sqrt{n}}\,F(t)\,, \qquad (7.46)$$

$$F(t) = \left[4n'!\,(2\pi n)^{1/2}\right]^{-1}$$

$$\times\left\{\left[1+\zeta'\,(1+n'/a)^{-1/2}\right]\left[n'D^2_{n'-1}(t) - D^2_{n'}(t)\right] - \frac{2}{t}\,\frac{d}{dt}D^2_{n'}(t)\right\},$$

where $\alpha = e^2/\hbar c$. Here the quantum number $\zeta' = \pm 1$ specifies the spin orientation of an electron in the final state along ($\zeta' = +1$)or opposite ($\zeta' = -1$) to the magnetic field direction $\boldsymbol{H}$, the probability of Eq. (7.46) in this case ($n >> 1,\ n' << n$) being independent of the initial spin orientation ζ. Eq. (7.46) is valid in particular for transitions to the low lying levels $n' = 0, 1, 2, \ldots \sim 1$.

We note that in a superstrong magnetic field $H \sim H_c$ ($a \sim 1$) in the final state with $n' \sim 1$ the transverse momentum turn out to be weakly relativistic: $p'_\perp = mc\,(n'/a)^{1/2} \sim mc$. At a small pitch-angle in the initial state ($p_z >> p_\perp >> mc$) the final longitudinal momentum remains relativistic ($p'_z >> mc$) since, according to Eqs (7.42) and (7.43)

$$\frac{E'}{mc^2} \sim \sqrt{\epsilon}\,\frac{E}{mc^2} \sim \frac{E}{mc^2}\left(1-\beta_0^2\right)^{1/2} = \frac{E}{E_\perp} >> 1. \qquad (7.47)$$

Upon integrating Eq. (7.46) over p_z (with relativistic accuracy extending integration limits to infinity) the total probability of transitions to states with fixed n' is found to be

$$w_{n'n} = \int_{-\infty}^{\infty} dp'_z\,\frac{dw_{n'n}}{dp'_z} = \alpha\,\frac{mc^2}{\hbar\sqrt{a}}\left(1-\beta_z^2\right)^{1/2}$$

$$\times \int_{\sqrt{q}}^{\infty} dtt \left(t^2 - q\right)^{-1/2} F(t) = \left(1 - \beta_z^2\right)^{1/2} w_{n'n}^{(0)} \tag{7.48}$$

where $q = 4(n' + a)$. Here $w_{n'n}^{(0)}$ is the probability in the reference frame, where $p_z = 0$. It is related to $w_{n'n}$ by the Lorentz transformation as it should be. Integration over t in Eq. (7.48) is carried out by means of the relation

$$\int_{\sqrt{q}}^{\infty} dtt \left(t^2 - q\right)^{-1/2} D_{n'}^2(t) = (-1)^{n'} n'! \left(\frac{\pi}{a}\right)^{1/2} \int\limits_q^{\infty} dx I_{n'n'}(x), \tag{7.49}$$

where, according to Eq. (7.38),

$$I_{n'n'}(x) = \exp(-x/2) L_{n'}(x).$$

Combining Eqs (7.48) and (7.49) yields

$$w_{n'n} = \frac{\alpha}{4a} \frac{\left(mc^2\right)^2}{\hbar E} (-1)^{n'} \tag{7.50}$$

$$\times \left\{ I_{n'n'}(q) - \frac{1}{4}\left(1 + \varsigma' (1 + n'/a)^{-1/2}\right) \int\limits_q^{\infty} dx \left(I_{n'n'}(x) + I_{n'-1,\, n'-1}(x)\right) \right\}.$$

for the probability of transitions $(n >> 1) \rightarrow (n' << n)$.

In particular the probability of transition to the ground state $(n' = 0, \quad \zeta' = -1)$ reads

$$w_{n'n} = \frac{\alpha}{2\hbar} \frac{\left(mc^2\right)^2}{E} \frac{H}{H^*} \exp\left(-\frac{H^*}{H}\right). \tag{7.51}$$

This particular result was also obtained (for $p_z = 0$) in Ref. [40] by means of a numerical method. We remark that in the ground state $p'_\perp = 0$, whereas the longitudinal momentum according to Eq. (7.47) can be large: $p'_z >> mc$.

Consider transitions to highly excited states $n' >> 1$, while $n' << n$ above. In this case the asymptotic formula is valid for the Laguerre function $I_{n'n'}(q)$ in Eq. (7.50) (assuming $n'/a = \left(p'_\perp/mc\right)^2 >> 1$) [36]

$$I_{n'n'}(q) \simeq (-1)^{n'} (2n')^{-1/3} Ai(z), \tag{7.52}$$

where

$$Ai(z) = \frac{1}{2\pi} \int\limits_{-\infty}^{\infty} dt \exp\left[i\left(zt + t^3/3\right)\right] \tag{7.53}$$

is the Airy function of the argument

$$z = 2^{2/3}(a - 1/2)(n')^{-1/3}. \tag{7.54}$$

With allowance for Eq. (7.52) upon summation over the spin numbers ς' in Eq. (7.50) we have

$$w_{n'n} = \alpha \frac{(mc^2)^2}{\hbar E} \frac{H}{H_*} (2n')^{-2/3} [-Ai'(z)], \tag{7.55}$$

where $Ai'(z) = dAi(z)/dz$.

In the case when $n' >> |a - 1/2|^3$ the argument $z << 1$, and one can put

$$Ai'(z) \simeq Ai'(0) = -\frac{3^{1/6}}{2\pi}\Gamma(2/3),$$

whence Eq. (7.55) takes the form

$$w_{n'n} = \frac{1}{2} \overset{\bullet}{w} \left(\frac{n}{n'}\right)^{2/3}, \tag{7.56}$$

$$\overset{\bullet}{w} = \frac{3^{1/6}}{\pi}\Gamma(2/3)\,\alpha\,\frac{mc^2}{\hbar}\left(1-\beta_z^2\right)^{1/2}\left(\frac{H}{H_*}\right)^{5/3}\left(\frac{m}{p_\perp}\right)^{7/3}.$$

Hence in the region of large quantum numbers n' the probability decreases with growing n' according to the power law: $w_{n'n} \sim (n')^{-2/3}$, i.e. very slowly. We emphasize that the result of Eq. (7.56) holds only for $n' << n$. In the case when $1 << n' \lesssim n$ Eq. (7.56) is replaced by a well-known formula for the spectral probability of SR in the semiclassical region [2,36]

$$\frac{dw}{dy} = \frac{\sqrt{3}}{2\pi}\alpha\,\frac{mc^2}{\hbar}\left(1-\beta_z^2\right)^{1/2}\frac{H}{H_*}(1+u)^{-2} \tag{7.57}$$

$$\times\left[\int_y^\infty dx K_{5/3}(x) + \frac{u^2}{1+u}K_{2/3}(y)\right],$$

where a discrete variable n' was replaced by a continuous variable y according to the relations

$$\xi y \equiv u = \left(\frac{n}{n'}\right)^{1/2} - 1 = \frac{p_\perp}{p'_\perp} - 1, \tag{7.58}$$

$$\xi = \frac{3}{2}\chi = \frac{3}{2}\frac{p_\perp}{mc}\frac{H}{H_*}, \quad dn' = \frac{H}{H_*}\left(\frac{p_\perp}{mc}\right)^2 \frac{\xi dy}{(1+u)^3}.$$

The variable y is related to the photon energy

$$\hbar\omega = E\frac{u}{1+u} \tag{7.59}$$

It should be remarked that on multiplying Eq. (7.57) by $\hbar\omega$ the spectrum of SR follows (see Eq. (7.25)).

When $n' >> a^3$, i.e. $\chi' = (p'_{\perp}/mc)(H/H^*) >> 1$, the argument y in Eq. (7.57) is small

$$y = \frac{2}{3}\left(\frac{1}{\chi'} - \frac{1}{\chi}\right) << 1,$$

and one can use a well-known asymptotics of the Macdonald function $K_{\nu}(y)$ [37]. We end up with the following asymptotics for spectral probability

$$\frac{dw}{dy} = \frac{\sqrt{3}}{2^{1/3}\pi}\Gamma(2/3)\,\alpha\,\frac{mc^2}{\hbar}\left(1-\beta_z^2\right)^{1/2}\frac{H}{H^*}y^{-2/3}(1+u)^{-3}\left(1+u+\frac{u^2}{2}\right) \qquad (7.60)$$

Here we may change y for a discrete variable n' according to Eq. (7.58)

$$w_{n'n} \doteq w\; u^{-2/3}\left(1+u+\frac{u^2}{2}\right). \qquad (7.61)$$

In the limiting case $n' << n$ we have $u \simeq (n/n')^{1/2} >> 1$, and Eq. (7.61) is transformed to Eq. (7.56). Hence for $1 << n' << n$ the validity domains of Eqs (7.56) and (7.61) overlap.

In the opposite limiting case, when $\nu \equiv n - n' << n$ ($u \simeq (\nu/2n) << 1$), it follows from Eq. (7.61) that

$$w_{n'n} \doteq w\;(2n/\nu)^{2/3}, \qquad (7.62)$$

which gives a well-known classical result for the low frequency region of the spectral radiation power [2]

$$W_{n'n} = \hbar\omega w_{n'n} \sim \omega^{1/3} \sim y^{1/3}.$$

We emphasize that Eq. (7.62) is valid for $\nu >> 1$ (but $\nu << n$). In the region $\nu \sim 1$ the radiation probability is expressed in terms of Bessel functions $J_{\nu}(\nu\beta_0 \sin\vartheta_0)$ (see Ref. [2]) by virtue of a well-known asymptotic relation for the Laguerre functions of Eq. (7.38)

$$\lim_{n\to\infty} I_{n,\,n-\nu}\left(\frac{x^2}{4\pi}\right) = J_{\nu}(x)\,.$$

Therefore at $\nu = n - n' = 0$ the probability $w_{nn} = 0$, as it should be.

Consider now the total power of radiation and estimate the contribution of transitions to the low lying Landau levels. From Eqs (7.50) and (7.51) the probability of the transition to the levels $n' = 0, 1, 2... \sim 1$ in weak fields $H << H^*$ can be estimated as exponentially small. This is why the formula for the spectral probability of SR is uniformly applicable throughout the whole interval $0 \leq y < \infty$ (for the high energy electron $E >> mc^2$ in the field $H << H^*$). The total radiation power W is related to the probability $w_{n'n}$ by the formula

$$W = W\;(\chi\,,\,f_1) = \sum_{n'=0}^{n-1} W_{n',n}, \qquad W_{n',n} = \hbar\omega w_{n'n}, \qquad (7.63)$$

where $\omega = \omega_{n'n}$ is the transition frequency of Eq. (7.36), and W is an invariant function of two invariant parameters [41]

$$\chi = \frac{e\hbar}{m^3c^4}\left[-\left(F_{\mu\nu}p^{\nu}\right)^2\right]^{1/2} = \frac{p_{\perp}}{mc}\frac{H}{H^*}, \tag{7.64}$$

$$f_1 = \frac{e\hbar}{m^2c^3}\left|\frac{1}{2}F_{\mu\nu}F^{\mu\nu}\right|^{1/2} = \frac{p_{\perp}}{mc}\frac{H}{H^*},$$

so that

$$\chi = (2n)^{1/2} f_1^{3/2}. \tag{7.65}$$

For $H << H^*$ and $p_{\perp} >> mc$ ($f_1 << 1$, $\chi >> f_1$) the main contribution to the sum of Eq. (7.63) comes from transitions to levels $n' >> 1$. Therefore replacing summation over n' by integration over y (see Eq. (7.58)) and taking into account Eq. (7.59)), we obtain

$$W\ (\chi\,,\ f_1)|_{f_1<<1} \simeq W\ (\chi\,,\ 0) \equiv W\ (\chi\,) = E\int_0^{\infty} dy\,\frac{u}{1+u}\,\frac{dw}{dy}, \tag{7.66}$$

where dw/dy is defined in Eq. (7.57). At $p_z = 0$ W yields Eq. (7.14). From here for $\chi >> 1$ using the asymptotics of Eq. (7.60), we arrive at (see Eq. (7.26))

$$W\ (\chi\,) \simeq W^*\ (\chi\,) \equiv \left(\frac{2}{3}\right)^5 \Gamma\,(2/3)\,\frac{\alpha m^2c^4}{\hbar}\,(3\chi)^{2/3}.$$

Note that as it has already been mentioned above the main contribution to the power in Eq. (7.66) is provided by the spectral region $u \equiv \xi y \sim 1$ ($n' \sim n$) and therefore up to a numerical factor of the order of unity the result of Eq. (7.26) immediately follows from Eq. (7.63) where we have to put (see Eqs (7.61) and (7.59)) $w_{n'n} \sim w^*$, $\hbar\omega \sim E$: $W \sim nEw^*$. The contribution of the region $n' << n$ ($u >> 1$) in this case is small, as according to Eq. (7.60) we have here $dw/dy \sim y^{-5/3}$.

Consider now superstrong fields $H \gtrsim H^*$. In this case the semiclassical formula of Eq. (7.57) correctly describes only the initial part of the SR spectrum ($y << 1$) and reduces itself to Eqs (7.60), (7.61), as (see Eq. (7.58))

$$y = \frac{2}{3}\left(\frac{H^*}{H}\right)^{3/2}\left(\frac{1}{\sqrt{2n'}} - \frac{1}{\sqrt{2n}}\right) << 1$$

for $n' >> 1$. The hard limit of the spectrum $\left(y \gtrsim 1\right)$, which is formed by transitions from the state with small quantum numbers ($n' = 0, 1, 2, \ldots \sim 1$), is described by Eq. (7.50). It should be once again emphasized that Eqs (7.57) and (7.50) are valid in the regions that overlap (see the sequence of Eqs (7.50), (7.56), (7.57), (7.61)).

The total radiation power is determined by the sum of Eq. (7.63) and in the general case can not be reduced to the integral of Eq. (7.66), as contribution of transitions to the states with $n' \sim 1$ can not be neglected. However, for $n >> 1$ and $H \gtrsim H^*$ Eq. (7.65) gives $\chi >> f_1$, hence Eq. (7.26) provides the main term of the asymptotic expansion of $W\ (\chi\,,\ f_1)$ in powers of the parameters $1/\chi$ and $1/n = f_1^3/\chi^2$ in Eq. (7.63). Next terms of the expansion are given in Ref. [42]

$$W\ (\chi\,,\ f_1)\,/W^*\ (\chi) = 1 - \frac{27\sqrt{3}}{8\pi}\,\Gamma\,(1/3)\,(3\chi)^{-2/3} \tag{7.67}$$

$$+\frac{55\sqrt{3}}{8\pi}\,\Gamma^2\,(1/3)\,(3\chi)^{-4/3}\left(1+\frac{4}{275}\,f_1^2\right).$$

We note that the complete asymptotic series in powers of χ for $W\ (\chi\ )$ in Eq. (7.66) has been obtained already in Ref. [11], the first two terms of the series being described by Eq. (7.67) at $f_1 = 0$. It follows from Eq. (7.67) that consideration for the field parameter f_1 for $f_1 \lesssim 1$ leads to an unessential modification (less than 1.5%) of the second term of the asymptotic expansion $W\ (\chi) \equiv W\ (\chi\,,\ 0)$. Hence at $\chi >> 1$ and $n >> 1$ the simple formula of Eq. (7.26) can be used. Now we estimate the contribution of the states with $n' \sim 1$ to the total power W with the help of the relation

$$R_{n'n} = W_{n'n}/W \tag{7.68}$$

which is considered as a function of n' and n being fixed for various values of the magnetic field intensity. Corresponding plots for the function of Eq. (7.68) made by means of Eqs (7.50) and (7.61) at $n = 10^3$ are depicted in Fig. 7.4. One can see

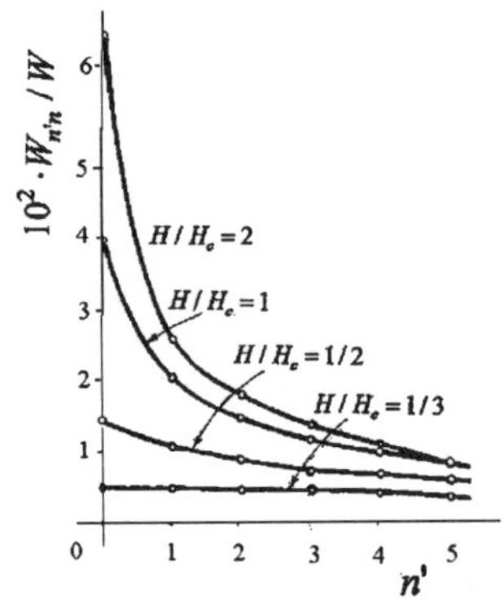

Fig. 7.4. Relative contribution of the states with $n' \sim 1$ to the total power of SR.

that the contribution $W_{n'n}$ of transitions to the low lying level $n' \sim 1$ increases with growing H and at $H \gtrsim H^*$ amounts to about 10% of the total contribution of highly excited states, $W^* \simeq W$ (see Eq. (7.67)). For example, for the ground state $(n' = 0)$

at $H = H^*$, $n = 10^3$ ($E_\perp = 23$ MeV) we obtain $R_{0n} = 0.04$ and at $H = 2H^*$, $n = 10^2$ ($E_\perp = 10$ MeV) we find $R_{0n} = 0.16$. In Fig. 7.5 the evolution of the normalized spectrum $W_{n'n}/W_{0n}$ with changing field intensity is depicted.

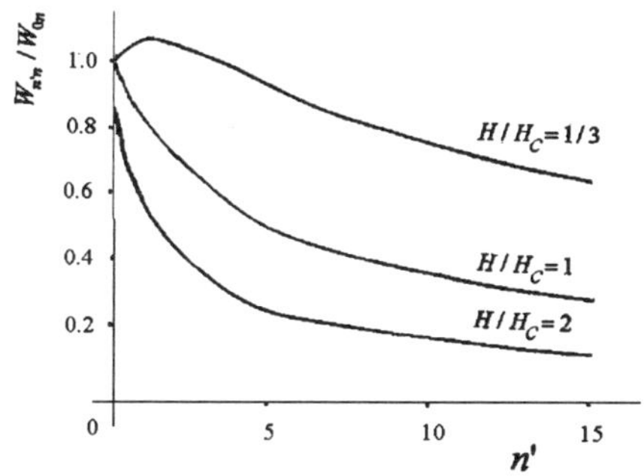

Fig. 7.5. Evolution of the normalized spectrum of SR with changing magnetic field sthrength.

As one can see a monotone spectral function for $H/H^* \gtrsim 1$ is gradually transformed into a characteristic for $H/H^* \gtrsim 1$ spectrum with a well expressed maximum in the region $n' >> 1$.

We note that radiative corrections were not taken into account in the above presentation. However, even in a superstrong magnetic field they turn out to be small. For instance radiative shift of electron energy in the ground state in the field $H >> H^*$ has the form (see e.g. Refs[41,42])

$$\Delta E \simeq \frac{\alpha}{4\pi} mc^2 \ln^2 \left(\frac{2H}{H^*}\right).$$

At $H/H^* = 10^2$ the estimate follows $\Delta E/mc^2 \simeq 0.02$. For the high energy electron ($p_\perp >> mc$) for $\chi >> 1$ we have [41]

$$\frac{\Delta E}{E} = \frac{7}{27\sqrt{3}} \Gamma\left(2/3\right) \left(1 - i\sqrt{3}\right) \alpha \left(3\chi\right)^{2/3} \left(\frac{mc^2}{E}\right)^2,$$

which at $H = H^*$, $n = 10^3$, $p_z = 0$ gives $|\Delta E|/E \simeq 4 \cdot 10^{-5}$. Therefore disregard of the radiative corrections in the above presentation is fairly justified. One of the processes characteristic for the neutron stars is the process inverse to that considered above, i.e. excitation of electrons from low lying Landau levels to much higher levels at absorption of a hard gamma-quantum (see e.g. Ref. [34]). This process in a superstrong field $H \gtrsim H^*$ was for the first time studied in Ref. [31], where in particular for the cross-section of electron excitation from the ground state ($n = 0$, $\zeta = -1$) by a gamma-quantum with the energy $\hbar\omega >> mc^2$ the following expression was obtained

$$\sigma_0 = 2\pi^{3/2} \alpha \frac{\hbar}{m\omega} \left(\frac{H^*}{H}\right)^{1/2} \exp\left(-\frac{H^*}{H}\right). \tag{7.69}$$

The cross-section as a function of H reaches its minimum at $H = 2H^*$. Compare it with the cross-section of the ordinary Compton scattering of a gamma-quantum on an electron at rest in the absence of a field

$$\sigma_c = 2\pi\alpha^2 \left(\frac{\hbar}{m\omega}\right)^2 \frac{1}{\kappa}\left(\ln\kappa + \frac{1}{2}\right), \quad \kappa = \frac{2\hbar\omega}{mc^2} >> 1.$$

This and the relation of Eq. (7.69) allows the following ratio to be written

$$r = \frac{\sigma_0^{\max}}{\sigma_c} \simeq \left(\frac{2\pi}{e}\right)^{1/2} (\alpha \ln\kappa)^{-1},$$

which at $\hbar\omega = 1$ GeV is equal to $r \simeq 25$.

Thus, a relativistic electron moving in a superstrong magnetic field $H \gtrsim H^*$ has a fairly large probability to emit a hard gamma-quantum and make a transition from the highly excited level $(n >> 1)$ to the low lying one $(n' = 0, 1, 2, ... \sim 1)$. As a result a hard discrete end of the SR spectrum is formed, that makes an appreciable contribution to the total SR power. The cross-section of the inverse process of photoexcitation of an electron in the field $H \sim H^*$ is also fairly large. As it was already mentioned above the processes in question play an important role in the astrophysics of neutron stars.

7.3. Dynamical Nature of the Anomalous Magnetic Moment of an Electron

We first of all consider briefly the question why it is so important to study deviation of the value of the anomalous magnetic moment (AMM) from the Bohr magneton: $\mu = -\mu_0 (1 + a_e)$, $a_e = (g - 2)/2$. Undoubtly important here is the strong, and sometimes decisive, influence of the AMM on the spin dynamics of a particle in an external field. This follows directly from the Bargmann-Michel-Telegdi spin evolution equation. In particular, under the influence of the AMM there arise depolarization resonances capable of destroying the transverse polarization of an electron beam acquired while they move in a storage ring. At the same time, if the spin precession frequency is a multiple of the frequencies characteristic of the orbital motion (frequencies of the betatron radial, ω_r, vertical, ω_z, and phase, ω_φ, oscillations) transverse polarization is destroyed and because of the depolarization the radiative self-polarization effect does not lead to a preferred electron spin orientation. The resonance condition [43]

$$a_e \frac{eH}{mc} = N_1\omega_r + N_2\omega_z + N_3\omega_\varphi,$$

(N_i are integers) depends strongly, as can be seen from the expression, on a_e.

The AMM has an even stronger influence on the longitudinal polarization of an electron — spin projection of a particle onto the direction of its motion. As we have just shown, as a result of this influence the longitudinal polarization is no longer conserved in time. Physically, this is due to the fact that the frequency of the gyration of a particle in a magnetic field,

$$\dot{\boldsymbol{P}} = (ec/E)\,[\boldsymbol{H}\boldsymbol{P}]\,,$$

and the frequency of the spin precession,

$$\dot{\boldsymbol{\zeta}} = (ec/E)\left(1 + a_e E/mc^2\right)[\boldsymbol{\zeta}\boldsymbol{H}]\,,$$

are not equal, and this leads to the expression

$$\frac{\boldsymbol{\zeta}\boldsymbol{P}}{|\boldsymbol{P}|} = \cos\left(a_e \frac{eH}{mc} t\right). \tag{7.70}$$

As we have already noted, the vacuum AMM of an electron also plays an important role in the stability of electromagnetic vacuum. The solution of the generalized Dirac-Pauli equation

$$\left\{ i\gamma^{\mu}\left(\frac{\partial}{\partial x_{\mu}} + ieA_{\mu}^{Ext}\right) - m + a_e\,\mu\,H/m \right\}\Psi = 0, \tag{7.71}$$

where $a_e = \alpha/2\pi$ is an anomalous part of the magnetic moment, leads to the solution in which the spin degeneracy is lifted [44]

$$E = \pm mc^2\left[1 + \frac{H}{H^*}\,(2n + \zeta + 1)^{1/2} - \zeta a_e\,\mu\,H/mc^2\right].$$

(We have omitted here the motion of a particle along the field.) If, now, one takes the Schwinger value of the anomalous moment, then in the ground state ($n = 0, \quad \varsigma = -1$) the vacuum becomes unstable if $(\alpha/4\pi)\ (H/H^*) = 1$, since

$$E = \pm mc^2\left(1 - \frac{\alpha}{4\pi}\,\frac{H}{H^*}\right).$$

However, a more detailed analysis shows that in a strong magnetic field ($H \to H^*$) the energy gap between the states $E = \pm mc^2$ not merely does not decrease but actually increases. The fact is that the electron AMM manifests its dynamical nature and $a_e = f\ (E, H)$becomes dependent on the magnetic field - in an ultrastrong magnetic field the electron AMM decreases strongly.

The anomalous magnetic moment of an electron is also of great interest in connection with the development of a high-precision measuring technique, opening up the possibility of investigating not only the terms of the order of $\alpha = e^2/c\hbar$ but also higher approximations in α

$$\mu = -\mu_0\left(1 + a_1\left(\frac{\alpha}{\pi}\right) + a_2\left(\frac{\alpha}{\pi}\right)^2 + a_3\left(\frac{\alpha}{\pi}\right)^3 + \ldots\right).$$

The dynamical nature of the AMM may lead to the contribution accessible to observation, since, the influence of a magnetic field is appreciable on the background of high powers of the expansion with respect to the square of the charge. In the single-loop approximation, the electron AMM was first calculated by Schwinger [45], but in general case the magnetic moment of an electron is rather complicated function that depends on the strength of an external magnetic field. Attention was first drawn to this circumstance by Gupta [46], who considered in the first order of perturbation theory with respect to α not only linear but also higher terms in the expansion of the magnetic moment with respect to the characteristic parameter a^{-1}, where $a = H^*/2H$, and H^* is the critical field.

It should be emphasized that in all the early studies [45,46] the nonrelativistic approximation of the problem was considered. This comes out especially clearly in the work of Luttinger [47] in which an electron in an external magnetic field is taken in a ground state in which the energy of the orbital motion of a particle and the energy of the interaction of the magnetic moment with the external magnetic field compensate each other.

In this connection,one can expect that in the case of relativistic motion of an electron this magnetic moment is not only a function of the field strength but may also depend on the electron energy; finally, we note that in all the cited studies only the case of a weak field, with the strength H restricted to the interval $0 < H << H^*$, was considered. In the light of recent discoveries of extremely strong magnetic fields near pulsars, and also under laboratory conditions (as we have already briefly mentioned), it is of interest to investigate the AMM value in a strong magnetic field.

We now consider the problem of the anomalous magnetic moment of an electron in the first order of perturbation theory in α by a method that is free of the limitations just specified. We note that the problem of dynamical nature of the electron AMM was posed and solved for the first time in our collaboration with Bagrov and Bordovitsyn [10].

We base our study of the electron anomalous magnetic moment on the integrodifferential Dirac-Schwinger equation

$$\left\{i\gamma^{\mu}\left(\frac{\partial}{\partial x_{\mu}} + ieA_{\mu}^{\text{ext}}\right) - m\right\}\Psi(x) = \int M(x,y)\,\Psi(y)\,d^4y, \tag{7.72}$$

the right-hand side of which contains the interaction of an electron with electromagnetic vacuum and includes self-energy effects and corrections associated with vacuum polarization by an external field. Mass operator in the first order in e^2 of perturbation theory for quantum electrodynamics,

$$M(x,y) = -i\,e^2\gamma^{\mu}\,S^{c}(x,y)\,\gamma^{\nu}D_{\mu\nu}^{c}(y-x)\,,$$

contains the electron propagator $S^{c}(x,y)$ with allowance for an external magnetic field,

$$\left\{i\gamma^{\mu}\left(\frac{\partial}{\partial x_{\mu}} + ieA_{\mu}^{\text{ext}}\right) - m\right\}S^{c}(x,y) = -\delta^{4}(y-x)\,,$$

and also the photon propagator $D^{\,c}_{\mu\nu}\,(y-x)$.

We shall use the Feynmann representation for the electron and photon propagators

$$S^{\,c} = \begin{cases} -i \sum\limits_{E_n > 0} \Psi_n\,(\boldsymbol{r})\,\bar{\Psi}_n\,(\boldsymbol{r}')\exp\left[-iE_n\,(t-t')\,/\hbar\right], & t > t', \\ i \sum\limits_{E_n < 0} \Psi_n\,(\boldsymbol{r})\,\bar{\Psi}_n\,(\boldsymbol{r}')\exp\left[-iE_n\,(t-t')\,/\hbar\right], & t < t', \end{cases}$$

where $\bar{\Psi}_n = \Psi_n^+\gamma^0$,

$$D^{\,c}_{\mu\nu}\,(x-x') = -i\frac{g_{\mu\nu}}{(2\pi)^2}\int \exp\left[i\boldsymbol{\kappa}\,(\boldsymbol{r}-\boldsymbol{r}') - i\omega\,|t-t'|\right]\frac{d^3\kappa}{\omega}.$$

In these expressions, $\Psi_n = \Psi_n\,(\boldsymbol{r})\exp\left[-iE_nt/\hbar\right]$ are stationary electron functions satisfying the Dirac equation unperturbed by radiative corrections. The sum over n for $t > t'$ includes states with only positive energy, while for $t < t'$ the causal function is equal to the same sum with the opposite sign, taken over states with negative energy (positron states).

Integration over time leads to the time-independent equation

$$\left(E-\widehat{H}\right)\Psi = \int K\ (\boldsymbol{r},\boldsymbol{r}')\,\Psi\ (\boldsymbol{r}')\,d^3\,x' = \widehat{R}\,\Psi\ (\boldsymbol{r}), \tag{7.73}$$

where $K\ (\boldsymbol{r},\boldsymbol{r}')$ the kernel has the form

$$K\ (\boldsymbol{r}-\boldsymbol{r}') = -\frac{e^2}{4\pi^2}\sum_{n',\epsilon}\int\frac{d^3\kappa}{\kappa}\,\frac{1}{K_n - \epsilon\,(K_{n'}+\kappa)}$$

$$\times\exp\left(i\boldsymbol{\kappa}\boldsymbol{r}\right)\alpha^{\mu}\,\Psi_{n'}\ (\boldsymbol{r})\,\Psi^+_{n'}\ (\boldsymbol{r})\exp\left(i\boldsymbol{\kappa}\boldsymbol{r}\right)\alpha_{\mu},$$

$E = c\hbar K$ is the electron energy, and $\epsilon = \pm 1$ is the sign of energy in the intermediate states, with $\epsilon = 1$ for electron states and $\epsilon = -1$ for positron states, $\alpha^{\mu} = \gamma^0\gamma^{\mu} = (I,\boldsymbol{\alpha})$.

We mention here that the simplest way of calculating a mass operator is the linear approximation in the magnetic field strength and nonrelativistic approximation. Then one can readly find that

$$\int M\ (x,y)\,\Psi\ (y)\,d^4y = -\frac{\alpha}{2\pi}\mu_0\sigma^{\,\mu\nu}F_{\mu\nu}\Psi\ (x),$$

and we arrive at the Dirac-Pauli equation (7.71). In this equation, additional interaction of the magnetic moment with the field is manifested locally, while the dynamical nature of μ remains hidden. To reveal the dynamical origin of the electron AMM, and to reveal the nonlocality of the interaction that generates it, we must return to Eq. (7.73) in its full form, giving up the limit of a quasistatic field. In this way we will also establish the limit of applicability of the Dirac-Pauli equation for an electron in the approximation linear in a field, since sometimes the dynamical nature of the electron AMM is ignored and the AMM is automatically included in the Dirac-Pauli equation even in the case of a strong electromagnetic field. This then leads to difficulties in explaining vacuum stability.

We return to the Dirac-Schwinger equation (7.73). Regarding the right-hand side of this equation as a perturbation, we can see that radiative corrections to the energy are completely determined by the matrix elements of the operator $\widehat{R}$, which characterizes the effective energy of the vacuum interaction

$$W_{\zeta\zeta'} = \int \Psi^+_{n\zeta'}(\boldsymbol{r})\,\widehat{R}\,\Psi_{n\zeta}(\boldsymbol{r})\,d^3x,$$

where the indices $\zeta, \zeta' = \pm 1$ characterize the initial and final spin orientations. We must consider the spin-dependent field correction to an electron mass, since this part of the mass arises because an electron possesses a vacuum magnetic moment [10]. As a result of calculations, we obtain

$$\Delta m^{\rm sp} = W^{\rm sp}_{\zeta\zeta'} = \frac{e^2}{4\pi} \sum_{n,\,\epsilon=\pm 1} \int_0^\infty \int_0^\pi \frac{\kappa d\kappa \sin\vartheta d\vartheta}{K - \epsilon\,(K' + \kappa)} \left(D_{-1}D'_{-1} - D_1 D'_1\right) \Phi, \tag{7.74}$$

where D are spin coefficients [2] determined by wave functions that are exact solutions of the Dirac equation separated with respect to the spin states by means of the operator $\widehat{\Sigma}_3$

$$\widehat{\Sigma}_3\Psi = \zeta\Psi, \quad \zeta = \pm 1,$$

$$\Phi = \frac{k_0}{K}\left[\left(1 - \epsilon\frac{K}{K'}\right)\frac{n' - n}{y} - \epsilon\frac{K}{K'}\right]\left[I^2_{nn'}(y) - I^2_{n-1,n'-1}(y)\right],$$

and the Laguerre function is related to the Laguerre polynomials $Q_n^{n-n'}(y)$ by

$$I_{nn'}(y) = \frac{1}{\sqrt{n!n'!}} \exp(-y/2)\; y^{(n-n')/2}\, Q_n^{n-n'}(y).$$

The part of the vacuum operator $\widehat{R}$ associated with the anomalous magnetic moment is usually replaced in nonrelativistic approximation by the operator $\widehat{R'} = \mu\rho_3(\boldsymbol{\sigma H})$, where constant $\mu = -\mu_0\alpha/2\pi$ is interpreted as the anomalous magnetic moment of an electron. As Pauli noted, such a generalization of the Dirac equation is covariant, and therefore it can be assumed that in general case the operator replacement $\widehat{R} \to \widehat{R}'$ remains valid as well, though now μ may depend on the magnetic field strength and on the electron energy. This means that the anomalous magnetic moment of an electron is of dynamical nature.

Comparing further matrix elements of the operator $\widehat{R}'$,

$$W'_{\zeta\zeta'} = \int \Psi^+_{\zeta'}(\boldsymbol{r})\,\widehat{R}'\Psi_\zeta(\boldsymbol{r})\,d^3x = \mu H\left(D_{-1}D'_{-1} - D_1D'_1\right),$$

with the energy $\Delta m^{\rm sp}$ of the vacuum interaction [see Eq. (7.74)], we may conclude that the operator substitution $\widehat{R} \to \widehat{R}'$ is indeed possible if μ is chosen in the form

$$\mu = \frac{\mathrm{Re}\,\Delta m^{\rm sp}}{H\left(D_1D'_1 - D_{-1}D'_{-1}\right)} = -\frac{\alpha}{2\pi}\,\mu_0\,\zeta\, f\,(n, a). \tag{7.75}$$

Summing now in Eq. (7.74) over the sign factor $\epsilon = E_n/|E_n| = \pm 1$, we obtain the expression [10] for the function $f\ (n,a)$

$$f\ (n,a) = -8a \sum_{n'} \int_0^\infty x\, dx \int_0^\pi \frac{\sin\vartheta\, d\vartheta}{\left(\sqrt{\eta + x^2 \cos^2\vartheta} + x\right)^2 - 1} \tag{7.76}$$

$$\times \left\{1 + \frac{\eta - 1 + x^2 \sin^2\vartheta}{x \sin^2\vartheta \sqrt{\eta + x^2\cos^2\vartheta}}\right\} \left[I^2_{nn'}(z) - I^2_{n-1,n'-1}(z)\right],$$

where $a = k_0^2/4\gamma = H^*/2H$, $\eta = (n'+a)/(n+a)$, $z = (n+a)\, x^2 \sin^2\vartheta$. This expression for the function $f\ (n,a)$ does not contain divergences, is finite in the complete range of variation of energy of an external field, and is exact in the framework of Dirac's theory.

7.3.1. Weak Field

We consider the case of a weak magnetic field, when $a = H^*/2H \gg 1$. At the same time, we can also assume that $n + a \gg 1$, by virtue of which we can make an expansion with respect to $(n+a)^{-1}$ and, integrating and summing in Eq. (7.76), we obtain [10]

$$f\ (n,a) = 1 - \frac{7}{3a^2}\left(\ln\ a - \frac{576\ \ln\ 2 - 83}{420}\right),$$

as a result of which the electron AMM acquires a correction to the Schwinger value in the form of a dependence that is quadratically logarithmic in the field

$$\mu = -\mu_0\,(1 + a_e)\,, \quad a_e = \frac{\alpha}{2\pi}\left[1 + \frac{28}{3}\left(\frac{H}{H^*}\right)^2 \ln\left(\frac{H}{H^*}\right) + \ldots\right]. \tag{7.77}$$

However, as it was shown in Ref. [10], this is valid only for not too strongly excited quantum states of an electron, when $n \ll a$. This corresponds to the non-relativistic approximation because, since the principal quantum number is given by $n = a\,(p_\perp/mc)^2$, the requirement $n \ll a$ is equivalent to the case $\beta \ll 1$. We note that the case considered here corresponds to the approximation $\chi = \frac{1}{2}\,(n/a^3)^{1/2} = (p_\perp/mc)\,(H/H^*) \ll 1$, but under the condition of small numbers n (discrete variation of the parameter χ). In this case, the corrections in the field do indeed have a quadratically logarithmic nature and do not depend on electron energy. Nevertheless, even in this simplest case the dynamical nature of the AMM is manifested. If, however, the field remains weak ($a \gg 1$) but the electron is relativistic, $n/a = (p_\perp/mc)^2 \gg 1$, then the behavior of the anomalous magnetic moment is strongly changed. Indeed, let us return to the original expression of Eq. (7.76) and analyze it under the assumption $a \gg 1$ and $n \gg 1$. The main contribution in this case will be made by transitions to excited intermediate levels with $n' \gg 1$, and therefore we can again go over from a summation over n' to an integration, though the variable of the integration must be chosen in the form $u = \left(\sqrt{n} - \sqrt{n'}\right)/\sqrt{n'}$, i.e., one must introduce

$$n' = n \,/\, (1+u)^2, \quad dn' = -2n\,du \,/\, (1+u)^3 .$$

At the same time, for the Laguerre functions $I_{nn'}(z)$ and $I_{n-1,n'-1}(z)$ we can use the approximation by means of the Macdonald functions $K_{1/3}$ and $K_{2/3}$, and then Re $f\,(n,a)$ can be represented in the form [2,41]

$$\mathrm{Re}\, f\,(n,a) = 2\int_0^\infty \frac{du}{(1+u)^3}\int_0^\pi dt\,\sin\left(t + \frac{\chi^2}{3u^2}\,t^3\right),$$

where χ is the invariant dynamical parameter. The approximate integration leads to the results

$$a_e = \begin{cases} \frac{\alpha}{2\pi}\left[1 - 12\chi^2\left(\ln\frac{1}{\chi} + \ldots\right)\right], & \chi \ll 1, \\ \alpha\Gamma(1/3)\,/\left[9\sqrt{3}\,(3\chi)^{2/3}\right], & \chi \gg 1 \end{cases} \tag{7.78}$$

(see also Refs[41,48]). Thus, in the region $\chi \ll 1$ (quasiquantum region) the corrections to the Schwinger value of the AMM are small as above, but for the parameter $\chi \gg 1$ there is a quite different behavior of the AMM - its value decreases sharply with increasing energy of an electron and is very different from the Schwinger value $\alpha/2\pi$ (ultraquantum case).

7.3.2. Ultrastrong Magnetic Field

An entirely new situation with regard to the field corrections to the anomalous magnetic moment arises in the region of a superstrong magnetic field $H \gg H^*$. We return to Eq. (7.76) and, substituting the variable $x \to x/\sin^2\vartheta$ and integrating exactly over ϑ, reduce it to the form

$$f\,(n,a) = \frac{a\ln a}{n+a} - 8a\int_0^\infty \sum_{n'=0}^{\infty} \frac{x^2(1+\eta) - (1-\eta)^2}{x\sqrt{s}}$$

$$\times\left\{I_{nn'}^2\left[(n+a)\,x^2\right] - I_{n-1n'-1}^2\left[(n+a)\,x^2\right]\right\}\Phi(x,\eta)\,dx,$$

$$\eta = \frac{n'+a}{n+a}, \quad s = \left[(x-\sqrt{\eta})^2 - 1\right]\left[(x+\sqrt{\eta})^2 - 1\right],$$

and the discontinuous function Φ is defined by

$$\Phi(x,\eta) = \begin{cases} \tan^{-1}\left[\sqrt{s}\left[\left(x+\sqrt{\eta}\right)^2 - 1\right]^{-1}\right] & \text{for } \left|\sqrt{\eta}-1\right| \le x \le \left|\sqrt{\eta}+1\right|, \\ \frac{1}{2}\ln\left|\left(x^2+\eta-1+\sqrt{\eta}\right)\left(2x\sqrt{\eta}\right)^{-1}\right| & \text{for others } x. \end{cases}$$

In the asymptotic limit $a \to 0$, the only term $n' = 0$ remains from the complete sum (localization of the Green's function), and its integration gives

$$f\ (1, a) = -2a\left[\ln(1/a) - 1.08\right], f\ (n, a) = 2a \ln(a/n), \quad a_e = (a/2\pi)\ f. \qquad (7.79)$$

Thus, in the limit $H \gg H^*$ intermediate states with $n' = 0$ make a real contribution to the electron AMM. This circumstance emphasizes once more that both the real and virtual states in a superstrong magnetic field are strongly localized in the directions at right angles to the magnetic field with localization within a circle defined by $\pi R^2 \cong \pi (\hbar/mc)^2 (H^*/H)$. As follows from Eq. (7.79), the function $f\ (n, a)$ in the limit $a \to 0$ is negative, and, therefore, the value of the AMM tends to zero with decreasing a (with increasing field) from the side of negative values. However,

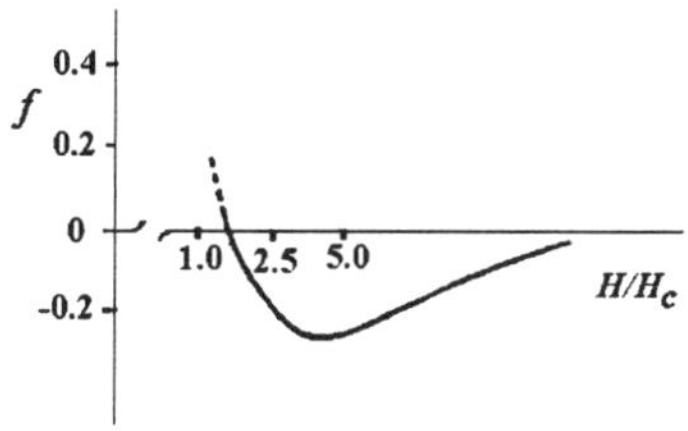

Fig. 7.6. Asymptotic behavior of the anomalous magnetic moment in an ultrastrong magnetic field for $n = 1$.

since for $a \gg 1$ the function $f\ (n, a)$ is positive, and, in addition, it is continuous in the complete range of its argument a (for fixed value of n), this region contains at least two characteristic points, at one of which $f\ (n, a)$ vanishes, while at the other it reaches its minimum value, which, near $a \sim 1$, i.e., $H \sim H^*$. Fig. 7.6 shows the graph of the function $f\ (n, a)$ for the state $n = 1$ calculated in accordance with Eq. (7.79). The change in the sign of the electron AMM was first noted in the Ternov group study [10], and later it was noted in Ref.[49]

Thus, the electron anomalous magnetic moment manifests a clear dynamical nature, being a nonlinear function of the particle energy and of the magnetic field strength.

We have shown that the motion of relativistic electrons in a strong magnetic field is accompanied by a number of fundamental effects: quantum fluctuations of powerful SR lead to broadening of quasiclassical macroscopic orbits of electrons in cyclic accelerators and storage rings, result in primary spin polarization of high energy electrons contrary to the direction of an external magnetic field. In a superstrong magnetic field the SR spectrum differs in a radical way from the classical one, transitions to the low lying Landau levels providing substantial contribution to its formation. Moreover, electromagnetic properties of an electron itself are substantially altered: its AMM depends in a nonlinear way on the field strength and on electron energy. Generally speaking, we may state that the theory of quantum processes in strong magnetic fields is a very rapidly developing direction of physical investigations [29,50], where I. M. Ternov's contribution is fundamental.

Bibliography

[1] I.M. Ternov and V.V. Mikhailin, *Synchrotron Radiation. Theory and Experiment* [in Russian](Energoatomizdat, Moscow, 1986).

[2] A.A. Sokolov and I.M. Ternov, *Radiation from Relativistic Electrons* (AIP, New York, 1986).

[3] V.V. Vladimirskii, *Zh. Eksp. Teor. Fiz.* **18** (1948) 392.

[4] J. Schwinger, *Phys. Rev.* **70** (1949) 1912.

[5] A.A. Sokolov, *Doklady Akad. Nauk SSSR* **67** (1949) 1013.

[6] A.A. Sokolov and I.M. Ternov, *Zh. Eksp. Teor. Fiz.* **25** (1953) 698.

[7] V.B. Beresteskii, E.M. Lifshitz, and L.P. Pitaevskii. *Quantum Electrodynamics*, 2nd ed. (Pergamon Press, Oxford, 1982).

[8] I.M. Ternov, in *History and Methodology of the Natural Sciences* [in Russian] (Moscow State University Press, Moscow, 1985), p. 3.

[9] A.A. Sokolov and I.M. Ternov, *Sov. Phys. Dokl.* **8** (1963) 1203.

[10] I.M. Ternov, V.G. Bagrov, V.A. Bordovitsyn, and O.F. Dorofeev, *Sov. Phys. JETP* **28** (1969) 1206.

[11] N.P. Klepikov, *Zh. Eksp. Teor. Fiz.* **26** (1954) 19.

[12] A.A. Sokolov and I.M. Ternov, *Synchrotron Radiation* (Akademie-Verlag, Berlin, 1968).

[13] I.M. Ternov, *"Development of the Theory of the Motion of Fast Elementary Particles with Allowance for the Effects of Radiation"*, Candidate's Dissertation [in Russian] (Moscow, 1954).

[14] A.A. Sokolov, N.P. Klepikov, and I.M. Ternov, *Zh. Eksp. Teor. Fiz.* **23** (1952) 632.

[15] J. Schwinger, *Proc. Natl. Acad. Sci. USA* **40** (1954) 132.

[16] V.N. Korchuganov, G.N. Kulipanov, N. A. Mezentsev et al., Preprint 77-83 [in Russian], Institute of Nuclear Physics of the Siberian Branch of the USSR Academy of Sciences, Novosibirsk (1977).

[17] N.N. Bogolyubov and S.V. Tyablikov, *Izv. Akad. Nauk Uzb.SSR* **5** (1946) 10.

[18] M. Sands, *Phys. Rev.* **97** (1955) 470.

[19] F.A. Korolev and A.G. Ershov, *Sov. Phys. Dokl.* **5** (1960) 1011.

[20] I.M. Ternov, *"Investigations in the Quantum Theory of Radiating Electrons"*, Doctoral Dissertation [in Russian]. (Moscow, 1961).

[21] A.A. Sokolov and I.M. Ternov, *Izv. Vissh. Uchebn. Zav.* **No. 10** (1967) 66.

[22] A.A. Sokolov, A.V. Borisov, D.V. Galtsov, and V.Ch. Zhukovsky, *Izv. Vissh. Uchebn. Zav.* **No.12** (1974) 5.

[23] I.M. Ternov, *Sov. J. Part. Nucl.* **17** (1986) 389.

[24] A.A. Sokolov and I.M. Ternov, *Zh. Eksp. Teor. Fiz.* **31** (1956) 473.

[25] A.A. Sokolov, I.M. Ternov, V.G. Bagrov, and R.A. Rzaev, in *Synchrotron Radiation* [in Russian] (Nauka, Moscow, 1966).
[26] V.G. Bagrov and O.F. Dorofeev, *Vestn. Mosk. Univ., Fiz. Astron.* **No. 2** (1966) 87.
[27] I.M. Ternov, V.G. Bagrov, and O.F. Dorofeev, *Izv. Vissh. Uchebn. Zav.* **No.10** (1968) 63.
[28] V.G. Bagrov, O.F. Dorofeev, Yu.N. Kargin, G.F. Kopytov, and G.K. Razina, *Numerical Analysis of the Spectral Distribution of Synchrotron Radiation* Paper No. 8002-V87 deposited at VINITI, Moscow (1987).
[29] I.M. Ternov, V.Ch. Zhukovsky, and A. V. Borisov, *Quantum Processes in a Strong External Field* [in Russian] (Moscow University Press, Moscow, 1989).
[30] V.M. Lipunov, *Astrophysics of Neutron Stars* [in Russian] (Nauka, Moscow, 1987).
[31] A.A. Sokolov, V.Ch. Zhukovsky, and N.S. Nikitina, *Vestn. Mosk. Univ., Fiz. Astron.* **15** (1974) 548.
[32] I.G. Mitrofanov, *Space Sci. Rev.* **74** (1995) 417.
[33] S.P. Kelner and Yu.D. Kotov, *Pisma v Zh. Eksp. Teor. Fiz.* **41** (1985) 200.
[34] J.M. Hameury, J.P. Lasota, S. Bonazzola, and J. Heyvaerts, *Astrophys. J.* **293** (1985) 56.
[35] V.V. Usov, *Mon. Notices Roy. Astron. Soc.* **267** (1994) 1035.
[36] A.A. Sokolov, A.V. Borisov, and V.Ch. Zhukovsky, *Izv. Vyssh. Uchebn. Zav.* **No.12** (1975) 51.
[37] I.S. Gradshtein and I.M. Ryzhik, *Tables of Integrals, Sums, Series and Products* (Academic Press, N.Y., 1965).
[38] I.S. Shklovsky, *Astrophys. J.* **159** (1970) L77.
[39] A.Z. Dolginov, Yu.N. Gnedin, and N.A. Silantyev, *Propagation and Polarization of Radiation in Cosmic Media* [in Russian] (Nauka, Moscow, 1979).
[40] D. White, *Phys. Rev.* **D9** (1974) 868.
[41] I.M. Ternov, V.R. Khalilov, and V.N. Rodionov, *Interaction of Charged Particles with a Strong Electromagnetic Field* [in Russian] (Moscow University Press, Moscow, 1982).
[42] V.N. Baier, V.M. Katkov, and V.M. Strakhovenko, *Zh. Eksp. Teor. Fiz.* **98** (1990) 1173.
[43] M. Froissart and R. Stora, *Nucl. Instrum. Methods* **7** (1960) 297.
[44] I.M. Ternov, V.G. Bagrov, and V.Ch. Zhukovskii, *Vestn. Mosk. Univ., Fiz. Astron.* **No. 1** (1966) 30.
[45] J. Schwinger, *Phys. Rev.* **73** (1948) 416.
[46] N.D.S. Gupta, *Nature* **163** (1949) 686.
[47] J.M. Lattinger, *Phys. Rev.* **74** (1948) 893.
[48] V.I. Ritus, *Tr. Fiz. Inst. Akad. Nauk SSSR* **111** (1979) 5.
[49] V.N. Baier, V.M. Katkov, and V.M. Strakhovenko, *Sov. J. Nucl. Phys.* **24** (1976) 197.
[50] V.R. Khalilov, *Electrons in Strong Electromagnetic Fields: an Advanced Classical and Quantum Treatment* (Gordon and Breach, Amsterdam, 1996).

Chapter 8

Synchrotron Radiation in Astrophysics

G.S.Bisnovatyi-Kogan [a]
Space Research Institute, Profsoyuznaya 84/32, Moscow 117810, Russia.

8.1. Introduction

Importance of SR in astrophysics was understood more then 40 years ago, when it was used for explanation of the origin of diffuse Galactic radiation and nonthermal spectra of some objects, most important of which was a supernova remnant Crab nebula. This big step in realization of SR origin of different astronomical objects of star-like and diffuse nature is connected with names of H. Alfven and N. Herlofson,[1] K.O. Kiepenheuer[58], V.L. Ginzburg[35], G.G. Getmatsev[33], I.S. Shklovski[87,88] and I.M. Gordon[45], whose pioneering works gave birth to large number of papers, where synchrotron, and later synchrocompton radiation was used for modeling of radiation of active galactic nuclei (AGN), quasars, pulsars and many other traditional and exotic objects in the Universe (see Refs[89,39]).

The main indication to synchrotron nature of radiation coming from space is its nonthermal spectrum, which is formed when relativistic electrons are injected into a region with a magnetic field. In conditions of a rarefied plasma when collisions are negligible, arbitrary distribution of relativistic electrons is rapidly transforming into a power-law distribution due to SR energy losses, which in turn leads to formation of power-law distribution of emitted radiation.[33,56] Another distinguished feature of SR is its polarization, which was discovered in the optical radiation of Crab nebula[29,101] after its theoretical prediction.[36,45]

Discovery of pulsars[50] and their identification with rapidly rotating neutron stars with huge (10^{12} Gs and larger) magnetic field occurs almost simultaneously with an appearance of the idea of radiation of very strong electromagnetic wave by a

[a]This work was supported in part by RFFI grant 96-02-16553, Astronomy Programm of RMSTP topic 3-169 and NSF grant AST 93-20068.

neutron star due to its rotation as a magnetic nonaligned dipole.[78] Electrons are accelerated up to ultrarelativistic energies in an electrical field of this wave and loose energy moving in its magnetic field emitting radiation on the frequency much higher then the frequency of the order of Hz of the original strong wave. This radiation is resulted from interaction of electrons with electromagnetic wave and may be called as Compton radiation. But due to huge number of coherent photons in the strong wave this radiation may be treated classically and has all features of SR. It is realized now that plasma around pulsars may prevent formation of such strong electromagnetic wave, so this question is not clear yet.

Here we consider different manifestations of SR in astrophysics with main attentions to X-ray sources, supernova, supernova remnants, AGN and quasars where most violent processes are going on and which are in the centre of interest of modern astrophysics.

8.2. SR Spectra of Extraterestial Sources

A charged particle (electron) moving with a velocity **v** in a magnetic field **B** with an angle θ between **v** and **B** follows[41,43] a spiral trajectory with a following radius of the orbit projection on a plane perpendicular to **B**

$$r_B = \frac{v\sin\theta}{\omega_B} = \frac{mcv\sin\theta}{eB\sqrt{1-\frac{v^2}{c^2}}}. \tag{8.1}$$

An angular frequency of rotation ω_B of an electron with an energy E in the magnetic field B is equal to

$$\omega_B = \omega_0\sqrt{1-\frac{v^2}{c^2}} = \frac{eB}{m_e c}\frac{m_e c^2}{E}, \quad E = \frac{m_e c^2}{\sqrt{1-\frac{v^2}{c^2}}}, \quad \omega_0 = \frac{eB}{m_e c}. \tag{8.2}$$

Electron moving in the magnetic field is emitting discrete spectrum multiple to the frequency[43]

$$\omega_B^* = \frac{\omega_B}{\sin^2\theta}. \tag{8.3}$$

Maximum of energy is emitted at frequency[64]

$$\omega_{max} \sim \omega_B\left(\frac{E}{m_e c^2}\right)^3 = \frac{eB}{m_e c}\left(\frac{E}{m_e c^2}\right)^2. \tag{8.4}$$

Energy losses of the electron I(ergs/s) in a magnetic field are determined by

$$I = \frac{2e^4 B_\perp^2}{3m_e^2 c^5}\frac{v^2}{1-\frac{v^2}{c^2}} = \begin{cases} \frac{2e^4 B_\perp^2}{3m_e^2 c^3}\left(\frac{E}{m_e c^2}\right)^2 & \text{at } E \gg m_e c^2; \\ \frac{2e^4 B_\perp^2}{3m_e^2 c^3}\left(\frac{v^2}{c^2}\right) & \text{at } v \ll c. \end{cases} \tag{8.5}$$

One may distinguish 2 (sometimes 3) regimes of radiation depending on difference between ω_B and $\omega_{\max}$. For nonrelativistic electrons with $v \ll c$, $E \approx mc^2$ these frequencies coincide, and practically all energy is emitted as a line at frequency $\omega_0 = \frac{eB}{m_e c}$. This frequency called cyclotron as well as the regime of radiation. For ultrarelativistic electrons with $E \gg m_e c^2$ we have $\omega_{\max} \gg \omega_B$ and this discrete spectrum may be approximated as a continuous, synchrotron spectrum of SR itself. We deal sometimes in astrophysics with a situation when magnetobremstrahlung radiation is produced by electrons with $E \sim m_e c^2$. A spectrum of such radiation is discrete, consisting of many harmonics, what differs from both previous cases. This type of radiation is called sometimes as gyrosynchrotron radiation.[52] In what follows we use the term "cyclotron" for this case also.

A spectrum of the circularly moving with velocity v electron consists of harmonics at the frequency ω_B from Eq. (8.2), with the following distribution of energy over harmonics[64]

$$I_n = \frac{2e^4B^2}{m_e^2c^2v}\left(1-\frac{v^2}{c^2}\right)\left[n\frac{v^2}{c^2}J'_{2n}\left(\frac{2nv}{c}\right) - n^2\left(1-\frac{v^2}{c^2}\right)\int_0^{v/c} J_{2n}(2n\xi)d\xi\right]. \tag{8.6}$$

In a cyclotron limit $nv \ll c$ we get a following distribution over harmonics, using[26] asymptotics of a Bessel function J_l

$$I_n = \frac{e^4B^2}{m_e^2c^2v2^{2n}\Gamma(2n+1)}\left(1-\frac{n}{2n+1}\right)\left(\frac{2nv}{c}\right)^{2n+1}. \tag{8.7}$$

For very small v practically all radiation goes in the first harmonics at frequency ω_0. At $n = 1$ Eq. (8.7) coincides with a nonrelativistic limit of Eq. (8.5). The case $nv \gg c$ is considered in Ref.[64] The CR limit of a relativistic electron is the most interesting. Here a difference between harmonics is so small that approximation of continuous spectrum is used with a frequency distribution over quanta energy as[64]

$$dI = d\omega\frac{\sqrt{3}}{2\pi}\frac{e^3B}{m_ec^2}\frac{\omega}{\omega_c}\int_{\omega/\omega_c}^{\infty} K_{5/3}(z)dz. \tag{8.8}$$

Here

$$\omega_c = \frac{3eB}{2m_ec}\left(\frac{E}{m_ec^2}\right)^2 \tag{8.9}$$

is proportional to $\omega_{\max}$ from Eq. (8.4). Macdonalds function $K_l(z)$ has the following asymptotics[4]

$$K_{5/3}(z) = \begin{cases} \sqrt{\frac{\pi}{2z}}e^{-z} & \text{at } z \gg 1, \\ \frac{1}{2}\Gamma\left(\frac{5}{3}\right)\left(\frac{z}{2}\right)^{-5/3} & \text{at } z \ll 1. \end{cases} \tag{8.10}$$

For an ensemble of electrons with an energy distribution $n(E)$ spectrum of SR is determined by integration over an ensemble

$$J_\omega = \int_0^\infty J_{\omega E} n(E) dE, \ \ J_{\omega,E} = \frac{dI}{d\omega} \tag{8.11}$$

Note, that maximum of SR of the ultrarelativistic electron corresponds to $\omega_m \approx 0.29\omega_c$. We consider here only isotropic distributions $n(E)$, and do not take into account anisotropy implied by presence of large scale magnetic field. In the last case the value $B_\perp \approx \sqrt{2/3B}$ could be used for estimations.

Distribution of electrons with concentration N_e in sources of nonthermal radiation is often follows a power law in an energy interval $E_{\min} < E < E_{max}$

$$n(E) = 4\pi p^2 f(p) = A_e E^{-\gamma_e}, \text{ with } \gamma_e > 1; \ \frac{1}{A_e} = \frac{E_{\min}^{1-\gamma_e} - E_{max}^{1-\gamma_e}}{\gamma_e - 1} \frac{1}{N_e}; \tag{8.12}$$

$$E^2 = p^2c^2 + m_e^2 c^4.$$

Using Eqs (8.8), (8.12) in Eq. (8.11) we get a SR spectrum of power distributed electrons. To simplify the calculations we may use with good precision for spectra at $\omega < y_2(\gamma_e)\omega_c(E_{\max}) = y_2(\gamma_e)\omega_{\text{cmax}} \equiv \omega_{\max}$ and $\omega > y_1(\gamma_e)\omega_c(E_{\min}) = y_1(\gamma_e)\omega_{\text{cmin}} \equiv \omega_{\min}$ the integration over dE from 0 to ∞ beyond the interval $(E_{\min}, E_{\max})$. It is possible, because of small amount of electrons with $E > E_{\max}$ and small input into the radiation of electrons with $E < E_{\min}$. A spectrum of SR at $\omega_{\min} < \omega < \omega_{\max}$ has a power law form[41]

$$J_\omega = \frac{A_e}{2\pi} \frac{\sqrt{3}}{\gamma_e + 1} \Gamma\left(\frac{3\gamma_e - 1}{12}\right) \Gamma\left(\frac{3\gamma_e + 19}{12}\right) \frac{e^3}{m_e c^2} \left(\frac{3e}{m_e^3 c^5}\right)^{\frac{\gamma_e - 1}{2}} B_\perp^{\frac{\gamma_e+1}{2}} \omega^{-\frac{\gamma_e - 1}{2}}. \tag{8.13}$$

For large frequencies $\omega \gg \omega_{\text{cmax}}$ we get a spectral distribution from Eq. (8.11), using Eq. (8.12) and asymptotic expressions given by Eq. (8.10)

$$J_\omega = \frac{A_e}{4} \sqrt{\frac{3}{2\pi} \frac{\gamma_e - 1}{\gamma_e + 1}} \left(\frac{e^3}{m_e c^2}\right) \left(\frac{3e}{2m_e^3 c^5}\right)^{1/2} B^{3/2} \left(\frac{E_{\max}}{m_e c^2}\right)^{2-\gamma_e} \omega^{-1/2} \exp\left(-\frac{\omega}{\omega_{\text{cmax}}}\right). \tag{8.14}$$

Eqs (8.13), (8.14) are valid for a uniform magnetic field. For chaotic magnetic field the flux J_ω is obtained after averaging over angles by expression[41]

$$J_\omega = 2A_e a(\gamma_e) \frac{e^3}{m_e c^2} \left(\frac{3e}{2m_e^3 c^5}\right)^{\frac{\gamma_e - 1}{2}} B^{\frac{\gamma_e+1}{2}} \omega^{-\frac{\gamma_e - 1}{2}}. \tag{8.15}$$

The coefficient $a(\gamma_e)$ is determined as

$$a(\gamma_e) = 2^{\frac{\gamma_e - 1}{2}} \sqrt{3} \Gamma\left(\frac{3\gamma_e - 1}{12}\right) \Gamma\left(\frac{3\gamma_e + 19}{12}\right) \Gamma\left(\frac{\gamma_e + 5}{4}\right) \times \left[8\sqrt{\pi}(\gamma_e + 1)\Gamma\left(\frac{\gamma_e + 7}{4}\right)\right]^{-1} \tag{8.16}$$

with $a(\gamma_e) =$	0.283,	0.103,	0.0742,	0.0725,	0.0922
$y_1(\gamma_e) =$	0.80,	1.8,	2.7,	3.4,	4.0
$y_2(\gamma_e) =$	0.00045,	0.032,	0.18,	0.38,	0.65
for $\gamma_e =$	1,	2,	3,	4,	5.

The synchrotron radiation spectrum of a relativistic Maxwellian plasma with distribution

$$n(E) = \frac{N_e}{2} \frac{E^2}{(kT)^3} \exp\left(-\frac{E}{kT}\right) \tag{8.17}$$

is[16,8]

$$J_\omega = \frac{\sqrt{3}}{2\pi} N_e \frac{e^2}{c} \omega_0 \left(\frac{2\omega}{3\omega_0}\right)^2 \left(\frac{m_e c^2}{kT}\right)^2 \int_0^\infty \gamma^{-3} K_{5/3}\left(\frac{\omega}{\omega_c}\right) \exp\left(-\gamma \frac{m_e c^2}{kT}\right) d\gamma, \tag{8.18}$$

where $\gamma = E/(m_e c^2)$. In two limiting cases the integral in Eq. (8.18) may be calculated, using saddle-point method[16] for high energy limit

$$J_\omega = \begin{cases} \frac{\sqrt{2}}{6} N_e \frac{e^2}{c} \omega_0 z \exp\left[-\left(\frac{9z}{2}\right)^{1/3}\right], & z \gg 1; \\ \frac{3^{1/6}}{\pi} \Gamma(4/3)\Gamma(5/3)\, N_e \frac{e^2}{c} \omega_0 z^{1/3}, & z \ll 1, \end{cases} \tag{8.19}$$

where

$$z = \frac{\omega}{\omega_0} \left(\frac{m_e c^2}{kT}\right)^2.$$

The SR emissivity

$$\varepsilon_B = \int_0^\infty J_\omega d\omega = \int_0^\infty I n(E) dE \tag{8.20}$$

of a Maxwellian plasma is obtained[16] by integration using Eq. (8.5)

$$\varepsilon_B = 2 \frac{N_e}{\rho} \frac{e^2}{c} \omega_0^2 \frac{kT}{m_e c^2} \approx 0.5 \frac{N_e m_p}{\rho} T B^2 \text{ erg g}^{-1}\text{s}^{-1} \tag{8.21}$$

for $kT \ll m_e c^2$ (nonrelativistic cyclotron),

$$\varepsilon_B = 8 \frac{N_e}{\rho} \frac{e^2}{c} \omega_0^2 \left(\frac{kT}{m_e c^2}\right)^2 \approx 3.2 \times 10^{-10} \frac{N_e m_p}{\rho} T^2 B^2 \text{ erg g}^{-1}\text{s}^{-1} \tag{8.22}$$

for $kT \gg m_e c^2$ (relativistic SR). Nonrelativistic Maxwell distribution is used in obtaining of Eq. (8.21)

$$n(E) = \frac{2}{\sqrt{\pi}} \frac{N_e}{(kT)^{3/2}} \sqrt{E}\, \mathrm{e}^{-E/kT}, \quad E = \frac{m_e v^2}{2}. \tag{8.23}$$

8.3. Accretion of Magnetized Gas into a Black Hole

Evolution of massive stars with $M \gg M_\odot$ goes through stages of a nuclear burning of hydrogen, helium, carbon, oxygen etc., until a core of the most stable iron group elements is formed in the centre. Subsequent elements transformation occurs with energy absorption, instead of energy production on previous stages, so intensive energy losses, mainly due to neutrino emission, are not compensated and in order to increase a pressure and to support a star against its self-gravity star compression accelerates. Such compression is transformed into a free-fall collapse, whose fate depends on a mass of a collapsing body.

If a mass of collapsing core exceeds a certain limiting mass, which lay approximately between 2 and 3 solar masses, the collapse cannot be stopped and a black hole emerges, that is, an object having a very strong gravitational field with a gravitational potential $\varphi_G \sim c^2$. The existence of black holes is connected to general relativity. The black hole properties are described in different books (see, for instance, Refs[105,76]). The most important observational property of a black hole consists in that it does not let any light to escape, so that in vacuum it can be detected only for making some known light source invisible. The space between stars and galaxies is however, filled in with a gas that falls on to the black hole, gets heated and emits radiation, making in principle the black hole observable. The most powerful accretion and best observational conditions occur when the black hole is incorporated in a binary, and the material flows from the neighbor star on to the black hole. Such are X-ray sources in binaries — candidates for black holes: CygX-1, LMCX-1, A0620-00, 4U1658-48(GX339-4) and others.[27]

We discuss below models of accretion on to black holes where magnetic field and SR radiation are important (see e.g. Ref.[11]). All of them are hydrodynamical, i.e. the mean free path of particles is assumed to be less than stellar radius. At strong ionization in interstellar medium, this is accounted for by entanglement of electron and ion trajectories by the magnetic field, while in binaries, where density is high, the mean free path is determined by Coulomb collisions and is short as well.

8.3.1. Spherically Symmetric Accretion.

In the case of a spherically symmetric accretion, the flow passes through the sound-velocity point at a saddle-like singularity of the system of hydrodynamical equations.[105] The condition for the flow to pass through the sound-velocity point determines unambiguously the mass flux $\dot{M}$ and all the flow properties for given T_∞

and ρ_∞. An adiabatic flow in the point-mass gravitational field can pass through the sound-velocity point only if the adiabatic index $\gamma_1 < 5/3$. The theory of adiabatic accretion is given in Ref.[105] We discuss hereafter the flows passing through the sound-velocity point and supersonic near the centre of gravity.

Studies of spherically symmetric accretion of interstellar gas on to black holes have shown that in the presence of magnetic field frozen in plasma, the effectiveness of the kinetic energy conversion into heat approaches $\eta \approx 10\%$,[92] while in the absence of field the bremstrahlung causes very small efficiency of accretion $\eta \approx 10^{-8}$. When the gas falls radially, the lines of magnetic force stretch along radius, $B_r \sim r^{-2}$, and the magnetic energy per unit volume $E_M \sim B^2 \sim r^{-4}$ increases more rapidly than the kinetic energy $E_{kin} \sim \rho v^2 \sim \dot{M}v/r^2 \sim r^{-5/2}$ ($\dot{M} = 4\pi\rho v r^2$ is the stationary mass flux, free-fall velocity $v \sim r^{-1/2}$). Since the energy E_M cannot physically exceed E_{kin}, it is assumed in Ref.[92] that an equipartition of energy $E_M \approx E_{kin}$ is supported by the dissipation of magnetic energy, the excess of which is consumed by plasma heating. This heating is taken into account in Ref.[20] and leads to an increase in efficiency η to 30% that may be considered under these assumptions as a realistic estimate.

If $E_M \sim r^{-4}$ is the energy of magnetic field with no dissipation, and $E'_M = E_{kin} \sim r^{-5/2}$ is the energy of magnetic field in the flow, then the increase in entropy per unit volume along radius due to the field annihilation in stationary flow is given by

$$\begin{aligned} Q_M &= \left(\rho T \frac{dS}{dr}\right)_M = \left(\frac{dE_M}{dr} - \frac{dE'_M}{dr}\right)_{E_M = E'_M} \\ &= -4\frac{E_M}{r} + \frac{5}{2}\frac{E_M}{r} = -\frac{3}{2r}\frac{B^2}{8\pi}. \end{aligned} \tag{8.24}$$

Consider separately the regions with non-relativistic

$$kT \ll m_e c^2, \quad \gamma_1 = 5/3,$$

and relativistic electrons. Note, that protons are always considered as non-relativistic

$$\begin{aligned} kT &\gg m_e c^2, \\ \rho E &= \frac{1}{2}\left(3P + \frac{3}{2}P\right) = \frac{9}{4}P = nP, \\ \gamma_1 &= 1 + \frac{1}{n} = 13/9. \end{aligned} \tag{8.25}$$

Here $P_e = P_p = 1/2P$, $\gamma_1 = \left(\frac{\partial P}{\partial \rho}\right)_S$ is adiabatic power,[63] n is the adiabatic index, and for simplicity we treat hydrogen plasma. From the energy balance equation

$$\frac{dE}{dr} - \frac{P}{\rho^2}\frac{d\rho}{dr} = \frac{Q_M}{\rho} - \frac{\epsilon_B}{v_r}, \tag{8.26}$$

where ϵ_B (erg g^{-1}s^{-1}) is the rate of magnetobremstrahlung losses of the Maxwellian plasma, with

$$\frac{B^2}{8\pi} = \frac{1}{2}\rho v_r^2, \quad v_r = \alpha v_{ff} = \alpha\sqrt{\frac{2GM}{r}}, \quad \overline{B_\perp^2} = \frac{2}{3}\overline{B^2}, \tag{8.27}$$

ε_B from Eq. (8.21) for $kT \ll m_e c^2$ (NR) and ε_B from Eq. (8.22) for $kT \gg m_e c^2$ (R), we get the equations for $T(r)$

$$\frac{3}{2}\frac{dT}{dr} + \frac{3}{2}\frac{T}{r} + \frac{3}{4}\frac{\alpha^2 2GM}{\mathcal{R}_g r^2} - 1.5\frac{T\dot{M}}{\mathcal{R}_g r^2} = 0, \quad \text{(NR)}; \tag{8.28}$$

$$\frac{9}{4}\frac{dT}{dr} + \frac{3}{2}\frac{T}{r} + \frac{3}{4}\frac{\alpha^2 2GM}{\mathcal{R}_g r^2} - 2.2\times 10^{-10}\frac{T^2\dot{M}}{\mathcal{R}_g r^2} = 0, \quad \text{(R)}. \tag{8.29}$$

Here $\mathcal{R}_g = 2k/m_p$ is the gas constant for ionized hydrogen. For specified values of ρ_∞, T_∞, M, the mass flux is determined by

$$\begin{aligned}\dot{M} = 4\pi\rho v_r r^2 &= \frac{10^{32}}{c^2}\left(\frac{M}{M_\odot}\right)^2\left(\frac{\rho_\infty}{10^{-24}\text{g cm}^{-3}}\right) \\ &\times \left(\frac{T_\infty}{10^4\text{ K}}\right)^{-3/2} \text{g s}^{-1}.\end{aligned} \tag{8.30}$$

Neglecting the radiation in Eq. (8.28) and adiabatic heating in Eq. (8.29), we find the solution in the form

$$T = 2\times 10^{12}x + 2.7\times 10^{12}x\alpha^2\ln\left(\frac{10^8 x}{T_\infty/10^4\text{ K}}\right), \quad \text{(NR)}; \tag{8.31}$$

$$T = \frac{m_e c^2}{k} + T_1\frac{e^{a(x-x_0)}-1}{e^{a(x-x_0)}+1}, \quad T_1 \gg \frac{m_e c^2}{k}, \quad \text{(R)}. \tag{8.32}$$

Here

$$\begin{aligned}x &= r_g/r = \frac{2GM}{rc^2} < 1, \\ T_1 &= 2.8\times 10^{12}\alpha\left(\frac{T_\infty}{10^4}\right)^{3/4}(M/M_\odot)^{-1/2}\left(\frac{\rho_\infty}{10^{-24}\text{gcm}^{-3}}\right)^{-1/2}, \\ a &= 1.3\,\alpha(M/M_\odot)^{1/2}\left(\frac{T_\infty}{10^4}\right)^{-3/4}\left(\frac{\rho_\infty}{10^{-24}\text{gcm}^{-3}}\right)^{-1/2}.\end{aligned} \tag{8.33}$$

The quantity x_0 depends slightly on T_∞, ρ_∞, and for various values of α is

α^2	1	1/3	1/10

$x_0 \quad 2 \times 10^{-4} \quad 5 \times 10^{-4} \quad 1.2 \times 10^{-3}$

When $x = x_0$, we have $T = m_e c^2/k$, and the solutions given by Eqs (8.31), (8.32) become fitted to one another. The luminosity and spectrum for this model have been calculated in Ref.[20] The luminosity due to magnetobremstrahlung is determined mainly by relativistic electrons from Eq. (8.32) and equals

$$L_B = \begin{cases} 2.7 \times 10^{31} \alpha^4 (M/M_\odot)^3 \left(\frac{T_\infty}{10^4}\right)^{-3} \left(\frac{\rho_\infty}{10^{-24}\text{gcm}^{-3}}\right), & a < 1; \\ 9 \times 10^{31} \alpha^2 (M/M_\odot)^2 \left(\frac{T_\infty}{10^4}\right)^{-3/2} \left(\frac{\rho_\infty}{10^{-24}\text{gcm}^{-3}}\right), & a \gg 1. \end{cases} \tag{8.34}$$

Comparing this expression with Eq. (8.30), we find that at a realistic value $\alpha^2 = 1/3$ the quantity $\eta = L_B/\dot{M}c^2 \leq 30\%$. An approximate emission spectrum L_ω ($L_B = \int_0^\infty L_\omega \, d\omega$) of a black hole with mass $M = 10 M_\odot$ is given in Fig. 8.1 from Ref.[11]

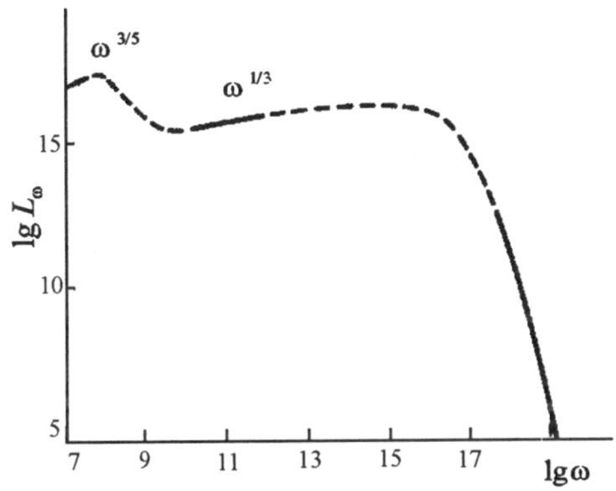

Fig. 8.1. Magnetobremstrahlung spectrum of a black hole of $M \doteq 10\, M_\odot$ for a spherically symmetric accretion and random magnetic field at $\rho_\infty = 10^{-2}$ g·cm^{-3}, at $T_\infty = 10^4$ K, $\alpha^2 = 1/3$. The solid lines represent asymptotic dependencies, dashed lines show extrapolation.

The range with $L_\omega \sim \omega^{3/5}$ is related to the emission of non-relativistic electrons; at $kT \gg m_e c^2$, $\hbar\omega \ll kT$, $L_\omega \sim \omega^{1/3}$, while at $\hbar\omega \approx \hbar\omega_{B,\max}(kT_1/m_e c^2)^2 \sim 10$ eV for $\rho_\infty = 10^{-24}$ g cm^{-3}, $T_\infty = 10^4$ K, $B_{\max} \sim 10^5$ Gauss an exponential cutoff occurs. The visible magnitude m_V for such a black hole under assumption of a plane spectrum is[20]

$$\begin{aligned} m_V &= 4.8 - 2.5 \lg L/\odot L + 5 \lg(R/10 \text{ pc}) \\ &\approx 14.1 - 7.5 \lg M/M_\odot \\ &- 2.5 \lg \left[\left(\frac{T_\infty}{10^4}\right)^{-9/4} \left(\frac{\rho_\infty}{10^{-24}\text{gcm}^{-3}}\right)^{3/2} \right] + 5 \lg \left(\frac{R}{10 \text{ pc}}\right). \end{aligned} \tag{8.35}$$

At a high luminosity and high density the role of bremstrahlung increases and the interaction of the accretion flow with outgoing flow of radiation becomes essential, thereby changing Eqs (8.28)– (8.35).

The above discussion assumes the equipartition to set in at any point in the flow. In reality the field annihilation may occur discontinuously and lead to the shock formation.[23] This is a reason for the emission spectrum to differ even at low luminosity from Fig. 8.1 that should be treated as a rough estimate.

8.3.2. Accretion at an Ordered Magnetic Field

If the characteristic scale of non-uniformity of magnetic field greatly exceeds the accretion radius

$$r_a = \frac{GM}{v_s^2}, \quad v_s \text{ is the sound velocity in the gas,} \tag{8.36}$$

the flow is no longer spherically symmetric. In the case of a uniform magnetic field the accretion symmetry is cylindrical. For a black hole in rest, a stationary pattern of magnetic lines sets in, the material flows along them and forms a disk in the plane of symmetry. A qualitative picture of the flow is shown in Fig. 8.2 from Refs[20,21]

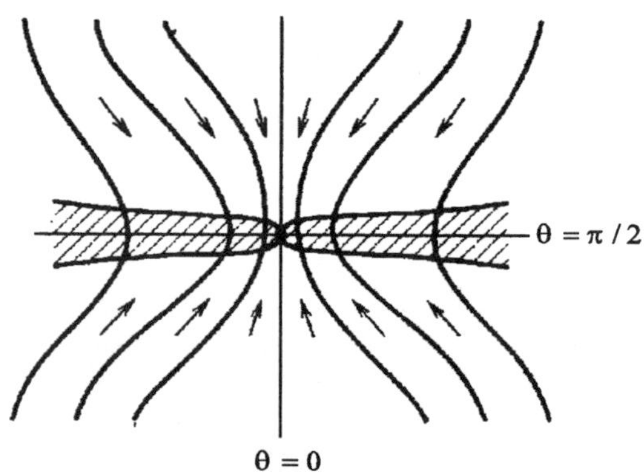

Fig. 8.2. Schematic pattern of lines of magnetic force in the ambient matter of a black hole for a field uniform at infinity, with the inclusion of distortions due to disk currents. The arrows indicate the direction of the gas flux, hatched is the region filled in by the dense disc.

At a large but finite conductivity the disk material infiltrates slowly through lines of magnetic force towards the black hole. The formation process has been studied, the structure of a disk supported by magnetic field and its radiation have been calculated in Refs[21,10]

Consider approximately the stationary disk structure. Equilibrium of a non-rotating disk is determined by balance between magnetic forces and gravity:

$$\frac{GM\Sigma}{r^2} = \frac{1}{c} B_\theta I_\varphi \approx \frac{2\pi}{c^2} I_\varphi^2. \tag{8.37}$$

Here $\Sigma = 2h\rho$ is the surface density, ρ is the mean matter density, I_φ is the circular current propagating through an element of disk area with unit length along r and height $2h$. We have roughly[13]

$$B_\theta \approx B_r \approx \frac{2\pi}{c} I_\varphi. \tag{8.38}$$

Equilibrium along vertical is supported by the pressure gradient

$$\frac{dP}{dz} = -\frac{\rho GM}{r^2} \frac{z}{r}, \quad h \approx \left(\frac{r^3}{GM} \frac{P}{\rho}\right)^{1/2}. \tag{8.39}$$

Taking into account the conversion into heat of the energy of material inflowing to the disk at a velocity close to the free-fall velocity, we have for the energy flux from unit disk surface

$$F = \frac{GM\dot{M}}{4\pi r^3} \left[1 + \frac{1}{2}\left(\frac{r}{R}\right)^{3/2}\right]. \tag{8.40}$$

From the mass conservation we have for the radial velocity

$$v_r = -\frac{\dot{M}}{2\pi r \Sigma} \left[1 - \left(\frac{r}{R}\right)^{3/2}\right]. \tag{8.41}$$

Here $\dot{M}$ is determined by the values at infinity in Eq. (8.30), $R \approx r_a$ from Eq. (8.36). The disk undergoes an ohmic dissipation that causes the material to infiltrate through the magnetic field and determines most of the released energy

$$\frac{GM\dot{M}}{4\pi r^3} \left[1 - \left(\frac{r}{R}\right)^{3/2}\right] = \frac{I_\varphi^2}{4\pi\sigma}, \tag{8.42}$$

where σ is the conductivity. If the disk is opaque for radiation, the energy is carried to its surface by radiative heat conduction, so that we have roughly

$$acT^4 = \kappa \Sigma F, \tag{8.43}$$

$T(r)$ is the mean disk temperature. For a transparent disk

$$2F = \Sigma(\epsilon_{ff} + \epsilon_B). \tag{8.44}$$

This relation includes plasma bremstrahlung ϵ_{ff} and magnetobremstrahlung ϵ_B. Eqs (8.37)–(8.44) with known functions P,κ, σ, ϵ_{ff} and ϵ_B give the structure of a non-rotating disk with magnetic field around the black hole.

In a laminar approximation the disk is always optically thick, electrons are non-degenerate and non-relativistic, the pressure is determined mainly by the ionized gas

$P = \rho \mathcal{R} T$, $\mathcal{R}$ is a gas constant. The conductivity due to Coulomb collisions is given by[11]

$$\sigma = 3.31 \cdot 10^7 \frac{6}{Z} \frac{T^{3/2}}{\Lambda} \mathrm{c}^{-1}, \tag{8.45}$$

where $\Lambda \sim 10$ is a Coulomb logarithm Two regions may be distinguished in such a disk: an outer region, where bremstrahlung processes with

$$\kappa_{ff} \approx 6 \cdot 10^{22} \frac{\rho}{T^{7/2}} \qquad \text{(NR)} \tag{8.46}$$

for nonrelativistic electrons (Kramers formula), and[21,11]

$$\kappa_{ff} \approx 2 \cdot 10^{18} \frac{\rho}{T^3} \ln \frac{kT}{m_e c^2} \qquad \text{(UR)} \tag{8.47}$$

for ultrarelativistic electrons predominate in the opacity, and an inner one, with predominance of the opacity due to magnetobremstrahlung absorption of non-relativistic electrons with[21,11]

$$\kappa_B \approx 2 \cdot 10^2 \frac{B^2}{T^3}. \qquad \text{(NR)} \tag{8.48}$$

Note, that approximate relation for magnetobremstrahlung opacity, as well as for ultrarelativistic free-free case were obtained[21] from averaged Kirghoff law, giving relation between emissivity ε_i and corresponding opacity κ_i in the form

$$\varepsilon_i = A \sigma T^4 \kappa_i, \tag{8.49}$$

where σ is the constant of a radiation density. The constant A was taken in Ref.[21] from the well known relation for free-free processes with nonrelativistic electrons,[38] giving $A \approx 170$. The free-free emissivity of electrons is determined by relations[38]

$$\varepsilon_{ff} = \begin{cases} 6 \cdot 10^{20} \rho T^{1/2} & \text{(NR);} \\ 2 \cdot 10^{16} \rho T \ln \dfrac{kT}{m_e c^2} & \text{(UR).} \end{cases} \tag{8.50}$$

Using the same relation we get from Eq. (8.22) for SR opacity of ultrarelativistic electrons

$$\kappa_B \approx 3 \cdot 10^{-8} \frac{B^2}{T^2} \qquad \text{(UR).} \tag{8.51}$$

As the Coulomb conductivity is small, the material infiltrates slowly through lines of magnetic force, and in a stationary case the disk mass turns out to be large: for a black hole of $M = 100 M_\odot$ the mass of the stationary disk is $M_d \approx 0.2\, M_\odot$. Most of the disk mass is accumulated in its outer layers. The disk interiors are the main source of radiation, the temperature in them attains $10^8 - 10^9$ K, the magnetic field $10^{10} - 10^{12}$ Gauss. The disk thickness nowhere exceeds 0.01 of the radius.

In the turbulent disk approximation the dissipation proceeds much more rapidly owing to twisting of magnetic field lines, and[21]

$$\sigma \approx \sigma_T \approx \frac{c^2}{\tilde{\alpha} 4\pi h \sqrt{P/\rho}}, \quad \tilde{\alpha} = 0.1 - 0.01. \tag{8.52}$$

The outer regions of such a disk are transparent for radiation, electrons are non-relativistic there, the gas pressure predominates, and contributions of bremstrahlung with ε_{ff} from Eq. (8.50) and magnetobremstrahlung with ε_B from Eq. (8.21) losses are of the same order. At $\frac{M}{M_\odot}\frac{r_g}{r} = mx \approx 100\alpha^2$ the electrons become relativistic and $\varepsilon_B \gg \varepsilon_{ff}$ from Eqs (8.22) and (8.50). The zone of relativistic electrons is narrow by radius because the optical thickness increases rapidly and the disk becomes opaque with decreasing radius. In the disk interiors at $10 < mx < 1000$ the gas pressure and electron scattering predominate with the opacity

$$\kappa_{es} = 0.2(1 + x_H + \frac{4n_+ m_p}{\rho}) f(x_0). \tag{8.53}$$

Here x_H is a hydrogen mass part, and the last term in brackets takes into account a presence of positrons with a concentration n_+, calculated in Ref.[11] A function $f(x_0)$ approximately takes into account Klein-Nishina corrections to Thomson scattering in the form[11]

$$f(x_0) = \frac{3}{4}\left[\left(1 - \frac{4}{x_0} - \frac{8}{x_0^2}\right)\ln(1 + x_0) + \frac{1}{2} + \frac{8}{x_0} - \frac{1}{2(1+x_0)^2}\right]\frac{1}{x_0} \tag{8.54}$$

$$= \begin{cases} 1 - x_0 & \text{at } x_0 \ll 1, \\ \frac{3}{4x_0}\left[\ln(1 + x_0) + \frac{1}{2}\right] & \text{at } x_0 \gg 1. \end{cases}$$

Here x_0 is related to the temperature T by[11]

$$x_0 = 2\beta \frac{kT}{m_e c^2}\left[\beta \frac{kT}{m_e c^2} + \sqrt{\left(\beta \frac{kT}{m_e c^2}\right)^2 + 1}\right] \tag{8.55}$$

with a numerical coefficient $\beta \approx 4$. At $mx > 1000$ the predominant role in pressure belongs to radiation. The mass of a turbulent disk remains always small because of a large dissipation and fast infiltration of material through magnetic field. Note again the non-monotonical character of the function $T(r)$ arising from the transformation of a transparent disk into opaque one with decreasing r. Relations describing the distribution of disk parameters are given in Ref.[21] for various disk regions.

As the gravitational energy has always time to convert into heat during stationary disk accretion, the black hole luminosity with a minimum radius equal to $1.5\, r_g$ is

$$L = \frac{1}{3}\dot{M}c^2 \quad \text{with} \quad \dot{M} \quad \text{from Eq. (8.30).} \tag{8.56}$$

The emission spectrum of an opaque disk is related to its effective temperature determined by

$$\frac{ac}{4}T_{ef}^4 = F \tag{8.57}$$

locally at any point of the disk surface under condition that the shock energy be completely thermalized. Hence, using Eq. (8.40), we have for the disk interiors with $r \ll R$, where most of the energy is released,

$$T_{ef} \sim r^{-3/4}. \tag{8.58}$$

The distribution of Eq. (8.58), upon being integrated over the disk surface, yields the spectrum

$$L(\omega) \sim \omega^{1/3} \tag{8.59}$$

with an exponential cutoff at $\hbar\omega \sim kT_{\max}$, $T_{\max} \approx 7 \times 10^5$ K for $M = 10M_{odot}$. So, despite a substantial differences in flow patterns for random and ordered magnetic fields, for the case of a laminar disk their emission spectra are similar, and Eq. (8.35) for m_V is valid in such a case as well. A turbulent disk has a large transparent region with magnetobremstrahlung in the infrared range that may be comparable by power with the ultraviolet and soft X-ray emission of opaque disk interiors.

If the accreting material has an intrinsic angular momentum, the magnetized disk generates electric fields giving rise to the formation of relativistic particles.[14] Such a mechanism is likely to act in Cyg X-1, galactic nuclei, and is analogous to unipolar machine proposed for explaining the pulsar radiation.[47]

Observations from satellites lead to discovery of sources which main radiation is emitted in the X-ray region. These sources are usually binary systems, consisting of a compact object, neutron star or a black hole, and a normal star.

8.4. SR in Radio-Galactics, Quasars and Extragalactic Jets

Discrete sources of radio-emission were identified on the Galactic background due to their flickering. First identification of the most powerful radio-sources Virgo A and Centaurus A with nearby galaxies M 87 and NGC 5128 have been done in 1949 year.[24] After identification of the strong radio-source Cygnus A with a galaxy with a visible magnitude 15^m, having a red shift $r \approx 0.06$, the extragalactic nature of discrete sources was firmly established. The main extragalactic sources of radio-emission are active nuclei of galaxies (AGN) and quasars. The last could be predecessors of AGN, representing most active stage of their evolution.

Point sources in quasars and AGN are often connected with extended directed jets, which are visible in radio, as well as in optics. These jets, which are often visible as one-sided, are believed to be originated as a result of outbursts from central powerful source. The luminosity of jets, nevertheless, is comparable with the luminosity of the central "parent" source, so jets could be considered as tools, by which central source in several times increase its luminosity.

Radiation of AGN, quasars and jets has two features, which indicate on its SR origin: nonthermal, sometimes power-law, spectra and strong degree of polarization. While a central sources has a powerful thermal component in optical, UV and X-ray region, the radiation of jets is usually nonthermal in all spectral regions, indicating to a presence of a strong acceleration mechanisms, by which electrons gain very large energies. Discovery of very hard (up to tens of GeV) gamma emission from many AGN[98,49] indicates that electron energies in jets could be much larger. There are many reviews of properties of ANG, quasars and jets (see e.g. Refs[59,3,68]). Here we consider properties of SR radiation in these objects, which are important for construction of their physical model, and a problem of stability and particle acceleration in jets.

8.4.1. SR Self-absorption and Estimation of a Size of a Source

In powerful sources SR spectral intensity of radiation S_ω may reach values, when its reabsorption is becoming important. At low frequencies, when the photon energy is much smaller then the energy of emitting electron, it is possible to write the following expression for frequency dependent absorption coefficient for isotropically distributed electrons, using Einstein relations for spontaneous and induced radiation[41]

$$\mu_r = -\frac{1}{S_\omega}\frac{dS_\omega}{dx} = -\frac{4\pi^3 c^2}{\omega^2}\int \frac{df(p)}{dp} J_{\omega E} p^2 dp. \tag{8.60}$$

For ultrarelativistic electrons momentum distribution is simply connected with the energy distribution $4\pi f(p)p^2dp = 4\pi f(E)E^2dE$, and for energy distribution $n(E) = 4\pi E^2 f(E)$ we get from Eq. (8.60)

$$\mu_r = -\frac{\pi^2 c^2}{\omega^2}\int E^2 \frac{d}{dE}\left(\frac{n(E)}{E^2}\right) J_{\omega E} dE. \tag{8.61}$$

For power-law energy distribution given by Eq. (8.12) we get,[41] using Eqs (8.8), (8.11) the absorption coefficient

$$\mu_r = g(\gamma_e)\frac{2\pi e^3}{m_e}\left(\frac{3e}{m_e^3 c^5}\right)^{\frac{\gamma_e}{2}} A_e B^{\frac{\gamma_e+2}{2}} \omega^{-\frac{\gamma_e+4}{2}}, \tag{8.62}$$

where for the uniform magnetic field $\mathbf{B}$ in Eq. (8.62) must be taken $B = B_\perp$, and

$$g = g_1(\gamma_e) = \frac{\sqrt{3}}{4}\Gamma\left(\frac{3\gamma_e+2}{12}\right)\Gamma\left(\frac{3\gamma_e+22}{12}\right), \tag{8.63}$$

with $g_1(\gamma_e) = 0.96,\ 0.70,\ 0.65,\ 0.69,\ 0.83$
for $\gamma_e = 1,\ 2,\ 3,\ 4,\ 5.$
In a catholic magnetic field we get after averaging over angles the following expression for $g(\gamma_e)$ [43]

$$g = g_2(\gamma_e) = \frac{\sqrt{3\pi}}{8}\frac{\Gamma\left(\frac{\gamma_e+6}{4}\right)}{\Gamma\left(\frac{\gamma_e+8}{4}\right)}\Gamma\left(\frac{3\gamma_e+2}{12}\right)\Gamma\left(\frac{3\gamma_e+22}{12}\right), \tag{8.64}$$

with $g_2(\gamma_e) = 0.69,\ 0.47,\ 0.40,\ 0.44,\ 0.46$
for $\gamma_e = 1,\ 2,\ 3,\ 4,\ 5.$
With account of reabsorption the spectral flux S_ω from the uniform layer with a thickness d is proportional to

$$S_\omega \simeq \int_0^d J_\omega\, e^{-\mu_r r} dr = \frac{J_\omega}{\mu_r}[1 - \exp(-\mu_r d)]. \tag{8.65}$$

For optically deep layer $\mu_r d \gg 1$ the spectrum of the outgoing radiation will be of the form

$$S_\omega = \frac{J_\omega}{\mu_r} \sim \omega^{-\frac{\gamma-1}{2}}\omega^{\frac{\gamma+4}{2}}B^{-1/2} = \omega^{5/2}B^{-1/2}. \tag{8.66}$$

This dependence is different from Plank relation for equilibrium black-body spectrum, where $S_\omega \sim \omega^2$. Using Eq. (8.61) for calculation of the absorption coefficient μ_{rm} of ultrarelativistic Maxwellian electrons given by Eq. (8.17) we get

$$\mu_{rm} = \frac{\pi^2 c^2}{\omega^2}\frac{J^m_\omega,}{kT} \tag{8.67}$$

where J^m_ω for ultrarelativistic Maxwell is given in Eq. (8.18). Using Eq. (8.66) for estimation of low frequency spectrum with reabsorption we get

$$S^m_\omega \sim \frac{J^m_\omega}{\mu_{rm}} \sim \omega^2, \tag{8.68}$$

similar to equilibrium Planckian spectrum. Deviation of Eq. (8.66) from equilibrium spectrum is connected with nonequilibrium, power-law distribution of electrons in Eq. (8.12).

Using Eq. (8.18), the integral in Eq. (8.61) may be integrated by parts leading to the following expression

$$\mu_r = \pi c \frac{4\sqrt{3}}{9}\frac{e^2}{\omega_0}(m_e c^2)^2 \int_0^\infty \gamma^{-3} K_{5/3}\left(\frac{\omega}{\omega_c}\right)\frac{n(E)}{E^2}d\gamma, \tag{8.69}$$

which is positive for any distribution $n(E)$. This lead to important conclusion, that negative absorption (maser-like emission) is impossible for SR emission of isotropically distributed particles[77] in vacuum. Influence of plasma on the wave propagation and (or) nonisotropic distribution of emitting particles may overcome this restriction.

Note, that we may use Eq. (8.61) for calculations of reabsorbed spectrum of free-free radiation. Approximating free-free spectrum of an electron with an energy E by[5]

$$J_{\omega E}^{ff} = K_{ff}\theta(E - \hbar\omega), \tag{8.70}$$

where

$$\theta(x) = \begin{cases} 1, & \text{for } x \geq 0; \\ 0, & \text{for } x < 0 \end{cases}$$

is a Heviside function, and using Eqs (8.17), (8.70) in Eq. (8.61) we get

$$\mu_{rm}^{ff} = \frac{\pi^2 c^2}{\omega^2} \int_{\hbar\omega}^{\infty} K_{ff} \frac{N_e}{2} \frac{E^2}{(kT)^4} \exp\left(-\frac{E}{kT}\right) dE \tag{8.71}$$

$$= \frac{\pi^2 c^2}{\omega^2} \frac{K_{ff}}{kT} N_e \left[1 + \frac{\hbar\omega}{kT} + \frac{1}{2}\left(\frac{\hbar\omega}{kT}\right)^2\right].$$

At low frequencies $\hbar\omega \ll kT$ we have $\mu_{rm}^{ff} \sim \omega^{-2}$. So, for ultrarelativistic Maxwell distribution we get, similar to Eq. (8.67),

$$\mu_{rm}^{ff} = \frac{\pi^2 c^2}{\omega^2} \frac{J_{\omega}^{ffm}}{kT}, \tag{8.72}$$

what gives, using Eq. (8.65), $S_{\omega}^{ffm} \sim \omega^2$ in according with equilibrium spectrum. The absorption coefficient of free-free radiation of power-law distributed electrons given by Eq. (8.12) is written, using Eq. (8.61) as

$$\mu_{r}^{ff} = \frac{\pi^2 c^2}{\omega^2} \frac{\gamma_e + 2}{\gamma_e} K_{ff} A_e (\hbar\omega)^{-\gamma_e} \sim \omega^{\gamma_e - 2}. \tag{8.73}$$

Free-free emissivity of power-law distributed electrons is written, using Eq. (8.11) as

$$J_{\omega}^{ff} = \frac{A_e K_{ff}}{\gamma_e - 1} (\hbar\omega)^{1-\gamma_e} \sim \omega^{1-\gamma_e}, \tag{8.74}$$

so we get for a low frequency spectrum with reabsorption $S_{\omega}^{ff} \sim J_{\omega}^{ff}/\mu_{r}^{ff} \sim \omega^3$. This spectrum has also a nonequilibrium shape because of nonequilibrium electron distribution.

When a SR spectrum of a source has a low-energy cutoff, produced by SR self-absorption, it is possible to estimate its angular size[93] using observational data at higher frequencies. Let ω_* is a frequency at a maximum of a spectra, corresponding to a unity of an optical depth for SR self-absorption. If d is a size of a source, then $\mu_{r*}\, d = \mu_r(\omega_*)\, d = 1$ and

$$S_{\omega *} = (1 - \frac{1}{\mathrm{e}}) \frac{J_{\omega *}}{\mu_{r*}} = (1 - \frac{1}{\mathrm{e}}) d\, J_{\omega *}. \tag{8.75}$$

Let $\omega_2 > \omega_*$ is a frequency, where self-absorption is unimportant with

$$S_{\omega 2} = d\, J_{\omega 2}, \tag{8.76}$$

and $\omega_1 < \omega_*$ is a frequency in the region of a cutoff where a spectrum is determined by Eq. (8.66) as

$$S_{\omega 1} = \frac{J_{\omega 1}}{\mu_{r1}}. \tag{8.77}$$

If an angular size of a SR source, situated at a distance R is $\theta = \frac{d}{R}$, then observed fluxes at different frequencies are determined as

$$\begin{aligned}
F(\omega_*) &= (1 - \frac{1}{\mathrm{e}})\frac{J_{\omega *}}{\mu_{r*}}\frac{\theta^2}{4} = (1 - \frac{1}{\mathrm{e}})d\, J_{\omega *}\frac{\theta^2}{4},\\
F(\omega_2) &= d\, J_{\omega 2}\frac{\theta^2}{4},\\
F(\omega_1) &= \frac{J_{\omega 1}}{\mu_{r1}}\frac{\theta^2}{4}.
\end{aligned} \tag{8.78}$$

Here J_ω and μ_r are given in Eqs (8.13), (8.62). Assuming that magnetic field in the source is uniform, we find from Eqs (8.78) and (8.13)

$$\begin{aligned}
(1 - \frac{1}{\mathrm{e}})F(\omega_1) &= F(\omega_*)\left(\frac{\omega_1}{\omega_*}\right)^{5/2},\\
F(\omega_*) &= (1 - \frac{1}{\mathrm{e}})[F(\omega_2)\omega_2^{\frac{\gamma_e - 1}{2}}]\omega_*^{-\frac{\gamma_e - 1}{2}},\\
F(\omega_1) &= [F(\omega_2)\omega_2^{\frac{\gamma_e - 1}{2}}]\omega_1^{5/2}\omega_*^{-\frac{\gamma_e + 4}{2}}.
\end{aligned} \tag{8.79}$$

Substituting $F(\omega_1)$ by use of Eqs (8.13), (8.62), (8.63) into the last expression in Eq. (8.79) we find[93]

$$\omega_*^{\frac{\gamma_e + 4}{2}} = [F(\omega_2)\omega_2^{\frac{\gamma_e - 1}{2}}]R(\gamma_e)B^{1/2}\frac{(2\pi)^{7/2}}{\pi\theta^2}, \tag{8.80}$$

where

$$R(\gamma_e) \tag{8.81}$$

$$= \frac{(\gamma_e + 1)(3\gamma_e + 10)}{3\gamma_e + 7}\frac{\Gamma\left(\frac{3\gamma_e + 2}{12}\right)\Gamma\left(\frac{3\gamma_e + 10}{12}\right)}{\Gamma\left(\frac{3\gamma_e - 1}{2}\right)\Gamma\left(\frac{3\gamma_e + 7}{12}\right)}\left(\frac{e}{m_e^3 c}\right)^{1/2}\frac{1}{4}\sqrt{\frac{6}{\pi}}.$$

The value $R(\gamma_e) \sim 10^{31}$ for different γ_e is tabulated in Ref.[77] as a coefficient c_{14}.

For a power-law spectrum in hard region $\omega > \omega_2$ the value in square brackets does not depend on frequency and is determined rather good together with a power γ_e. A weak dependence of the angular size θ on the magnetic field value ($\sim B^{1/4}$) gives a possibility to get a plausible evaluation of θ .

8.4.2. SR in Extragalactic Jets; Magnetic Collimation and Particle Acceleration

Extragalactic jets are often observed near point-like extragalactic sources, like quasars and AGN. Jets are seen at very different scales, from parts of parsec in very long base interferometric (VLBI) observations, up to hundreds of kiloparsecs, visible in optical and radio images. It is now generally believed, following,[67] that energy emitted by quasars and AGN comes from accretion of matter into super-massive black hole (SBH) with mass $M = 10^6 - 10^9\ M_\odot$. Such SMB could be formed in quasars and AGN at the end of evolution of a very dense stellar cluster.[9]

Matter, falling into SBH is taken from stars moving around it, which are disrupted by tidal interaction with SBH or by collisions between themselves. This matter has a large angular momentum and is holding by centrifugal force, forming a disk around SBH. Due to rapid cooling the disk becomes thin, where pressure gradient is small compared to gravitational and centrifugal forces. Turbulent friction between disc layers, rotating with a Keplerian speed, leads to loss of angular momentum and accretion of matter into SBH. Such a picture of a disc accretion is observed also in strong galactic X-ray sources, where disc accretion occurs into a stellar mass black hole (SMBH) or a neutron star in a close binary system.

The efficiency of a transformation of a gravitational energy into radiation during disc accretion onto a BH varies from about 6% for a nonrotating (Schwarzschild) BH to about 40% for a rapidly rotating (Kerr) BH.[105] At large accretion rates luminosity becomes so large that radiation force balance a gravitational one. This happens when a luminosity determined by accretion L_a is equal to

$$L_a = L_{Edd} = \frac{4\pi c G M_{bh}}{\kappa}. \tag{8.82}$$

The luminosity L_{Edd} is called Eddington critical luminosity. As a rule, it cannot be exceeded during accretion and approaches it when matter supply from outside increases. An extra matter flux $\dot{M}$, which cannot reach BH, changes direction of its motion and flows out. So at supercritical $\dot{M}$ luminosity approaches L_{Edd} and matter outflow from the planes of the disc take place simultaneously with a critical disc accretion.[85,14,15] In absence of a magnetic field this outflow from a geometrically thin disc is almost spherical and does not lead to a formation of collimated jet. The situations is changing if a matter falling into a BH is magnetized having large scale ordered magnetic field. It was described in the section 3.2 the picture of accretion of matter with an ordered magnetic field, but without rotation. In reality we may expect presence of both factors. In this case the picture if magnetic disc accretion from section 3.2 becomes more complicated. Matter rotating in the disc is twisting magnetic field lines and rotation in magnetic field is leading to appearence of an electrical field like in an unipolar generator. This magnetic field has a component along a direction of the magnetic field, so charged particles are accelerated along the magnetic field lines, leading to formation of a collimated gas flow in a form of jet. When the induced electrical field is strong enough, the charged particles are accelerated up to very high energies, and moving along the curved magnetic field lines give birth to an electron–positron pair cascade. So jets may contain large amount of positrons, which are not annihilated during an active phase of a jet because of a very low density in them. Particle acceleration in a disc accretion of a magnetized matter into a BH was considered in Ref.[14], and formation of a collimated jets in such a picture was described in Ref.[69], see also Ref.[3] The general picture of formation of magnetically collimated jet is shown in Fig. 8.3.

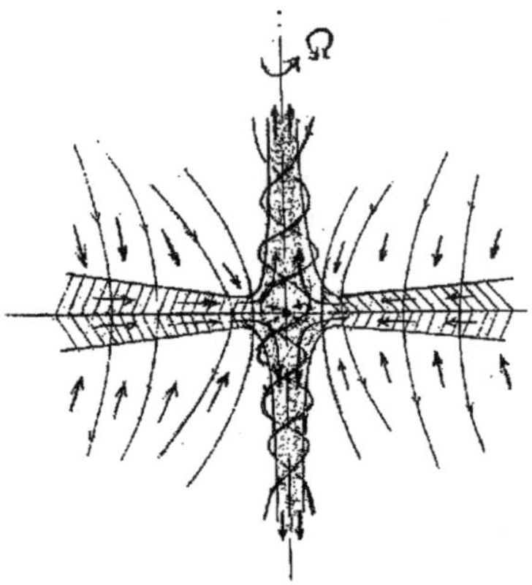

Fig. 8.3. A schematic picture of force lines of the magnetic field and matter motion, indicated by arrows, in the magnetized accretion disk with rotation around a black hole. Rotational axis coincides with the plane of symmetry of the magnetic field. Far from the star a magnetic field is directed along z-axis. The falling matter forms a disk in the symmetry plane and settles slowly to the black hole through magnetic field lines because of dissipation due to finite conductivity, turbulent from Eq. (8.52), and turbulent viscosity. In the central parts of the disc, where rotation is close to Keplerian and magnetic field lines are twisted by rotation, magnetically and centrifugally driven mass outflow, collimated by magnetic field, takes place. Additional driving force for the mass outflow comes from the radiation pressure when a local radiation flux l_r ergs/cm^2/s approaches a critical value $l_{rc} = L_{Edd}/4\pi r^2$, see Eq. (8.82).

Jets are represented by a long luminous structures, coming our from a central source, visible like a bright point source, but these structures are not uniform and are divided in a sequence of bright quasispherical blobs and dark intervals between them. The most studied jets in the nucleus of the galaxy M 87 [51,83] and quasar 3C 273 [81] show such a structure, see also Ref.[25]

Assuming a plausible hypothesis, that extragalactic jets are formed as a result of matter outflow from the accretion disc collimated by a large scale magnetic field, we need to explain two basic properties of these jets: their stability against expansion and dispersion; and long time luminosity of jets, indicating to continuous particle acceleration inside them. Luminosity of extragalactic jets in wide spectral region, from radio to X-ray, has a synchrotron nature. This radiation has all properties of SR: nonthermal, power-law spectrum with a low energy cutoff, and strong polarization on the level of tens of percents, impossible in the thermal radiation.[51,81]

Outflow from the rotating magnetized accretion disc implies twisting of the magnetic field lines and creation of azimuthal component of the magnetic field. It was suggested in Ref.[18] that azimuthal magnetic field in jets is responsible for their stability, preventing them from expansion and dispersion, so jets may be considered as **magnetically collimated.** Presence of the azimuthal magnetic field is connected with an electrical current I_z along a jet axis. In addition to electrodynamic origin of this current, connected with a rotational twisting of magnetic field lines, formation of an axial current may be connected also with a process of origin of a jet, which may be connected with a radiation pressure, acting on electrons much stronger, then on protons (ions). This pressure is proportional to the Thomson scattering cross-section $\sigma_T = \frac{8\pi}{3}\left(\frac{e^2}{mc^2}\right)^2$ which depend on mass as $\sim m^{-2}$. In a spatially restricted jet of a radius R a motion of electrons with respect to protons produces an azimuthal magnetic field on the surface of the jets $B_\phi \sim I_z/cR$, which is supporting a jet against dispersion all along it up to large distance from a central engine. A strength of an axial magnetic field falls down with a distance from a centre,[21] so soon azimuthal field B_ϕ becomes larger than B_z. In such conditions a jet with axial current is becoming unstable against constriction (pinch instability[102,55]) and jet is divided in blobs, visible in observations. Such blobs may be separated from each other and be self-supported by electrodynamic oscillations inside them.

In the spatially restricted body an axial current leads to collection of an electrical charge whose electrical field stops the current, and then reverse it. If a blob is born with initial charge separation, or (and) initial axial current, it gives birth to large scale electromagnetic oscillations like in capacitance-inductance system.[18] If Q is an initial charge separation and R is a characteristic size of a blob, then electrostatic energy of the blob is of the order of $\varepsilon_{el} \sim E^2R^3 \sim Q^2/R$, where E is a strength of an electrical field. Magnetic energy in presence of axial current I_z is of the order $\varepsilon_{mag} \sim B^2R^3 \sim I_z^2R/c^2$, $I_z = dQ/dt$. During oscillations we have $Q = Q_0 e^{i\omega_b t}$. Balance of forces is violated during oscillations, what implies oscillational hydrodynamical motion in the blob, which size and form are oscillating with the same frequency. To estimate approximately a period of these oscillations we neglect kinetic energy of a motion which due to its origin does not exceed electromagnetic energy, and consider the energy conservation law in the form $\frac{d}{dt}(\varepsilon_{el} + \varepsilon_{mag}) = 0$, what gives the following relations for a frequency ω_b and period T_b of the blob's oscillations

$$\omega_b^2 \sim c^2/R^2, \quad T_b \sim 2\pi R/c. \tag{8.83}$$

Optical observations of blobs in the nucleus of the galaxy M 87 have shown strong linear polarization in their radiation with a polarization angle changing from one blob to another, with a difference of about $\pi/2$ between blobs.[51] This fact is explained in the picture of magnetically collimated lets divided in blobs due to action of pinch instability. Electromagnetic oscillations take place in blobs in presence of a common axial magnetic field $B_z \sim (B_\phi)_{\max}$. If in neighboring blobs oscillations occur in antiphase, then the direction of the field B_ϕ is opposite, and the direction of the total magnetic field $\mathbf{B}(0,\, B_z,\, B_\phi)$ has an angle $\pi/2$ between blobs, what may be a reason of the observed behavior of the polarization angle in M 87.[18]

Ultrarelativistic electrons moving preferentially along magnetic field lines have a polarization angle uniquely connected with a direction of a magnetic field.[41] The level of polarization Π for isotropic power law distributed electrons depends only on the power γ_e and equal to[41]

$$\Pi = \frac{\gamma_e + 1}{\gamma_e + \frac{7}{3}} = 75\%;\;\; 69\% \qquad \text{for} \quad \gamma_e = 3;\;\; 2 \qquad \text{respectively.} \tag{8.84}$$

One observational conclusion following from this picture is a variability of the polarization angle with time. The period of such variability in Eq. (8.83) $T_b \sim 500(R/30 \text{ pc})$ years, what may be few hundred years for M 87. This prediction needs observational check, which is very important for confirmation of the picture of magnetically collimated jets, but is difficult because of long expected periods and poor precision in measurements of a polarization angle.

Jets are extended to the distances of kiloparsecs and more and are sufficiently bright all along their length. The life time of a jet τ_j, ejected from the nuclei is estimated by its length

$$\tau_j \geq L/c \geq 3.3 \times 10^3 (L/1\,\text{kpc}) \;\; \text{years}, \tag{8.85}$$

and reaches $\sim 2.6 \times 10^5$ years $\approx 8 \times 10^{12}$ s for the jet in the quasar $3C$ 273. Here the angular size is[81] about 20 arcsec, the redshift $z = 0.158$, what corresponds to a distance 10^3 Mpc at the Hubble constant $H = 50$ km/s/Mpc. That gives a length of the jet about 80 Kpc, leading to estimation of an lower boundary of its age, because the speed of motion of the jet may be considerably less then the light speed. Let us compare the time τ_j with the life time of relativistic electrons, in order to make conclusion about necessity of an acceleration of electrons in a jet. Taking for energy losses of an ultrarelativistic electron from Eq. (8.5)

$$I = \frac{2e^4 B_\perp^2}{3m_e^2 c^3}\gamma^2 \approx 1.6 \times 10^{-15}\gamma^2 B_\perp^2 \;\; \text{ergs/s}, \tag{8.86}$$

we find the life time of this electron τ_{sr} from the estimation

$$\tau_{sr} \approx \frac{\gamma m_e c^2}{I} = \frac{3m_e^2 c^3}{2e^4 B_\perp^2 \gamma} \approx \frac{3 \times 10^8}{B_\perp^2 \gamma} \;\; \text{s}. \tag{8.87}$$

A maximum of the spectrum of an ultrarelativistic electron corresponds to frequency[41]

$$\omega_m \simeq 0.29\omega_c \approx 8 \times 10^6 B_\perp \gamma^2 \ \mathrm{s}^{-1}, \tag{8.88}$$

with ω_c from Eq. (8.9). From observations we know the frequency of a high energy cutoff, which characterize a life time of a most energetic electrons. It corresponds to $\omega \sim 10^{14}$ for the jet in quasar $3C$ 273.[74] Taking ω_m as a high energy cutoff frequency, we obtain for τ_{sr} from Eqs (8.87), (8.88)

$$\tau_{sr} \simeq \frac{3m_e^{5/2}c^{9/2}\kappa^{1/2}}{2e^{7/2}B_\perp^{3/2}\omega_m^{1/2}} \approx \frac{8.1 \times 10^{11}}{B_\perp^{3/2}\omega_m^{1/2}} \ \mathrm{s}. \tag{8.89}$$

Here $\kappa = 0.29 \times 1.5 = 0.435$ is a coefficient, connecting ω_m and ω_0 from Eq. (8.2): $\omega_m = \kappa\omega_0\gamma^2$. Using conventional value for $B = 10^{-4}$ Gs. and $\omega_m = 10^{14}$ s^{-1} in Eq. (8.89) we obtain an estimation $\tau_{sr} \approx 8 \times 10^{10}$ s, what is at least two orders of magnitude less then the observed life time of the jet in $3C$ 273. Such a value of magnetic field corresponds to very high energy of electrons with $\gamma \approx 4 \times 10^5 = 2 \times 10^{11}$ eV from Eq. (8.88). Larger magnetic field would demand less energetic electrons and would give smaller value for τ_{sr}. So, discrepancy between τ_j and τ_{sr} may be explain only in a model where there is a continuous acceleration of electrons in the jet itself.

The mechanism of shock acceleration of particles, often considered[30] seems to be an uncertain explanation of the acceleration of electrons in extra-galactic jets. This is because of unknown efficiency of shock acceleration, which is referred to as the injection problem,[32] and unknown partition of energy between electrons and protons.[30] Further, it is unlikely that shock acceleration can give a fairly uniform brightness jet as observed in some cases. The shock acceleration may be expected mainly from the region near the bow-shock of the jet. The mechanisms of magnetic field reconnection[80] and plasma turbulence acceleration[30] are also highly uncertain.

A new acceleration mechanism for charged particles in extragalactic jets was proposed,[19] which considers emission of intense long-wavelength electromagnetic wave accompanying the electromagnetic oscillations in blobs, discussed above. The electromagnetic wave amplitudes envisioned are sufficient to give *in situ* acceleration of electrons to the very high energies observed.

When the plasma density in the surrounding medium is small enough, the electromagnetic wave generated by the nonpotential plasma oscillations of the confined body is emitted outside and can accelerate single particles. It is clear that when the emitted wave is strong enough it washed out the medium around and the density can become very small, consisting only of the accelerated particles. In that sense the action of the oscillating knot is similar to the action of the pulsar, as inclined magnetic rotator. Both emit strong electromagnetic waves, which could effectively accelerate particles.[78,48]

The solution of the whole problem needs consideration of the dynamical behavior of the confined knots embedded into the large scale elongated magnetic field and producing the toroidal field, connected with the electromagnetic oscillations. The whole problem can be solved by self-consistent calculations of the knot oscillations, using together the hydrodynamical and complete Maxwell equations, and obtaining as a result not only the time dependencies of the distributions of the density and the electrical current but also the power and distributions for the emitted radiation.

In order to imagine the properties of the long wave radiation of the oscillating knot we solve instead the idealized problem having the analytical solution. The model presented here is even simpler then the one considered in Ref.[19], where there was a vacuum layer, surrounding a cylindrical jet, and outside the ambient plasma was involved. Here[12] a plasma cylinder is embedded into a vacuum. Consider linear oscillations of the infinitive long uniform cylinder with a constant cross-section. The problems of such kind have been intensively studied for plasma waveguides.[61] The main difference in our problem are another boundary conditions which suggest vacuum state around the cylinder. When we consider linear electromagnetic oscillations in the static plasma cylinder, only Maxwell equations are needed

$$\nabla \times \mathbf{B} = \frac{4\pi}{c}\mathbf{j} - \frac{i\omega}{c}\mathbf{E}, \tag{8.90}$$

$$\nabla \times \mathbf{E} = \frac{i\omega}{c}\mathbf{B}. \tag{8.91}$$

Time dependence is taken in the form $\sim \exp(-i\omega t)$ and the background constant field $B_z = B_0$ is adopted. Dependence if j on $\mathbf{E}$, $\mathbf{B}$ and B_0 can be obtained, using the expression for the dielectric permeability

$$\begin{aligned} \epsilon_{ij} &= \delta_{ij} + \frac{4\pi i}{\omega}\left[\sigma_{ij}(e) + \sigma_{ij}(p)\right], \\ j_i &= \left[\sigma_{ij}(e) + \sigma_{ij}(p)\right] E_j. \end{aligned} \tag{8.92}$$

We consider for simplicity pure hydrogen plasma. The components for $\sigma_{ij}(\alpha)$ in the case of perfect conductivity are[61]

$$\sigma_{11}(\alpha) = \sigma_{22}(\alpha) = \frac{i}{4\pi}\frac{\omega_{p\alpha}^2\,\omega}{\omega^2 - \omega_{B\alpha}^2}, \tag{8.93}$$

$$\sigma_{12}(\alpha) = -\sigma_{21}(\alpha) = -\frac{1}{4\pi}\frac{\omega_{p\alpha}^2\,\omega_{B\alpha}}{\omega^2 - \omega_{B\alpha}^2}, \quad \sigma_{33}(\alpha) = \frac{i}{4\pi}\frac{\omega_{p\alpha}^2}{\omega},$$

where $\omega_{p\alpha}$ and $\omega_{B\alpha}$ are plasma and Larmor frequencies of electrons (e) and protons (p)

$$\omega_{p\alpha} = \frac{4\pi n_0 e_\alpha^2}{m_\alpha}, \qquad \omega_{B\alpha} = \frac{e_\alpha B_0}{m_\alpha c}. \tag{8.94}$$

In the cylindrical coordinates $(r,\ \phi,\ z)$ with $\frac{\partial}{\partial\phi} = 0,\ \frac{\partial}{\partial z} = ik$ we have from Eq. (8.90)

$$-ikB_\phi = \frac{4\pi}{c}j_r - \frac{i\omega}{c}E_r, \quad ikB_r - \frac{dB_r}{dr} = \frac{4\pi}{c}j_\phi - \frac{i\omega}{c}E_\phi,$$

$$\frac{1}{r}\frac{d}{dr}(rB_\phi) = \frac{4\pi}{c}j_z - \frac{i\omega}{c}E_z, \tag{8.95}$$

and from Eq. (8.91) we have

$$-ikE_\phi = \frac{i\omega}{c}B_r, \quad ikE_r - \frac{dE_z}{dr} = \frac{i\omega}{c}B_\phi, \quad \frac{1}{r}\frac{d}{dr}(rE_\phi) = \frac{i\omega}{c}B_z. \tag{8.96}$$

We are interested in the long wave oscillations with $\omega \ll \omega_{pe}, \omega_{pp}, \omega_{Be}, \omega_{Bp}$, so approximately

$$j_r = -\frac{i}{4\pi}\frac{\omega_{pp}^2\omega}{\omega_{Bp}^2}E_r, \quad j_\phi = -\frac{i}{4\pi}\frac{\omega_{pp}^2\omega}{\omega_{Bp}^2}E_\phi, \quad j_z = -\frac{i}{4\pi}\frac{\omega_{pe}^2}{\omega}E_z. \tag{8.97}$$

Substituting Eq. (8.97) into Eqs (8.95), (8.96) we can see that two types of long waves, corresponding to different polarizations exist independently: electric-type (E) waves with $B_z = 0$ and nonzero (E_r, E_z, B_ϕ), and magnetic-type (B) waves with $E_z = 0$ and nonzero (B_r, B_z, E_ϕ). The equations for $E-$ wave has a form

$$\frac{d}{dr}\left[\frac{1}{r}\frac{d}{dr}(rZ_E)\right] + \left(\frac{\omega_{pe}^2}{\omega^2} - 1\right)\left(\frac{k^2c_A^2}{c^2} - \frac{\omega^2}{c^2}\right)Z_E = 0, \tag{8.98}$$

$$\frac{1}{r}\frac{d}{dr}\left(r\frac{dE_z}{dr}\right) + \left(\frac{\omega_{pe}^2}{\omega^2} - 1\right)\left(\frac{k^2c_A^2}{c^2} - \frac{\omega^2}{c^2}\right)E_z = 0, \tag{8.99}$$

where $Z_E = E_r$ or B_ϕ, and

$$c_A^2 = \frac{c^2}{1 + \frac{\omega_{pp}^2}{\omega_{Bp}^2}} = \frac{c^2}{1 + \frac{4\pi\rho c^2}{B^2}} \tag{8.100}$$

is a speed of the Alfwen waves. The equations for a $B-$ wave has a form

$$\frac{d}{dr}\left[\frac{1}{r}\frac{d}{dr}(rZ_B)\right] + \left(\frac{\omega^2}{c_A^2} - k^2\right)Z_B = 0, \tag{8.101}$$

$$\frac{1}{r}\frac{d}{dr}\left(r\frac{dB_z}{dr}\right) + \left(\frac{\omega^2}{c_A^2} - k^2\right)B_z = 0, \tag{8.102}$$

where $Z_B = B_r$ or E_ϕ. Outside the cylinder Eqs (8.98), (8.99), (8.101), (8.102) retain their form after substituting c everywhere instead of c_A. The solution of these equations is written in the form of Bessel functions:

$$Z_B = Z_{B0}J_1(\kappa_B r), \quad B_Z = B_{Z0}J_0(\kappa_B r),$$
$$Z_E = Z_{E0}J_1(\kappa_E r), \quad E_Z = E_{Z0}J_0(\kappa_E r), \tag{8.103}$$

where

$$\kappa_B^2 = \frac{\omega^2}{c_A^2} - k^2, \qquad \kappa_E^2 = \left(\frac{\omega_{pe}^2}{\omega^2} - 1\right)\left(\frac{k^2 c_A^2}{c^2} - \frac{\omega^2}{c^2}\right) \tag{8.104}$$

inside the cylinder. Outside the same solutions given by Eq. (8.103) are valid with

$$\kappa^2 = \frac{\omega^2}{c^2} - k^2, \tag{8.105}$$

instead of κ_E^2 or κ_B^2. The relations which select discrete values of κ_B and κ_E and give the dispersion equation are obtained from the boundary conditions on the surface of the cylinder with the vacuum.

For $B-$ wave the components B_r and E_ϕ are continuous on the boundary and $B_z = 0$ is taken at the boundary $r = r_0 - 0$. That leads to the relations

$$\kappa_{Bn} r_0 = \lambda_{0n}\ , \quad \omega_n^2 = c_A^2\left(k^2 + \frac{\lambda_{0n}^2}{r_0^2}\right). \tag{8.106}$$

For $E-$ wave E_z is continuous and $E_r = 0$ at the boundary $r = r_0 - 0$, so we have

$$\kappa_{Ei} r_0 = \lambda_{1i}, \quad \omega_i^2 = c_A^2 k^2 \left(1 + \frac{\lambda_{1i}^2 c^2}{r_0^2 \omega_{pe}^2}\right)^{-1}, \tag{8.107}$$

where λ_{0n} and λ_{1i} are zero's of the Bessel functions $J_o(x)$ and $J_1(x)$. We have then for the wave vectors outside the cylinder

$$\kappa_n^2 = \frac{c_A^2}{c^2}\frac{\lambda_{0n}^2}{r_0^2} - k^2\left(1 - \frac{c_A^2}{c^2}\right) \qquad (B),$$

$$\kappa_i^2 = -k^2\left(1 - \frac{c_A^2/c^2}{1 + \frac{\lambda_{1i}^2 c^2}{r_0^2 \omega_{pe}^2}}\right) \qquad (E). \tag{8.108}$$

It is clear from Eq. (8.100) that $\kappa_i^2 < 0$ for $E-$ waves , but for each k there is corresponding n_* such that $\kappa_n^2 > 0$ for $n > n_*$ and Alfven $B-$ wave is transforming into electromagnetic wave radiating into vacuum. The outer value of $B_z(r_0+0)$ is found from the last equation in Eq. (8.96). The jump $\Delta B_z = B_z(r_0+0)$ on the boundary determines the induced surface current i_ϕ.

Very long electromagnetic wave emitted with the amplitude E_ϕ could accelerate particles up to energies $\epsilon \approx e E r_0 \approx \alpha 10^3$ ergs for $r_0 = 1$ pc and $E = \alpha \cdot 10^{-6}$ in CGS. Real energy of the accelerated particles would be much less due to radiation losses by synchrotron and synchrocompton mechanisms, because a particle acceleration takes place in presence of quasistationary magnetic field which is much stronger, then the magnetic field of an electromagnetic wave. It is clear that the expected energies are comparable with the energies of the particles accelerated by the pulsar.

Similarity of the acceleration mechanisms of the particles in the pulsars and relativistic jets could be a reason of the similarity of the high energy radiation around 100

Mev which have been observed by EGRET in number of radio-pulsars; [31,97] quasars and AGN,[98,49] but in no other compact objects.

8.5. Magnetized Neutron Stars: Radio and X-ray Pulsars

Neuron stars, representing the end point of evolution of massive stars with $M \geq (8-10)\, M_{\odot}$[105,11] have been expected to be seen as X-ray sources due to accretion, but first firm evidence of existence of neutron stars came from radio observations, when radio pulsars have been discovered.[50]

8.5.1. Radio Pulsars; Coherent SR

This name was given to point-like radio-sources, emitting signals with very precise periodicity, corresponding to periods ≤ 1 seconds. It became clear soon,[46] that pulsars are rotating neutron stars with enormous magnetic field $B \sim 10^{12}$ Gs, in which SR must be very important. Now more then 700 radio pulsars are known which periods vary from 1.5 ms up to several seconds.[96,72] Neutron stars (see Ref.[65]) are very small ($R \approx 10$ km) and very dense ($\rho_c \geq 10^{15}$ g/cm^3) stars with masses of the order of $1.5 M_{\odot}$. Observed pulsational periods correspond to periods of rotation, and pulses of radiation are determined by strong magnetic field, responsible for a beaming.

A period of a rigidly rotating star cannot exceed a value when a centrifugal force on the surface of equator is equal to the gravitational force $\Omega^2 R < GM/R^2$. It follows that $\Omega < (GM/R^3)^{1/2}$ and rotational period $P > 0.54 R_6^{3/2} M_s^{-1/2}$ ms, where $R_6 = R/(10^6\text{cm})$, M_s is scaled by Solar mass $M_{\odot} = 2 \times 10^{33}$ g. The most compact stars after neutrons are white dwarfs whose radii exceed 10^8 cm and masses are less then $1.2 M_{\odot}$. Their rotational periods must exceed ~ 0.5 seconds. The pulsar in the Crab nebula has a period $P_{crab} = 0.033$ s, what can belong only to a neutron star.

Long time observations have shown, that periods of pulsars increase slowly, in good accordance with a picture, where rotational energy is a main energy source of a pulsar radiation. A neutron star rotating with a period P has rotational energy $E_{rot} = \frac{1}{2} I_s \Omega^2$, where stellar angular velocity is $\Omega = \frac{2\pi}{P}$, and stellar axial inertia momentum $I_s = bMR^2 \approx 10^{45}$ g cm^2 for $M = 1.5 M_{\odot} = 4.5 \times 10^{33}$ g, $R = 10^6$ cm, $b = \frac{1}{4.5}$. If rotational energy is the main source, then luminosity of a pulsar L_p cannot exceed a derivative of a rotational energy $L_p < -\dot{E}_{rot} = -I_s \Omega \dot{\Omega}$. Observations have shown that this limit is never reached and radio luminosity of pulsars is $10^2 - 10^5$ less then $(-\dot{E}_{rot})$. It means that main energy losses are spent for particle acceleration, what is observed in the Crab nebula. Radio luminosity of pulsars has a very large brightness temperature T_b, which determine a source with a Reighley-Jeans flux on the frequency ω

$$F(\omega) = \frac{\omega^2}{2\pi c^2} k_B T_b \left(\frac{R_e}{l_p}\right)^2, \quad T_b = \frac{2\pi c^2 F(\omega)}{k_B \omega^2} \left(\frac{l_p}{R_e}\right)^2, \tag{8.109}$$

where k_B is a Boltzmann constant. For the observed $F(\omega)$, radius of emission $R_r \simeq 10^8$ cm and estimated distances to pulsars l_p it follows enormous brightness temperature in radio region $T_b \geq 10^{25}$ K. This indicates to a nonthermal coherent mechanism of a pulsar radio emission, because no existing source may be heated up to such a huge temperature. The brightness temperature of a pulsar in the Crab nebula, observed also in optical, X-ray and gamma-ray regions is much less $\leq 10^9$ K, indicating a possibility of noncoherent emission, while their luminosities in these regions are 3 - 5 orders of magnitude larger, than in radio. Most other radio pulsars are not seen in optical and X-ray regions, but some of them seen in hard ($E_\gamma \geq 100$ MeV) gamma region,[31,97] what may be an indication of a powerful particle acceleration, taking place continuously in radio pulsars.

Observation of a pulsed radiation is possible only if an axis of a magnetic field is inclined with respect to rotational axis. The inclined rotator emit electromagnetic radiation, which for magnetic dipole with a field B_0 on the surface of the star at the pole, has a power L_d[64,78]

$$L_d = \frac{2\Omega^4}{3c^2} R^6 B_0^2 \sin^2\beta, \tag{8.110}$$

where β is an angle between rotational and magnetic axis. Other magnetic momenta (quadrupole et al) are at least $(v/c)^2$ less effective[64] and usually are not considered. Assuming that loss of rotational energy is determined by magnetodipole radiation, we may estimate magnetic field of a neutron star as

$$L_d = -\dot{E}_{rot} = -I\Omega\dot{\Omega} = \frac{4\pi I\dot{P}}{P^3}, \quad B = \left(\frac{3I_sP\dot{P}}{8\pi^2R^6}\right)^{1/2}. \tag{8.111}$$

For $I_s = 10^{45}$ g$\times$cm^2, $R = 10^6$ cm and P in seconds we get

$$B \approx 3.2 \times 10^{19}(P\dot{P})^{1/2} \text{ Gs}. \tag{8.112}$$

From observations of P and $\dot{P}$ the values of B lay[71,66] between 2×10^8 and 2×10^{13} Gs, while the typical value is 10^{12} Gs. We have taken $\sin\beta = 1$ in Eq. (8.110) when making estimations of the magnetic field strength. This is not because in all pulsars the axis of rotation and magnetic field are perpendicular to each other. Rotating magnetized neutron star is acting like an unipolar generator, in which a strength of an electrical field on the surface $E_u = \frac{1}{c}\mathbf{B} \times (\mathbf{\Omega} \times \mathbf{r})$ is enough to accelerate particle up to very high energies. Energy losses due to ejection of these particles, estimated in Ref.[47], coincides with Eq. (8.110) with $\sin\beta = 1$ for aligned rotator with parallel axis. So, Eq. (8.110) with $\sin\beta = 1$ is valid for any angle β. It follows from Eqs (8.110), (8.111) the relations

$$L = A\Omega^4 = I_s\Omega\dot{\Omega}, \quad \dot{\Omega} = \frac{A}{I_s}\Omega^3, \quad \ddot{\Omega} = 3\frac{A}{I_s}\Omega^2\dot{\Omega}, \quad n = \frac{\Omega\ddot{\Omega}}{\dot{\Omega}^2} = 3. \tag{8.113}$$

Measurements of $\ddot{\Omega}$ for the pulsar in Crab nebula gave a value of breaking index $n \approx 2.5$[65] in satisfactory accordance with the theory.

In a huge field of pulsars the cyclotron frequency ω_0 corresponds to X-ray quanta, $\hbar\omega_0 = 12(B/10^{12}\text{Gs})$ keV. The energy losses in such magnetic fields are so high (see sec.4.2), that electrons occupy only a basic and few excited quantum levels in the magnetic field, called Landau levels. This quantization is related only to the motion in a plane, perpendicular to magnetic field direction. A motion along the direction of a magnetic field is not quantized and may be arbitrary.

The mechanism of radio radiation of pulsars is coherent and connected with a magnetic field, but consistent theoretical picture of it is not constructed. Two possible coherent mechanisms are considered. The first one, called antenna mechanism is based on a radiation of moving charged drops with a charge n times larger then the electron charge. The radiation of a charged drop with n electrons is $\sim n^2$ and n times larger then the radiation of n chaotically moving electrons. Taking characteristic temperature $T \sim 10^{13}$ K, corresponding to $\gamma \sim 10^3$ we need for production of a brightness temperature 10^{25} K drops with $n \sim 10^6$. Note, that coherent emission takes place only[37] for waves with a length $\lambda > d_d$, where d_d is a size of the drop. Such drops may be seeded near the surface of the neutron star, where motion of electrons along the curved magnetic field lines leads to emission of gamma quanta, which give birth to electron–positron pairs in the magnetic field and so on, what finally results in a birth of a pair cascade.[95,82] It is expected, that two stream instability will lead to formation of strongly charged drops, which emit in radio range, moving farther along magnetic field lines. As was mentioned above, in a large magnetic field most electrons occupy a basic Landau level and their emission is connected with a motion along a curved magnetic field lines. This type of radiation[79] called **curvature radiation** has all properties of SR, if we substitute Larmor radius of a rotation in a magnetic field r_B by from Eq. (8.1 a curvature radius of a magnetic field line r_{cur}, taking everywhere $B = \frac{E}{er_{cur}}$. For example, the energy losses of an ultrarelativistic electron in curvature radiation are determined from Eq. (8.5) as

$$I_{cur} = \frac{2e\gamma^3}{3m_e c^3 r_{cur}}, \tag{8.114}$$

and the frequency, where maximum energy is going, is determined from Eq. (8.4) as

$$\omega_{\max}^{cur} \sim \frac{c\gamma^3}{r_{cur}}. \tag{8.115}$$

Observations of a hard gamma radiation from radio-pulsars may be considered as an evidence in favor of this model. They may be explained by curvature and annihilation radiation of particles in pair cascade near the neutron star having $\gamma \sim 10^4 - 10^5$. During a motion from the neutron star γ and magnetic field drop, so a power of radiation becomes very low when characteristic frequencies reach X-ray or optical range. The curvature radiation is becoming stronger again in radio region, when development of instability lead to formation of charged drops and strong coherence of emission. It follows from observations, that efficiency of hard gamma radiation from pulsars is higher for older and fainter ones. It may be connected with the fact, that generated electrical field in old pulsars is smaller and accelerate particles to smaller γ. That is leading to more rapid pair annihilation,[5] and increase efficiency of hard gamma radiation. In young pulsars, like Crab, pair cascade is strong enough to give also detectable optical and X-ray radiation in smaller magnetic field at distances $R_e \approx 10^7 - 10^8$ cm.

Second possible coherent mechanism is based on maser-like radiation, connected with distribution of charged particles in a phase space or nonequilibrium excitations in plasma surrounding pulsars. Coherent mechanism, connected with amplification of plasma waves with farther transformation into radio radiation was considered as a more suitable for pulsars, the coherent SR. Nevertheless, the last one is also possible.

Let us consider coherent SR, which appears in the medium with nonzero refraction coefficient due to plasma influence. Refraction coefficient n_r for electromagnetic waves in "cold" plasma is written as

$$n_r^2 = 1 - \frac{\omega_p^2}{\omega^2}, \tag{8.116}$$

where an electron plasma frequency ω_p is

$$\omega_p^2 = \frac{4\pi N_e e^2}{m_e} = 5.64 \times 10^4 \sqrt{N_e}. \tag{8.117}$$

It was shown,[41,42,107] that even in the case of sufficiently high-frequency waves, for which $1 - n \ll 1$, plasma influence is important when $\gamma^2 \frac{\omega_p^2}{\omega^2} \geq 1$. With account of plasma an emissivity of one electron $J_{\omega,E}^{(p)}$, instead of Eqs (8.10), (8.11) is written as[41,42,107]

$$J_{\omega,E}^{(p)} = \frac{\sqrt{3}\, e^2 \omega_0}{2\pi \, c} \frac{\omega}{\omega_c'} \left(1 + \frac{\omega_p^2}{\omega^2}\gamma^2\right)^{-1/2} \int_{\omega/\omega_c'}^{\infty} K_{5/3}(z) dz. \tag{8.118}$$

Here

$$\omega_c' = \omega_c \left(1 + \frac{\omega_p^2}{\omega^2}\gamma^2\right)^{-3/2}, \quad \omega_c = \frac{3}{2}\omega_0 \gamma^2. \tag{8.119}$$

Absorption coefficient of SR in presence of plasma $\mu_r^{(p)}$ for isotropic distribution of electrons, instead of Eq. (8.62), is written with account of Eq. (8.118) as

$$\mu_r^{(p)} = -\frac{\pi^2 c^2}{\omega^2} \int_0^\infty E^2 \frac{d}{dE}\left(\frac{n(E)}{E^2}\right) J_{\omega,E}^{(p)} dE = -\frac{\pi^2 c^2}{\omega^2} \int_0^\infty \gamma^2 \frac{d}{d\gamma}\left(\frac{n(E)}{\gamma^2}\right) J_{\omega,E}^{(p)} d\gamma \tag{8.120}$$
$$= \frac{\pi e^2 c}{\sqrt{3}\omega} \left(\int_0^\infty 2\frac{n(E)}{\gamma^3} z K_{5/3}(z) d\gamma - \frac{\omega_p^2}{\omega^2} \int_0^\infty \frac{n(E)}{\gamma}\left[z K_{5/3}(z) - 2\int_z^\infty K_{5/3}(x)dx\right] d\gamma\right).$$

Here (see Eqs (8.9), (8.118))

$$z = \frac{\omega}{\omega_c'} = \begin{cases} \frac{\omega}{\omega_c}, & \text{at } \gamma^2 \frac{\omega_p^2}{\omega^2} \ll 1; \\ \frac{2}{3} \frac{\omega_p^3 \gamma}{\omega_0 \omega^2}, & \text{at } \gamma^2 \frac{\omega_p^2}{\omega^2} \gg 1. \end{cases} \tag{8.121}$$

The expression inside square brackets in Eq. (8.120)

$$\Psi(z) = zK_{5/3}(z) - 2\int_z^\infty K_{5/3}(z)dz \tag{8.122}$$

changes sign. It is negative at small z, changes sign at $z \approx 1.35$, reaches a maximum negative value $|\Psi|_{\max} \approx 0.1$ at $z \approx 2$ and with account of Eq. (8.10) has an asymptotics $\Psi(z) \approx \sqrt{\frac{\pi}{2z}}(z-2)e^{-z}$ at $z \gg 1$. So, there are distributions $n(E)$ and for which some frequencies have negative absorption and may be amplified in maser-like way.

For monoenergetic electrons with $\gamma = \gamma_0$ we get[107] from Eq. (8.120)

$$\mu_r^{(p)} = N_e \frac{\pi e^2}{\sqrt{3}\omega m_e c}\left(2\gamma_0^3 z_0 K_{5/3}(z_0) - \frac{\omega_p^2}{\omega^2}\frac{1}{\gamma_0}\left[z_0 K_{5/3}(z_0) - 2\int_{z_0}^\infty K_{5/3}(x)dx\right]\right). \tag{8.123}$$

Here $z_0 = z(\gamma_0)$. For very high frequencies the plasma influence is negligible and we get a positive absorption, determined by first term in square brackets. When $\gamma^2 \frac{\omega_p^2}{\omega^2} \gg 1$, but $z_0 < 1.35$ absorption is also positive. The negative absorption happens when $\gamma^2 \frac{\omega_p^2}{\omega^2} \gg 1$ and $z_0 >\sim 1.35$. Here $\Psi(z)$ is positive and first term in brackets is negligible, so we have a maser-like amplification. At $z_0 \gg 1$ we get in this limit

$$\mu_r^{(p)} = N_e \frac{\pi e^2}{\sqrt{3}\omega m_e c}\left[\frac{2}{\gamma_0^3} z_0 - \frac{\omega_p^2}{\omega^2}\frac{1}{\gamma_0}(z_0 - 2)\right] K_{5/3}(z_0) \tag{8.124}$$

with a negligible first term.

For power-like distribution of Eq. (8.12) we get[42] for the case $\gamma^2 \frac{\omega_p^2}{\omega^2} \gg 1$, when first term in Eq. (8.120) may be neglected

$$\mu_r^{(p)} = -A_e \frac{\pi\sqrt{3}e^2 c\omega_0}{2\omega\omega_p}\left(\frac{2\omega_p^3}{3\omega^2\omega_0}\right)^{\gamma_e+1}(m_e c^2)^{-\gamma_e}\left(1 + \frac{2}{\gamma_e}\right)\int_0^\infty x^{-\gamma_e} K_{5/3}(x)dx. \tag{8.125}$$

Eq. (8.125) is valid only for $\gamma_e < -2/3$, what is connected with the condition of convergence.[107] It follows from Eq. (8.125), that negative absorption takes place at $\gamma_e < -2$. As amplification of both linear polarized waves takes place in such a way, that an amplified signal has a strong circular polarization, it can serve as an observational test for a maser-like action.

Another situation, in which negative absorption of SR is possible, happens in presence of strong anisotropy of an electron distribution function.[108] Let us consider monoenergetic electrons with energy E, moving in the plane, perpendicular to magnetic field, so the distribution function $n(\mathbf{p})$ is written as

$$n(\mathbf{p}) = n(E,\phi) = \frac{N_e}{2\pi}\delta(E - E_0)\delta(\phi - \frac{\pi}{2}), \tag{8.126}$$

$$\text{with a normalization} \quad 2\pi\int_0^\pi \sin\phi\, n(E,\phi)\, d\phi dE = N_e.$$

Here ϕ is an angle calculated from the direction of the magnetic field. For waves moving almost perpendicular to the magnetic field the absorption coefficients for extraordinary ($\mu_r^{(1)}$) and ordinary ($\mu_r^{(2)}$) waves are written as[108]

$$\mu_r^{(1)} = \frac{8e^2N_e}{3cm_e\omega_0}(\psi^2 + \frac{1}{\gamma^2})^2 K_{2/3}(y)\left[2yK_{1/3}(y) + \frac{K_{2/3}(y)}{3}\right], \tag{8.127}$$

$$\mu_r^{(2)} = \frac{8e^2N_e}{3cm_e\omega_0}\psi^2(\psi^2 + \frac{1}{\gamma^2})K_{1/3}(y)\left[2yK_{2/3}(y) - \frac{K_{1/3}(y)}{3}\right].$$

These expressions have been obtained in neglecting of plasma effects for $n_r = 1$ in Eq. (8.116); $\psi \ll 1$ is an angle between a direction of wave propagation and a plane, perpendicular to the direction of a magnetic field,

$$y = \frac{\omega\gamma^2}{3\omega_0}\left(\psi^2 + \frac{1}{\gamma^2}\right)^{3/2}. \tag{8.128}$$

Second relations in Eq. (8.127) is valid at restrictions

$$\psi \ll \frac{\omega_0}{2\omega} \quad \text{at} \quad y \geq 1, \qquad \omega \ll \omega_0\gamma \quad \text{at} \quad y \ll 1. \tag{8.129}$$

The following conclusions follow from Eq. (8.127). There is no absorption of the ordinary wave in the plane, perpendicular to the direction of a magnetic field. There is no negative absorption of the extraordinary wave without plasma. Negative absorption of the ordinary wave is possible when $y < y_* \simeq 0.1$ and second restriction from Eq. (8.129). It is shown in Ref.[108] that in presence of plasma both waves may have a negative absorption.

8.5.2. Accretion into Neutron Stars; Strong X-ray Sources

X-ray sources had been discovered in the rocket flight,[34] but their firm identification with neutron stars had been done only after launch of the X-ray satellite UHURU (see review in Ref.[65]). Its observations had shown that many strong X-ray sources occur to be X-ray pulsars, which existence had been predicted before theoretically.[91] X-ray pulsars as well as most other strong galactic X-ray sources are neutron stars in close binary systems. Mass flow from the companion star falls into the neutron star and a gravitational energy of the falling gas is transformed into a radiation. Accretion into a neutron star differs from the accretion into a black hole by existence of a boundary of a neutron star, due to what gravitational energy of a falling gas is not lost under the gravitational radius, but is distributed between radiation and rotational energy of a neutron star. So an efficiency of transformation of the gravitational energy into heat cannot be small and varies between ~ 10 and $\sim 20\%$.

When a neutron star has a large magnetic field a mass flux is stopped by a pressure of a magnetic field when $\rho v_f^2 \approx B^2/8\pi$. For a free-fall velocity $v_f = (2GM/r)^{1/2}$ density of the matter in the stationary flow with a mass flux $\dot{M}$ varies as

$$\rho = \frac{\dot{M}}{4\pi r^2 v_f} \approx \frac{\dot{M}}{4\pi(2GM)^{1/2} r^{3/2}}. \tag{8.130}$$

For a dipole magnetic field of the neutron star $B = B_0 \left(\frac{R}{r}\right)^3$ the Alfven radius, where the flux is stopped by a magnetic field, is determined as

$$\frac{\dot{M}\sqrt{2GM}}{4\pi r^{5/2}} = \frac{B_0^2}{24\pi}\left(\frac{R}{r}\right)^6, \qquad r_A = \left(\frac{B_0^2 R^6}{6\dot{M}\sqrt{2GM}}\right)^{2/7}. \tag{8.131}$$

This radius reaches a surface of a neutron star $r_A = R$ at

$$B_0 = B_* = \left(\frac{6\dot{M}\sqrt{2GM}}{R^{5/2}}\right)^{1/2} \approx 10^8 L_{37}\ \text{Gs}, \tag{8.132}$$

for $L_{37} = L/10^{37}$ ergs/s, $L = GM\dot{M}/R$, $GM/R = 1.8 \times 10^{20}$ cm^2/s^2, corresponding to the efficiency $\eta = 20\%$, and $R = 10^6$ cm. At $B \leq B_*$ the accretion flow cannot be stopped by a magnetic field and falls directly to the surface of the neutron star.

A gas in a binary system has large angular momentum and forms a disc around a neutron star. Due to a rapid cooling a pressure in the disc falls down and it becomes geometrically thin and rotating with a Keplerian velocity. Due to a friction between layers, which has a turbulent nature, matter in the disc is loosing its angular momentum and is moving to a neutron star. If $B \leq B_*$ the inner edge of the disc reaches a surface of the neutron star and matter from the disc is joining the neutron star. For slowly rotating neutron star a boundary layer is formed at the surface, where considerable amount of energy is produced. For nonrotating star half of the gravitational energy of the accreting gas is emerged in the boundary layer and half in the accretion disc.

Matter joining the neutron star has a large angular momentum and is accelerating a rotation of the neutron star. When accretion disk reaches a surface of the neutron star with $B \leq B_*$ its angular velocity with time reaches a critical value, when centrifugal force is balancing a gravitational one on the equator. After the end of evolution of a companion and its transformation into a white dwarf or another neutron star, accretion is ceased and rotationally powered radio pulsar is borning, which is called "recycled pulsar". Such pulsars may have very small periods, reaching the order of millisecond.

When $B \gg B_*$ radial or disc accretion is stopped at the Alfven radius given by Eq. (8.131) and farther motion of the matter is canalized by the magnetic field, because at $r < r_A$ a magnetic pressure for a dipole field increases as $P_B \sim r^{-6}$, what is much faster then the dynamical pressure $\rho v_f^2 \sim r^{-5/2}$. Due to Rayleigh-Taylor instability matter penetrates inside the magnetosphere of the neutron star, moves along magnetic field lines and falls into the neutron star near its magnetic poles. Only magnetic poles of the neutron star are heated during such accretion, and when angular velocity axis does not coincide with the magnetic one the radiation of such star is modulated with the period of rotation. It is easy to estimate to what temperature is heated the surface of the neutron star, assuming a spectrum close to black body one. We get

$$T_{eff} = \left(\frac{L}{4\pi R^2 \sigma}\right)^{1/4} \approx 1.1 \times 10^7 L_{37}^{1/4}\ K, \tag{8.133}$$

for $R = 10^6$ cm. At this temperature the maximum of radiation falls into an X-ray region. In full accordance with a theory observations have shown that accretion powered X-ray pulsars are decreasing their periods with time, opposite to rotationally powered radio pulsars. As in the case of the accretion into a black hole, the luminosity during accretion cannot exceed the value of the order of a critical Eddington luminosity given by Eq. (8.82), which for neutron stars and opacity, determined by a Thomson scattering $\kappa = 0.4$ is equal to $L_{Edd} \approx 1.3 \times 10^{38}(M/\dot{M})$ ergs/s. The observed luminosity of most strong X-ray sources are of the order of a critical, but most stationary X-ray sources have luminosity which are several times less, then a critical, so the estimation of an effective temperature in Eq. (8.133) is close to the observed one. Stress, that it is related only to an effective temperature, characterizing the luminosity of the star, because local energy of particles may be much higher then kT_{eff}, and spectra of the X-ray sources are often far from the black-body ones.

Supposing that an angular size θ_s of the hot spot at the pole of the neutron star is determined by the pattern of a magnetic field line, arriving to the pole from the Alfven surface, it is possible to write an estimation[62]

$$\sin\theta_s \approx \sqrt{\frac{R}{r_A}} \approx 0.25 B_{10}^{-2/7} L_{37}^{1/7}, \tag{8.134}$$

where $B_{10} = B/(10^{10}$ Gs). Even in the case of a small angular size the beam from an X-ray pulsar would modulated not more, then $\sim \sin(2\pi t/P)$, but observations show much more narrow and more complicated forms of the beam. That needs a formation of an anisotropic radiation near the poles of the neutron star.

Two possible variants of a formation of a beam have been discussed in Ref. [44], and both are connected with SR. First mechanism is using a property of higher harmonics $s \geq 5$ of a magnetobremstrahlung radiation of mildly relativistic electrons with $T_e \sim 10^9$ K, for which an angle of a beam does not exceed 30°. A width of a resonant cyclotron line $\delta\omega$ is about $(kT_e/m_e c^2)^{1/2}\omega$ with $\omega = s\omega_0$, $s = 1, 2....$ Radiation spectrum $J_{\omega s}^l$ of a thermal plasma near the resonance with a number s may be written as[44]

$$J^l_{\omega s} = \mu^l_{rs}(\omega)\mathcal{B}(\omega, T_e), \tag{8.135}$$

where

$$\mathcal{B}(\omega, T_e) = \frac{\hbar\omega^3}{4\pi^3 c^2}\left[\exp\left(\frac{\hbar\omega}{kT_e}\right) - 1\right]^{-1}$$

is a Planckian function, $l = 1$ for extraordinary and $l = 2$ for ordinary waves, respectively. An absorption coefficient of bremstrahlung in a region of a resonance with a number s is written as[106]

$$\mu^l_{rs}(\omega) = \sqrt{2}\pi^{3/2}\frac{s^{2s}}{2^s s!}\frac{e^2}{m_e c\omega}N_e\left(\frac{kT_e}{m_e c^2}\right)^{s-\frac{3}{2}} \tag{8.136}$$

$$\times\frac{1}{|\cos\theta|}f^l_s(\theta)\exp\left[-\frac{m_e c^2}{2kT_e}\frac{(1-\frac{s\omega_0}{\omega})^2}{\cos^2\theta}\right].$$

An angular dependence of radiation is determined by a function

$$f^l_s(\theta) = [1 - (-1)^l\cos\theta]^2\sin^{2s-2}\theta, \tag{8.137}$$

where θ is an angle between the direction of the magnetic field and quantum trajectory. Eqs (8.136), (8.137) are valid in neglecting plasma effects, for $2 < s \leq s_m$, where s_m is determined by a relation

$$\left(\frac{kT_e}{m_e c^2}\right)s_m^2\sin^2\theta = 1. \tag{8.138}$$

At

$$\frac{\omega_p}{\omega} \equiv \left(\frac{4\pi e^2 N_e}{m_e\omega^2}\right)^{1/2} \ll \left(\frac{kT_e}{m_e c^2}\right)^{1/4}$$

Eq. (8.136) is valid also for $s = 1$. At $s > s_m$ the ultrarelativistic limiting formula is valid

$$\mu_{rs}(\omega_0 s) \approx \mu_{rs_m}\sqrt{s - s_m}\exp\left[-\frac{2}{3}\frac{(s - s_m)}{|\sin\theta|}\right], \tag{8.139}$$

where μ_{rs_m} is determined by Eq. (8.136). In the case $s = 1$ the sum of two polarized waves gives a form of a beam as $\overline{f_1}(\theta) = 1 + \cos^2\theta$, what means an almost isotropic beam.

An optical depth to magnetobremstrahlung absorption τ_s in first resonances is large, $\tau_s = \mu_{rs}d > 1$ due to large μ_{rs}, so their intensity in thermal Maxwellian plasma is equal to Rayleigh-Jeans one, $J_{\omega s} \sim \omega^2$. Maximum of radiation corresponds to larger s, where beaming of radiation may be substantial. Estimations above have been done for isotropic distribution of electrons in the magnetic field, while in a strong field this distribution may be anisotropic due to rapid loss of a perpendicular momentum. Besides, observations have shown, that the X-ray pulsars have a beam close to a pencil-like form (hollow-cone model), instead of the knife-like diagram, characteristic for higher harmonics, considered above.

From that point of view more attractive seems to be another model of a creation of a narrow beam, based on a possibility of a formation of anisotropic distribution of relativistic electrons with $T_{\|} \sim 10^{11}$ K and $T_{\perp} \ll T_{\|}$. In this case the beaming is connected with a relativistic motion along a magnetic field leading to a formation of a pencil beam due to aberration. A simplified model of formation of such a beam was considered in Ref.[7]

For an observed beam angle equal 30° we may get from Eq. (8.134) restriction on the value of magnetic field

$$B_{10} \geq 0.1\sqrt{L_{37}} = B_{1*}. \qquad (8.140)$$

The value of B_{1*} is about 10 times larger then B_* from Eq. (8.132) at which Alfven radius reaches a surface of a neutron star. We consider a model, when accretion flow, canalized to magnetic poles is stopped at the surface of the neutron star and a kinetic energy of a motion is transformed into heat in a noncollisional shock wave.[16] When electrons and protons, moving with the same velocity, are stopped by the shock, protons are heated to much higher energies, and as a result of relaxation protons transfer their energy to electrons. We suggest, that plasma is turbulent behind the shock, so energy exchange goes faster, then in binary coulomb collisions.

Let λ is a characteristic length at which protons transfer all their energy to electrons due to development of a beam instability. In general a heating of electrons in strong magnetic field may occur anisotropically, so we distinguish between $\lambda_{\|}$ and $\lambda_{\perp}$. A part of energy $\lambda/2\lambda_{\|}$ goes to increase an energy $E_{\|}$, parallel to magnetic field, and part of energy $\lambda/2\lambda_{\perp}$ goes to increase an energy $E_{\perp}$ in the plane perpendicular to magnetic field, so that $\lambda = 2\lambda_{\|}\lambda_{\perp}/(\lambda_{\|} + \lambda_{\perp})$. Simultaneously $E_{\perp}$ is lost due to SR, and anisotropy in the electron distribution implies a development of anisotropic instability leading to relaxation between parallel and perpendicular energy, so that a ratio between the increments of beam and anisotropic instability $\lambda_3/\lambda \sim 1$.

Define values $v_{\perp}$ and $E_{\perp} = m_e c^2/\sqrt{1 - \frac{v_{\perp}^2}{c^2}}$ in a coordinate system, where electron is rotating in the magnetic field without a motion along it, and values $v_{\|}$ and $E_{\|} = m_e c^2/\sqrt{1 - \frac{v_{\|}^2}{c^2}}$ in a coordinate system of an observer, where a centre of a Larmor circle is moving with a velocity $v_{\|}$. The choice of $E_{\|}$ and $E_{\perp}$ is done in such a way that magnetobremstrahlung is decreasing only $E_{\perp}$ and does not influence $E_{\|}$. In the system of the observer the total electron energy is written as $E = E_{\|}E_{\perp}/m_e c^2$. A simplified system of equations, describing a change of $E_{\|}$ and $E_{\perp}$ along the radius of the neutron star, is written as[7]

$$\frac{dE_{\parallel}}{dx} = E_0 \frac{\lambda_{\perp}}{\lambda} \frac{m_e c^2}{\lambda_{\perp} E_{\perp} + 2\lambda_{\parallel} E_{\parallel}} - \frac{2E_{\parallel} - E_{\perp}}{\lambda_3} \frac{m_e c^2}{E_{\perp}}, \tag{8.141}$$

$$\frac{dE_{\perp}}{dx} = 2E_0 \frac{\lambda_{\parallel}}{\lambda} \frac{m_e c^2}{\lambda_{\perp} E_{\perp} + 2\lambda_{\parallel} E_{\parallel}} + \frac{2E_{\parallel} - E_{\perp}}{\lambda_3} \frac{m_e c^2}{E_{\parallel}} \tag{8.142}$$

$$- \frac{m_e c^2}{E_{\parallel}} \frac{2e^2 \omega_0^2}{3cw_{\parallel}} \left[\left(\frac{E_{\perp}}{m_e c^2} \right)^2 - 1 \right].$$

This system is valid at $x < \lambda$, where protons transfer energy to electrons. At $x > \lambda$, where only magnetobremstrahlung and anisotropic relaxation occur, the system is written as

$$\frac{dE_{\parallel}}{dx} = - \frac{2E_{\parallel} - E_{\perp}}{\lambda_3} \frac{m_e c^2}{E_{\perp}}, \tag{8.143}$$

$$\frac{dE_{\perp}}{dx} = \frac{2E_{\parallel} - E_{\perp}}{\lambda_3} \frac{m_e c^2}{E_{\parallel}} - \frac{m_e c^2}{E_{\parallel}} \frac{2e^2 \omega_0^2}{3cw_{\parallel}} \left[\left(\frac{E_{\perp}}{m_e c^2} \right)^2 - 1 \right].$$

Here $w_{\parallel} \simeq 0.5c$ is an accretion speed at the surface of the neutron star; $E_0 = 0.2 m_p c^2$ is a proton energy near the surface of the star. Coefficients at $E_{\parallel}$ and $E_{\perp}$ in Eqs (8.141)—(8.143) are connected with number of degrees of freedom $k = 1$ in parallel and $k = 2$ in perpendicular directions. The general expression for magnetobremstrahlung cooling from Eq. (8.5) is used here.

Equation for change of total energy E of electrons in the observer's system may be obtained from Eqs (8.141)—(8.143) as

$$\frac{dE}{dx} = \frac{E_0}{\lambda} - \frac{2e^2 \omega_0^2}{3cw_{\parallel}} \left[\left(\frac{E_{\perp}}{m_e c^2} \right)^2 - 1 \right] \quad \text{at} \quad x < \lambda, \tag{8.144}$$

$$\frac{dE}{dx} = - \frac{2e^2 \omega_0^2}{3cw_{\parallel}} \left[\left(\frac{E_{\perp}}{m_e c^2} \right)^2 - 1 \right] \quad \text{at} \quad x > \lambda.$$

The solution of the system of Eqs (8.141)—(8.143) may be easily obtained only in some limiting cases. Consider estimations, following from Eqs (8.141)—(8.143) for a fully transparent case with additional account of the energy of protons.

Take in estimations $\lambda_1 \simeq \lambda_2 \simeq \lambda_3 \simeq \lambda$. During excitation of plasma oscillations protons loose their parallel energy, increasing a parallel energy of electrons. So we may expect $\lambda_1 < \lambda_2$, what increases a level of anisotropy, obtained below. Following cases may be distinguished depending on relation between heating by plasma oscillations and cooling by magnetobremstrahlung.

1. Strong magnetic field, corresponding to inequality

$$\omega_0^2 > \frac{3E_0 c w_{\parallel}}{8\lambda e^2}, \quad \text{with} \quad B_{10} > \sqrt{\frac{6 \text{ cm}}{\lambda}}. \tag{8.145}$$

This case corresponds to a very strong anisotropy, where $E_{\parallel} \gg m_e c^2$, and $E_{\perp} - m_e c^2 \ll m_e c^2$. The estimations with account of the proton energy give

$$E_{\parallel} \simeq \sqrt{0.5 m_e c^2 E_0} \simeq 14 m_e c^2, \tag{8.146}$$

$$\frac{E_{\perp}}{m_e c^2} \simeq 1 + \frac{3}{8}\frac{E_0}{\lambda}\frac{c w_{\parallel}}{e^2}\frac{1}{\omega_0^2} \simeq 1 + \frac{6}{\lambda\ \mathrm{cm}}\frac{1}{B_{10}^2}.$$

Electrons with $E_{\perp} \ll 2 m_e c^2$ radiate in the system of a standing Larmor circle mainly on frequency $\omega_0 \simeq 2 \times 10^{17} B_{10}$, see Eq. (8.2). Due to relativistic elongated motion the observer accept a strongly Doppler shifted frequency, which is $\frac{2E_{\parallel}}{m_e c^2}$ larger, so the frequency of maximum radiation corresponds to

$$\omega_{\max} = \omega_1 \simeq 5 \times 10^{18} B_{10}, \quad E_{\omega,\max} \simeq (3B_{10})\ \mathrm{keV}. \tag{8.147}$$

When a maximum of nonthermal component of the X-ray source, identified with $E_{\omega,\max}$, lay at 3 or 10 keV we get an estimation of magnetic field $B_{10} \simeq 1$ or 3. With account of Eq. (8.145) we get that this case may be realized at $\lambda > 6$ or 2/3 cm.

2. Middle magnetic field corresponds to inequalities

$$\frac{3}{4\lambda}\frac{c w_{\parallel}}{e^2} m_e c^2 < \omega_0^2 < \frac{3 E_0 c w_{\parallel}}{8 \lambda e^2}, \quad \text{or} \quad \sqrt{\frac{0.04\ \mathrm{cm}}{\lambda}} < B_{10} < \sqrt{\frac{6\ \mathrm{cm}}{\lambda}}. \tag{8.148}$$

This case corresponds to a mild anisotropy, where both $E_{\parallel}$ and $E_{\perp}$ are larger then $m_e c^2$. We have

$$E_{\parallel} \simeq \sqrt{0.5 m_e c^2 E_0} \simeq 14 m_e c^2, \tag{8.149}$$

$$\frac{E_{\perp}}{m_e c^2} \simeq 1 + \sqrt{\frac{6}{\lambda\ \mathrm{cm}}\frac{1}{B_{10}}}.$$

Here the frequency of maximum radiation for the observer corresponds to

$$\omega_{\max} = \omega_2 \simeq \frac{5 \times 10^{18}}{B_{10}(\lambda/6)}, \quad E_{\omega,\max} \simeq \frac{3}{B_{10}(\lambda/6)}\ \mathrm{keV}. \tag{8.150}$$

Here for $E_{\omega,\max} = 3$ or 10 keV we get the values of $B_{10}(\lambda/6) = 1$ or $1/3$. It follows from Eqs (8.147) and (8.150) that the boundary between two cases, corresponding to $B_{10} = \sqrt{\frac{6\ \mathrm{cm}}{\lambda}}$, has the following extremal properties. For a given λ and changing B_{10} the value of $\omega_{\max} = \omega_1$ increases with increasing B_{10} due to increase of a cyclotron frequency ω_0, and with decreasing B_{10} the value of $\omega_{\max} = \omega_2$ increases due to increasing of a perpendicular energy $E_{\perp}$. So, the boundary value of B_{10} corresponds to the minimal value of $\omega_{\max}$ at the given λ.

From the other side at a fixed B_{10} increasing of λ does not change $\omega_{\max} = \omega_1$, which is of the order of the cyclotron frequency, and decreasing of λ leads to increasing of $\omega_{\max} = \omega_2$ also because of increasing of $E_\perp$. So, for a given B_{10} the boundary value of $\lambda = \frac{6\text{ cm}}{B_{10}^2}$ also corresponds to the minimal value of $\omega_{\max}$.

We see then, that the same value of $\omega_{\max}$ in the nonthermal part of a spectrum of the X-ray source may correspond to two different combinations of parameters. For example, for $\omega_{\max} = 10^{19}$ we may expect $B_{10} = 2$ and $\lambda > 3/2$ cm from Eqs (8.145), (8.147), or another combination at $B_{10}\lambda = 3$ and $\lambda < 3/2$ cm from Eqs (8.148), (8.150). In both cases we expect a pencil-like beam with an angle $\theta = 2\theta_s + (m_e c^2/E) \simeq 2\theta + 0.07$ radian.

3. The third case correspond to the limit of "weak" field, when

$$\omega_0^2 < \frac{3}{4\lambda}\frac{cw_\parallel}{e^2} m_e c^2, \quad \text{or} \quad B_{10} < \sqrt{\frac{0.04\text{ cm}}{\lambda}}. \tag{8.151}$$

In this case the energy exchange between degrees of freedom is so rapid that almost isotropic distribution of electrons with $2\overline{E_\parallel} \simeq \overline{E_\perp}$ is reached and radiation ceased to be beamed.

It is rather difficult to estimate the lengths of energy relaxation between protons and electrons λ, λ_1, λ_2,, as well as the length λ_3, at which isotropisation of electrons happens. There are several characteristic lengths in the turbulent magnetized plasma, connected with plasma and Larmor frequencies given by Eq. (8.94)

$$\frac{c}{\omega_{pe}}, \quad \frac{c}{\omega_{pp}}, \quad \frac{c}{\omega_{Be}}, \quad \frac{c}{\omega_{Bp}}, \quad \frac{m_p}{m_e}\frac{c}{\omega_{pp}}. \tag{8.152}$$

Taking the largest length, the last in Eqs (8.152), we get an estimation

$$\lambda \simeq \frac{m_p}{m_e}\frac{c}{\omega_{pp}} \simeq \frac{4 \times 10^{10}}{\sqrt{N_e}}. \tag{8.153}$$

Plasma parameters near the surface of the poles of the neutron star were estimated in Ref.[62] The electron density N_{ep} and the optical depth to Thomson scattering τ_p are estimated as

$$N_{ep} \simeq 10^{19} L_{37}^{5/7} B_{10}^{4/7}, \quad \tau_p \simeq 4 L_{37}^{5/7} B_{10}^{4/7}. \tag{8.154}$$

The beaming of X-ray pulsar is possible only at $\tau_p < 1$, what leads to restrictions

$$B_{10} < 0.1 L_{37}^{-5/4}, \quad N_{ep} < 2.5 \times 10^{18}. \tag{8.155}$$

It follows from Eq. (8.153) the inequality $\lambda > 25$ cm. At $L_{37} = 0.3$ we get a restriction $B_{10} < 0.4$. Comparing with Eqs (8.145) and (8.148) we find that this combination of B_{10} and λ corresponds to the case 2, for which $0.04 < B_{10} < 0.4$. On the boundary $B_{10} = 0.4$ we get $\omega_{\max} \simeq 3 \times 10^{18}$, $E_{\max} \simeq 2$ keV. If the maximum is observed at higher energy it could correspond to smaller magnetic field.

anisotropic electrons radiate along a magnetic field in both directions, so half of the beam radiation illuminate the surface of the poles of the neutron star, creating hot spots on the poles giving a black body radiation at softer energy region. So the form of the beam in this model is expected to be dependent on the energy band, and the width of the beam will be narrower in hard region where the input of the anisotropic magnetobremstrahlung prevails. Such kind of relations between the beams in hard (1-30) keV, and in soft (0.02-2) keV regions were observed in radiation of the X-ray pulsar Her X-1.[100]

The model of magnetobremstrahlung of ultrarelativistic electrons with strongly anisotropic distribution may explain also a feature in the hard radiation of Her X-1.[99,103,75] This feature is usually interpreted as a cyclotron line in emission with $E_\gamma \simeq$ 50 keV, or in absorption with $E_\gamma \simeq 34$ keV. It is related to $B \simeq 5\times 10^{12}$ Gs in the case of absorption, or $B \simeq 3\times 10^{12}$ Gs in the case of emission. Both values are rather high for an old neutron star with an age $\sim 10^8$ years. It is more important that such a high magnetic field is strongly contradicting to selfconsistent model of this object, which explains satisfactory all complex of its properties, including the form of the beam at different phases of the orbital and 35-day cycle, spectral changes, change of the orbital light curve during 35 day cycle etc.[86] The value of the magnetic field following from this model is restricted by 10^{11} Gs, because the radius of the Alfven surface given by Eq. (8.131) cannot be situated too far from the star. If to interpret the observed feature as magnetobremstrahlung emission in the model, developed above, then the estimation of the magnetic field decreases $2E_{||}/m_e c^2 \simeq 30$ times, making the interpretation of this feature in consistency with the model from Ref.[86]

8.6. SR from Supernovae and Supernova Remnants

Supernova explosions are the most spectacular events in the universe of stars. The major part of energy releases in a time interval (< 1 s), small relative to not only stellar but also human lifetime, whereas its quantity exceeds by an order of magnitude and more the energy emitted by the star over its total lifetime reaching $\sim 10^{10}$ years for Sun-like stars. During the neutron star formation most of the energy releases in the form of scarcely observable neutrinos.

A supernova explosion is the end of the life of most massive stars with $M > 8M_\odot$. The flash itself results either from the thermal instability development in a degenerate core, or from gravitational and partly nuclear energy release during collapse which leads to the neutron star formation. The rotation and magnetic field may play an important role in conversion of gravitational energy into energy of observable flash. A small part of stars (the most massive ones) seem to end their life with collapse and black hole formation. The collapse in this case may be "silent" and not lead to supernova explosion.

From observational point of view, supernovae are divided into Type I supernovae (SN I) with no hydrogen lines in their emission spectrum and numerous absorption lines of various heavy elements instead, and Type II supernovae (SN II) with much hydrogen and nearly normal (solar) chemical composition. SN II have been established to be a result of the evolution of massive stars, while SN I have less massive stars for progenitors; many SN I are likely to occur in binaries.

The pulsars are thought to be produced by SN II explosions. The common feature between SN II and some SN I of subclasses Ib and Ic is a nonthermal radioemission visible during several years after explosion. This is a real indication, that SN Ib,c also give birth to pulsars. Contrary, SN of subclass Ia never show such kind of radioemission, so it is suggested that they are produced by nuclear explosions, leading to total disruption of the star, while other types of SN result from hydrodynamical collapse, leading to formation of a neutron star. In SN II the collapsing core is surrounded by a large hydrogen envelope, while in SN Ib,c the collapsing core is almost naked, presumably originated from less massive stars, which have lost its hydrogen-rich envelope in preceding evolution, may be in a binary. There are some spectral differences between 3 types of SN I, while they all do not show hydrogen lines.

The presence of hydrogen in the SN II spectra gives evidence that the explosion occurred before the star has lost its hydrogen envelope. The shape of the light curve shows that prior to the explosion this envelope could be extremely extended, $(10^3 \div 10^4)R_\odot$.[53] Observation in the Large Magellanic Cloud of SN1987A, the first visible by the naked eye for more than 300 years supernova, provide evidence for a SN II explosion in a fairly compact blue supergiant. The SN1987A light curve may be reproduced only in the case of a prolonged energy release after the main explosion.

The absence of hydrogen in the SN I spectra leads to the conclusion that the star loses its hydrogen envelope during its pre-explosion evolution and that the presupernova is compact. The shape of SN I light curve gets an explanation if the emitted energy is accounted for by the radioactive decay ${}^{56}_{28}\mathrm{Ni} \rightarrow {}^{56}_{27}\mathrm{Co} \rightarrow {}^{56}_{26}\mathrm{Fe}$; to maintain the luminescence energetically, it must be in this case $M_{^{56}\mathrm{Ni}} = 0.3 \div 1.1\, M_\odot$. The slow energy release during two—twenty days makes the maximum light phase prolonged (10–20 days) for a compact stellar model.[53] The same radioactive decay accounts for the character of the SN1987A light curve at $t \geq 120$ days after the explosion; here it is necessary to have initially $0.078 M_\odot$ of ^{56}Ni.[54]

At the endpoint of the evolution of massive stars iron cores with mass $M_{\mathrm{Fe}} \geq 1.4\, M_\odot$ are formed, which lose their stability through the iron dissociation directly leading to a rapid collapse. SN II are supposed to appear as a result of a reversal of this collapse and explosion after formation of a stable neutron star. The iron core forms in stars with initial mass $M_i > 10 M_\odot$, while for $M > 13 M_\odot$ all stages of nuclear burning proceed smoothly. As all evolutionary calculations yield significant uncertainty in the relationship $M_{\mathrm{Fe}}(M_i)$, the stability loss due to the iron dissociation is certain for single stars with $M_i > 13 \pm 3 M_\odot$.

Hydrodynamical calculations of iron core collapse have been performed with account of the effects of electron and muon neutrinos, the role of neutrino deposition leading to the envelope heating and probable ejection, and the effect of burning of thermonuclear fuel ^{12}C, ^{16}O remaining around the iron core. Reflection of the infalling matter from the surface of stable neutron star and formation of a shock wave (bounce) may also be important for producing supernova explosion.

Numerous calculations to date (see reviews in Refs[53,11]) have revealed sensitivity of the results to the equation of state of nuclear matter, quantity of remaining thermonuclear fuel, probable convection. The results are strongly influenced by adopted methods for including neutrinos at transparent and opaque stages. Self-consistent calculations in spherically symmetric approximation with most accurate account of neutrino processes and neutrino transport had shown that collapse is reversed very little, so that explosion does not happen, or its energy is much smaller then it is necessary for explanation of a SN explosion. This results from the structure of the new born hot neutron star which is opaque for neutrino. At the level of neutrinosphere, from which neutrino escape free, the temperature and neutrino energy are rather small. While the cross-section of neutrino interaction is proportional to square of its energy, heating of matter due to neutrino deposition is not enough for giving hydrodynamical explosion.

The only possibility to get an explosion by neutrino deposition is to increase the energy of outgoing neutrino which may heat a matter with higher efficiency. This may be done if the matter of a hot neutron star is convectively unstable. Then convective overturns bring a hot matter to the neutrinosphere and energetic neutrino have a chance to make a supernova by deposition. Calculations have shown, that convective instability is really present, but the final result depends on its treatment. Numerical calculations, especially in 3-dimension necessary for neutrino convection, cannot be done presently with enough precision to get a definite answer. Physically it is clear, that the critical parameter is the scale of the neutrino convection: the larger is the scale, the larger is the energy of neutrino on the neutrinosphere brought from deeper layers, the easier is to obtain a strong explosion. It is impossible to determine this parameter otherwise but in numerical simulations which are not enough reliable. So even in presence of some calculations resulting in strong enough explosion, the problem of "neutrino supernova" still does not have a definite answer.

Discovery of radio pulsars had shown that new born neutron star are rapidly rotating and have a huge magnetic field. This led to the idea of a **magnetorotational supernova.**[6]

When other mechanisms of explosion prove to be inefficient, the magnetic field may convert the rotational energy of a neutron star resulting from collapse into kinetic energy of the envelope and thus ensure a supernova explosion. Observations of radio emission at a relatively early period of the flash, roughly one year after the light peak, provide an indirect evidence for an essential role played by the magnetic field in supernova explosion. A radio emission has been detected in SN II, where a pulsar birth is assumed, and in those types of SN I (SN Ib,c), where a neutron star birth is suspected as well.

Numerical calculations for this model have been made in cylindrical approximation, spherically symmetric approximation and in a simplified two-dimensional formulation. The results of these calculations are in qualitative agreement and give a conversion of $\sim 3\%$ of the rotational energy into kinetic energy of the outburst. For $E_{rot} = 10^{53}$ erg we have the energy $E_{kin} = 3 \times 10^{51}$ erg sufficient to account for a supernova explosion.

The mechanism of magnetorotational supernova explosion may be formulated as follows. When a rapidly rotating presupernova collapses, it leads to formation of a rapidly rotating neutron star surrounded by a deferentially rotating envelope in which the centrifugal forces are comparable with gravitational ones. The differential rotation twists the lines of the magnetic field, thereby causing the magnetic field with initial energy $\epsilon_M \ll \epsilon_G$ to increase linearly in time. When the energy of the field in the envelope approaches $\epsilon_M \sim \epsilon_G$, the magnetic pressure pushes the material outward. The arising wave of compression propagates over medium with falling density, gets enhanced and transforms into shock to result in a powerful explosion. As the compression wave and subsequent shock move outward, their energy keeps increasing, maintained by rotational energy supplied by magnetic field. The magnetic field serves also to transfer to outer layers an essential part of the total angular momentum. Twisting of magnetic field lines has been used earlier[57] to account for the energy transfer from the neutron star to the Crab nebula and maintaining its luminescence.

The envelope ejection leaves behind a young pulsar that keeps supporting the envelope acceleration by pressure of its radiation in the form of high-energy particles and dipole electromagnetic radiation. The action of the magnetohydrodynamical rotational mechanism of envelope ejection and subsequent activity of the young pulsar could affect the explosion and light curve formation of the supernova SN1987A.

8.6.1. Radio Emission From Supernovae

The radio emission detected in young SN was a distinct evidence for importance of magnetic field and SR in these giant events. Discovery of radio emission from SN in distant galaxies has became possible after the completion of the Very Large Array (VLA), which gave enough sensitivity on the level of submillijansky flux density levels with only a few minutes time of integration.[104] The unit of radio flux power 1 Jansky $\equiv$ 1 Jy= 10^{-26} Watt/m^2/Hz= 10^{-23} ergs/cm^2/Hz. Besides, VLA gave an angular resolution on arcsecond level, allowing radio-optical positional agreement to be established even in presence of the effects of confusing sources and large scale structure of the host galaxy. Radio emission from SN was first discovered in April 1980 by observations of the optical position of the SN1979C in the galaxy NGC 4321 (M100). This SN was observed in average once per month from 1980 to 1983, and measured $\sim$ 4 times per year since 1984. The radio light curves and spectral index evolution are shown in Fig. 8.4. SN1979C belongs to SN II type, and these radio light curves are thought to be typical of that class, when radio emission is present.

The emission brightens first at high frequencies and later at progressively lower frequencies. Following the initial sharp rise, the flux density decays slowly over a period of months or years, longer then the optical light curve with decay time not exceeding 1-2 months. After the emitting region becomes optically thin, the radio spectral index is constant and reasonably typical for a nonthermal synchrotron source ($S \sim \omega^{-0.7}$). The radio flux density of SN1979C has remained sufficiently high that

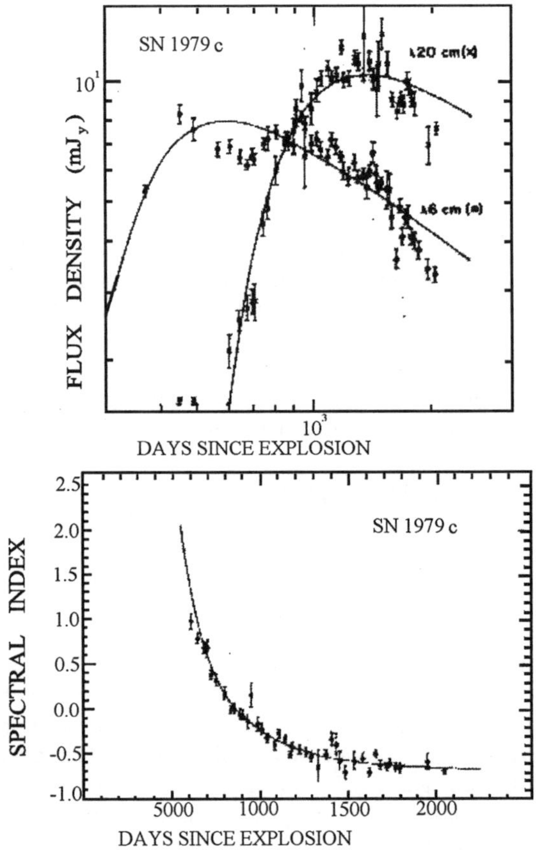

Fig. 8.4. (*top*) Radio light curves at 6- and 20- cm wavelength for Type II supernova SN 1979C in NGC 4321 (M100) and (*bottom*) spectral index a ($S \sim \omega^{+a}$) evolution between 20 and 6 cm (from Ref.[104]).

Very Long Base Interferometry (VLBI) measurements have been carried out to establish its structure on milliarcsecond scale. The diameter of the radio source was measured and its expansion was established. At 6-cm wavelength the diameter increased from 1.05 milliarcsec in December 1982 to 1.43 milliarcsec in December 1983.

After successes with SN1979C and later SN1980K, regular observations were begun with the VLA of all newly discovered, bright SN. First SN Type I discovered in radio was SN1983N in NGC 5236(M83). The radio radiation from this supernova was detected at 6 cm 11 days before optical maximum. The radio emission evolved extremely rapidly, and while SN Type II have remained detectable in radio for many years, in less then one year SN1983N could no longer be detected to the sensitivity limit of the VLA. Optically thin spectral index here was significantly steeper $\sim \omega^{-1}$ than that found for SN Type II. SN Type I with radioemission belong to special class Type Ib, where neutron star is probably formed in the remnant.

Traditionally, all SN have first been found by their optical emission, but few SN have been discovered by their radio emission, including SN1981K in NGC 4258 and SN1986J in NGC 891.[104] Both belong to Type II, the explosion itself were missed in optics, but they were identified on optics later by a search of optical photographic plates. The object received the designation SN1986J probably exploded in 1982 or 1983 and remained in 1988 the most luminous of the radio SN. Its absolute radio luminosity at 6 cm was about 1000 times larger then that of Cas A, the brightest of radio SN remnants (SNR) in the Galaxy, and over four times the peak radio luminosity of SN1979C, the next most luminous radio SN. The properties of SN1986J in different bands suggest that a pulsar may be an important source of energy in this SN at an age 4 years.

Radio emission from SN1987A was detected with the first radio observations made at 0.843 and 1.4 GHz only two days after discovery. Observations one day later at 2.3 GHz also detected the supernova but with a flux density that was already declining rapidly. The peak radio flux from SN1987A was > 100 mJy, but Magellanic Clouds are so close, that its absolute radio luminosity was more then 3 orders of magnitude less then that of SN1979C. Also, the radio emission from SN1987A declined extraordinary rapidly and was undetectable in few days, except at 0.843 GHz. If this SN would be in at a normal distance for extragalactic SN, such as Virgo Cluster or even the Local Group it certainly would not have been detected.

Following an initial radio outburst, SN1987A remained quiescent at radio frequencies for about two years. In July 1990 the source again became a radio emitter. Since that time it shows steady increase of flux density reaching a level of 5 mJy at 8.6 GHz in April 1993.[94] This behavior is different from other radio supernovae, where after 10 - 1000 days the overall flux density suffers a power-law decline. Moreover, this SN is much less luminous in radio then usual, about 10^6 times then SN1986J. The reason for this difference appears to be that radio supernovae have red supergiant progenitors which have massive outflows, or winds. The interaction of the shock front with the wind produces copious radio emission. On the other hand, the progenitor of SN1987A was blue supergiant with a fast tenuous wind which produces little radio emission when hit by a shock. This wind appears to be responsible for sweeping away the former red supergiant wind into an equatorial ring of radius $\sim 6 \times 10^{17}$ cm. Therefore, the radio emission which is seen now may well be the initial development of a classical SNR, similar to Cassiopea A.

Radio SN show steep radio spectra and their angular sizes imply extremely high brightness temperatures, what indicate to a nonthermal origin of their radio emission. Usually, SR is considered, so models of RSN have to describe the origin and evolution of the relativistic particles and magnetic fields. Two classes of models have been developed, which may be designated[104] as "minishells" and "miniplerions". The last ones utilize the rotational energy losses by a rapidly spinning, rapidly slowing pulsar left over by the SN explosion. This model has serious difficulties. Since the source of the energy is a pulsar buried deeply within the large amount of mass ejected in a supernova, the radiation generated by a pulsar and its surroundings needs to escape to a radius where the density of ionized matter is low enough to be optically thin in the radio emission. This problem could be solved in the model, where mass ejected by SN clumps very early into dense filaments between which the miniplerion can be seen and the radio emission can escape. Another difficulty is the large size of RSN measured by VLBI, which is difficult to explain. For these reasons miniplerion model is not considered usually as an appropriate model for RSN, et least for the very young SNe.

The minishell models utilize the Reyleigh-Taylor unstable region between the supernova shock, which propagates into the preexisting circumstellar shell of dense material that originated by a stellar wind from the red supergiant presupernova star, and reverse shock, which propagates back into the supernova envelope to excite turbulence. This gives the magnetic field enhancement and particle acceleration for radio synchrotron radiation. Since the circumstellar shell is ionized and generally optically thick at radio frequencies, emission becomes observable only after the shock wave has passed far enough through the shell that the optical depth for radio emission approaches unity. The change in the remaining optical depth produces the observed rapid turn-on at progressively lower frequencies, and the density gradient of the SN envelope provides a declining radio light curve of the form $S \sim \omega^a t^b$ similar to that observed. Both intensity and time scale of radio emission are dependent on the characteristics of the circumstellar material and are thus indicate to the mass-loss rate of the supernova progenitor.

The observed flux results from a convolution of the absorption process with the process responsible for the intrinsic emission, which itself depends on the way in which electrons are accelerated and on the evolution of the magnetic field within the source. It was suggested in Ref.[28] that the intrinsic radio emission from RSN could be understood by assuming that a fixed fraction of about 2% of the flux energy through the shock front is divided equally between relativistic electrons and the magnetic field. Approximating the emission region as a thin shell behind the shock, and including the effect of free-free absorption in a stellar wind with density falling off with the square of the distance from the progenitor star, it was predicted[28] that the radio flux density

$S(\omega, t)$ should be described by five parametric model. This model occurs to be good enough for several RSN, however it failed to fit one of the brightest and best-observed RSN: SN1986J.

This SN, first discovered as a radio source, belongs to a "peculiar" Type II SN with an extremely massive $M \sim 20 - 60\ M_{\odot}$ progenitor. The radio emission rose much more slowly that is compatible with the rather rapid drop in absorption implied in the standard model. This led to the suggestion that there might be an additional contribution to the free-free absorption due to the thermal material which is mixed with the nonthermal synchrotron emitting material, giving a significantly better fit to the data from early epochs. The model based on the diffusive acceleration of the electrons at the expanding supernova shock is considered in Ref.[2] This model was applied earlier for the unique second phase of radio emission from SN1987A, but requires for its understanding a detailed treatment of the environment of the progenitor and the reaction of accelerated particles on the hydrodynamics of the explosion. For the case of SN1986J such complication can be neglected, and it is possible to fit a simple form of the model, including a combination of internal and external free-free absorption, to the radio data. The free parameters are simply related to physical properties of the explosion, enabling estimation of quantities such as the compression ratio of the shock front, the magnetic field and the effective diffusion coefficient of the electrons.

The model assumes diffuse particle acceleration at a shock expanding at constant speed with a shock radius $r_s \sim (t - t_e)$ and magnetic field dependence. For a predominantly azimuthal magnetic field in the wind of the progenitor a time dependence of the magnetic field at the shock front is determined as $B(r_s) \sim (t - t_e)^{-1}$, where t_e is the time of the explosion. If the magnetic field encountered by shock is predominantly radial, then $B(r_s) \sim (t - t_e)^{-2}$. The general situation was considered in Ref.[2] by setting $B(r_s) \sim (t - t_e)^{-n}$, where n is a free parameter. The model also assumes that the rate of injection of electrons into the acceleration mechanism is a constant. In these conditions a solution of equation, describing a particle acceleration gives a solution for intrinsic (unabsorbed) flux density as a function of time and frequency

$$F(\omega, t) \sim \omega^{-\alpha}(t - t_e)^{1-n(1+\alpha)}. \tag{8.156}$$

Here the parameter α is connected with the distribution of the accelerated particles $N(p)$ over their momentums p, so that $N(p) \sim p^{-2\alpha-1}$. Account for an intrinsic absorption leads to more rapid decrease of the radio luminosity with time, fitting observations, but at large times and shock radiuses when absorption is negligible, dependence given by Eq. (8.156) must take place.

The points of the best-fit diffusive model to the radioluminous SN1986J and SN1988Z are (i) a magnetic field dependence $B \sim t^{-2}$ and (ii) a compression ratio not too different from the value 4 expected at a shock front which is not modified by acceleration particle pressure. Such magnetic field dependence could arise if progenitor's were very slowly rotating and had the field close to radial, or such dependence was established by turbulent processes within the wind itself. If there is a substantial shock deceleration, then even dipole field with $B \sim r^{-3}$ could give a time dependence $B \sim t^{-2}$. The value of compression indicate to small input of accelerated particles to the pressure. It may be connected with sharp fall-of of the magnetic field. The pressure in relativistic particles is dominated by particles of TeV energy or higher, and a weak magnetic field will strongly increase the time needed to accelerate particles to such energies. The magnetic field at the shock position 1000 days after explosion is found to be about 3×10^{-5} Gs, significantly stronger then the ambient interstellar field. This is consistent with the picture in which the field is related to the progenitor. Extrapolating back to the stellar surface implies an enormously high surface field, but the model does not require the t^{-2} behavior to apply earlier then about 500 days after explosion. High stellar magnetic field resulting from this model may indicate also to the importance of magnetic forces in the explosion itself and to validity of the magnetorotational mechanism. The effective diffusion coefficient derived in Ref.[2] is equivalent to a mean free path of an electron of 10^{14} cm, some four orders of magnitude greater then gyroradius of a radio emitting electrons. This could be possible if diffusion is mainly directed along the magnetic field. The value of mass-loss rate of the progenitor is determined in this model with large errors.

The diffusive acceleration model also provides adequate fits to the radio emission from other less luminous RSN, including Type II and Type Ib SN. In contrast to SN1986J and SN1988Z the fits to these other SN imply a magnetic field dependence $B \sim t^{-1}$, like in the modeling of the second phase of radio emission from SN1987A. The reason of this difference may be connected with a properties of a progenitor stars. The diffuse model has eight parameters for fitting observations, so its validity would be confirmed if the same set of parameters would give a best fit for several RSN.

Different modifications of a synchrotron model cannot be directly applied to the first phase of radioemission of SN1987A, registered very soon ($\sim$ 2 days) after the explosion. There is a very short time for acceleration of electrons needed in SR model. Alternative model developed in Ref.[17] avoid this problem and may be applied even at closer start of radio emission by this SN. The model is based on acceleration of positrons up to energy $\sim$ 15 MeV by a neutrino flux in the outer transparent layers in the reaction

$$\tilde{\nu}_e + p \rightarrow n + e^+. \tag{8.157}$$

Rapid positrons excite plasma oscillations at frequency given by Eq. (8.117), which conversion into radiation at the same and doubled frequency gives explanation for the radio emission. The small duration of the observed radio emission may be connected with annihilation of the positrons and damping of plasma oscillations if no other energy supply is present. By this mechanism radio emission could start in several seconds after explosion, and the drift of the maximum frequency radiation in time to smaller frequencies is connected with outflow of rapid positrons and excitation of plasma oscillations in regions with falling density

8.6.2. Radio Emission of Supernova Remnants

Extragalactic SNR are the only SNRs for which reasonably reliable distances, or at least relative distances, can be obtained, but they are generally small, poorly resolved, and faint, so much more information may be obtained from Galactic SNRs. About 150 SNRs are known in the Galaxy, but their list is incomplete, what is confirmed by regular discovery of new SNRs. This incompleteness is related mainly to young bright remnants and older remnants with diameter $\geq$ 15 pc. Galactic SNRs are identified as having a shell or partial shell form and nonthermal radio spectrum. On the basis of more detailed study four categories of SNRs have been suggested:[73] – Balmer-dominated, oxygen-rich, plerionic-composite, and evolved. In Ref.[104] a possible fifth class was added – centrally influenced SNR.

1. Balmer-dominated SNR are defined, as ones whose filamentary shells are strong in the Balmer lines of hydrogen but absent or very weak in oxygen [O III] and sulphur [S II]. These are shell- or partial shell-shaped radio remnants, in which high-velocity, nonradiative, collisionless shock encounters the partially neutral gas of the surrounding ISM. These SNRs are thought to be young SN Ia remnants, resulting from explosion and total disruption of the low-mass ($\sim 1.4\,M_\odot$) progenitor. Two most recent ralactic SNRs, SN1604 (Kepler's SN) and SN1572 (Tycho's SN) belong to this class.

2. Oxygen-rich SNR are generally located near H II regions and are thought to be the result of explosion of a massive star ($\leq 25\,M_\odot$), where the observed oxygen-rich eject a consist of processed material from deep within the stellar interior. These SNRs are powered by shock energy, but the shock is interacting more with the circumstellar material lost by the progenitor star in the last stages of evolution before explosion then with the mater swept from ISM. The youngest ($\sim$ 1670) and brightest radio source on the sky SNR Cassiopea A belongs to this class.

3.Plerionic-composite SNRs contain plerions, filled-centre remnants in combination with a surrounding shell. The plerions derive their energy from the rotational energy of the central neutron star – radiopulsar. The reason for existence of shells around plerions was connected with the properties of ISM, so that pure plerions are formed in the more tenuous ISM. Review of properties of plerionic SNRs is given in Ref.[84] The most prominent member of this class is the Crab nebula, remnant of SN1054, where the bright pulsar is observed. The action of the pulsar fills the plerion with relativistic particles, which emit SR in the broad band, from radio to X-rays.

The X-ray synchrotron nebula occupies a small volume in the centre of the nebula, close to the pulsar which appears as an unresolved source. The smaller size of the X-ray nebula indicates that the lifetime of the relativistic electrons producing X-rays is shorter then the time it takes these electrons to diffuse to the edge of the Crab. In the X-ray band the Crab is the brightest SNR in our galaxy, Ninety-five percent of the soft X-ray luminosity comes from the synchrotron nebula. The unique features of the Crab are the presence of a rapid, most luminous pulsar, contributing to the brightness of the nebula, and the low expansion velocity of the optical filaments, which confine the deposited energy to a relatively small volume with a consequent high energy density being a contributing factor to the unique brightness of the Crab. The optical and radio continuum are confined to a volume bounded by the filaments. Radio emission is brightest close to some of the brightest filaments, indicating a compression of the nebular magnetic field caused by the filaments blocking the flow of energy from the pulsar vicinity. Optical spectra indicate that the filaments are photoionized, so even the power in optical emission lines from the filaments originates in the pulsar.

The remnant SNR 0540-69.3 has been called the "twin" of the Crab nebula. It contains a 50-millisecond pulsar, visible in X-ray and optics and losing rotational energy at 1/3 rate of the Crab Pulsar. This remnant is situated in Large Magellanic Cloud (LMC) and shows synchrotron continuum from radio to X-rays similar to Crab. The difference from Crab is in the emission from the rings, which goes mainly in oxygen lines [O II] and [O III]. This may indicate that the progenitor star was more massive then that of Crab Nebula. Twenty-three percent of the observed X-rays from this remnant is pulsed, what is much higher then 4% observed from the Crab Nebula. This may be due to a more favorable orientation of the rotating neutron star.

Another famous representative of this class is the Vela SNR, containing a second young rapid pulsar in the Galaxy. The pulsar is clearly visible through radio, optical and hard ($>$ 30 MeV) gamma rays, where it is the brightest source in the sky. It is surrounded by a small X-ray emitting nebula at least $4'$ in diameter. The X-rays from the "point source" are not pulsed with an upper limit 1% in the 0.2-4 keV energy band.

There are 18 SNR which X-ray morphology indicate existence of neutron stars in their centres. Among them 6 SNR contain radio pulsars with the X-ray luminosity, produced by SR. Another SNR indicate an existence of central source of energy by their radio morphology. Thus, approximately 25 remnants have, or show evidence for, internal neutron stars.

4. Evolved SNR are characterized by a shell-shaped form with [S II] to Hα ratio greater then 0.7 and shock waves with modest velocities of 50-200 km s^{-1}. The emission arises in shocked cloudlets of ISM. It is likely that this category represents the final evolution of all of the three other types of SNRs. Balmer-dominated and oxygen-rich SNRs are already shell shaped even when young, and composite SNRs are expected to resemble pure shell-type SNRs when the internal plerion fades away. In all cases, energy has been put into ISM, and once the properties of the individual explosions become unimportant, all remnants are observable as faint shells generated by shock wave before final fading.

5.Centrally influenced SNR are unusual remnants having a general shell shape, which is significantly altered by the activity of central, compact, peculiar active object. The most striking example is W50 (G39.7-2.0), containing inside the object SS 433 showing strong sub-relativistic jets with strong influence on the remnant.

Detailed investigation of SNRs[60] had shown that the majority of them are "barrel-shaped". The explanation of this fact may be connected with a toroidal supernova outburst, expected in magnetorotational supernova, or with a density distribution around, connected with the anisotropic wind from the presupernova, or another previous history. To explain some particular SNR shapes theoretically by numerical modeling, both factors were taken into account.[22]

Analysis of sample of SNR in our and nearby galaxies gave following relations between their properties.

1.The radio surface brightness Σ_R and the X-ray surface brightness Σ_X are strongly correlated ($\Sigma_R \sim \Sigma_X^{0.69}$), indicating, that heating and farther acceleration of thermal electrons, giving X-rays, to relativistic energies (to produce radio synchrotron) takes place in common by the shock wave in SNR of different ages.

2. SNRs evolve adiabatically at nearly constant Σ_R, followed by a steep decrease near the maximum observable diameter when the remnant enters the radiative phase.

3. The observed variation in Σ_R is connected with variation in the explosion energy E_0 and ambient plasma density n_0, but not by are and evolutionary effects.

4. Σ_R and D_R (radio diameter) are related to the ambient density by $\Sigma_R \sim n_0^{1.37}$ and $D_R \sim n_0^{-0.39}$, what is valid for all shell-type SNR.

5. There is a weak correlation between increasing radio spectral index a ($S \sim \omega^{+a}$) with increasing D_R.

Applying the theory of strong point explosion to SN explosion in ISM gives[90] the time dependence of SNR diameter D on the adiabatic stage as

$$D = (2.02E_0/\rho_0)^{0.2} t_{(s)}^{0.4} \text{ cm}. \tag{8.158}$$

Synchrotron origin of SNR radioemission confirmed by their power-law radio spectra and strong polarization, permits to obtain for them some quantitative relations (see Ref.[70]). Energy losses by SR change the shape of distribution function of electrons. In presence of energy losses and input of electrons from outside $Q(E,t)$ we may write a kinetic equation describing the evolution of the electron distribution function $n(E,t)$ as[59,40]

$$\frac{\partial n(E,t)}{\partial t} + \frac{\partial}{\partial E}\left[\frac{dE}{dt} n(E,t)\right] = Q(E,t), \tag{8.159}$$

where the energy loss rate given by Eq. (8.5) for UR electrons may be written as

$$\frac{dE}{dt} = -gB_\perp^2 E^2, \quad g = \frac{2e^4}{3m_e^4 c^7}. \tag{8.160}$$

Eq. (8.160) is a characteristic equation for (8.159), its solution, starting from $E = E_0$ at $t = 0$ is

$$E = \frac{E_0}{1 + gB_\perp^2 t E_0}, \quad \text{or} \quad E_0 = \frac{E}{1 - gB_\perp^2 t E}. \tag{8.161}$$

It follows from Eq. (8.161) the characteristic time of the energy losses due to SR given by Eq. (8.87), the place of the break of the electron spectrum in absence of the outside supply E_b

$$E_b(t)\ \text{eV} \approx \frac{8.3\times 10^6}{B_\perp^2 t\ (\text{years})}, \tag{8.162}$$

and the place of the break in the radiation spectrum at ω_b

$$\omega_b(t)\ \text{sec}^{-1} \approx 0.29\omega_c(E_b) \approx \frac{2.1\times 10^9}{B_\perp^3 t^2\ (\text{years})}. \tag{8.163}$$

The time-dependent spectrum in presence of SR without the electron supply is obtained by substitution $E_0(E,t)$ from Eq. (8.161) into initial distribution of Eq. (8.12), taking into account $dE_0 = \frac{dE}{(1-gB_\perp^2 tE)^2}$

$$n(E,t) = A_e E^{-\gamma_e}(1-gB_\perp^2 tE)^{\gamma_e-2}. \tag{8.164}$$

The time-dependent spectral intensity of radiation is obtained after substitution Eq. (8.164) into Eq. (8.11) and integration. For low frequencies before the break $\omega \ll \omega_b$ the main input into the spectrum is given by electrons with $E \ll E_b$ and we get the spectrum of Eq. (8.13). At frequencies after the break the main input is given by electrons near the break with $E \simeq E_B$ and for $\omega \gg \omega_b$ the exponential in $Z = \omega t^2 B_\perp^3$ cutoff takes place. Another conclusion was obtained in Ref.[56], where power-law decrease in Z was claimed. In the stationary regime with continuous injection of the electrons with the distribution given by Eq. (8.12) we get from Eq. (8.159) at $\frac{\partial n}{\partial t} = 0$ the distribution

$$n(E) \sim E^{-\gamma_e - 1}, \tag{8.165}$$

what, similar to Eq. (8.13), gives the spectrum $J_\omega \sim \omega^{-\frac{\gamma_e-1}{2}}\omega^{-\frac{1}{2}}$. This distribution is established only at $\omega > \omega_b$, while at smaller frequencies the distribution given by Eq. (8.13) is valid. So, in vicinity of ω_b there is a break of the spectral index $a = -\frac{\gamma_e-1}{2}$, equal to $\Delta a = -0.5$, $J_\omega \sim \omega^a$.

Adiabatic expansion leads to decrease of the radiation flux without a change of the spectral slope. The energy of ultrarelativistic electrons and their distribution (see Eq. (8.12)) in the total volume with account of conservation of the adiabatic invariant are changing at expansion as

$$E = E_0\left(\frac{R_0}{R}\right), \quad n(E,R) = \left(\frac{R_0}{R}\right)^3 n_0(E,R)\frac{dE_0}{dE} = A_{e0}\left(\frac{R_0}{R}\right)^{\gamma_e+2} E^{-\gamma_e}. \tag{8.166}$$

At constant magnetic flux magnetic field is changing as $B = B_0\left(\frac{R_0}{R}\right)^2$ and total spectral flux from the remnant with power-like distributed electrons (see Eq. (8.13)) is

$$L_\omega \sim R^3 J_\omega \sim R^{-\gamma_e+1} B^{\frac{\gamma_e+1}{2}}\omega^{-\frac{\gamma_e-1}{2}} \sim R^{-2\gamma_e}\omega^{-\frac{\gamma_e-1}{2}}. \tag{8.167}$$

The conclusion that SR radiation from SNRs is decreasing was confirmed by observations.

Let us finally describe a relation between SN and SNR types. Different arguments lead to conclusion that plerionic and composite SNRs result from the explosion of a Type II SN, where presence or lack of a surrounding shell depend on the density of the local ISM.[104] It is generally accepted the connection between Type 1a SNe and Balmer-dominated, shell-type SNRs with no remaining stellar remnant. The most questionable is the origin of massive oxygen-rich SNRs. The class of SN Type Ib seems to be remaining for it, but there are more inclinations to consider SN Ib progenitors as not massive. If the massive star origin for Type Ib SNe is proven correct, this may be the solution to a number of long-standing problems.[104]

8.7. Cosmic Rays and SR

SR application to astrophysics was first done for explanation of the origin of the diffuse radio emission of the Galaxy.[1,58,35,36] The existence of cosmic rays was known to that time, so cosmic ray electrons were used for this explanation. Electrons in the cosmic ray flux consist only about 1% of all particles, but because of low rest mass only they give the recorded diffuse SR in a radio region. There are different points of view on the origin of the proton- nuclear component. There are evidences, that cosmic rays, except its most energetic tail with $E_p > 10^{17}$ eV, are originated in the Galaxy. It is definitely related to the electronic component, where particle energy does not exceed[40] $\sim 2 \times 10^{12}$ eV, $\gamma_e < 4 \times 10^6$. The characteristic length of SR energy losses by relativistic electron is about $\lambda_{sr} = c\tau_{sr}$, what gives, using Eq. (8.87),

$$\lambda_{sr}^{(e)} \approx \frac{10^{19}}{\gamma_e B^2} \text{ cm}. \tag{8.168}$$

To estimate a characteristic length for inverse Compton losses we need to substitute $B^2/8\pi$ in Eq. (8.168) by aT^4. The characteristic length for Compton losses for electrons due to interaction with a background 3 K radiation is $\lambda_{ic} \approx 10^{30}/\gamma_e$ cm, what for energetic tail of electrons with $\gamma_e \approx 4 \times 10^6$ gives $\lambda_{ic} \approx 2.5 \times 10^{23}$ cm ≈ 82 kpc. This distance is very small compared to cosmological distances and excludes the extragalactic origin. Another characteristic length for cosmic rays in the Galaxy is the Larmor radius of particles given by Eq. (8.152)

$$\lambda_{lar}^{(e)} \approx \frac{1.7 \times 10^3 \gamma_e}{B} \text{ cm}, \quad \lambda_{lar}^{(p)} \approx \frac{3.1 \times 10^6 \gamma_p}{B} \text{ cm}. \tag{8.169}$$

For $E_p > 10^{17}$ eV, $\gamma_p > 10^8$ the Larmor proton radius in the characteristic Galactic field $\sim 3 \times 10^{-6}$ Gs is about 10^{20} cm ≈ 30 pc. For retaining of cosmic rays in the galaxy the Larmor radius must not exceed the characteristic length of nonuniformity of the magnetic field, which is between 3 and 100 pc. So cosmic ray protons with $E > 10^{17}$ eV do not retain in the magnetic field of the Galactic disc.[40]

The electrons and positrons, which part in the cosmic ray flux near the Earth is about 10% of the electron flux, are originated in primary and secondary processes. To secondary processes are related the processes of interaction of the proton component of cosmic rays with a galactic gas, giving origin to $\pi^{\pm}$ pions, which produce electrons by decay in the channel $\pi^{\pm} \to \mu^{\pm} \to e^{\pm}$. About 10% of electrons and all positrons of the cosmic rays are probably born in the secondary processes, but some positrons may be originated by radiopulsars. Cosmic ray electrons of different energies are originated by objects in the Galaxy, mainly in the galactic disc, instead of the visible isotropy of their distribution. The isotropization is produced by chaotic magnetic field of the galactic disc, changing and mixing the trajectories of the electrons. Low energy cosmic ray electrons with $E < 1$ GeV are mainly produced by in solar flares and by activity of low-massive stars. The most active are red dwarf stars, called also "flaring stars" which masses are less then the solar mass.

SR from the galactic electrons may be observed from the Earth in the limited frequency interval from 10 MHz ($\lambda = 30$ m) to 10 GHz ($\lambda = 3$ cm). The lower boundary is connected with an absorption in the regions of ionized interstellar hydrogen, and the upper frequency limit is determined by the microwave background radiation prevailing at higher frequencies. For the galactic magnetic field $B_G \sim 3 \times 10^{-6}$ Gs we may estimate the energetic interval of cosmic ray electrons, available for detection by their radio radiation (see Eqs (8.2), (8.9), (8.88)).

$$\omega_0 \approx 1.76 \times 10^7 B, \;\; \omega_m \approx 7.7 \times 10^6 B \gamma_e^2, \tag{8.170}$$

$$\gamma_{emin} \approx 1.6 \times 10^3, \;\; E_{emin} \approx 0.8\,\mathrm{GeV}, \;\; \gamma_{emax} \approx 5.2 \times 10^4, \;\; E_{emax} \approx 30\,\mathrm{GeV}.$$

The radio background of the Galaxy is formed, in addition to SR radiation of the electrons, by the contaminated metagalactic radio sources. It is estimated now, that in the direction of minimum background the metagalactic input is between 25 and 50% of the observed flux. For analysis of distribution of relativistic electrons in the Galaxy the directions are used where there are no nonuniformities with enhanced radio emission, probably old remnants of the SN explosions. Such directions are the Galactic anticentre ($l = 180°$ - Galactic longitude, and $b = 3°$ - Galactic latitude); and halo-minimum direction ($l = 180°, \;\; b \approx 60°$). Spectral index of radio emission is $a = -0.5 \div -0.6$ at low frequencies less then 100 MHz. At larger frequencies its absolute value is gradually increasing up to $-0.8 \div -1.0$. Such a break in the spectral index is probably connected with the energy losses of electrons, see Eq. (8.165). Note that low-frequency part of the radio background is determined by electrons, which galactic spectrum cannot be measured near the Earth due to a modulation in the circumstellar space.

Spectral index a is changing with a direction. Its absolute value increases with increasing of the angle between the galactic plane and direction of observations (with increasing of b).The difference in spectral index a between high and low latitudes is about $0.1 \div 0.2$. This effect takes place for the low-frequency part of the spectrum ($10 \div 100$ MHz). At higher frequencies the opposite dependence of a is observed, its absolute value is decreasing with the latitude. Such anomalous behavior of a may be connected with the radio radiation of nearby SNR, having relatively more energetic electrons.

Our position in the Galaxy does not permit to make direct observations of the Galactic halo and to estimate its extend. Theoretically it is possible to calculate from the model, describing injection and propagation of the electron cosmic rays in the galaxy, but the model contains free parameters: spectral index of the energy distribution of the galactic electrons near the Earth γ_e and of injected electrons γ_{e0}, relative input of the metagalactic electrons into the background. For commonly accepted $\gamma_e = 2.7$ the properties of the radio halo in the model depend substantially on other input parameters, which are not firmly determined.

Indirect evidences concerning the Galactic radio halo are obtained from radio observations of nearby galaxies. The main problem with the nearest giant galaxy M 31 (Andromeda nebula) is its position at low galactic latitudes, where radio emission from the Galaxy is intensive. Increase of sensitivity of radiotelescopes permitted to investigate more distant galaxies, especially those, visible from the edge of the disc. Extended radio halo with a noticeable flux at distances of several kiloparsecs from the galactic plane were detected in several galaxies, properties of NGC 891 are listed in Ref.[40]

Besides solar-like and flaring stars and pulsars, the most important source of the cosmic ray electrons are SN and SNR. Powerful nonthermal radioemission of synchrotron origin indicates, that continuous electron acceleration takes place in SNR. Mechanisms of acceleration of cosmic rays are different is SNRs and pulsars. In the last acceleration takes place in the regular electrical field. In SNRs the acceleration happens in the region of the shock front: electrons are scattered on the plasma nonuniformities and many times cross the surface of the shock. Each crossing is accompanied by a gain of the energy. The effective masses of nonuniformities are very large, what corresponds to very high temperatures. In the process of scattering the electron temperature tends to the "temperature" of nonuniformities due to establishing of thermodynamic equilibrium. This mechanism of acceleration based on scattering of elementary particles on macroscopic bodies was proposed by E. Fermi, who considered a scattering of particles by interstellar magnetized clouds. Scattering of clouds may accelerate protons to much higher energies then electrons because of much lower SR and other energy losses during a time between scatterings. Different problems connected with cosmic rays are considered in Ref.[40]

Bibliography

1. Alfven, H. and Herlofson, N., *Phys. Rev.* **78** (1950) 616.
2. Ball, L. and Kirk, J.G., Preprint Astron Dep. Univ. of Sydney No.IOA R95-29 (1995).
3. Begelman, M. and Blandford Rees M., *Rev. Mod. Phys* **56** (1984) 255.

4. Erdelyi, A. et al., *Higher transcendental functions* (McGraw-Hill, N.Y., 1953).
5. Beresteskii, V.B., Lifshitz, E.M. and Pitaevskii, L.P., *Quantum Electrodynamics* (2nd ed., Pergamon Press, Oxford, 1982).
6. Bisnovatyi-Kogan, G.S., *Astron. Zh.* **47** (1970) 813.
7. Bisnovatyi-Kogan, G.S., *Astron. Zh.* **50** (1973) 902.
8. Bisnovatyi-Kogan, G.S., *Equilibrium and stability of stars and stellar systems* (in Russian) (Doctor Thesis, Moscow, IKI 1977)
9. Bisnovatyi-Kogan, G.S., *Astron. Zh.* Lett. **4** (1978) 130.
10. Bisnovatyi-Kogan, G.S., *Rivista Nuovo Cimento* **2** (1979) No.1.
11. Bisnovatyi-Kogan, G.S., *Physical Problems of the Theory of Stellar Evolution* (Nauka, Moscow, 1989).
12. Bisnovatyi-Kogan, G.S. in *Cosmoparticle Physics.* Proc. "COSMION-94", eds M.Yu. Khlopov et al., (Editions Frontiers, 1996) 333.
13. Bisnovatyi-Kogan, G.S. and Blinnikov, S.I., *Astrophys. Space Sci.* **19** (1972) 93.
14. Bisnovatyi-Kogan, G.S. and Blinnikov, S.I., *Sov. Astron. Lett.* **2** (1976) 191.
15. Bisnovatyi-Kogan, G.S. and Blinnikov, S.I., *Astron. Ap.* **59** (1977) 111.
16. Bisnovatyi-Kogan, G.S. and Fridman, A.M. *Astron. Zh.* **46** (1969) 721.
17. Bisnovatyi-Kogan, G.S., Illarionov, A.F. and Slysh, V.I., *Proc. School Plasma Astrophys.* ESA SP-311 (1990) 289.
18. Bisnovatyi-Kogan, G.S., Komberg, B.V. and Fridman A.M., *Astron. Zh.* **46** (1969) 465.
19. Bisnovatyi-Kogan G.S. and Lovelace, R.V.L., *Astron. Ap.* **296** (1995) L17.
20. Bisnovatyi-Kogan, G.S. and Ruzmaikin, A.A., *Astrophys. Space Sci.* **28** (1974) 45.
21. Bisnovatyi-Kogan, G.S. and Ruzmaikin, A.A., *Astrophys. Space Sci.* **42** (1976) 401.
22. Bisnovatyi-Kogan, G.S. and Silich, S.A., *Rev. Mod. Phys.* **67** (1995) 661.
23. Bisnovatyi-Kogan, G.S. and Syunyaev, R.A., *Astron. Zh.* **48** (1971) 881.
24. Bolton, J.G., Stanley, G and Slee, B., *Nature* **164** (1949) 101.
25. Bridle, A.H. and Perley R.A., *Ann. Rev. Astron. Ap.* **22** (1984) 319.
26. Bronshtein, I.N. and Semendyaev K.A., *Handbook of mathematics (in Russian)* (GITTL, Moscow, 1953).
27. Cherepashchuk, A.M. ed. *Catalog of close binary systems on late stages of evolution.* (MGU, Moscow, 1988).
28. Chevalier, R.A., *Astrophys. J.* **258** (1982) 790.
29. Dombrovsky, V.A., *Dokl. Akad. Nauk SSSR* **94** (1954) 1021.
30. Eilek, J.A. and Hughes, P.E., in *Astrophysical Jets*, ed. P.E. Hughes (Cambridge University Press, 1990) 428.
31. Fiero, J. et al., *Astrophys. J.* **413** (1993) L27.
32. Fedorenko, V., in *Extragalactic Radio Sources From Beams to Jets*, eds J. Roland, H. Sol, and G. Pelletier (Cambridge University Press, 1992) 1.
33. Getmansev, V.V., *Dokl. Akad. Nauk SSSR* **83** (1952) 557.
34. Giacconi, R., Gursky, H., Paolini, F.R. and Rossi B.B., *Phys. Rev. Let.* **9** (1962)

439.
35. Ginzburg, V.L., *Dokl. Akad. Nauk SSSR* **76** (1951) 377.
36. Ginzburg, V.L., *Uspekhi Fiz. Nauk* **51** (1953) 343.
37. Ginzburg, V.L., *Pulsars (theoretical consideration)* (1970) Preprint No.52, FIAN.
38. Ginzburg, V.L., *Theoretical physics and astrophysics* (Nauka, Moscow, 1975).
39. Ginzburg, V.L., in *Early Years of Radio Astronomy* ed. W. Sullivan, (Cambridge Univ. Press, 1984) p.289.
40. Ginzburg, V.L., ed. *Astrophysics of cosmic rays* (Nauka, Moscow, 1990).
41. Ginzburg, V.L. and Syrovatsky, S.I., *Uspekhi Fiz. Nauk* **87** (1965) 65.
42. Ginzburg, V.L. and Syrovatsky, S.I., *Ann. Rev. Astron. Ap.* **7** (1969) 375.
43. Ginzburg, V.L., Sazonov V.N. and Syrovatsky, S.I., *Uspekhi Fiz. Nauk* **94** (1968) 63.
44. Gnedin, Yu.N. and Syunyaev, R.A., *Astron. Astrophys.* **25** (1973) 233.
45. Gordon, I.M., *Dokl. Akad. Nauk SSSR* **94** (1954) 813.
46. Gold, T., *Nature* **218** (1968) 731.
47. Goldreich, P. and Julian, W., *Astrophys. J.* **157** (1969) 869.
48. Gunn, J. and Ostriker J., *Astrophys. J.* **165** (1971) 523.
49. Hartman, R. et al., *Astrophys. J.* **407** (1993) L41.
50. Hewish, A., Bell, S.J., Pilkington, Scott, P.F., Collines *Nature* **217** (1968) 709.
51. Hiltner, W.A., *Astrophys. J.* **130** (1959) 340.
52. Hughes, V.A., *Astrophys. J.* **383** (1991) 280.
53. Imshennik, V.S. and Nadyozhin, D.K., *Itogi Nauki i Tekhniki, Astronomiya* **21** (1982) 63.
54. Imshennik, V.S. and Nadyozhin, D.K., *Uspekhi Fiz. Nauk* **156** (1988) 561.
55. Kadomtsev, B.B., *Problems in Plasma Theory* **2** (1963) 132.
56. Kardashev, N.S., *Astron. Zh.* **39** (1962) 393.
57. Kardashev, N.S., *Astron. Zh.* **41** (1964) 807–813
58. Kiepenheuer, K.O., *Phys. Rev.* **79** (1950) 738.
59. Kellerman, K.I., in *Galactic and Extra-Galactic Radio Astronomy* (Springer-Verlag, 1974).
60. Kesteven, M.J. and Caswell, J.L., *Astron. Ap.* **183** (1987) 118.
61. Kondratenko, A.N., *Plasma Waveguides* (Atomizdat, Moscow, 1976).
62. Lamb, F.K., Pethick, C.J. and Pines, D., *Astrophys. J.* **184** (1973) 271.
63. Landau, L.D.and Lifshiz, E.M., *Statistical Physics* (Nauka, Moscow, 1964).
64. Landau, L.D. and Lifshitz, E.M., *The Classical Theory of Fields* (Pergamon, New York, 4th edition, 1975).
65. Lipunov, V.M., *Astrophysics of neutron stars* (Nauka, Moscow, 1987).
66. Lorrimer, D.R., Nicastro, L., Lyne, A.G. et al., *Astrophys. J.* **439** (1995) 933.
67. Lynden-Bell, D., *Nature* **223** (1969) 690.
68. Longair, M.S., *High energy astrohysics* (Cambridge Univ. Press, Cambridge, 1981).
69. Lovelace, R.V.E., *Nature* **262** (1976) 649.
70. Lozinskaya, T.A., *Supernovae and stellar wind. Interaction with a G alactic gas*

(Nauka, Moscow, 1986).
71. Manchester, R.N. and Taylor, J.H., *Pulsars* (W.H.Freeman and company, San Francisco, 1977).
72. Manchester, R.N., Lyne A.G., D'Amico N. et al., *Month. Not. RAS* **279** (1996) 1235.
73. Mathewson, D.S, Ford, V.L., Dopita, M.A., Tuohy, I.R., Mills, B.Y. and Turtle A.J., *Astrophys. J. Suppl.* **55** (1984) 189.
74. Meisenheimer, K., Röser, H.-J, Hiltner, P.R., Yates, M.G., Longair, M.S., Chini, R. and Perley, R.A., *Astron. Ap.* **219** (1989) 63.
75. Mihara, T., Makishima, K., Ohashi, T. et al., *Nature* **346** (1990) 250.
76. Novikov, I.D. and Frolov, V.P., *Physics of black holes* (Nauka, Moscow, 1986).
77. Pacholczyk, A.G., *Radio Astrophysics* (W.H.Freeman and Company, San Francisco, 1970).
78. Pacini, F., *Nature* **216** (1967) 567.
79. Radhakrishnan, V. and Cocke, D.J., *Astrophys. Lett.* **3** (1969) 225.
80. Romanova, M.M. and Lovelace, R.V.E., *Astron. Ap.* **262** (1992) 26.
81. Röser, H.-J., Conway, R.G. and Meisenheimer, K., *Astron. Ap.* **314** (1996) 414.
82. Ruderman, M.A. and Sutherland, P.G., *Astrophys. J.* **196** (1975) 51.
83. Schlötelburg, M., Meisenheimer, K., and Röser, H.-J., *Astron Ap.* **202** (1988) L23.
84. Seward, F.D., *Astrophys. J. Suppl.* **73** (1990) 781.
85. Shakura, N.I. and Syunyaev, R.R., *Astron. Ap.* **24** (1973) 337.
86. Sheffer, E.K., Kopaeva, I.F., Averintsev, M.V. et al., *Astron. Zh.* **69** (1992) 82.
87. Shklovsky, I.S., *Astr. Zh.* **30** (1953) 15.
88. Shklovsky, I.S., *Dokl. Akad. Nauk SSSR* **90** (1953) 983.
89. Shklovsky, I.S., *Radioastronomy* (GITTL, Moscow, 1955).
90. Shklovsky, I.S., *Astr. Zh.* **39** (1962) 209.
91. Shwarzman, V.F., *Astr. Zh.* **48** (1971) 438.
92. Shwarzman, V.F., *Astr. Zh.* **48** (1971) 479.
93. Slysh, V.I., *Nature* **199** (1963) 682.
94. Staveley-Smith, L. et al., *Nature* **366** (1993) 136.
95. Sturrock, P.A., *Astrophys. J.* **164** (1971) 529.
96. Taylor, J.H., Manchester, R.N. and Lyne, A.G., *Astrophys. J. Suppl.* **88** (1993) 529.
97. Thomson, D.J. et al., *Nature* **359** (1992) 615.
98. Thomson, D.J. et al., *Astrophys. J.* **415** (1993) L13.
99. Trümper, J., Pietsch, W., Reppin, C., Voges, W., Staubert, R. and Kendziorra, E., *Astrophys. J.* **219** (1978) L105.
100. Trümper, J., Kahabka, P., Ögelman, H., Pietsch, W. and Voges, W., *Astrophys. J.* **300** (1986) L63.
101. Vashakidze, M.A., *Astron. Tsirk.* No.147 (1954) 11.
102. Vedenov, A.A., Velikhov, E.P. and Sagdeev, R.Z., *Uspekhi Phys. Nauk.* **73** (1961) 701.

103. Voges, W., Pietsch, W., Reppin, C. Trümper, J., Kendziorra, E. and Staubert, R., *Astrophys. J.* **263** (1982) 803.
104. Weiler, K. and Skamek, R.A., *Ann. Rev. Astron. Ap.* **26** (1988) 295.
105. Zeldovich, Ya.B. and Novikov, I.D., *Gravitational theory and stellar evolution* (Nauka, Moscow, 1971).
106. Zheleznyakov, V.V., *Radio radiation of the Sun and planets* (Nauka, Moscow, 1964).
107. Zheleznyakov, V.V., *Electromagnetic waves in space plasma. Generation and propagation* (Nauka, Moscow, 1977).
108. Zheleznyakov, V.V. and Suvorov, E.V., *Astrophys. Space Sci.* **15** (1972) 24.

Index